AF598415

SEPARATION METHODS FOR WASTE AND ENVIRONMENTAL APPLICATIONS

Environmental Science and Pollution Control Series

1. Toxic Metal Chemistry in Marine Environments, *Muhammad Sadiq*
2. Handbook of Polymer Degradation, *edited by S. Halim Hamid, Mohamed B. Amin, and Ali G. Maadhah*
3. Unit Processes in Drinking Water Treatment, *Willy J. Masschelein*
4. Groundwater Contamination and Analysis at Hazardous Waste Sites, *edited by Suzanne Lesage and Richard E. Jackson*
5. Plastics Waste Management: Disposal, Recycling, and Reuse, *edited by Nabil Mustafa*
6. Hazardous Waste Site Soil Remediation: Theory and Application of Innovative Technologies, *edited by David J. Wilson and Ann N. Clarke*
7. Process Engineering for Pollution Control and Waste Minimization, *edited by Donald L. Wise and Debra J. Trantolo*
8. Remediation of Hazardous Waste Contaminated Soils, *edited by Donald L. Wise and Debra J. Trantolo*
9. Water Contamination and Health: Integration of Exposure Assessment, Toxicology, and Risk Assessment, *edited by Rhoda G. M. Wang*
10. Pollution Control in Fertilizer Production, *edited by Charles A. Hodge and Neculai N. Popovici*
11. Groundwater Contamination and Control, *edited by Uri Zoller*
12. Toxic Properties of Pesticides, *Nicholas P. Cheremisinoff and John A. King*
13. Combustion and Incineration Processes: Applications in Environmental Engineering, Second Edition, Revised and Expanded, *Walter R. Niessen*
14. Hazardous Chemicals in the Polymer Industry, *Nicholas P. Cheremisinoff*
15. Handbook of Highly Toxic Materials Handling and Management, *edited by Stanley S. Grossel and Daniel A. Crowl*
16. Separation Processes in Waste Minimization, *Robert B. Long*
17. Handbook of Pollution and Hazardous Materials Compliance: A Sourcebook for Environmental Managers, *Nicholas P. Cheremisinoff and Madelyn Graffia*
18. Biosolids Treatment and Management: Processes for Beneficial Use, *edited by Mark J. Girovich*

19. Biological Wastewater Treatment: Second Edition, Revised and Expanded, *C. P. Leslie Grady, Jr., Glen T. Daigger, and Henry C. Lim*
20. Separation Methods for Waste and Environmental Applications, *Jack S. Watson*

Additional Volumes in Preparation

SEPARATION METHODS FOR WASTE AND ENVIRONMENTAL APPLICATIONS

JACK S. WATSON
Oak Ridge National Laboratory
Oak Ridge, Tennessee

MARCEL DEKKER, INC. NEW YORK • BASEL

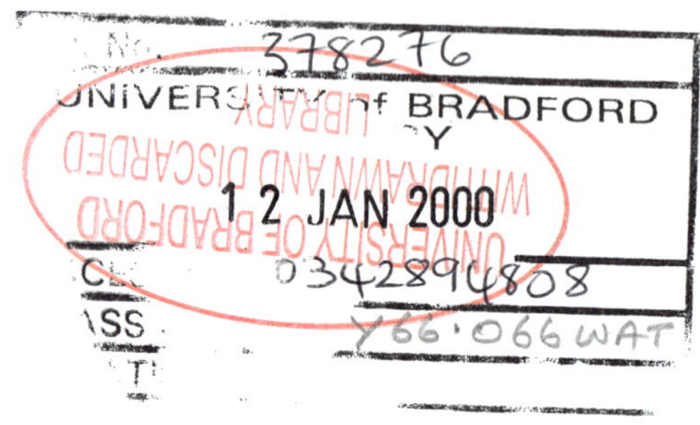

ISBN: 0-8247-9943-7

This book is printed on acid-free paper.

Headquarters
Marcel Dekker, Inc.
270 Madison Avenue, New York, NY 10016
tel: 212-696-9000; fax: 212-685-4540

Eastern Hemisphere Distribution
Marcel Dekker AG
Hutgasse 4, Postfach 812, CH-4001 Basel, Switzerland
tel: 41-61-261-8482; fax: 41-61-261-8896

World Wide Web
http://www.dekker.com

The publisher offers discounts on this book when ordered in bulk quantities. For more information, write to Special Sales/Professional Marketing at the headquarters address above.

Current printing (last digit):
10 9 8 7 6 5 4 3 2 1

PRINTED IN THE UNITED STATES OF AMERICA

To my wife Patricia
whose patience and encouragement
helped me to complete this work

Preface

The purpose of this book is to bring together information and concepts needed by those concerned with selection and/or design of separation process equipment for treating wastes or environmental streams. In treating a waste stream or a contaminated portion of the environment there are usually only two or three options: separation of the contaminant from the stream of material, destruction of the contaminant, or isolation of the material to prevent its release or spread in the environment. Separation processes are often the best choice, but their roles may not be obvious because these processes are sometimes called by different names such as *removal* of contaminants, *concentration* of contaminants, or *purification* of waste material. These terms all refer to separation processes. *Enriching*, *stripping*, *clarification*, and *benefication* are a few of the additional terms used to describe separation processes, but they have come to have more specific meanings.

It hardly seems necessary to explain how or why treatment of waste streams and contaminated soils or groundwater has become a topic of increasing concern and importance. The past few decades have seen a growing awareness of the important effect of environmental pollution on the health of people, plants, and animals. This increased awareness was generated, in part, by individuals, by organizations such as the Sierra Club, and by popular publications such as *The Silent Spring*. The concern has generated a large increase in research on the effects of chemicals and other pollutants on both our health and the health of the environment. This research on environmental effects made industry, the government, and the technical community more aware of the impact of many waste effluents on the environment, and the understanding of this impact has reached the public at large. Many chemicals and materials that were in common use only a decade or so ago are now known to be hazardous, and in some cases, the manufacture of those materials has ceased or is scheduled to cease. The list of known toxic substances has grown significantly during this period. This increased research and identification of more toxic materials has also increased the concern of the public, and that has stimulated additional research.

Despite the increased recognized risk and the tighter regulations, it is often necessary to use hazardous materials. Although the use of hazardous materials makes industrial operations more difficult and more costly, it is possible to operate industrial systems safely using hazardous materials without releasing significant quantities of toxic wastes, and industrial firms are learning to do just that. Use of a toxic material does not have to be equated with release of significant quantities of that material. However, industrial and consumer activities in the past, when less effective containment methods were used, have left us with a legacy of contaminated soil, buildings, equipment, etc.; this legacy has had a negative effect on public confidence in some companies, and even some entire industries. This legacy of hazardous materials must be handled when "remediating" sites. There is a need for better ways to both handle and contain hazardous materials during current operations and to remove contaminants from water, soil, equipment, and anywhere contamination from past operations has spread.

Separation methods play important roles in solving all of these problems, and their roles are expected to increase. Although enormous sums have been spent on contaminated sites during the past decade, the number of sites that have been completely remediated has been disappointing to many. To accelerate the remediation of contaminated sites, it will be necessary to utilize the most cost-effective methods. It is important that those working on cleaning up environmental problems and designing or modifying process systems to reduce future spreading of hazardous contamination be well versed in most aspects of separation methods, including the selection of the best methods for a problem, the best way to apply methods, and the limitations of each method. This book is an effort to help with these needs.

Separation Methods for Waste and Environmental Applications

The separation methods that are or are likely to become most important in solving waste and environmental problems include a number of chemical and physical separation methods that have traditionally not been covered in any one university curriculum or in a single textbook. Since this book is aimed at applications in a small, but rapidly growing, segment of the process industries—the waste and environmental industries—coverage is based on the importance of the methods to this segment. Separation methods for any application should be based on the problem to be solved and not influenced by limitations in the background of the engineer or others assigned to solve the problem.

Most separation processes are concerned directly with meeting regulations or discharge limits, mostly for water and air. The processes are also important in meeting solid waste disposal regulations, especially those involving the concentration of contaminants in the solid waste. Separations can also be used in other, earlier, parts of a process system to prevent pollutants from reaching the effluent streams. For instance, pollutants that result from undesired components in the feed stream could be removed from the feed stream as well as from the effluent stream. It is common practice to remove hydrogen sulfide from natural gas so it will not enter the environment when the gas is burned or processed in any other way (usually to make "syngas"). This book focuses principally on removal of contaminants from effluent streams, but the examples include several cases in which separation systems can be applied elsewhere to meet the same regulations.

Potential Readers

The complexity of the material in this book varies, but most of it should be understandable to graduate engineers or scientists, graduate students, and upper-division undergraduate students. No exceptional mathematics background is required, and most of the subjects covered are not exceptionally complex. A basic understanding of general chemistry is needed to understand many sections, but most engineers and scientists working on environmental problems should have a year, or more, of undergraduate study in chemistry. Attempts are made in the book to explain the few concepts that are less likely to be covered in basic chemistry courses. Those who will be most interested in the material are environmental engineers, chemical engineers, and other engineers and scientists who have some experience in the field, and they will probably have more than basic understanding of the chemical concepts needed. A brief discussion of the chemistry involved in a concept is sometimes presented, but this is not principally a chemistry book. In most cases, the additional details may not be absolutely necessary to use the concepts presented.

This book is intended to be used by some readers to review separation methods and by others to become acquainted with methods with which they are not sufficiently familiar. Civil and environmental engineering textbooks cover some of the same topics, but the coverage is usually different. In most respects the organization is more like that of a chemical engineering book, with the focus on individual separation processes, and the treatments of water and air streams using the same separation principles are discussed together. Information on the separation methods most important in environmental and waste problems is emphasized. There is also an attempt to provide insight into the concepts and principles that must be understood for selection of separation methods for new problems. Traditional books on separations for chemical engineers usually describe each separation method well but do not give much attention to comparison of the different methods. That is partially because the traditional books do not focus on a constrained group of applications—environmental and waste problems—and such comparisons would be difficult to make for a wider group of applications. Of course, even for the relatively narrow range of applications in this book, judgment is often needed until detailed performance and economic studies are completed. Nevertheless, at least the introduction to selective criteria is expected to be helpful to the readers. The general comparisons are intended to guide the reader by suggesting which separation methods should be considered for a given application and conditions, not as a final judgment on which method to use.

Separation processes are usually a strong part of chemical engineering practice, and chemical engineering textbooks probably provide more information on fundamentals of separation processes than books written for other fields. Because the chemical and process industries are involved in such a wide variety of materials, there is less temptation to seek a few approaches that will satisfy most separation needs than one might find in books for fields with less varied needs. This book is also expected to be useful to the trained chemical engineer because it differs significantly from most chemical engineering books by including more detail on separation methods important in waste and environmental treatment.

Differences Between This Book and Most Separation Texts

The strong focus on separation methods for waste and environmental treatment is intentional. The most obvious sign of this focus is probably in the examples given, which are mostly drawn from environmental or waste problems. However, the most important aspect of the focus is reflected in the selection of the separation methods covered and the space devoted to each separation method. This is most evident when one compares the topics and the space allotted to each topic with the coverage in other separation books. This is important because many separation methods that are important in environmental and waste problems are not covered extensively in standard separations texts, usually texts aimed at chemical engineers working in the process industries; this is especially notable in texts used at the undergraduate level.

Of course, the selection of the methods to emphasize in this book comes from my own experiences and views. The career of each engineer or scientist working on waste and environmental problems is different; so there can be no claim that this book is custom-tailored for every engineer and scientist in the field. The reader will find adsorption covered in far more detail than in separation texts that attempt to cover all separation methods. Adsorption is especially important for removing the last traces of contaminants from both aqueous and air effluents. Adsorption can be the principal method for removing a contaminant or it can be a "polishing" method for removing traces of contaminants after other methods have been used to remove the bulk of the contaminants.

Other separation methods that are especially important for waste treatment are gas absorption and gas stripping. Some separation methods are covered in more detail in this book than they have been covered in most "classic" texts on separations because of their high potential for future applications in waste and environmental processing. Perhaps the most notable example is membrane separations. Applications for membranes are currently growing rapidly, and this growth is not limited to waste and environmental applications. The chapter on membrane processes focuses on reverse osmosis and removal of hydrophobic contaminants from effluent water and/or air emissions rather than on some of the other exciting new applications of membranes in gas separations.

A similar focused (or limited) treatment is given for liquid-liquid (solvent) extraction. Although the inherent solubility and entrainment that occur in all liquid-liquid extraction systems introduce traces of organic compounds to the treated stream that could be considered a second contaminant, there are places where liquid-liquid extraction can remove some contaminants more effectively than other methods. The removal of metal ion contaminants from aqueous solutions is a notable example. The chapter on liquid-liquid extraction thus focuses on systems in which a highly insoluble extractant can be used, and especially on the removal of metal ion contaminants.

The most important separation method in the process industries is distillation, which involves, by far, the largest capital investment and energy consumption of any separation method. Despite its great importance to the process industries, distillation is not given prominent treatment in this book, principally because it is not expected to have such great importance in waste and environmental problems. A chapter is devoted to distillation because it can still be important in some waste and environmental applications, but readers familiar with other separation texts are likely to find the treatment of distillation more brief and less detailed than they expected. Distillation is generally most useful in recovering products at higher concentrations than those usually found in waste and environmental problems.

Crystallization is another separation method that is covered in less detail than one may find in other separation texts. There is no separate chapter on crystallization, but some aspects of crystallization are included in several chapters. Many aspects of crystallization, such as the slow growth of crystals to produce high-quality and relatively pure crystal products, are not likely to be issues in waste and environmental operations, which are far more likely to deal with very dilute solutions. However, precipitation is simply another name for crystallization, a name that is more likely used with "crystallization" of highly insoluble materials, usually initiated by a change in pH or concentration of a reactant. Precipitation of insoluble materials is discussed along with adsorption of contaminant on precipitants (or incorporation of contaminants into precipitants) and the removal of precipitants from solutions. However, these are rapid "crystallization" operations in which the principal interest is in the chemistry to form highly insoluble components and remove the precipitated solids. When the precipitated solids have extremely low solubilities, it is less likely that the crystal growth can be controlled very well, and that is usually not a key goal of waste/environmental precipitation operations.

Readers who are familiar with other separations texts will also note a stronger emphasis on physical separations such as filtration and sedimentation. Physical separations are often covered in separate texts from chemical separations. Here they are covered with chemical separations, because they are already important in waste and environmental systems, and their importance is expected to grow. Filtration and sedimentation are well established and used almost universally, because almost all waste and groundwater contain solids that must be removed. Even if the solids are not contaminants (there are limits on solids in discharge waters), they often adsorb "soluble" contaminants, and discharge limits often cannot be met without removal of the majority of the solids.

Solid-fluid separations are also often required in chemical separation systems because so many chemical separations cannot function well in the presence of high solids concentrations. Filtration is usually required on any fluid stream going to adsorption or other packed-bed separation equipment; so it is needed even when it does not remove a contaminant. Filtration and sedimentation are also important steps in precipitation processes, and the physical separations may be the most difficult aspect of such systems.

Other physical separation methods that currently play some, but limited, roles in waste and environmental separations are believed to have potential for increased importance in future waste and environmental treatment systems, and those methods are discussed at least briefly. The coverage is largely descriptive rather than quantitative. These methods are grouped together into a single chapter. The reader should have some familiarity with these methods, but their current importance does not warrant more than the brief summaries given in this book.

Descriptions of Design Procedures

All the major chapters on separation methods that are expected to be most important to waste and environmental problems include at least some discussion of the science involved in the separation method and design procedures that allow the reader to size equipment or to estimate the performance of given equipment. Throughout these chapters, the reader is reminded that computer packages are available that perform many of the calculations described. However, it is important that the reader understand what is going on in such calculations. There are assumptions and limitations in all such calculations that users should understand. Also, in some cases it is as easy to do calculations manually as it is to set them up for the computer. The descriptions focus on relatively simple cases because they aim to provide insight into the design method; often the insight into the simpler cases is all that is needed to understand the principles that are used in the more complicated cases that almost certainly should be evaluated using powerful computer codes and packaged computer analyses.

Essentially all the design procedures given in this book are available in standard textbooks on separations. The only significant exception is the description of a generalized approach to design of adsorption beds with constant pattern breakthrough curves, and even this approach is based on principles that are dispersed throughout other textbooks. The unique feature of this book is the inclusion of the particular collection of methods presented. Thus, one would need to look at several texts to find the information presented here, but in looking at several texts, one would find far more detail on some separation methods, perhaps more detail than was wanted or than would be needed.

Example Applications

Example applications are usually given in the major chapters in two places. Early in each major chapter there is a brief mention of a few of the most important general classes of the method, giving an immediate sense of how a separation method could be used to address a waste or environmental problem. However, most examples are presented near the end of the chapter, usually following a general description of conditions that make the method appropriate for a particular application.

There is no practical way to include all applications that could be of interest, so those given should be viewed only as examples. Some of the most important applications are mentioned, but most are simply interesting. There are applications that include combinations of separations and applications that are only under study and could be important in the future. In some cases, the applications are mentioned because they illustrate a point about conditions that make a particular separation method attractive. The lists of applications are not complete. That would be impractical because of the constant generation of new ideas and new applications for separation methods.

Jack S. Watson

Contents

1
Introduction

The treatment of industrial wastes and contaminated soils and groundwater is a rapidly growing industry in itself and now consumes a substantial portion of the activity of many firms. Several new firms have been organized during the last two decades whose business is dedicated entirely or largely to solving waste and environmental problems. Some of these firms are now large and have operations throughout the United States and even overseas. Although the rate of growth of this industry appears to have slowed in the last few years, it is still significant and appears to be significantly higher than the growth of industry as a whole. There has been a accompanying increase in interest in the academic community in environmental issues and research on ways to treat environmental problems. Ecology programs and environmental engineering programs have grown greatly on many campuses, and environmental activities have become important parts of other academic programs from engineering to law and business.

This increase in activity and business concerned with environmental and waste problems has resulted from the increased awareness of the public to the hazards of many industrial wastes. Research during the last few decades illustrates the importance of pollution in air, water, and soil to the public health and to environmental quality. The results of these studies have raised the public awareness and concern with industrial pollution, and the public concern has resulted in new regulations that affect waste and environmental treatment. Environmental problems such as acute spills or gradual environmental damage have been recognized by the press and electronic media to be of interest to the public, and this has enhanced the rate at which environmental news reaches the general public.

The increased awareness and importance of environmental issues have not escaped the attention of lawmakers and regulators. Essentially everyone is aware of the great increase in laws and regulations that have come into effect during recent years. This growth in regulations was crowned by the formation of a major agency, the Environmental Protection Agency (EPA), and the formation of related programs within other federal agencies. Every state has its own state version of the EPA, and in some cases the regulations from the state agencies have been stricter than those from the EPA. This increased regulation of environmental problems and issues has not been limited to the United States; essentially every country has set up similar environmental agencies, and in some cases the laws or enforcement is even stricter than in the United States. The changes in the regulatory environment during the past two decades have greatly affected most industrial operations. Almost everyone who has worked in an industrial company has seen these changes, and when hazardous materials are involved in industrial operations, the changes can be profound.

These three changes—in our awareness of the seriousness of environmental problems, in new research on environmental and health problems, and in environmental regulations—are closely connected; the activities in each area have increased because of results from the other changes. Despite some newer voices in Congress with complaints about federal environmental regulations, it appears more likely that improved understanding of the effects of industrial emissions on the population and the environment will continue to require strict regulation, perhaps in some cases regulations even stricter than those currently in place.

The recent history of environmental regulation has shown an almost continuous reduction in discharge limits allowed for a growing number of materials. Although it would be simplistic to expect the regulations to continue to be tightened at the same rate forever, there is still a reasonable expectation that environmental regulations will continue to become more restrictive during the coming decades, but perhaps at a less rapid pace. There are powerful forces currently attempting to reduce the burden (or effectiveness) of some regulations that seem (to them) to be very costly and to play few roles in protecting the public or the environment. There also are clear cases where significant parts of the public object to environmental regulations, usually because the regulations or their enforcement affect or threaten jobs, but there is little evidence that the general public favors a significant reduction in environmental regulations or enforcement. Fear of hazardous materials and wastes appears to still be strong in the general public, and a significant relaxation in regulations does not appear to be likely as long as that fear remains. Current complaints are focused on some of the regulations and regulatory practices that appear to be least justified, especially the liability regulations that hold each contributor to a contaminated site potentially liable for remediation of the entire site. However, there is no reason to expect a major pullback of regulations except in these few specific areas where regulations may have been applied incorrectly. Overall the environmental regulations may continue to become stricter, hopefully in ways that will improve health and environmental quality efficiently as well as effectively. Waste treatment and environmental restoration is now a large industry in the United States and is still growing rapidly with expectations for considerable future growth.

SEPARATION PROCESSES IN WASTE AND ENVIRONMENTAL TREATMENT

Separation operations are currently important in almost every aspect of waste and environmental treatment, but the separation treatment methods may go under other related names. These names include "removal" of a contaminant, "decontamination" of a material or effluent stream, "concentration" of a contaminant, "purification," etc. There are generally only three options that can be used to treat waste or environmental streams: separation, destruction (reaction), or immobilization. Of course, in many systems, all three options can be and are used. A stream of waste materials could be treated by each option for remediation problems associated with different contaminants. Separations could be used to concentrate contaminants for eventual destruction or immobilization and to remove contaminants and concentrate or purify them to the point where they can be used again and no longer be considered "contaminants."

The reduction or elimination of waste streams is likely to become an increasingly important aspect of environmental protection; so one should not think of waste treatment entirely in terms of treatment of effluent streams. The addition of process systems to reduce toxic emissions from existing industrial facilities is currently the principal focus of waste and processing, and that is also the principal focus of the illustrations in this book. However, one should be prepared to view waste problems in a more global manner, including modifications to the industrial operations themselves to produce less waste and to utilize more of the wastes that must be produced. This projected shift in waste treatment techniques is one reason why readers are encouraged to seek a general understanding of the separation methods themselves and not focus too strongly on individual applications that could become less important in the future. Applications of separation methods are expected to increase as efforts are made to reduce waste volumes, reuse contaminants rather than discharging them, to reduce the concentration of contaminants in inlet streams even further, or to remove contaminants from soils and groundwater. If the future approaches to waste management involve a greater emphasis on changes in the manufacturing processes and rely less on effluent treatment, the role of separations is not expected to decline significantly, but the roles of separations could shift.

"GREEN MANUFACTURING"

There has been considerable interest lately in "green manufacturing." One "driving force" for seeking green manufacturing is the Pollution Prevention Act of 1990 [1]. The term "green manufacturing" can have different meaning to different people. In some cases, it can mean manufacturing with zero or minimal release of toxic materials. Note that the call may be for zero release, but "zero" may not have the meaning in these cases that some people may intend. It can simply mean that the presence of the toxic material is not detected; so better chemical analyses could change this meaning of the term. Of course, in the ultimate sense, if we had suitable chemical analyses, we probably could detect a molecule or a few molecules of essentially anything in any waste, product, or emission. Thus the practical definition of green manufacturing would specify that the emissions would be acceptable (within specified limits) and minimized. That is, one would reduce the emissions to levels as far below the accepted limits as possible or as practical (again another chance to hold different definitions of the term).

Others define green manufacturing as minimizing or eliminating the use of toxic materials as raw materials or intermediates in manufacturing. Although one can argue that only the emissions count and that consumed or recycled intermediates do not contribute to environmental pollution, a case can also be made for minimizing the presence of the toxic material where ever possible. Even the use of toxic materials leaves the possibility that there will be accidental releases, and the absence of toxic materials eliminates the possibilities that there will be toxic releases, either accidental or undetected. There is also the advantage that eliminating the use of toxic substances will eliminate the exposure of workers to those materials, even if the toxic materials are not released.

This seems to be one common definition used in recent workshops and publications on the subject. The major concern is usually the use of chlorine and chlorine compounds in the synthesis of organic chemicals. These processes leave the possibility for release of some of the organic intermediates and for formation of trace quantities of other toxic compounds, such as dioxins. Intermediates such as isocynates are also targets for elimination. There have even been calls by a few people for total elimination of chlorine in chemical manufacturing. A total elimination of chlorine would be difficult and would eliminate the production of many useful products, including important pharmaceutical products. A general discussion of the subject can be found in a brief article by Browner [2]. An example of the results of workshops and technical session can be found in the proceedings of a symposium on clean chemical manufacturing organized by the American Chemical Society [3].

Since most of the separation processes discussed in this book are concerned with removing toxic or other materials from solid, liquid, or gaseous wastes, it may appear that separation methods are concerned with the more conventional approaches of using or producing toxic compounds but trapping the toxic components so that they cannot reach the environment. In most cases, that is a correct assessment. Increased use of green manufacturing would eliminate the presence of toxic materials in many waste streams and the subsequent need for separation methods. As desirable as these approaches are, they are not expected to eliminate or even reduce the growth in the need for more and better separation methods. There are several reasons why separation methods will remain among the primary tools in reducing pollution of the environment. Although a reduction of the use of toxic components in the manufacture of nontoxic products will be appropriate and is expected, the total elimination of the use of toxic intermediates is not expected. Note that several "useful" products are toxic, and to eliminate those toxic materials would eliminate the products themselves. Most insecticides and herbicides are toxic, at least slightly, to humans or the environment to a limited (but accepted) degree. Rather than eliminate these products, the public is likely to call for more restricted uses of these materials and the development of less toxic alternatives where uses are necessary. Operations that handle highly toxic materials are increasingly likely to require separations equipment on the off-gas and ventilation systems to prevent release of contaminants during accidental or normal operating conditions.

Finally, one should note that there have already been significant "spills" or releases of toxic materials, and continuing efforts are needed to remove the contaminants from the environment. All of these activities are likely to require separations operations to provide better containment of the toxic materials that truly need to be produced and/or used and to decontaminate soil, water, equipment, and buildings that have been contaminated by earlier operations. To this author, the need to increase the use of separations operations for waste and environmental problems is likely to develop more quickly than the ability to decrease the use of toxic materials.

ENVIRONMENTAL REGULATIONS

Because the regulatory restrictions are currently being assessed by Congress and regulatory agencies, and the approaches to regulations may be changed in the near future, a detailed description of the regulations will not be given here. Even if the approach to regulations is changed, the need for reducing the release of contaminants is likely to continue, perhaps with no more than selected limited changes in the allowable release of selected contaminants. Numerous laws are concerned with the release and disposal of contaminants, including the Research Conservation and Recovery Act (RCRA) which regulates the storage and disposal of solid wastes, the Clean Air Act (CAA), the Clean Water Act (CWA), and the subsequent amendments to these acts. Government facilities are also covered by the National Environmental Protection Act (NEPA), and some sites and facilities are covered by the "Superfund" laws. Those facilities dealing with radioactive wastes (hospitals, universities, electric utility companies, and many industrial concerns, as well as the Department of Energy facilities, etc.) must also deal with the Atomic Energy Act and regulations from the Nuclear Regulatory Commission. Most states have created their own regulatory agencies and regulations that must be obeyed by organizations operating in those states, and state regulations can be different, and sometimes more restrictive, than the federal regulations.

No attempt will be made here to provide details on the current regulations on waste and concentrations in effluent streams, and there is no way to know how those regulations will change in the near future. However, the reader should read the regulations or explanations of the regulations.

One likely change in these acts in the near future appears to include giving more consideration to the "risk" of a contaminant. Such a change appears to be particularly interesting to Congress. Under risk analyses, the release limits could become different for different forms of a contaminant and even different under different local conditions that would affect the likelihood of the contaminant reaching the public or sensitive parts of the environment. Costs of remediation or release reduction could also become a factor in setting acceptable release rates and concentrations. That is, cost-benefits may become important to the regulations. (The "benefit" part of the analyses is likely to benefit society, not necessarily a particular company.) Although the allowed levels of contaminant release are almost certain to remain sufficiently low to present no more than accepted risk, there is a growing recognition that it is not possible to always achieve zero release or zero risk of release. Under these real conditions, there is a practical level of effort that should be expended to "minimize" release or to bring the releases to lower and lower levels. It may be more effective to use the remaining funds to reduce further the emission of other pollutants that may be posing more risk. Risk analysis, if used properly, can suggest the most appropriate way to expend our efforts and resources to reduce dangers to human health and the environment. Risk analysis can, in principle, determine when it is more appropriate to start expending more effort on other problems rather than trying to reduce an individual emission much further.

Such risk-based regulations could make it possible for some facilities to release more of some contaminants, but there is no assurance that such analyses will not make the release of other contaminants even more restrictive. The merits of risk-based regulations are the opportunites to base all release regulations on comparable bases. Perhaps the greatest potential benefit from expanded use of risk analysis could be the increased focus of regulations and waste and environmental treatment efforts on the most important problems, problems that were not necessarily obvious from less rigorous approaches. Risk analysis will have to be at least somewhat specific to the site, and requirements for risk analysis are likely to complicate the selection of waste treatment methods for some contaminated sites.

The availability of reliable risk analysis methods will be critical to the success of any such changes in regulations. Risk analyses always predict a probability of damage to the public or the environment, but it is also important to know how accurate and reliable those probabilities are. If more reliance is to be placed on risk analyses, we will certainly need reliable methods and information to make the risk assessments; otherwise, there could be an increase in "risk" by greater use of risk analyses.

Of course, the present regulatory system is ultimately based upon a concept of risk, but the regulators have approached the problem in a more generic way to set release restrictions that will apply to all sites, and the pathways through which contaminants endanger human health or the environment appear to be the most conservative available. Regulations appear to have been developed by separate individual studies that do not necessarily use the same assumptions and techniques. It is also not clear that political considerations (usually through the level of public concern) do not sometimes override scientific evaluations of the problems. The proper use of risk analyses could put environmental regulations on a more rational basis and could provide better regulation by highlighting problems with specific sites and at reduced cost, if it is possible to reduce the restrictions on some sites. On the other hand, risk-based regulations are likely to restrict releases of contaminants at certain sites that are particularly sensitive because of large local populations, endangered specie, or unfavorable geology. The restrictions at those sites could become significantly more difficult to meet than the current regulations. Of course, if the regulations are based upon valid risk analyses, such restrictions would be justified.

There is also some consideration in Congress to change the liability basis for environmental cleanup. Although there is no way to be sure what changes, if any, will result, the complaints with current regulations appear to focus on the practice of holding each contributor to a contaminated site potentially liable for the entire cleanup effort and cost, regardless of the amount of material they added to the site or the toxicity of the material they added. This clause is viewed as grossly unfair to many, but any changes in regulations to limit any organizations responsibilities to the portion of the contamination they contributed will certainly leave some difficult questions to answer. How does one determine how much each organization is responsible? Who is responsible for the contamination for which no source is identified. There are also questions of who will pay for cleanup of contamination created by organizations that no longer exist or that cannot possibly (with their total assets) pay their share of the cleanup.

Resource Conservation and Recovery Act

The Resource Conservation and Recovery Act (RCRA) is usually the principal regulation for solid wastes. Although the act is intended to encourage conservation and recovery of resources and does do that, its regulation of the storage and disposal of solid wastes is often the major reason for its importance. "Hazardous" wastes fall into one or more of four categories based upon their ignitability, corrosiveness, reactivity, or toxicity. The RCRA regulates both the disposal and storage of hazardous wastes. Its objectives include prevention of long-term on-site storage that could eventually result in a legacy of wastes stored at abandoned industrial or waste storage sites. Waste must be treated properly and removed to approved storage facilities or disposal facilities in a timely manner.

The RCRA classifies wastes as hazardous if they come from certain sources (listed wastes) of if they have certain characteristics (wastes that are hazardous by characteristics). A waste can be hazardous in any of several ways.

Ignitable wastes are liquids with a flash point lower than 60°C (as determined by the Pensky-Martens closed-cup test or an approved equivalent test), gases that can be ignited, wastes that are strong oxidizers, and solid wastes that are capable of starting a fire through friction or from spontaneous chemical changes at standard temperature and pressure.

Corrosive wastes are usually liquids. They can have a pH less than 2 or greater than 12.5. To be noncorrosive, the liquid must also not corrode steel (SAE 1020) at a rate greater than 6.35 mm/yr at a temperature of 55°C.

Reactive wastes are those that are unstable enough to undergo violent changes (even without detonation), react violently with water or release toxic gases upon reaction with water, or capable of detonation from an initializing force.

Toxic wastes are those that can be leached in the environment and release toxic components in concentrations greater than specified limits. Environmental leaching is simulated by a standard leach test specified in Appendix II of 40CFR261 or by an equivalent and approved method. The allowable concentration limits in the leach liquor from the standard test are given in Table 1 of 40CFR261.24. The concentrations allowed in the leach liquor are usually tied to the drinking water standards associated with the Clean Water Act (Table 1). The toxicity standards appear to be the most important aspects of most solid wastes, and this book will focus somewhat on the toxic materials in liquid and solid waste streams.

In the RCRA, the EPA also defined certain wastes as hazardous simply because of their source and/or the presence of certain components. These are called "listed" wastes because there is a "list" of such wastes. This is explained in Appendix VIII of 40CFR261. There are four "lists" of hazardous waste, each designated by an F, K, U, or P. List F includes hazardous waste from nonspecific sources but with specific hazardous components. List K includes wastes from specific sources. See 40CFR261.32. Lists P and U include discarded products, off-specification products, and spilled or stored residue from these sources. List P includes the most acutely hazardous wastes, and list U includes the others. Table 1 gives some of the wastes that fall into each of these lists. Details of the listed wastes are given in the regulations (40CFR261.32).

If a waste is not included in any of these lists, it can be nonhazardous or hazardous by characteristics. If a waste is not "listed," it can be still be hazardous if it fails one or more of the criteria given earlier for defining hazardous wastes (hazardous waste "characteristics"). For instance, the waste could fail the standard environmental leach test. If a waste is listed, it is declared hazardous regardless of its characteristics or performance in any of the specified tests. There is often confusion and doubt about the wisdom of declaring wastes toxic even if they do not show characteristics of hazardous wastes (although most listed wastes probably will show such characteristics). Although the intent of all aspects of regulations is not necessarily obvious, it is likely that the use of listed wastes is intended to place the burden of proof that a waste is not hazardous on the generator of the waste and to prevent the use of dilution as a tool to make a toxic waste nontoxic.

TABLE 1 RCRA "Listed" Wastes

F-Series: Hazardous wastes from nonspecific sources, except those excluded by other regulations such as 40CFR260.20 or 260.22)

- Spent selected halogenated solvents and solvents used in degreasing that contained initially 10% or more by volume of halogenated components (F001)
- Other halogenated solvents that contain 10% or more by volume of the halogenated solvents and still bottoms from the recovery of those solvents (F002)
- Solvent waste that contain before use 10% or more of selected nonhalogenated solvents such as xylene, acetone, ethyl benzene, MIBK, ethyl acetate, ethyl ether, etc. (F003)
- Solvents and residue that contain before use 10% or more of selected solvents such as cresols, cresylic acid, or nitrobenzene before use (F004)
- Solvents and residue that contain before use more than 10% of selected solvents such as benzene, toluene, methyl ethyl ketone, carbon disulfide, isobutanol, or pyridine (F005)
- Wastewater treatment sludges from electroplating operations, except for selected plating operations such as tin plating, anodizing of aluminum, zinc plating, aluminum plating, and some cleaning/stripping operations associated with plating these materials (F006)
- Spent cyanide plating bath solutions (F007)
- Residues from the bottom of plating baths (F008)
- Spent stripping/cleaning bath solutions from processes where cyanide is used (F009)
- Residues from oil-quenching baths from metal treating where cyanides were used (F010)
- Spent cyanide solutions from salt bath cleaning from heat treatment of metals (F011)
- Wastewater from quenching after heat treatment of metals where cyanide was used (F012)
- Wastewater treatment sludges from chemical conversion coating of aluminum (F019)
- Wastes (except for wastewater and spent carbon from HCl purification) of tri- or tetrachlorophenol used in the production of pesticides or their derivatives (F020)
- Wastes (except for wastewater and spent carbon from HCl purification) from production of pentachlorophenol or intermediates (F021)
- Wastes (except for wastewater and spent carbon from HCl purification) from production of tetra-, penta-, or hexachlorobenzenes under alkaline conditions (F022)
- Wastes (except for wastewater and spent carbon from HCl purification) from production of tri- or tetraphenols (F023)

(continued)

TABLE 1 *(Continued)*

Wastes from processes using free-radical-catalyzed processes such as the production of chlorinated aliphatic hydrocarbons (F024)
Wastes (except for wastewater and spent carbon from HCl purification) that were produced in equipment used previously in the production of tetra-, penta-, or hexachlorobenzene under alkaline conditions (F026)
Discarded formulations using tri-, tetra-, or pentaphenols or materials derived from these compounds (F027)
Residues from the incineration or thermal treatment of soil contaminated by hazardous wastes of the type F020, F021, F022, F023, F026, or F027

K-Series: Solid wastes from specific sources
Sediments from treatment of wastewaters from wood-preserving processes that use creosote and/or pentachlorophenol (K001)
Wastewaters from production of any of several inorganic pigments (K002 through K008)
Distillation and stripper bottoms from the production of several hydrocarbons and halogenated compounds (K009 through K020, K022 through K027, K029, K030, K083, K085, K103, and K136)
Certain spent aqueous and solid catalysts (K021 and K028)
Certain wastewaters, condensate, and adsorbents (K111 through K118)
Brines and inorganic sludges (K071, K073, and K106)
Various wastes from the production of pesticides (K031 through K042, K97 through K99, and K123 through K126)
Wastes from the manufacture of explosives (K044 through K047)
Sludges, floated material and emulsions, and heat-exchange cleaning residue from petroleum refining (K048 through K052)
Furnace exhaust dust and pickle liquor from the steel industry (K061 and K062)
Dust or leach liquor from lead treatment (K069 and K100)
Wastes from the production of veterinary pharmaceuticals (K084, K101, and K102)
Wastes from ink production (K086)
Ammonia still lime sludge and decanter tank tar sludge from coking operations (K060 and K067)

P-Series: Discarded commercial chemicals, container residues, spill residue, and off-specification chemical products
More than 100 chemicals are listed by the EPA, including organic and inorganic compounds.

U-Series: Commercial chemical products, off-specification products, and chemical intermediates
A long list of chemicals and intermediates is given by the EPS, most of which are organic.

The EPA provides a procedure for "delisting" a listed waste [4]. Delisting is not a simple procedure and is only worth attempting when large volumes of waste are involved. Delisting, of course, involves showing that the waste does not fail any of the tests for hazardous wastes. Because the waste has been considered hazardous initially, delisting requires proof that the waste does not fail any test and that the waste used in the tests is truly representative of the waste being delisted. One certainly should seek delisting of any waste that is truly not hazardous, but one should not expect the delisting to be a quick procedure.

Drinking Water Standards

Drinking water standards were issued under the Safe Drinking Water Act (40CFR141) and are important to waste and environmental treatment

TABLE 2 National Primary Drinking Water Standards

Compound	Allowable concentration (mg/L)
Arsenic	0.05
Barium	1.0
Cadmium	0.01
Chromium	0.05
Fluoride	4.0
Lead	0.05
Mercury	0.002
Nitrate (as nitrogen)	10.0
Selenium	0.01
Silver	0.05
Endrin	0.0002
Lindane	0.004
Methoxychlor	0.1
Toxaphene	0.005
1,4-D (2,4-Dichlorophenoxyacetic acid)	0.1
2,4,5-Tp (Silvex or 2,4,5-trichlorophenoxypropionic acid)	0.01
Radium	5 pCi/L
Gross alpha radioactivity	15 pCi/L
Turbidity	1/turbidity unit
Coliform bacteria	1/100 mL

problems because they affect the concentrations that need to be reached by treatment of water effluents and groundwater and because the concentrations of toxic materials in the standard environmental leach test approved for the RCRA are tied to the drinking water standards. Allowable concentrations in discharge waters will not always be required to be as low as the drinking water standard because of subsequent degradation or dilution of contaminants before it is reasonable for anyone to drink the water, but the allowable release concentrations are still likely to be affected by the drinking water standards. There are even a few cases where it may be necessary to reduce contaminant concentrations to levels below the drinking water standards, such as when surface discharge of the water would be more harmful to fish than to humans. In those cases, one can think that the discharge standards will be set by standards other than the drinking water standards.

The drinking water standards are subject to change as more materials are found to be toxic, but the National Primary Drinking Water Standards are shown in Table 2. This list tends to include some of the more common acute toxic materials. There is also a Secondary Drinking Water Standard shown in Table 3 (see 40CFR143.3). Note that the secondary standards include less specific materials, such as those that cause color or odor, total dissolved solids, corrosiveness, and all foaming agents. Several materials are included that are common in most waters, but whose concentration should still be limited.

TABLE 3 National Secondary Drinking Water Standards

Material	Allowable concentration (mg/L)
Chloride	250
Color	15 color units
Copper	1
Fluoride	2
Foaming agents	0.5
Iron	0.3
Manganese	0.05
Sulfate	250
Zinc	5
Total dissolved solids	500
Color	15 color units
Corrosiveness	noncorrosive
Odor	3 × threshold odor number
pH	6.5–8.5

NEW TECHNOLOGIES FOR WASTE AND ENVIRONMENTAL TREATMENT

In addition to growing in size and financial importance, the waste and environmental industries are still growing in scientific and technical ways. Considerable research and development efforts are devoted to waste and environmental treatment, to find new treatment methods and change the way the industry works. Greater use of separation methods is likely to occur as new technologies are developed to handle specific wastes and waste streams, and these new technologies may include significant innovations. There is some concern that the research and development in the industry are geared too closely to testing and demonstration and are not encouraging participation from companies that are most proficient in developing high technologies. On the other hand, there is also concern that many "high-tech" approaches will prove too costly for many applications, especially those that involve large volumes of soil and groundwater. This particular concern may be relevant to only some types of "high-tech" concepts, not to all concepts offered by large and/or advanced companies.

In some ways, the current approaches to environmental regulation have not encouraged development of new and innovative regulations. Regulations that require extensive validation before a technology can be accepted can also slow the development and application of new technologies; such requirements can be justified or not justified. Any regulation that slows the deployment of new technologies is likely to slow the growing importance of new separation methods. Essentially all industrial companies want to meet all environmental regulations and even to protect the environment and the health of their workers and neighbors. However, there is room to debate how much effort and, most importantly to businesses, how much money should be expended for these purposes. There is even a degree of fear in some industrial communities that as new technologies are developed they may be forced to spend considerable funds to install the technologies at their facilities even if the technologies at their facilities are, in their opinion, adequate. This fear comes from the requirements to use the "best available technologies." Although there are merits in installing new technologies, no company is pleased to find its newly installed "best available technology" no longer the "best" and must be replaced.

This concept of requiring the best technology is reasonable in a static industry, but because new waste and environmental technologies are constantly being developed, it can become impractical to replace its effluent control equipment too often if better technologies are developed rapidly, especially if the improvements are relatively small. The merits of installing new emission technologies depend upon the current environmental and health effects from the current technology and the degree of improvement that would result from installation of the new technology. It is possible that there will be little incentive for the user of the technology to develop improved waste treatment methods for existing facilities as long as the currently installed technology is viewed as adequate; the principal incentives to develop better technologies may lie with the government and equipment suppliers who would benefit from selling the new technology. Most companies will be more likely to consider the very best technologies for new facilities than for replacing equipment at older facilities. Even if it is somewhat more costly to install a newer technology, improved performance may reduce the likelihood that the company would have to make changes soon to meet new regulations. There are several notable cases where companies found that they could reduce emissions by recovering and reusing components. They sometimes actually reduced costs, and the changes were profitable as well as environmentally desirable. These cost reductions probably incorporate the cost of waste disposal that has risen so steeply during the past decade. However, companies may be less interested in installing better technologies to replace equipment that they and the regulators once (or even recently) viewed as adequate.

The difficulties in "proving" that a new technology is superior can also be a deterrent to development and application of better technologies. This problem is not unique with the waste and environmental industries; every technology developer must convince the customers that they have developed a truly improved product. The special difficulty for the waste and environmental industry is the need to convince both the customer companies and the regulatory agencies that a new technology is significantly better. The motives of these two groups can be different. It may not be difficult to convince regulators that a new technology with costs comparable to or greater than the existing technology is better and worth using, but it may be more difficult to convince the companies. On the other hand, it may be easy to convince the companies that a new technology that is less costly but gives comparable performance is an improvement, but regulators may be more skeptical.

According to some the regulatory agencies, and especially the EPA, stress enforcement of the laws over protection of the environment. They contend that it may be better to work more closely with some industrial groups such as those developing waste and environmental technologies. That would help bring new technologies into regulatory approval and thus speed the rate at which better technologies for treating waste and contaminated environment are implemented. However, that would also create the possibility for the agency to become a "captive" of industries. Then the public and the environment could suffer. Critics of closer association of the EPA with industry cite other regulatory bodies that play important roles in promoting the industry they regulate.

There is a fundamental difference between the EPA and most other industrial regulatory agencies that could make it less likely that the agency would become captive to industries. Most other regulatory agencies are responsible for single industries or groups of industries. In those cases, it may be in the interest of the agency personnel to have the industries they regulate flourish. Complete decline of the industry would eliminate the need for the agency, and thus the need for the regulators. The EPA, however, is responsible for essentially all industries as well as governmental agencies, individuals, and others who do or could pollute the environment. It may be easier for the EPA to remain independent of each individual industry while taking a more active role in helping the industries develop and test better environmental treatment methods. Although the EPA does play important roles in technology development and testing, its focus does appear to be much stronger on the enforcement part of its mission.

Of course, enforcement of good regulations does protect the environment, but the agency also serves the public interests when it properly helps industries to improve their protection of the environment through innovation and apply enforcement activities only when other approaches fail. It would be difficult to prove that the EPA and other regulatory agencies are approaching their tasks in the wrong way, but the reports that companies involved in programs such as the Superfund projects are spending more on legal activities than on solving the problems do suggest that something is preventing the most efficient use of the funds from either the government or industry for waste and environmental cleanup. However, such reports do not clearly show whose policies are at fault, and one should not quickly blame the EPA or some other government agency. There may be tendencies for companies to defend their position legally rather than solving the problems. It is difficult to determine if such practices actually save money in the long term. This could be just another example of the apparent trend in the United States to go to court first and to think or evaluate later.

The personnel decisions of regulatory agencies are also criticized by some who claim that regulatory agencies have too many legal staff and too few scientists and engineers who could analyze new technologies reliably and confidently approve new technologies. Again this criticism reflects different judgments of the regulators' roles of enforcement and assisting in improving environmental quality. Since most readers of this book are likely to be engineers or scientists, even the readers may hold some bias on this issue.

Nevertheless, the need for new and improved approaches to the treatment of wastes and environmental problems is so clear that new technologies are essentially certain to be sought and adopted. Otherwise, the cost of waste and environmental treatment will continue to rise at a rate greater than the gross national product and will become an increasing economic burden unless we become willing to accept less control and greater environmental hazards. New separation technologies will play major roles in the use of new technologies.

ORGANIZATION OF THIS BOOK

There are two major parts of the book. The larger part covers chemical separation methods, such as adsorption, gas absorption (and stripping), liquid-liquid extraction, membrane processes, etc. The second part covers physical separation methods, such as filtration and sedimentation. The chemical methods separate materials as individual molecules, while physical methods separate large groups of molecules such as solid particles or liquid droplets. Although the careful reader will discover that there is no clear distinction between several of the separation methods, and not even a clear distinction between some chemical and physical separations when macromolecules are involved, the topics will be covered in chapters that are most appropriate.

Although the molecular separation methods believed to be of most importance to environmental and waste applications are covered by complete chapters, a few other separation methods could have some applications and do deserve some mention. These methods are grouped with related separation methods into a separate chapter with other separation methods that are likely to have more limited applications in waste and environmental processing. In these cases, much less detail is given. In most cases, there will be no quantitative description of design methods for the separation methods that are covered only briefly, but the similarities to design methods for other separation methods are mentioned. The aim of these descriptions is to give the reader an understanding of how the method works and for what problems the method could be considered, not to prepare the reader to design systems using these less common methods. Even in these short descriptions of the less common separation methods, there is an effort to describe the limitations of those methods, especially the limitations for their use in waste and environmental processing. In some cases, the limitation on the method may be the inability of the method to compete economically with other methods at the moderate-to-high throughput usually required for most waste and environmental processes. Several separation methods are available that are effective, but are currently far too costly to use on high-throughput systems, such as many industrial effluent systems. However, with further development and larger-scale deployment, the costs for some of these methods may decline sufficiently that larger-scale applications to waste and environmental problems become practical.

The second part of the book is devoted to physical separations, i.e., separations of different phases. They could be separation of solid particles from gases or liquids, separation of different liquid phase, or separation of different solid phases. The best way to divide the separation methods between molecular and physical separations is not always completely obvious, and examples are pointed out where there can be difficulty in deciding whether a separation is molecular or physical. In those cases, the rationale for placing the separation method in one group or the other is described, but other authors may choose to divide the topics differently.

The largest part of the physical separations part of the book is devoted to filtration methods. Filtration is so common in most process industries, including waste and environmental processing, that it deserves first billing in this part of the book. There are two sections in the chapter on filtration. The first is devoted to "surface" filters, i.e., filters in which the filtration takes place largely outside the filter medium. The principal filtration occurs at the surface of the filter medium or in a filter cake that builds up on the filter medium. These are also sometimes called "cake filters" because the filtered particles accumulate on the filter surface and form a "cake." This is probably the most common form of filtration and certainly the most easily recognized to most people.

The second filtration section is devoted to "deep-bed" filters. Those are filters in which the particles are removed largely within the filter medium, usually within a bed of fibers or granules. There are many differences in filters covered in both of these sections; so each chapter is further divided to describe some of the variations in each broad class of filters. The design methods or considerations for each class of filter are different, but the design considerations for different variations of filters within each class are usually similar, and many of the design procedures will be the same or similar for all filters within each class. As with the molecular separation chapters, a few general considerations are given on conditions under which each class and variation of filter is most likely to be favored.

Within each part of the book, chapters cover the separation methods most common or most important for waste and environmental problems. The amount of detail provided in each chapter comes from the current and/or expected importance of the method to waste and environmental problems. The relative importance of the separation methods in waste and environmental problems is not expected to be the same as their importance in the current process industries, so the reader often will find methods covered in either more or less detail than is usually given in books of similar length that do not focus on waste and environmental problems. Of course, to distribute the pages in the book among the different separations methods, it was also necessary to predict which methods will grow in importance so they can be covered in more detail.

CONTENT OF THE CHAPTERS

Each major chapter describes a particular separation method, the "science" or basis for the separations, some of the more common types of equipment used for this separation method, the principles used in "process" design of the equipment (generally sizing of equipment and estimation of the performance of equipment), examples where the method can be used in waste and environmental problems, and a general discussion of the competing separation methods and the conditions under which this method is more likely to be the most practical.

Each major chapter begins with a general description of the separation method and a few waste or environmental applications. This early mention of one or more applications is intended to give the reader an immediate idea of the potential benefits of the method. This is followed by a brief description of the types of equipment that can be used and a description of design methods. Then the reader is better equipped to understand the conditions under which the method is most likely to be the preferred one to use. The chapters then usually close with mention of a variety of applications of the method to waste and environmental problems. The list of applications is not complete (and would not be likely to remain complete if it were), but it is intended to give the reader a chance to think of the conditions under which the method could be used. The list and description include applications in active use and potential applications that have been studied. Each chapter also includes a description of the most likely competing separation methods and the conditions under which the method being studied is believed to most likely be practical.

LEVEL OF DIFFICULTY AND USE OF THE BOOK

The reader is expected to be a graduate or an upper-level undergraduate in engineering or one of the physical sciences. Chemical engineers and chemists are likely to find the book easier reading. However, in many ways the book may be of more use to environmental engineers and others working on waste or environmental projects who have had less background in separation methods. It is important that the reader have some basic knowledge of chemistry, but no advanced chemistry is needed. An introductory course, such as freshman chemistry, will be adequate for most of the book, but there will be advantages in have a somewhat stronger chemistry background. Those with very strong chemistry backgrounds may find some of the discussions oversimplified, but that is a risk that must be taken if the book is to have a sufficiently wide appeal. The mathematics involved should not challenge those with engineering or physical science backgrounds. Generally, the standard undergraduate calculus background will be fully adequate; in most cases, basic algebra will be adequate. There are references to more advance mathematical methods, but it is usually not necessary to understand the more complicated details to use the results.

Readers are expected to have an interest in waste and environmental problems, and often may have the responsibility to select or design separation treatment methods for the problems. It will be best if the book can be first studied as a whole since that is the only way to develop a basis for comparing separation methods and making selections. However, each chapter can be used as a reference book where the reader can study selectively those methods in which he or she is interested. Most chapters discuss competing methods when they are most likely to be serious competition to the method being discussed. Then readers can decide if they need to study more than one chapter for a given problem. It is anticipated that most readers would first read the book and then use it as reference material and review chapters pertaining to separation methods that may be of importance to the problems of concern at the moment. These cross references to competing methods may be one reason for using the book as a reference source. Many environmental engineers and others working on a variety of waste and environmental issues may find it difficult to remain expert in all of the separation methods of potential use and will want to refer to the appropriate chapters occasionally. The level of detail is considered appropriate for an overview, a review, or a "brush-up" on selected separation methods. The chapters are short enough to be reviewed relatively quickly.

The design procedures are given for several reasons. First, they provide insight into what is important to the size, cost, and performance of equipment. This insight is needed by those who select separation methods and by those who purchase equipment that others design. Some familiarity with the procedures is needed just to evaluate the work of others who may actually be designing and/or supplying the equipment. In many chapters, the description of design procedures is adequate for actual design of equipment. The reader designing separation equipment is encouraged to use computer procedures when available, but they are also encouraged to understand the procedures before relying upon the computer packages. Such understanding can be important in evaluating results of calculations and seeking alternative parameters, perhaps even other design specifications. Even when equipment is supplied by outside vendors, the buyer needs to understand how it works.

Those involved in research and development, consulting, or technical sales for a variety of separation methods may want to use the book in a similar manner, but those involved only in specific separation methods are likely to find that the chapters on those particular methods do not contain sufficient detail. Those people will need to be familiar with some of the most detailed specialty tests devoted to single separation methods or even specific aspects of those methods. Those involved on the leading edge of research and development need to remain aware of even the latest developments and should remain familiar with current journals and reports as well as current specialty books; no book can be or remain adequate to meet all of their needs. Such current information seldom can be obtained by books alone. However, even those actively involved in developing advances in individual separation methods may find the coverage of the other sections helpful to explore and evaluate alternatives to their methods. In developing any new technology, it is certainly useful to understand the competition as well as to know what standards the technology must reach.

REFERENCES

1. Pollution Prevention Act of 1990. *42 U.S.C. section 13101–13109* (1990).
2. Browner, C. M. *EPA J.*, *19*, 6 (1993).
3. Anastas, P. T. and T. C. Williamson (Eds.). *Green Chemistry: Designing Chemistry for the Environment*. ACS Symposium Series 626, Washington, DC (1996).
4. U.S. EPA, *Petitions to Delist Hazardous Wastes—A Guidance Manual*, EPA/530-SW-003 (1985).

2 Adsorption and Ion Exchange

Adsorption is the removal of a component from a fluid by physical or chemical attachment to a solid. Ion exchange is the substitution of one ion in or on a solid for an electrically equivalent number of ions from a solution. Although adsorption and ion exchange are different chemically, they are treated together in this volume because their operations are similar in so many ways. They both usually involve pumping a fluid through a bed or column packed with solid adsorbent or ion exchange materials. They are both important operations for removing toxic materials from liquid or gaseous wastes, especially for those components at low concentrations. Adsorption and ion exchange are becoming so important in waste and environmental processing that they are treated with considerably more detail in this book than in most other books on general separation topics, but, of course, even more details can be found in books dedicated entirely or primarily to adsorption operations. This importance of adsorption in these applications results largely from the ability of adsorption processes to remove components from very dilute fluids (gas or liquid) economically, and the importance results partially from the ability of adsorption beds to incorporate large numbers of separation stages (or transfer units) into modest size equipment. Adsorption can also be important in the removal of valuable materials from waste fluids. Even in cases where other methods are more economical for removing the bulk of the component(s), adsorption or ion exchange may be the best choice for a "polishing" step to remove the last traces of the component(s).

Adsorption and ion exchange are particularly attractive for removing dilute components, a common problem in pollution control. Usually, the adsorbed material, toxic or valuable, is recovered in a concentrated form for disposal or reuse. The solid adsorbent can often be regenerated for reuse, and the adsorbed or exchanged component(s) can be recovered in a concentrated form. When the toxic component(s) can be recovered alone or in a concentrated form, waste disposal costs are reduced because the volume of waste sent to disposal is reduced. The merits of adsorption depend upon its ability to reduce the concentration of the toxic components in effluent streams (often air or water) many-fold, sufficiently to permit release of the fluid, and upon its ability to concentrate the toxic materials sufficiently that disposal, destruction, or subsequent use of the materials will be less costly.

Current regulations generally give little financial incentive for removal of toxic components from waste streams unless the concentration is reduced sufficiently that the fluid can be released or that entirely different disposal methods can be used. This means that the concentration of the toxic component must be reduced below some specified value. Because adsorption and ion exchange are usually most effective with dilute solutions (liquid or gas) and can reduce concentrations many-fold, they are attractive for use as "polishing" steps to reduce the concentrations of contaminants to acceptable levels, perhaps after another approach has removed the bulk of the contaminant.

Adsorption also can be used to fix toxic materials in forms that are not hazardous simply because they are not easily removed from the solid. In those cases, the adsorbent may be part of a "waste form," and the adsorbent and toxic materials may be sent to disposal together. It still may be necessary to incorporate the adsorbent into an additional waste form such as concrete grout, but the adsorbent can help hold the contaminant in the waste form and thus meet EPA leaching restrictions. If the toxic material is combustible, it could be removed from the adsorbent and destroyed by incineration. The adsorbent could also be burned, or the adsorbent could be regenerated by burning off or otherwise destroying the contaminant. (Note that there is considerable public concern with incineration and often severe limitations on its use.) The adsorbent loaded with toxic contaminants could be sent to a landfill without further treatment if leaching of the toxic materials is sufficiently slow or if the landfill is protected to recover and prevent release of leached toxic materials. Otherwise, as noted, the adsorbent could be incorporated into a grout or another solid form with zero or extremely slow release rates, rates that pass EPA leach tests.

ADSORPTION EQUIPMENT

Adsorption operations usually involve passing the fluid over a static fixed bed of "adsorbent" particles until the bed is sufficiently loaded that the sorbate (solute or contaminant) can no longer be removed effectively

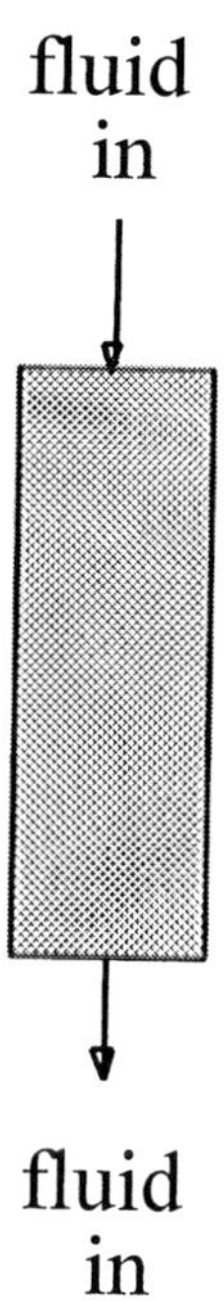

FIGURE 1 Adsorption bed showing the transient nature of a typical adsorption operation. Adsorbate accumulates in the bed with time.

(Figure 1). This transient type of procedure is only one way to operate an adsorption separation, but it is by far the most common procedure for bulk (high volume) processing operations such as those that are important in waste treatment. The feed stream is fed constantly to the adsorption bed, and the treated fluid exits the other end of the bed with a greatly reduced concentration of the adsorbed solute. There is thus an accumulation of the solute in the bed. The solute is adsorbed initially near the top (inlet) of the bed, and solute moves down the bed as the concentration on the adsorbent accumulates and approaches the concentration that is in equilibrium with the feed stream (Figure 2). Figure 2 shows the idealized accumulation for a "favorable" (to be defined later) adsorption. The upper diagram shows the accumulation in the bed at different positions for three times after adsorption begins. The diagram shows the concentration in the fluid, but this concentration (as will be seen) is approximately proportional to the loading of solute on the adsorbent. Similarly, after time 2, the front will have moved farther down the bed. The lower diagram

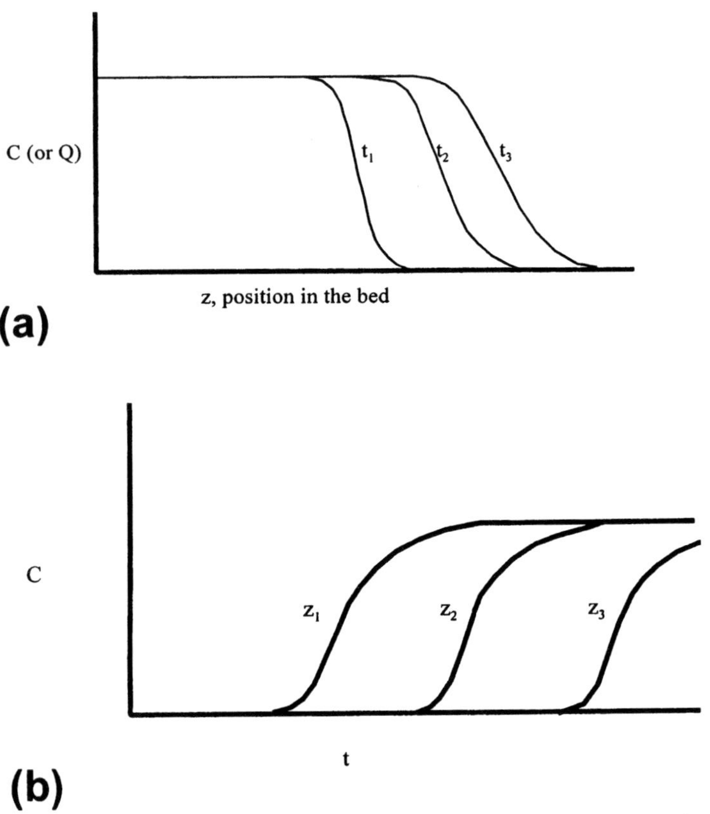

FIGURE 2 Movement of concentration fronts down an adsorption or ion exchange bed: (a) Concentration profile within the bed after different times; $t_1 < t_2 < t_3$. (b) Appearance of an adsorption front plotted as concentration seen as a function of time at fixed positions in the bed; $z_1 < z_2 < z_3$.

shows the concentration in the fluid exiting the bed as a function of time. Note that there is no solute in the fluid until breakthrough occurs. Then the concentration rises to the feed concentration. This is the type of operation covered most extensively in this book. In more realistic cases, the leading edge of the front in the bed and the breakthrough curve exiting the bed look like S-shaped curves rather than vertical steps.

Since adsorption operations usually are not steady-state operations, the adsorbent bed must be replaced or regenerated after it is used (sufficiently loaded). Steady-state operations with some solid adsorbents have been proposed and used, but they are not common and will be mentioned only briefly. When lower removal efficiencies can be tolerated or when the adsorbent has a particularly strong affinity for the toxic adsorbate, fluidized beds of adsorbent or even stirred tanks can be used, and the adsorbent can be removed and added continuously. These types of operations are generally limited to relatively low removal efficiencies because mixing (and fluid bypassing) allows some toxic materials to escape the adsorbent. One additional important merit of fluidized adsorption beds is their ability to handle small but significant quantities of solids in the feed; packed beds can tolerate only very small quantities of solids in the feed. Stirred tanks of ion exchange resins have been used to recover minerals from slurries of leached ores, where there are especially high concentration of solids in the feed.

The limited number of separation stages in fluidized or stirred beds of adsorbent or ion exchange material results from the mixing of both the fluid and the particles in a fluidized bed; this can severely limit the performance of some systems. However, one can increase the performance of fluidized beds in several ways. The fluidized bed volume can be divided into several smaller fluidized beds operating in series. In such cases, it may be possible to make each smaller bed perform approximately as a stage. A similar effect can be achieved by using adsorbent particles with two sizes. If the sizes of the particles differ sufficiently, the fluidized bed will separate the adsorbent particles with the smaller particles above the larger particles [1]. This makes the bed function approximately like two beds in series. It is also possible to add magnetic particles to the bed and to apply a magnetic field after the bed is fluidized. The magnetic field restricts or prevents motion of the magnetic particles and thus holds the other (adsorbent) particles in approximately stationary positions in the bed [2]. This can make the bed function somewhat like a "fixed bed" but with the bed expanded sufficiently by the fluidization to allow a lower pressure drop.

Continuous Operations

Other approaches to continuous adsorption or ion exchange involve continual or incremental movement of a packed bed of adsorbent or of an adsorbent fabricated in a different form such as a continuous belt, a disk, or a wheel. Such operations have found practical uses, but they are far less common than the simpler packed beds. Other "continuous countercurrent" systems have physically moved the adsorbent or ion exchange material. The solids can move as fixed beds as in the Higgins column, which has been used for continuous ion exchange processes, or can move as a series of fluidized beds which flow from one part of the system to another. Keller recently reviewed adsorption and ion exchange equipment [3]. In other continuous methods, the adsorbent is constructed as a porous monolith rather than as a bed of random spaced adsorbent particles, and the monolith can be moved mechanically in the direction countercurrent to the fluid. The porous monolith can be constructed as a "wheel" of adsorbent. Similar mechanical movement of solid adsorbents also can be achieved by transport of baskets or other porous particle containers countercurrent to the fluid. "Simulated" steady-state operations can be approximated by using several transient packed beds by periodically changing valving so the feed, elution, and product enter/leave different beds sequentially [4].

However, the following discussion will deal largely with the more common "fixed" bed operations which are far more likely to be used in waste and environmental treatment. There are the dominant systems in industry overall, and they are expected to remain so for a considerable time.

Elution Chromatography

Elution chromatography is a different way to operate an adsorption process. An adsorption/elution chromatographic method commonly used in analytical separations first involves placing a small sample of material to be separated in the upper portion of the column and then flushing (eluting) it down the bed. The different components of the sample are eluted down the bed at different rates and can be detected sequentially as they are eluted from the bed. Since elution chromatography is not commonly used in bulk waste treatment, it will not be discussed extensively. More details on elution chromatographic separations are usually found in textbooks on analytical chemistry. Gas chromatography is a form of elution chromatography that has become a standard method for analyzing multicomponent gas mixtures. Liquid chromatography is of growing importance in analyzing multiple solutes in liquid streams, especially solutions of biotech materials.

Elution chromatography is usually used to separate small quantities of materials, and it is usually limited to analytical chemistry and the separations of quite valuable materials such as those from biotech operations. Elution chromatography is a relatively slow operation that is not suitable for high volume and/or low value materials. In waste and environmental operations, elution chromatography is more likely to be used as an analytical chemistry tool than for bulk chemical separations.

TYPES OF ADSORBENTS

Materials used as adsorbents can be categorized in several ways. The adsorbents can be natural materials, usually with irregular granular shapes, or specially manufactured materials, in granular or specifically selected shapes. Even this simple categorization has problems because the "natural" materials can undergo various degrees of treatment before reaching their most effective form as adsorbents. Alternatively, adsorbents could be classified as organic materials or inorganic materials, as hydrophilic materials or hydrophobic materials, or in many other ways.

Perhaps the most common adsorbent materials are activated carbons, treated natural adsorbents. This is a high surface area carbon product usually prepared by charring a cellulose-based starting material. One of the oldest and most interesting applications of charcoal is in the removal of undesirable compounds from whiskeys and related alcoholic drinks. Scotch whiskey and U.S. bourbons are "aged" in partially charred barrels so undesirable compounds can be adsorbed by the charcoal on the sides of the barrel. Of course, other reactions also take place during the aging process, but the adsorption contributes to the desirable taste of the final product. The carbon can also be used in more conventional packed beds. One well-known domestic producer advertises its "trickle bed" adsorber used in manufacture of its "sipping whiskey." Other untreated natural materials such as coal, wood, and various minerals are also used as adsorbents. The most effective adsorbent materials are usually high surface area materials with either specific pore or cavity sizes or with specific active surfaces.

Numerous manufactured materials are used as adsorbents. Adsorbent materials produced in large volumes include silica gels, synthetic zeolites, and forms of cellulose. Note that cellulose-based adsorbents could be regarded as natural or manufactured, depending upon the amount of processing required. Many very effective, but relatively costly, adsorbent materials are entering the market for high value applications, principally for chemical analyses by elution chromatography. The financial opportunities in producing high quality adsorbents are high.

Adsorbents are also often categorized as physical adsorbents or as chemisorbents. This is simply a categorization based upon the strength of the bonding of the solute (the material being adsorbed and sometimes called the "adsorbate") with the adsorbent. Although a great many of the most important adsorption systems fall clearly into one of these two categories, some do not because the energy of adsorption may be between the values expected for physical adsorption and chemical adsorption. Physical adsorption is usually more common, especially for gas systems and when regeneration of the adsorbent by small changes in pressure or temperature is desired. Regeneration of chemically adsorbed materials may be difficult.

Ion exchange relies upon electrostatic forces to hold cations (or anions) on/within the solid "ion exchange material." However, the cations (or anions) in the ion exchange material can be replaced (exchanged) with other cations (or anions) in a surrounding solution.

EXAMPLES OF COMMON ADSORBENTS

Activated Carbon

Activated carbons, sometimes called granular activated carbons (GACs), are widely used for removing traces of nonpolar organic compounds such as oils and solvents from gas and water streams. Since control of hydrocarbon and solvent emissions and removal of oils from water discharges are common problems in waste and effluent management, these adsorbents have numerous applications in effluent treatment. They are able to reduce oil and solvent emissions in gas effluents to very low levels, and their applications are likely to increase further if emission requirements continue to be tightened. Carbon beds in air effluent streams (vent or off-gas streams) can also provide protection to the environment in situations when inadvertent discharges are made to the effluent stream.

The adsorption behavior of an activated carbon often results from the carbon surfaces and their normal hydrophobic (water repelling) nature. Essentially any nonpolar molecule will be adsorbed on an activated carbon, such as most common hydrocarbons, trichloroethylene, trichloroethane, dichloroethane, polychlorobiphenyls, etc., which are common pollutants. Activated carbons thus have numerous roles in environmental control because they remove organic pollutants from water as well as air effluent streams. The hydrophobic nature of carbon surfaces can be affected by reactions or chemical adsorption on the surfaces. Significant oxygen on carbon surfaces can decrease the hydrophobic properties. Even dissolved oxygen in water can affect the adsorptive properties of activated carbons for removal of certain organic contaminant, that is, contaminants that are likely to interact with oxidized sites. Nakhla and co-workers have studied the effects of dissolved oxygen on the removal of phenolic and creosolic compounds [5–8]. The capacity of the carbon surface for these contaminants is increased significantly by the presence of oxygen in the water. The increases could approach 50%. Heating the carbon in a vacuum can often remove the reacted or chemisorbed materials from the surfaces and restore the hydrophobicity or the original nature of the carbon surfaces.

Activated carbons are prepared by pyrolysis of natural carbon containing materials, usually woody materials. The woody material is heated to several hundred degrees Celsius in the absence of oxygen or in the presence of little oxygen. The material decomposes with only limited oxidation, and volatile compounds (water and other decomposition products) are driven off. The residue is a char that is mostly carbon. It will, of course, retain most of the inorganic mineral elements that were present in the original woody source material. Coconut shell is one material that pyrolyses to a particularly good activated carbon. Activated carbons may have surface areas of 100 to 1400 m^2/g and are especially useful for removing hydrocarbons, including benzene and toluene, from air [9]. The surfaces of activated carbons can be altered by treatment or even preadsorption of selected materials on the carbon. With preadsorption of some materials, the selectivity of the carbon for different adsorbates can be altered greatly. A useful recent review of activated carbon adsorption and mathematical equations for designing adsorption beds has been written by Crittenden et al. [10].

Activated carbons are sold in powdered and granular forms. The powdered form is used for single contact (single stage) operations, usually in some water treatment operations when high mass transfer rates are needed and when extremely high removal efficiencies are not needed. The powdered carbon is usually removed by filtration and discarded after a single use. Thus powdered activated carbon applications may be carried out much like precipitation operations, and even sometimes in conjunction with precipitation operations which remove inorganic contaminants while the carbon removes the organic contaminants.

Granular activated carbons most often come in size fractions between 8 and 30 mesh or between 12 and 40 mesh. Granular carbons can be regenerated and reused several times. For carbons used to remove organic contaminants, thermal regeneration may be sufficient, but it is common to introduce superheated steam or even some air to the regeneration furnaces/ovens. Both tray ovens and fluidized beds have been used to regenerate activated carbons. Temperatures in the regeneration ovens or beds are usually staged to do as much drying as possible at lower temperatures. The highest temperatures required depend upon the temperature needed to remove the particular contaminant(s) on the carbon. However, since so much of the activated carbon is regenerated in commercial furnaces at firms supplying regeneration services to several users, much of the carbon will be regenerated at rather high temperatures (near 1000°C) commonly used by commercial regenerators. Part of the carbon will be destroyed by the chemical treatment (essentially from burning), but attrition is probably the major cause of carbon loss during regeneration operations. Since regeneration usually takes place in separate ovens designed just for regeneration, it is necessary to "slurry" the carbon from the bed and transport the carbon to the regeneration ovens and dryers.

This physical handling of the carbon increases the fines content, which is most likely to be "burned off" or lost during regeneration. Regeneration loss may be as much as 10%, even with well-operated systems [11]. Somewhat higher losses have been reported for other systems.

Regeneration usually affects the properties of the carbon. As noted, part of the carbon is effectively burned, and regeneration is likely to increase the pore sizes and even the surface area of the carbon. These factors can enhance the performance of the carbon, and in many cases regenerated carbon may perform better than the original carbon. However, regeneration can affect surface chemistry of the carbon, producing unfavorable changes in carbon performance. Oxidation and reactions with steam can lower the carbon hydrophobicity, which usually needs to be restored before the carbon is put back into use. Restoration is usually done through a final stage of regeneration in which the carbon is heated in the presence of low concentrations of air (oxygen) or water. Generally, one can expect the capacity of carbons to be near 90% of the original capacity after regeneration.

As noted, regeneration can be performed in house, or the spent carbon can be sent to a commercial firm that specializes in regeneration of activated carbons. Such firms are often the original suppliers of the carbon. The choice of going off site to have the carbon regenerated is usually made when there is not enough supply of spent carbons on site to justify a separate regeneration system. One estimate gives approximately 100 kg/day as the regeneration rate when one is likely to find on-site regeneration more economical [12]. For smaller adsorption systems that require a lower rate of carbon regeneration, off-site regeneration is more likely to be the preferred and more economical choice. When considering on-site regeneration systems, remember that extensive off-gas treatment may be required for the regeneration ovens, and the capital costs may be relatively high. Although there may be several environmental/waste applications where one would want to regenerate carbon on site, there will probably be far more cases where carbon beds would operate on dilute streams for relatively long times between regenerations, and there will not be enough regeneration needed to justify an on-site regeneration facility.

Organic Polymers

Some organic polymers have been used as adsorbents for organic vapors or dissolved organic compounds. Many of them are polystyrene-divinylbenzene (DVB) copolymer spheres much like the basic support structure of ion exchange materials that will be discussed later. However, for adsorbing organic compounds it is not necessary to form the ionizable groups on the polymers that give the ion exchange properties. Some of the applications for polymer beads are similar to those for activated carbon. Because of the extremely high surface area of some activated carbons, polymer beads are more likely to have lower total capacities than many activated carbon adsorbents [13,14]. The principal advantage of polymer adsorbents is the ease of regeneration and the relatively small loss in adsorption capacity with each regeneration cycle. One study of chlorobenzene and benzoic acid removal from water using Dow Chemicals XUS resin beads showed capacities to a few hundred grams per liter of bed (determined from the 50% breakthrough curve, an approach to estimating adsorbent capacity that will be discussed later) when removing the contaminants from concentrations of 250 to 300 ppm in water. Both contaminants could be removed from the polymer with a small volume of methanol, approximately 10 to 20 times the volume of the adsorption bed.

Since polymer adsorbents are usually hydrophobic, they are often difficult to wet when they are used to remove organic contaminants from water streams. A pretreatment with methanol can allow the polymer adsorbent to be wet by water, and the methanol can be displaced by the organic contaminant that will be removed by the polymer subsequently. Note, however, that there will be a release of a small amount of methanol to the water effluent if this approach is used.

Equilibrium adsorption of organic contaminants by organic polymers can be complex and may not follow the behavior predicted by the more common isotherms used for more rigid adsorbents (Langmuir, Freundlich, etc.). The different behavior can be even more different in multicomponent systems. The different behavior results at least in part from swelling of the polymer particles as the organic contaminant loading increases. While the presence of a second adsorbate usually decreases the adsorption capacity of rigid adsorbents for the contaminant(s) of interest, the presence of a second adsorbate could swell the resin and increase the capacity of the polymer adsorbent for the contaminant of interest. This effect could be thought of as dissolution of the contaminant in the adsorbed second adsorbate. Some readers may notice that this could be called adsorption if the contaminant dissolves in the second adsorbate, or even in the polymer itself. In this book, this will still be considered adsorption since it involves removal of the contaminant to a solid, whatever the mechanism.

With suitable active groups attached to the polymers they can be used as ion exchange materials, but some of those materials can also be useful adsorbents. Polymers with amine groups can remove acid gases from effluent streams [15]. Of course, organic polymers are not stable at high temperatures, and they do not retain some acid gases at very high temperatures. Polystyrene-DVB polymers with amine groups can remove acid gases such as SO_x and NO_x at near ambient temperatures, and the gases can be removed by heating the adsorbent to temperatures approaching 100°C. Although SO_x can be removed from gas streams and regenerated effectively at modest temperatures with primary, secondary, or tertiary amines, NO_x is difficult to remove from resin adsorbents with either primary or secondary amines. Remember that carbon dioxide is also an acid gas and can be adsorbed on resins with basic groups such as amines. However, the adsorption of carbon dioxide on resins with tertiary amines is not great [16].

Carbon Molecular Sieves

Carbon molecular sieves are receiving increased attention, but their costs exceed those of the more common granular activated carbons. Like activated carbons, carbon molecular sieves are prepared by carbonization of carbon containing materials. The difference in the preparation is that the molecular sieves are prepared from chemical compounds (usually polymers) while activated carbons are prepared from natural materials. The homogeneity of the polymers results in a much more narrow range of pore sizes, and the pore size peaks at a significantly smaller size than the pores in activated carbons. The surfaces of carbon molecular sieves are hydrophobic, like those of activated carbons. Furthermore, since the carbon molecular sieves take the same shape as their polymer precursors, it is relatively easy to form them into a variety of shapes. Other than spherical or granular the most interesting shape is probably fibrous. Fibers can have small diameters and be woven into larger threads or even into fabrics.

The term "molecular sieve" describes the uniform and small pore sizes. The term was used earlier to describe zeolites and implies that the size distribution is narrow enough to "sieve" out larger molecules and thus permit only molecules under a given size to be adsorbed. However, the very small pores of carbon molecular sieves are still significantly larger than the pores of zeolites. Since many solutes have molecules significantly smaller than the pores of carbon molecular sieves, it is probably better to avoid thinking of them as sieves, but simply as high surface area carbon adsorbers with relatively uniform size and very small pores. Pore diameters can play important roles in the relative adsorption equilibrium of different contaminants and provide extremely high internal surface areas. For instance, one study of phenol compounds on fibrous carbon molecular sieves showed that 4-nitrophenol adsorbed more strongly than

4-chlorophenol, which adsorbed more strongly than phenol [17]. However, one earlier study [18] found the opposite order of adsorption preference. This could have resulted from a dependence of selectivity on pore diameter, but it is also possible that trace components or differences in the surface structure of the different materials also play a role.

Silica Gels

Silica gels are inexpensive adsorbents for removing water and other hydrophilic materials from gases. They are silica polymers dehydrated into an inorganic gel and are usually produced from alkali metal silicates. Silica gels have effective pore diameters of approximately 5 to 50 Å, usually with a peak in the size distribution near 20 Å. They are available in several particle sizes and are usually produced in approximately spherical shapes. The most common application for silica gels is in the removal of water, but that is not usually a major problem in environmental and waste processing.

With only modest difficulty and expense, however, other polar molecules can be attached to silica gels with polar or silane bonds. Such attached molecules can incorporate a variety of specific chemical groups which completely change the adsorptive properties of silica gels; some attached groups can even make the silica gel hydrophobic. Silica gels are also useful solid substrates for incorporating ligands that are highly selective for individual solutes such as metal ions. Many ligands can be easily attached to silica through silane bonds. These can become highly specific adsorbents or ion exchange materials capable of removing (largely) individual components or groups of components from complex mixtures. Some such treated silica particles are relatively expensive for large-scale operations, but with increased use and high production rates their cost could decline, perhaps considerably.

Zeolites

Zeolites are aluminosilicates with regular porous crystal structures [19]. Although zeolites occur naturally, a wider variety of zeolites are manufactured. All zeolites are constructed from a few crystal units, but the pore structures can differ considerably because the units can be assembled differently in the crystals. Aluminosilicate structures carry a net negative electrical charge that is neutralized by a cation. Further differences in zeolites can result when different cations are added to neutralize electrical charges on crystal unit surfaces.

The most common group of zeolites is type A. All type A zeolites have the same crystal structures, but they may have different cations on the unit surfaces. Type 5A zeolites have Ca^{2+} ions on the surfaces and pore sizes just under 5 Å. Type 4A zeolites have Na^+ ions on the surfaces and slightly smaller pores. The less common type 3A zeolites have K^+ ions on the unit surfaces and even smaller pores. Type X and type Y zeolites have more open structures and larger effective pore diameters of approximately 8 Å. More recently, more zeolite structures have been synthesized, usually with even larger pores. The initial use of a new zeolite is likely to be in catalysis if the costs are relatively high. However, if (or as) the costs of new materials decline, they may be considered for use as adsorbents.

Because zeolites have crystal structures and the pores are an integral part of those structures, the distribution of pore sizes is very narrow. Although each pore will be essentially like another, it is too simplistic to view the pores as straight cylindrical holes. A more realistic image of zeolite pores is a series of interconnected chambers with restricted openings. Although all chambers and openings may be identical, they are not described so neatly by a single pore diameter. The most important dimension is usually the size of the opening to the chambers. As noted, that dimension is affected when the cation that neutralizes the electric charge of the aluminate structure is changed. Different zeolite types have different pore dimensions. Type A zeolites have the smallest pores of the common zeolites, and type Y zeolites have the largest pores of the common zeolites. Some newer zeolites have significantly larger pores, but they are not commonly used as adsorbents. Effective pore size is determined by the structure (type) and the number of neutralizing cations in it, and the effective size of the pores can be altered by changing the neutralizing cations. For instance, one study found that the adsorption of some compounds by Na-Y zeolite could be altered significantly by preparing the zeolite with some sodium ions replaced by lithium ions [20].

Diffusion rates into the small pores of molecular sieves are usually very slow because of the small effective pore size, and it is necessary to use very fine zeolite particles to obtain satisfactory adsorption rates for most applications. However, small adsorbent particles cause an excessive pressure drop in packed beds when high processing (flow) rates are needed. To gain the advantages of both large adsorbent particles (a low pressure drop or high throughput) and small zeolite crystal sizes (high adsorption rates), commercially prepared zeolite adsorbents usually consist of very small zeolite crystals bonded into much larger overall adsorbent particles. The crystals are usually held together by a clay binder. Diffusion coefficients through the larger openings between the zeolite crystals are relatively high because the openings are much larger than the molecular dimensions of most adsorbing materials, and the pore walls do not greatly restrict molecular motion (diffusion). Diffusion coefficients within the zeolite crystals remain very small, but the diffusion paths are short because of the small crystal sizes. There will, of course, be a significant reduction in the effective diffusion coefficient even in the macropores between the crystals from the diffusion coefficients in the surrounding fluid simply because of the random direction of the pores and the presence of "dead end" pores. These effects are usually grouped into the "tortuosity factor," which decreases the effective diffusion coefficient approximately two- to six-fold below that observed in a free media.

Such adsorbents constructed with two size pores are often called "bidispersed" pore systems. That is, their behavior is governed by two pore sizes, namely the size of the opening between the zeolite crystals (held together with clay binder) and the size of the openings within the zeolite crystals. Under some conditions, the adsorption rates may be controlled by diffusion rates in the zeolite crystals (micropores); under other conditions, the rates may be controlled by diffusion through the larger (macro-pores) openings outside the crystals; and under still other conditions, the diffusion resistance in both pore systems may contribute to the adsorption rates. This will be discussed in more detail later.

The inorganic and ionic structures of zeolites make them particularly good adsorbents for polar materials, especially moisture (water) from gases and nonpolar liquids such as hydrocarbons and oils. They have such a high affinity for water that temperatures of approximately 200 to 400°C are required to desorb the water. The temperature and desorption time required depend upon the degree of desorption needed. For extremely effective regeneration, even higher temperatures and longer desorption times (hours) may be required. Such levels of desorption are needed when the zeolite is to be used for reaching extremely low moisture contents in the gas or liquid.

Since the strong affinity of zeolites for water makes desorption inherently costly in energy and time, zeolites are more likely to be used only where very high dryness is required; even then, they may be used only for the final polishing step. Type 5A and type 4A zeolites are especially effective in removing water.

As noted, drying is not one of the most common applications for adsorption in waste management and environmental management, but zeolites could also be used to remove other polar materials that contaminate gas or liquid effluents or products. However, remember that they are likely to remove water as well, so any moisture present will consume some of the adsorption capacity of the zeolite adsorbents operating on gaseous streams.

Zeolites can also be used as ion exchange materials because the cations neutralizing the net negative charge on the crystal faces can be exchanged. Use of zeolites in ion exchange will be described later.

Natural Products as Adsorbents

Several natural materials have been, or can be, used as adsorbents with little or no treatment. Some of the original adsorbents were natural materials, and synthetic materials became more commonly used as their properties improved. Eventually many of the synthetic materials became more effective than the original natural materials. As noted, the first zeolites used in adsorption were natural minerals, but synthetic zeolites have gained most of the high performance markets. Coal and plant materials have been used as adsorbents, and almost any biomass created by microorganisms can adsorb or undergo ion exchange with some metal ions. Useful adsorption by these materials can occur in situations when it is not even fully recognized. The adsorption of contaminants by the biomass in activated sludge digesters in municipal sewage plants is one example. There is a growing interest in making better use of the capabilities of low cost natural materials such as adsorbents, particularly for waste treatment.

In some ways it is difficult to draw a clear line between "natural" and synthetic adsorbents since many of the most effective and widely used adsorbents are "derived" from natural materials. As noted, activated carbon is prepared from wood or coconut shells. When a natural material undergoes significant processing or treatment, it is not clear if it should be considered as a natural or a synthetic adsorbent.

In many high performance applications, synthetic or highly treated adsorbents are likely to prove superior to untreated natural materials. However, when a very inexpensive natural material is sufficient, it should be considered. In recent years there has been interest in materials such as biomass and microorganisms (alive or dead) for removing materials such as heavy metals from aqueous wastes. As noted earlier, there has also been a growing recognition that adsorption on biomass can occur in other equipment, such as sludge digesters, which are not known principally as adsorbers, and affect its performance. Special biosorption equipment and operations also have been studied extensively, but few systems are known to be operating in the United States. However, there appears to be more interest in Europe [21]. It is difficult to know exactly why biosorption is not used more extensively in the United States. Expected difficulties with handling or disposing of the solid biomass after its use can be one problem. Incineration, rather than regeneration, appears to be an attractive way to destroy the relatively inexpensive spent adsorbent, but approval of incineration permits is difficult to obtain in the United States. The reliability (or lack of reliability) of living systems may also play a role in discouraging adoption of biosorption processes.

ION EXCHANGE MATERIALS

Solid ion exchange materials are insoluble solids with ionized groups located on the surface or in interior regions that can be reached by the fluid. For most practical purposes, the ion exchange systems discussed are used to exchange ions from aqueous (water) solutions. The ion exchange materials may be naturally occurring materials such as zeolites, a modified natural material such as sulfonated coal particles, or a synthetic material fabricated specifically for its ion exchange capabilities. Although the original ion exchange materials usually were derived from natural materials, the ability to synthesize materials with better properties has resulted in an industry dominated by the (usually) superior synthetic materials.

Synthetic ion exchange materials can be made of either organic or inorganic components. The most common materials are organic polymers (resins) with ion groups placed along the polymer chain. This permits the placement of numerous ions in the polymer and thus gives high "capacities" for ion exchange. The polymers are usually cross-linked sufficiently to make them essentially insoluble in water, even with the large number of ionized (polar) groups, and to give them structural strength. Without the ionizable groups, such polymers would be dense, and water would not penetrate the polymer particles significantly. However, when the ion groups are attached to the polymer, the particles swell from osmotic forces, and water penetrates the resin (polymer) particles. Water penetration permits the ions neutralizing the fixed charges on the polymer to move relatively freely and rapidly and thus to exchange places with other ions in the external solution.

It is common to think of exchangeable ions in the resin (polymer) being attached or associated with a specific charged group fixed on the polymer. We often write expressions to describe ion exchange that look like chemical reactions:

$$M_1^+ + M_2R \leftrightarrows M_2^+ + M_1R \tag{1}$$

where M_1^+ and M_2^+ are metal ions and R is a resin site with a charge opposite to that of the exchanging ions. (In this equation, both metal ions are monovalent.)

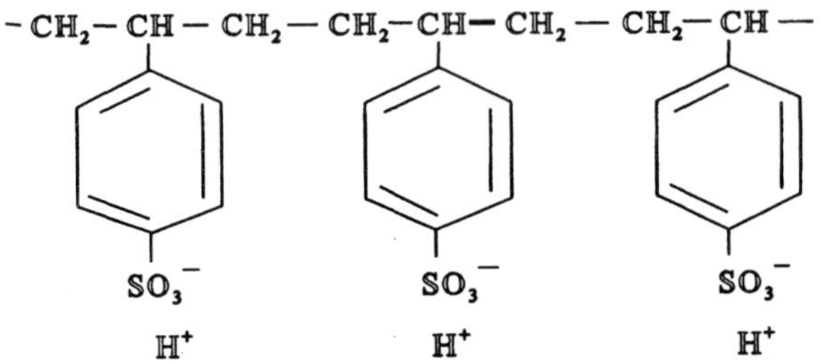

FIGURE 3 Chemical structure of a common polystyrene-based ion exchange resin with sulfonic acid groups, a strong acid cation resin.

In some cases, such as hydrogen ions neutralizing the charge of weak acid groups on the polymer, this may be a realistic view of the material: hydrogen ions are less free to "roam" in the water that penetrates the polymer. However, in many cases, especially those involving other neutralizing cations such as metal ions, the neutralizing (or exchangeable) ions may roam freely in the penetrating water. Then osmotic swelling is significant and limited only by the restraining force of the "stretched" polymer, which equals the osmotic pressure. In these cases, it is probably better to view the resin polymer as a concentrated solution with one set of ions free to migrate and the other fixed or restrained in spacial position because it is part of the polymer. Thus, a sodium ion in such a cation resin may have little more association with any particular negatively charged groups on the polymer than with any other negatively charged group on the polymer. Statistically, there will be a higher probability or concentration of mobile ions around the fixed ions of opposite charge, but individual mobile ions move randomly throughout the region by thermal motion.

An ion exchange material can be a cation exchanger (exchange cations) or an anion exchanger (exchange anions). Since the exchanging ions are mobile, a cation exchanger has negatively charged anion groups incorporated (fixed) in the polymer, and an anion exchanger has positively charged cation groups incorporated in the polymer.

One well-established and common group of synthetic organic polymer-based ion exchange materials is constructed of polystyrene polymers with divinylbenzene cross-linking. It is relatively easy to incorporate ionic groups into the polymer, usually on the benzene groups of the polystyrene (Figure 3). If sulfonic acid groups are incorporated, the material is a "strong acid" cation exchange material. The ion exchange capability results because the hydrogen ions from the acid are ionized, leaving the negatively charged sulfonate group fixed on the polymer. It is called a strong acid ion exchange material because the sulfonic acid is a strong acid and is, thus, essentially completely ionized over the entire pH range likely to be of interest, even at low pH (in acid solutions). On the other hand, if a carbolic acid group were incorporated in the polystyrene polymer, the material would be a "weak acid" cation exchange material. The hydrogen ions of carbolic acids can be exchanged, but with more difficulty. The hydrogen ions will not be completely ionized in low pH (acid) solutions. Thus, one basis for characterizing cation exchange materials is by the strength of the acid group incorporated in the polymer structure.

To create an anion exchange material, it is necessary to incorporate an ionizable basic group to the polymer. The ionized hydroxyl ion can then be exchanged with other anions in the solution. Amine groups are bases that are usually incorporated on polystyrene polymers. Again, the anion exchange materials may be classified as strong bases or weak bases, depending upon the strength of the base group incorporated in the polymer. Quaternary amines are strong bases, and the hydroxyl ions can be ionized at essentially any pH. Strong base anion exchangers are often classified as Type I or Type II materials. Both of these "types" of resins use quaternary amines and are considered "strong base" resins, but the two types use different components on the quaternary ammonium group. Type II resins are slightly weaker than Type I resins. However, if ternary amines (or even secondary amines) are built into the polymer, the material will be a "weak base" anion exchange material because these groups are weak bases. That means that hydroxyl ions will not be fully ionized at high (basic) pH.

The standard polystyrene-DVB resins and many other polymer resins are approximately homogeneous polymers that swell when placed in water or other aqueous solution. The degree of swelling depends upon the degree of cross-linking and the solution concentration. For weak acid or base resins, the swelling can also depend upon the concentration of hydrogen or hydroxide ion. For ions to exchange, they must diffuse through the swollen polymer structure. These diffusion rates can be relatively low because of the small spaces between polymer chains, that is, low compared to diffusion rates in water. Improved mass transfer rates can be achieved if larger pores can be created within the resin, and many of the more recent resins are constructed in this manner [22]. One common way to construct such resin particles involves first producing very small, and preferably uniform size, particles. These small particles can then be clustered into larger overall size particles. This results in two principal size pores: small pores within the small particles, and larger pores between the small particles within the clusters. This is similar to the structure of zeo-

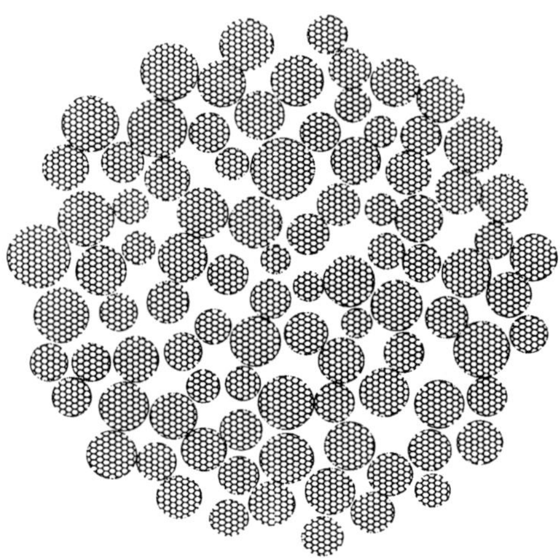

FIGURE 4 Micro-pore and macro-pore structures of adsorbents and ion exchange materials. Example structure of larger particles that are clusters of porous smaller particles. Micro-pores are within the smaller particles; macro-pores are spaces between the smaller particles.

lite adsorbents discussed earlier with binders holding together very small particles of zeolite crystals. Even greater porosity can be achieved by forming intermediate size clusters of the small particles and larger clusters of the smaller clusters. This results in at least three groups of pore sizes: pores within the small particles, pores in the small clusters between the small particles, and larger pores in the larger cluster between the smaller clusters (Figure 4). More recently there have been smaller quantities of relatively costly ion exchange resins constructed with the resin coating porous structures of rigid materials such as silica, but these materials are not likely to be useful for large-scale operations until their cost is reduced significantly.

Although the polystyrene-DVB-based polymers have been standard materials for several decades, manufacturers have made significant improvements in the structure of the materials and in the polymer and ionic groups used. Table 1 lists a few common organic-polymer-based ion exchange materials from major U.S. manufacturers. The type of resin refers to whether the resin is a strong acid, strong base, etc. All strong base resins are Type II unless specified in the table. Some of these materials may be used for a variety of ion-exchange-based separations, but others were developed for properties needed in specific applications.

TABLE 1 Some Exmples of Ion Exchange Resin Manufacturers and Resins

Manufacturer and resin name	Resin type	Comments
Dow Chemical		
Dowex 50	Strong acid cation resin Polystyrene DVB	Available in a variety of cross-linkage values and sizes
Dowex 1	Strong base anion resin Polystyrene DVB	Available in a variety of cross-linkage values and sizes
Dowex 2	Weaker base anion resin Polystyrene DVB	Available in a variety of cross-linkage values and sizes
Rohm and Haas		
Amberlite 200	Strong acid cation resin Polystyrene DVB	
Amberlite IRC76	Weaker acid cation resin Acrylic	
Amberlite IRC900	Strong base anion resin Polystyrene DVB	
Amberlite IRC93	Weak base anion resin Polystyrene DVB	
Sybron Chemicals		
IONAC C-249	Strong acid cation resin Polystyrene DVB	
IONAC CFP-110	Strong acid cation resin Polystyrene DVB	Macroreticular
IONAC CC	Weak acid cation resin Acrylic	
IONAC ASB-1	Strong base anion resin Polystyrene DVB	
IONAC A-641	Strong base anion resin Polystyrene DVB	Macroporous
IONAC AFP-329	Weak base anion resin Polystyrene DVB	
Purolite Resins		
Purolite C 100	Strong acid cation resin Polystyrene DVB	
Purolite C 105	Weak acid cation resin Acrylic	
Purolite A 400	Strong base anion resin Polystyrene DVB	
Purolite A 100	Weak base anion resin Polystyrene DVB	

This table should not be viewed as complete, and new resins are being developed annually. The length of the table should not be considered as an indication of the complete variety of resins sold by a vendor or the volume of resin they manufacture. The tables show examples and indicate the variety of ion exchange resins available. Before proceeding too far in developing an ion-exchange-based waste separation facility, it would be wise to consult the manufacturers to determine what new materials are available that may be more suitable for the application and to obtain data or run tests with the resins and solutions of interest.

Some resins come in a variety of particle sizes, but the most common size for larger-scale applications is about 0.5 mm in diameter, optimum for many applications. Smaller particles give higher mass transfer rates, but larger particles require less pressure drop for pumping the liquid through the beds. Smaller particles are likely to be desired for analytical chemistry and laboratory applications, especially for liquid chromatography. Large-scale industrial applications are likely to need somewhat larger particles to achieve high throughput.

One group of inorganic materials that have important ion exchange capabilities has been discussed as adsorbents, namely zeolites. In aqueous solutions, the cations can be exchanged; thus, zeolites can function as cation exchange materials. Remember, zeolite structures carry negative charges neutralized by metal cations at specific crystal faces. The different metal cations distinguish one type A zeolite from another. The size of the cation (and the number of cations required to neutralize the charge) determines the size of the cavity entrances through which molecules must pass if they are adsorbed. Because zeolite structures (crystals) have small openings, migration of ions through the structures to exchange ions can be much slower than diffusion in some organic-polymer-based ion exchange materials that can swell to give relatively large openings. The use of inorganic materials such as zeolites may, nevertheless, be favored over organic materials where slightly higher temperatures, oxidizing conditions, or radiation fields are required since those conditions can degrade organic materials rapidly. There are also cases where zeolites have better affinities (selectivities) for specific ions [23].

Several other inorganic materials have important ion exchange capabilities. One large class of materials is the hydrous oxides of metals such as aluminum or zirconium. These materials may be amphoteric and act as anion exchange materials at low pHs and cation exchange materials at high pHs. Even silica gel can have significant ion exchange capacity. A notable book that summarizes early work on inorganic ion exchange was written by Amphlett [24]. A review of inorganic ion exchange was edited

TABLE 2 Examples of Inorganic Compounds That Can Be Used as Adsorbents or Ion Exchange Materials

Alumina
Ammonium molybdophosphate
Antimonate acid
Barium sulfate (activated by calcium)
Cobalt hexacyanoferrate (with potassium hexacyanoferrate)
Copper oxide
Ferric hydroxide
Magnesium oxide
Manganese dioxide
Nickel hexacyanoferrate (with/without manganese dioxide or potassium hexacyanoferrate)
Silica
Sodium titanate
Tin antimonate
Titanium antimonate
Titanium dioxide
Titanium phosphate
Zinc oxide
Zinc titanate
Zirconium oxide
Zirconium phosphate

by Qureshi and Varshney [25]. This book focuses on applications in analytical chemistry, but there is a good review of properties that could be helpful in other applications.

Significant advances have been made in the development of inorganic ion exchange materials during the last three decades, and there is now a far wider range of materials studied and used. High selectivity can be achieved in inorganic ion exchange materials by selecting the geometry of the internal cavities of the material as well as the active groups that provide the negative or positive sites for attracting cations or anions that can be exchanged. A partial list of some of the minerals whose ion exchange properties have been studied is given in Table 2. One should not think of these as simple compounds because the structural properties of the materials can sometimes vary considerably with change in the preparation procedures. Layer spacing of clay or other layered materials or the pore size or other internal material shapes can affect the selectivity of the material for different ions. Commercial materials are now available that are prepared by proprietary methods to enhance their selectivity for ions

TABLE 3 Examples of Common Adsorbents

Adsorbent	Notable properties	Applications
Activated carbon	Hydrophobic surfaces Low cost	Removal of organic pollutants from air or water
Silica gel	Hydrophilic surface High capacity (for drying agents) Surfaces can be made hydrophobic	Drying gases or organic liquids
Activated alumina	Hydrophobic surface	Drying of gases or organic liquids
Zeolites (molecular sieves)	Very hydrophilic Uniform pores Lower capacities than silica gel or activated alumina	Drying of gases of liquids Separation of gas mixtures
Carbon molecular sieves	Hydrophobic surfaces Approximately uniform pores Moderately expensive	Gas separations
Organic solids (polymers)	Hydrophobic internals More expensive than carbon	Removal of organics from gases
Reactive (irreversible) adsorbent	Reactive surfaces Specific to some contaminants	Removal of trace contaminants Usually not cost effective for removing more than trace quantities Often used for strong acid or reactive contaminants

of importance. Clearfield discussed some of the more recently studied materials [26].

A new concept for adsorption and ion exchange systems is the use of monoliths of adsorbent material rather than randomly packed particles. Although such a material could be constructed of many different materials, some of the initial interest comes from the use of inorganic materials. Monoliths can have interesting properties, especially potentially low pressure drops, but more will be said about monoliths in the next section.

Structure of Ion Exchange Materials and Adsorbents

The basic shapes of organic-polymer-based ion exchange materials are usually spherical, but other shapes, such as thin membranes, have been prepared. The particles are usually formed in a continuous range of sizes, and selected "cuts" of different narrow size fractions are separated and marketed separately. Smaller particles are desirable to achieve rapid exchange rates because the distance for the exchanging ions to diffuse is shorter than for larger particles. However, smaller particles offer greater flow resistances, and ion exchange column design, like adsorption, involves selection of optimal particle sizes that permit acceptable exchange rates without an excessive pressure drop.

Two important physical properties of organic gel ion exchange materials are their tendency to swell or contract as the solution changes and their possible deformation under pressure. Swelling/shrinking occurs when the ionic strength of the solution around the resin is changed or, in weak acid/base ion exchange materials, when H^+ or OH^- ions are loaded or eluted from the resin. Such changes usually occur between the loading (operating) and unloading (regeneration) cycles. Ion exchange operations are likely to be carried out on dilute waste streams, and the loading cycle can be relatively long if the concentration of the ion to be removed is sufficiently dilute. However, it is usually desirable to recover the ions at higher concentrations. This means that more concentrated regeneration solutions are likely to be used. Gel-type ion exchange resins will shrink when regenerated with more concentrated eluate solutions and expand again when placed back in operation with a dilute feed stream. If a low ionic strength water wash is used between the regeneration and operating cycles, resin swelling could be even greater. In most cases, swelling and shrinking is not a major problem if sufficient excess bed volume is provided to accommodate for the swelling. Changes in the resin volume can generate some mixing of the resin between cycles, but that is not a major problem for all applications. If the resin does not expand/contract smoothly, variations in the void fraction and, thus, nonuniform flow could result. This could have serious effects on bed performance.

Since gel-like ion exchange resin particles are compressible, they can also deform in packed columns under pressure. The pressure is usually that imposed by resistance to flow through the column. The force on the resin particles is proportional to the pressure gradient, and resin deformation then increases as the flow rate is increased. Smaller particles can withstand higher pressure gradients better than larger particles, but, of course, smaller particles also require greater pressure gradients to sustain a given slow rate through the column. Deformation of the resin particles will cause the resin to "fill" part of the void volume in the bed and increase the flow resistance (or pressure drop) for a given fluid flow rate. Increasing the cross-linking of the gel polymer increases the strength of the particles against deformation as well as strength to resist swelling. When high flow rates and thus high pressure gradients are required, low cross-linked resins are not likely to be satisfactory. However, higher cross-linking also results in slower diffusion of ions within the resin and thus lower mass transfer rates. The best cross-linkage to use is determined from an optimization between resin strength (ability to withstand pressure gradients), mass transfer rates, and difficulties in bed packing that result from swelling/contracting during regeneration cycles.

Many newer resins have more complex internal structures than the gels just described. Usually the spherical external shape is retained, but several advantages can be achieved by altering the internal structure of the particles. Generally, the two most common reasons for developing more complex internal structures are to increase diffusion rates and to achieve stronger particles. In many cases, the resulting material makes improvements in both properties. Diffusion rates within ion exchange materials can be relatively slow, especially when high cross-linkage is needed for strength, and the most common approach to alleviate this problem is to prepare larger particles that are made from very small particles assembled into large structures, much like the synthetic zeolite-binder structures described earlier. Such structures are sometimes called "bidisperse" or "macro-reticular." The important feature is the large pores (usually the space between the smaller subparticles) and smaller pores within the "subparticles." As with the zeolites, diffusion rates into the smaller subparticles can be rapid because the diffusion paths are short, but the pressure gradient down the column is set by the size of the much larger aggregate particles, not by the size of the small crystals. A variety of inorganic and organic materials can be used to bind small "subparticles" into aggregate particles. Cellulose and polyacrylonitrile (PAN) [27–29] are examples of polymer gels used with considerable success.

Macro-reticular resins with rigid binders also are often stronger and less prone to swelling than "homogeneous" gel particles. However, some organic binding materials such as cellulose or PAN can also be "soft" and have no more than moderate strength. Organic resins can also be prepared on strong inorganic substrates of silica or alumina to give even greater strength. Many high performance ion exchange resins have been prepared for small analytical chemistry scale operations, but the costs for many of these resins are likely to be too high for use in high throughput wastewater treatment operations. Some such materials are more likely to be used on a preparative (production) scale in biotechnology and in the pharmaceutical industry, where high value products are involved and where the scale of operations is relatively modest. These industries also are often interested in large molecules that have especially low diffusion coefficients in conventional ion exchange resins. As the use and demand for high performance resins increase, the scale of operations used in their preparation will increase, and that could result in some cost reduction. As the merits of high performance resins become more evident, new and less costly preparation methods may be developed. However, large-scale waste handling operations are likely to remain the domain of relatively cheap conventional resins for at least several years.

Diffusion rates in inorganic materials can be particularly slow because the pores (or plate spacings) are often very small, so it is often necessary to keep the diffusion paths short and adopt a macro-reticular-type structure. Small particles of the ion exchange material also can be incorporated into larger agglomerates with the spacing between the small particles sufficiently large that the diffusion rates in the macro-pores are relatively high. It is especially desirable to incorporate smaller inorganic ion exchange materials into "engineered forms" with clusters of very small particles. Several approaches are used to "bind" the smaller particles together. Zeolites are often produced in spherical or granular (or other shapes) by using clay binders to hold the zeolite particles together.

As the macro-pores in a macro-reticular structure become sufficiently large, the effective apparent diffusion rate within the macro-pores can become very high and even appear to exceed the diffusion rates of the ions in water (and the same phenomena can occur with adsorption in macro-reticular adsorbents). This occurs because, as the pores become sufficiently large, ions begin to be transported through the pores by convection (flow) as well as by diffusion. These phenomena were discussed by Watson [30,31] and Carta [32]. The apparent macro-pore diffusion increases with fluid flow rates through the bed because the increasing pressure gradient forces fluid through the pores at increasing rates, and the contribution of convection to the mass transfer becomes increasingly important. Of course, the actual diffusion coefficient does not exceed that in water, but the apparent diffusion rate can exceed the rate calculated by ignoring the effects of convection within the particles. Adsorbents and ion exchange materials with very large pores and small micro-particles are likely to become more common, but the growth in their applications probably will begin in those cases where the value of the product can justify the use of new and, often, more expensive adsorbents and ion exchange materials.

Since the vast majority of adsorbents are spherical or granular shaped, and granular shapes are usually considered "near spherical," it is unusual to even mention other adsorbent shapes. However, this could be another topic where advances may be made in adsorbent development during the coming decades. The need for short diffusion lengths has already been discussed, and the coupling between short diffusion paths and particle size has been broken by developing composite particles with very small particles agglomerated into larger particles. However, the diffusion path can also be decoupled from the flow resistance (pressure drop) in other ways, but these have not all been explored fully. For instance, by using fiber-shaped adsorbents, the diffusion length is the radius of the fiber, and the pressure losses are related to the packing density of the fibers, not the diameter of the fibers. Packed beds of spheres or granules all have void fractions that are not far from 0.5, but with fiber beds the void fraction can be varied over a wide range by changing the number of fibers per unit volume. Very high void fractions, and thus very low pressure losses, can be achieved in fiber beds if needed. Although fiber beds are not common for adsorption, they are being studied [33].

Other adsorption shapes can be considered. For instance, solid monoliths with carefully formed flow channels and thin adsorbent "walls" between the channels may be constructed in the future to give low flow resistance, high mass transfer rates (short diffusion paths), and minimal flow dispersion. Straight channels through a monolith of a solid adsorbent will give lower pressure drops than packed beds of similar size particles. The mass transfer performance from monoliths, however, may not be as good as that of packed beds unless the size of the channels is reduced and the number of channels is increased appropriately. If monoliths are significantly more expensive than granular adsorbents, the initial applications of high performance adsorbent structures may be limited to high cost operations, usually the purification of high value products.

ADSORPTION EQUILIBRIA

To predict how an adsorber will perform, it is first necessary to understand how much of a toxic or valuable material will be adsorbed at a given concentration (or pressure) in the fluid phase and at a given temperature. If the adsorbent adsorbs more than one component from the fluid, the adsorption of each component may be a function of the concentration of any of several components in the fluid. Those are usually referred to as multi-component systems, meaning multiple adsorbable solutes. Of course, single solute systems are simpler and the most thoroughly studied. Multiple solute problems are not only more complex but are likely to be more specific for each application and to require more data

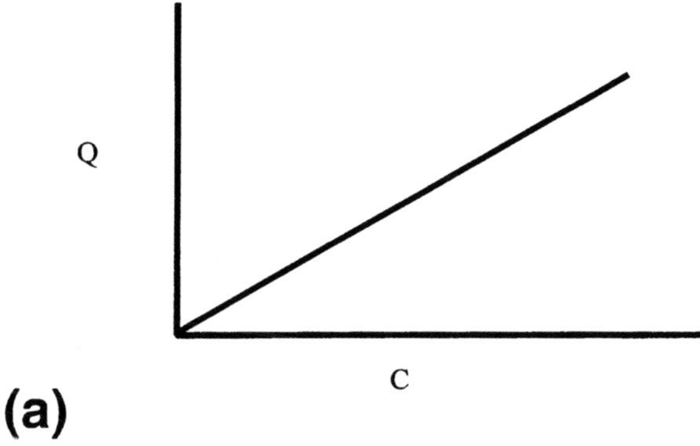

FIGURE 5 (a) Linear adsorption isotherm.

(information) to understand. Thus, each multi-component system may be of interest to only a few people.

The following discussion will begin with the single adsorbate systems. The equilibria will usually be discussed in terms of "isotherms." These are equilibrium curves for constant temperatures. The isotherm will usually give the amount of the adsorbate adsorbed per unit mass (or volume) of adsorbent as a function of the concentration or pressure of the adsorbate in the fluid. The term "isotherm" apparently arose from the frequent use of adsorption to remove components from gases where the equilibrium is strongly affected by the temperature and the pressure of the adsorbate in the gas. Although all adsorption phenomena are likely to be affected by temperature, liquid-solid adsorption systems are less likely to be operated over a significant range of temperatures and may show less dependence upon temperature. Nevertheless, "isotherm" is frequently used to describe adsorption equilibria for liquid as well as gas systems, and the term will be used in the following discussions.

Linear Isotherms

The simplest shape for equilibrium isotherms is a straight line, a linear isotherm. Actually this is a common shape of the isotherm for dilute systems. The linear isotherm corresponds to a constant distribution coefficient (the ratio of the concentration on the solid to the concentration in the fluid) (Figure 5a). It implies that a molecule of an adsorbate has a given probability of being adsorbed that is independent of the concentration of the adsorbate (solute). This is often the case when the loading of the adsorbate is small. The active sites, surface, or volume of the adsorbent are

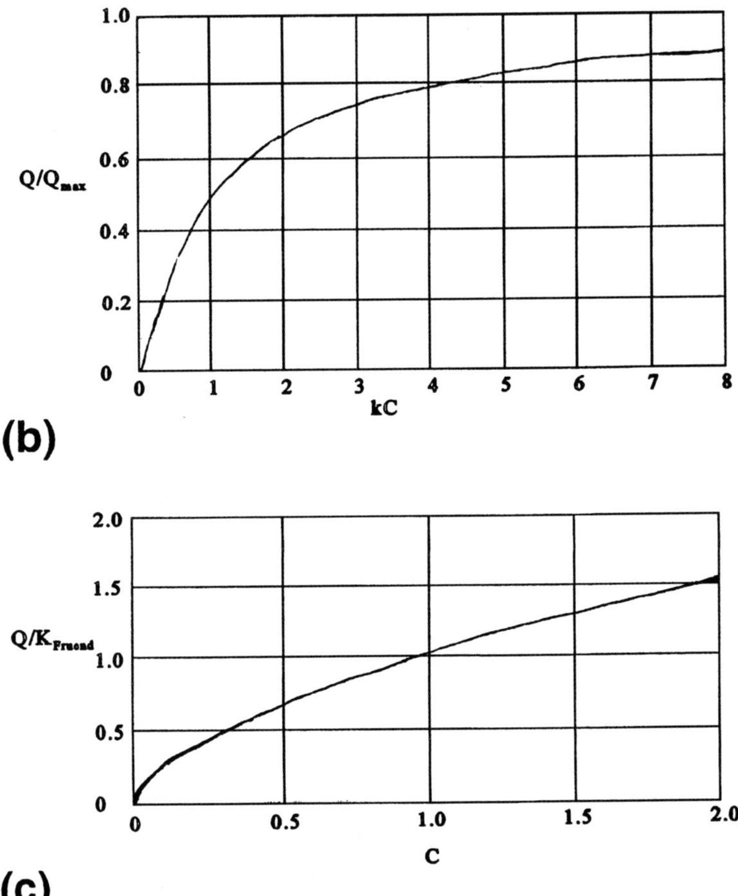

FIGURE 5 (b) Langmuir isotherm. This example illustrates the shape of Langmuir isotherms but is "normalized," showing the adsorbent loading in the vertical axis as a fraction of its maximum (total) capacity and the concentration normalized by multiplying by the Langmuir constant, k. (c) Freundlich isotherm. This example illustrates the shape of a Freundlich isotherm for a system with a Freundlich exponent of 0.6. The adsorbent loading is shown as the actual loading divided by the Freundlich constant, K.

largely unfilled with adsorbates, so the chemical activity of the adsorbent surface is approximately constant.

Even when an adsorption isotherm is not linear, it may be approximately linear over the concentration region of interest. That region is simply the range of concentrations from the minimum concentration up to the maximum concentration of the adsorbate expected. For the simple operation where a single adsorbate is removed from a fluid stream with a constant inlet composition, the maximum concentration is the inlet concentration. The minimum concentration is likely to be the concentration in the fluid that would be in equilibrium with the original adsorbent; if the adsorbent originally contains no solute (adsorbate), the minimum concentration will be zero. Even when the isotherm is not exactly linear over this range there may be merit in approximating the isotherm as linear because subsequent treatment of adsorption rates and bed behavior is less complex for linear isotherms.

Langmuir Isotherm

Deviations from the linear isotherm can result from a number of causes. Activity coefficients (fugacity coefficients for gases) may be dependent on the concentration, and that can cause some variation in the isotherm behavior. However, even with constant activity coefficients, deviations from the linear isotherm are likely to occur at high adsorbent loading because the available active sites (or surface or volume) will decrease. The Langmuir isotherm takes this into account:

$$q = \frac{kq_{\max}C}{1 + kC} \tag{2}$$

This equation assumes that all adsorption sites are identical, that the probability of a molecule adsorbing is proportional to the concentration of the adsorbate in the fluid and the number of adsorption sites (or surface area) available *without* adsorbate molecules, and that the rate at which molecules are being desorbed is proportional to the number of adsorbed molecules. The equation for this isotherm involves two constants rather than the single parameter required for the simpler linear isotherm. The shape of the isotherm is illustrated in Figure 5b. Note that the curve is linear for low concentrations (and thus low adsorption), and at higher concentrations the isotherm bends to form, eventually, a horizontal asymptote at high concentrations. The asymptote corresponds to the condition when all of the adsorption sites are filled with the adsorbate. The two constants required to describe a Langmuir isotherm can be written in terms of the initial slope (at low concentrations) and the horizontal asymptote (kq_{max} and q_{max}).

Although the Langmuir isotherm is very simple in form, it is not always easily incorporated into rate equations for adsorption bed dynamics. The Langmuir isotherm is based upon a simple and logical basis, but it clearly involves idealizations and approximations and is only an approximate model that cannot be expected to fit more than a few cases. As noted earlier, it does not account for variations in the activity coefficients in either the fluid or the solid with concentration or loading. It also assumes that all adsorption sites or surfaces are identical and not affected by the presence of other adsorbate molecules. These are not good assumptions for all systems.

Other Isotherms

It is not practical, or even possible, in this volume to describe all of the isotherms and isotherm shapes that can be important to adsorption operations. In some cases, it may be preferable simply to rely upon graphical presentations when data are not adequately described by a well-known isotherm equation. In other cases, it may be possible to use a portion of an experimental isotherm over the concentration range of interest, usually the low concentration region, and fit that portion of the isotherm to a simple equation, even if the data over an extended range clearly cannot be described by the equation.

There are many reasons why adsorption isotherms could differ from the relatively simple Langmuir form. Besides interactions between solute molecules, there can be significant variations in the sites on the adsorber itself. It is probably obvious that if an adsorbed molecule attracted more molecules to adjacent sites (or affected the adjacent sites in any way), there would then be more than one type of site. However, even when the solute is adsorbing on the surface of the adsorbent, there could be important differences in the surface. These differences could, of course, result from impurities on the surface, but even clean and uniform surfaces can look different to solute molecules. Remember that most adsorbents used in commercial separations involve pores and surface areas significantly greater than the external surface area of the particles. Furthermore, pore diameters usually vary considerably. Since solute molecules can be affected by all surfaces near the molecule, the adsorption affinity for solute molecules will depend upon the pore size. If all pores do not have the same diameter, the affinity of the solute molecules for the surface will not be uniform.

The Freundlich equation,

$$q = k_F C^n \tag{3}$$

is one commonly used empirical equation which contains curvature much like the Langmuir isotherm that can fit many sets of data. Its simpler form sometimes makes it easier to use than the Langmuir equation (Figure 5c). The Freundlich is another two-parameter isotherm. The exponent n is usually less than unity, and then the isotherm has negative curvature and resembles a Langmuir isotherm in general appearance. If data are available only over a limited range of concentrations or pressures and contain the usual experimental scatter, it may not be possible to determine if the data are fit better by a Langmuir or a Freundlich equation. The important differences between the Langmuir and Freundlich isotherms are evident at very low and at very high concentrations. Note that the Freundlich equation does not approximate the linear isotherm at low concentrations. In fact, when $n < 1$, the initial slope of the Freundlich isotherm is infinite at infinitely low concentrations. Although this is not usually realistic, the Freundlich equation is frequently used successfully when the extremely low concentrations are less important. Note also that the Freundlich equation does not approach an asymptote at high concentrations. This can also be a potential problem if there is a clear maximum adsorbent capacity. However, there may not necessarily be a maximum capacity, at least not in the simple sense of the Langmuir isotherm. Multi-layer adsorption can occur, and the clearly defined maximum capacities envisioned by the Langmuir isotherm may not occur.

The BET equation is frequently used to describe adsorption equilibria when interactions between adsorbed molecules are important. Adsorbed molecules can increase the probability that additional molecules will be adsorbed. This equation can describe the relative affinity of adsorbate molecules for active sites and for other adsorbed molecules, but it requires more parameters than the Langmuir or Freundlich isotherm equations.

It is important to remember that all of the equations used to describe isotherms involve significant assumptions or are entirely empirical. Thus, they are often viewed as approximate descriptions of the real isotherms. Data from real systems need not fit any of these relatively simple equations. To distinguish whether a given set of data agrees with (or is "fit" better by) one isotherm rather than another often requires considerable data taken over a proper range of concentrations, and if the system is not going to be operated over such a wide range of concentrations, the better fit by either equation may not be important. To distinguish the Langmuir isotherm from the Freundlich isotherm, one needs data from low concentrations where the Langmuir isotherm has a finite and constant slope (is linear) and/or at high concentrations where the slope of the Langmuir isotherm approaches zero and the curve approaches an asymptote, while the Freundlich isotherm continues to rise and does not approach an asymptote. Of course, it is possible for a real isotherm to have a linear slope at very low concentrations, but no asymptote at high concentrations. Such isotherms will not be fit well over the entire concentration range by either of these simple isotherm equations. One can try the BET equation for such systems. This can be evidence that solute-solute molecular interactions should not be ignored.

For most engineering applications, agreement with an equation is important principally because the equation can be used in mathematical models to describe and predict the behavior of adsorption beds. It is, however, important only that the isotherm expression describe the entire range of concentrations that will occur in the adsorption process. For a single solute (adsorbate), the maximum concentration that will occur during the adsorption period will be the concentration of the inlet (feed) fluid, so data at higher concentrations are not necessary. If very high removal efficiencies are required, as they are in many waste management applications, the isotherm will need to be known down to concentrations approaching zero. Remember that at low concentrations, the slopes of Langmuir and Freundlich isotherms can be significantly different. When data are taken principally at higher concentrations where chemical analyses may be more accurate, extrapolation to low concentrations using the wrong equation could cause significant errors (likewise extrapolation to very high concentrations would be risky). The role of the slope of the isotherm on column performance will be discussed later. Here it is sufficient to note that if one is interested in behavior at low breakthrough, accurate representation of the isotherm in that region may be necessary.

Some effective use has been made of a hybrid of the Langmuir and Freundlich isotherms. The form of the Langmuir isotherm is retained, but the concentrations are changed to concentrations raised to a power n, much as in the Freundlich isotherm:

$$q = \frac{K_1 C^n}{1 + K_2 C^n} \tag{4}$$

This is a three-parameter isotherm. By incorporating the empiricism of the Freundlich isotherm, there is little theoretical (or model) basis for this isotherm, but, if it is useful, it should be considered when needed.

For desoprtion, it would generally be desirable to know the isotherm at concentrations that correspond to the maximum loading that occurs during the adsorption period. That is, the isotherm needs to be known to the concentration at the higher temperature (or other difference in adsorption and regeneration conditions) that corresponds to the loading that would be in equilibrium with the adsorbent loaded at the inlet feed concentration at the adsorption conditions. Usually, desorption occurs at significantly different conditions; for gases this is often at higher temperatures or lower pressures. If the isotherm conditions strongly favor desorption, it may not be as common to analyze the desorption step as carefully as the adsorption period. This may be the case in many waste management and environmental applications. However, if energy requirements are high for desorption, if desorption is particularly difficult, or if very high solute removal is needed, it may be necessary to analyze the desorption more carefully, perhaps even more carefully than the adsorption conditions. Remember that some adsorbents, such as activated carbon, may be regenerated "off site," under relatively high temperatures, and by a commercial desorption service company, not by the user.

Multi-component Adsorption Isotherms

Multi-component adsorption is defined as an adsorption process in which more than one solute (adsorbate) is adsorbed. Obviously if there are so many adsorption patterns for single components, there must be far more patterns possible when a number of solutes are involved. Because the potential shapes of multi-component isotherms are so numerous, it is not practical to describe even a significant fraction of the potential isotherm shapes. Even for the experimentalist, multi-component isotherms present practical problems because so much data can be required to describe the equilibria. It may be necessary to describe the isotherm graphically.

Here, only one multi-component isotherm and one alternative form will be described—the multi-component Langmuir isotherm:

$$q_i = \frac{q_{\max} k_i C_i}{1 + \sum (k_j C_j)} \tag{5}$$

Note that this equation is a logical expansion of the single component Langmuir isotherm. The subscript i denotes the component of immediate interest. The equation differs from the single component isotherm only in the sum of the product kC for all components j rather than accounting only for component i. Thus this equation accounts for the portion of the adsorbent surface occupied by all solutes as well as the solute of immediate interest.

The multi-component Langmuir isotherm includes all of the assumptions of the single component isotherm along with two others. First, it assumes that all of the components adsorb with the conditions assumed in the Langmuir isotherm. Of course, that limits its strict applicability to components that adsorb in this manner, but there are such mixtures.

The second assumption is that the total or maximum capacity of the adsorbent is the same for all components. This assumption is contained in the use of the same constant q_{max} for all components. If the adsorption is by surface coverage, this assumes that the coverage per molecule is essentially the same for all components. For some systems, that means that the sizes of the adsorbed molecules are approximately the same and occupy the same adsorbent surface per molecule adsorbed. However, in some cases, adsorption occurs at "active sites" on the adsorbent surface, not on the entire surface. Then the assumption would only imply that all active sites adsorb the same number of molecules of any of the solutes. If a site adsorbs only one molecule, this may not be such a bad assumption.

One reason that so much multi-component adsorption data in the literature seem to be described by (or fit by) a multi-component Langmuir isotherm is the moderate number of parameters that are needed to describe the equilibria. Note that if there are n components being adsorbed, one needs to determine $n + 1$ parameters (q_{max} and n values of k_i) to describe the equilibria. Even evaluation of this number of parameters can require considerable experimental measurements as the number of components becomes large. Perhaps the best way to evaluate the parameters and test the multi-component Langmuir isotherm is to measure q_{max} and k_i independently with single component measurement and then test the predicted adsorption with results containing a few multi-component mixtures. Of course, the first signs of difficulty appear if the data for some components do not fit a Langmuir form and/or if the values for q_{max} are greatly different for the different components.

The use of other isotherm equations or graphical representations is likely to require more data. The data needs are made even greater because, in adsorption beds, individual components may adsorb sequentially and be displaced (eluted) by other components. Then the range of concentrations of individual components can vary greatly, and in parts of the beds the concentration of a component can even exceed the concentration of that component in the feed. It is not always easy to predict the range of concentrations that needs to be investigated for mixtures of many components.

CLASSIFICATION OF ISOTHERMS

The shape of the isotherm strongly affects the behavior of single component adsorption in beds. The adsorbent capacity at the feed concentration determines the minimum size of the bed required to remove the adsorbate for a specified period of time. As pointed out in the next section, the curvature of the isotherm usually has the most important (first order) effect on the spreading of concentration fronts that move through a bed of adsorbent particles. Thus, the curvature of the isotherm affects the additional volume or length of the bed required because of front spreading.

The isotherms discussed in the previous section should be considered only as examples; they do not cover all isotherms that could be observed in practice. A more generalized classification of isotherms was suggested by Brunauer et al. [34]. They proposed four types of isotherms that describe qualitatively essentially all observed adsorption behavior. These classifications are based upon the curvature of the isotherms, and no particular equations are suggested for any "type" of isotherms. However, we can identify which type includes each of the isotherms discussed earlier.

The Brunauer classifications are illustrated in Figure 6. Type I isotherms are concave downward; they have a negative curvature. That is, $d^2q/dc^2 \leq 0$ over the entire concentration range. The common equations discussed in the previous section (Langmuir and Freundlich) are both Type I isotherms. This is the most common type of isotherm, especially when the linear isotherm ($d^2q/dc^2 = 0$) is included in Type I. Although for engineering applications no mechanism is necessarily needed to explain the reasons why an isotherm falls into one type or another, it may be helpful to note that the negative curvature of a Type I isotherm implies that each increment of additional adsorption makes the next increment more difficult (or less likely). This may imply that there is a limited capacity for the adsorbent as assumed in the Langmuir model, and as the solute loading approaches that capacity it becomes increasingly more difficult to add more of the adsorbate.

Type II isotherms have a positive curvature; they are concave upward. These isotherms are less common than Type I isotherms, and the earlier section gave no equations specifically developed to describe Type II isotherms. However, the Freundlich isotherm equation can be made to have such behavior if the exponent n is greater than unity. Although not common, the Type II isotherm suggests that adsorption becomes more favorable as more of the adsorbate is adsorbed. That may imply that adsorbed adsorbate molecules have a greater affinity for other adsorbed molecules than for the adsorbent surface.

Types III and IV isotherms have both negative and positive curvatures in different concentration or pressure ranges. Type III isotherms have negative curvatures at low concentrations and positive curvatures at high concentrations, while Type IV isotherms have positive curvatures at low concentrations and negative curvatures at high concentrations.

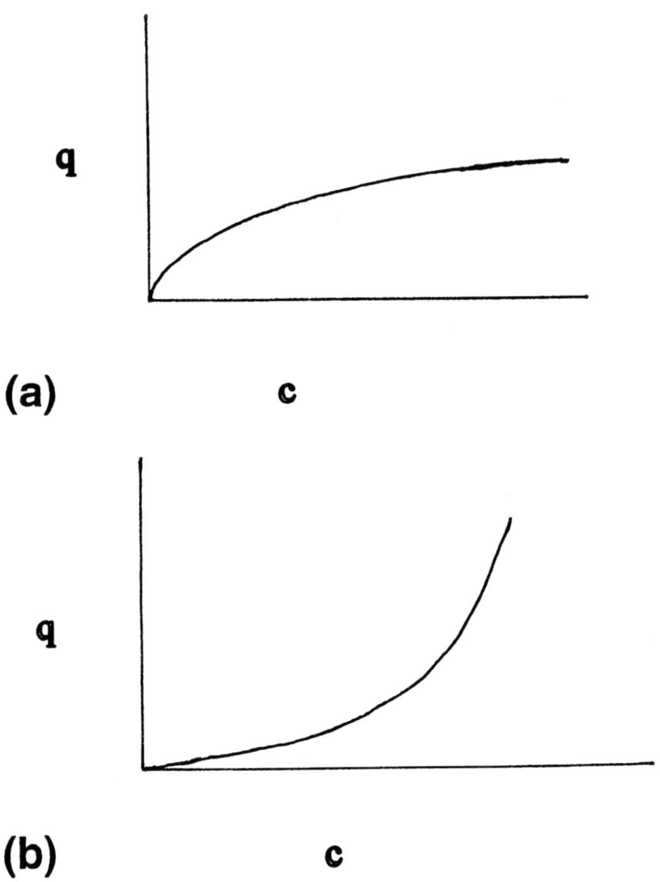

FIGURE 6 (a) Class I adsorption isotherm (negative curvature over entire isotherm range). (b) Class II adsorption isotherm (positive curvature over isotherm entire range).

These changes in the sign of the curvature suggest that there is a change in the nature of the adsorption. The BET isotherm is a Type III isotherm. At low concentrations adsorbate molecules are adsorbing largely on the adsorbent surface, but at higher concentrations the adsorption approaches a monolayer of adsorbed molecules, and further adsorption takes place largely on other adsorbed molecules, perhaps even filling the pores of the adsorbent. Note that at low concentrations Type III and Type IV isotherms can appear much like Type I and Type II isotherms, respectively.

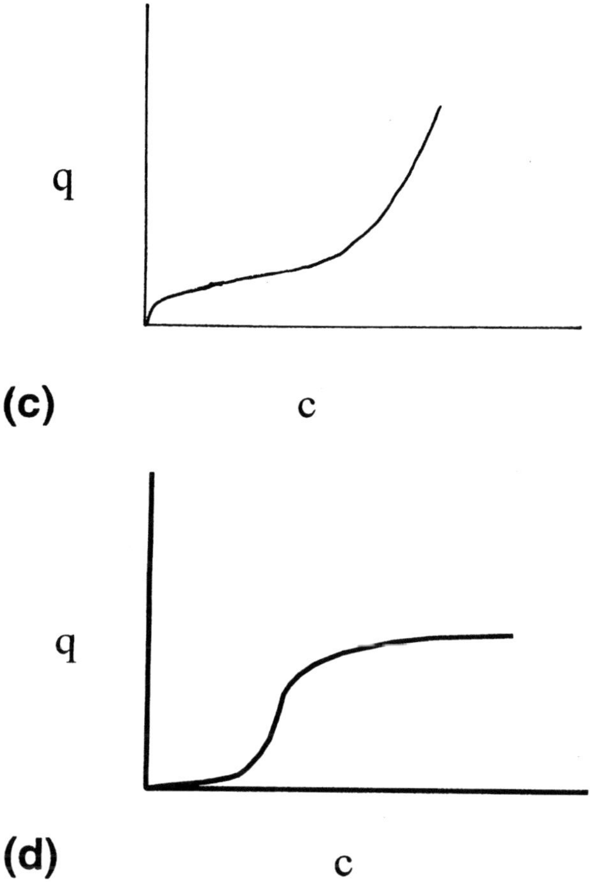

FIGURE 6 (c) Class III adsorption isotherm (negative curvature at low concentrations and positive curvature at higher concentrations). (d) Class IV adsorption isotherm (positive curvature at low concentrations and negative curvature at higher concentrations).

ION EXCHANGE EQUILIBRIUM

The exchange of ions between an ion exchange material and a solution can be described as follows:

$$\frac{1}{n_1}[M_1^{+n_1}] + \frac{1}{n_2}[\underline{M}_2] \rightleftarrows \frac{1}{n_1}[\underline{M}_1] + \frac{1}{n_2}[M_2^{+n^2}] \tag{6}$$

This expression describes the cation exchange of one metal ion in solution (M_1) for another metal cation within the resin (M_2). The valences of the two ions are n_1 and n_2, respectively. The underscore signifies that the ion is in the resin phase. This relation is sometimes written to look as if a chemical reaction is taking place with one ion in solution replacing an ion attached to the resin, but the formulation in (6) is preferred because it does not imply that ions in the resin are associated with any resin site. It is better to think of the resin and the solution as different phases. Although this equation was described in terms of cation exchange (all exchanging ions having positive charges), the valences could be negative, and the equation would then describe anion exchange.

Equilibrium can be written in terms of the mass action equation:

$$\frac{[\underline{M}_1]}{[M_1^{n_1}]} = K\{[\underline{M}_2][M_2^{n_2}]\}^{n_1/n_2} \tag{7}$$

K is an equilibrium "constant" that includes effects of the activity coefficients for both ions in both the aqueous and resin phases. If the activity coefficients are constant, then K is truly a constant. Otherwise, there will be some variation in K with solution or resin composition. In many cases, there will not be a large variation in activity coefficients over operation conditions, but one should take care when assuming a constant value of K, especially if a wide range of solution compositions is of interest.

This equation is written in terms of the ratio of the concentration of ion 1 in the resin to its composition in the solution. In describing adsorption equilibrium, this ratio was referred to as the "distribution coefficient." Note that for the simple exchange of two ions, the distribution coefficient of one ion is approximately constant only if the distribution of the other ion is approximately constant. During an operation when one ion is loading on the resin and another ion is being displaced, neither of these ratios will be approximately constant, even for dilute solutions. *Do not assume that at dilute concentrations ion exchange equilibria will be like linear isotherms in adsorption systems*. One should not expect a constant distribution coefficient (that is, a linear isotherm) for ion exchange, even from dilute solutions. The possible exception occurs when the ion of interest is dilute, but the other ion is concentrated in both the solution and the solid. Then the solution and the resin would remain largely loaded with the other component, so the distribution of the trace component could be approximately constant as the concentration of the trace component increases.

Another mistake that appears to be common is to assume that a dilute ion exchange system can be approximated by a Langmuir isotherm (or, as just mentioned, even a linear isotherm). There is an important difference in ion exchange and adsorption because even in a dilute system, a column initially saturated with component 2 can be completely loaded with sufficiently large volumes of even the most dilute solution of component 1. The resin will be saturated with component 1 regardless of the concentration. It is clear that it cannot be approximated well with a linear isotherm at low concentrations. The effective curvature in the isotherm (if one wants to think of ion exchange equilibrium in terms of an isotherm) depends upon the change in the ratio of concentrations of ions 1 and 2 in the solution and resin, not upon the actual concentration of either ions 1 or 2. There are similarities and differences between ion exchange equilibria and Langmuir isotherms that will be discussed in the next section.

Preferences of most polymer gel-type resins for different ions usually can be conveniently grouped as first order effects, based upon the valence (charge) of the ion, and second order effects, or special affinity of some ions for the resin phase. In only a few cases will these second order effects be as important as, or more important than, first order valence effects. The valence effect can be illustrated by using Equation (7) to calculate the separation factor for two ions between the resin and solution. The separation factor is the ratio of the distribution coefficients:

$$\alpha_{12} = \frac{[\underline{M}_1]/[M_1]}{[\underline{M}_2]/[M_2]} = K \left\{ \frac{[\underline{M}_2]}{[M_2]} \right\}^{n_2/n_1 - 1} \tag{8}$$

Note that $[\underline{M}_2] > [M_2]$ because the resin capacity is greater than the solution concentration (at least for dilute solutions), and $[\underline{M}_2]/[M_2] > 1$ (usually $[\underline{M}_2]/[M_2] \gg 1$). When the two ions have the same valence, $n_1 = n_2$, the exponent on the right side is zero and the separation coefficient is equal to K. In that case, there is no valence effect. The separation factor then is equal to K and reflects only second order activity effects. However, when $n_2 > n_1$, the exponent is positive and the separation factor is greater than unity unless the activity terms make K small enough to compensate for the valence effect. This is the case when an ion with a lower valance, n_1, is displacing an ion with a higher valance, n_2. This means that the resin favors the cation with the higher valence. For dilute solutions where $[\underline{M}_2] \gg [M_2]$, the resin may favor the higher charged ions strongly. Similarly, when $n_2 < n_1$, the exponent is negative and the resin favors ion 1 over ion 2 and still favors the ion with the higher charge.

Thus, most resins prefer to lose single charged ions, such as sodium or potassium ions, and gain ions from the solution, such as calcium or magnesium, that have double charges. When loading a resin with doubly charged ions such as calcium and regenerating with singly charged ions such as sodium or hydrogen, it is better to regenerate with high concentrations where the preference for the doubly charged ion will not be as great.

Second order selectivity can come from general behavior that follows a pattern for many ion exchange materials, or it can be specific for a few ions and a few resins. The "general" effect probably results because the interior of an ion exchange resin contains sufficient organic polymer that it is usually not very hydrophilic, despite the fact that the ionizable groups on the resin do bring large quantities of water into the polymer. This usually results in a selectivity for the less hydrophilic ions in each series, those ions that bring less water into the resin. For instance, in the alkali metal series, cesium is usually the most preferred ion in the series and lithium the least preferred. In most such series, the ions with the largest size (or molecular weight) are preferred by the resin, and they are usually the ions with the least waters of hydration. The degree of selectivity for ions in such a series can be altered by changing the organic content and type of organic components in the polymer.

Perhaps the most common examples where specific interactions or "second order" activity effects are extremely important are the weak acid cation exchange materials and weak base anion exchange materials. Weak acid resins can have strong affinities for hydrogen ions, and weak base resins can have high affinities for hydroxyl ions. Other ion exchange groups may show considerable preference or affinities for other specific cations or anions, but these effects are usually not as strong or as common. Although these special affinities may be important for selected applications, only a few resins have preferences for specific ions that are sufficient to overcome the valence effects.

Some of the other specific selectivities can result when a resin contains a group that interacts strongly with a specific cation, but that group may not be part of the ion exchange function. For instance, some ion exchange resins that contain thio-sulfur have high selectivity for mercury ions or for ions of most metals that form highly insoluble sulfides. Some selectivities resulting from these effects can be significant.

Comparison of Ion Exchange Equilibria and the Langmuir Isotherm

The relation between ion exchange and Langmuir adsorption can be illustrated best by considering exchange of ions with the same valance ($n_1 = n_2 = n$). Then the equilibrium loading of the first ion can be written as

$$\underline{\mathrm{M}}_1 = K\mathrm{M}_1\left[\frac{\underline{\mathrm{M}}_2}{\mathrm{M}_2}\right]^{n_1/n_2} = K\mathrm{M}_1\left[\frac{\underline{\mathrm{M}}_2}{\mathrm{M}_2}\right]$$

$$= K\mathrm{M}_1\left[\frac{\underline{\mathrm{C}} - n\underline{\mathrm{M}}_1}{\mathrm{C} - n\mathrm{M}_1}\right] \tag{9}$$

since

$$\mathrm{C} = n_1\mathrm{M}_1 + n_2\mathrm{M}_2 \qquad \text{and} \qquad \underline{\mathrm{C}} = n_1\underline{\mathrm{M}}_1 + n_2\underline{\mathrm{M}}_2$$

(That is, the total concentration of ions in the solution C, and the total concentration of ions in the resin $\underline{\mathrm{C}}$, is not changed by the exchange of ions.) Then

$$\frac{\underline{\mathrm{M}}_1}{\underline{\mathrm{C}}/n} = K\left[\frac{\mathrm{M}_1}{\mathrm{C}/n}\right]\frac{[1 - \underline{\mathrm{M}}_1/(\underline{\mathrm{C}}/n)]}{[1 - \mathrm{M}_1/(\mathrm{C}/n)]} \tag{10}$$

Here the concentrations in the resin and the solution are normalized by the total concentration of cations (or anions for anion exchange) in the resin and solution, respectively. Thus, the concentrations represent the fraction of the resin capacity and the fraction of the solution cations occupied by the reference ion, component 1. The loading of component 1 in the resin occurs in both sides of this equation; therefore, to get this into a form similar to the Langmuir form, the equation must be solved for $\underline{\mathrm{M}}_1/(\underline{\mathrm{C}}/n)$. Then

$$\frac{\underline{\mathrm{M}}_1}{\underline{\mathrm{C}}/n} = K\left[\frac{\mathrm{M}_1}{\mathrm{C}/n}\right] \Big/ \left[1 + \frac{(K-1)\mathrm{M}_1}{\mathrm{C}/n}\right] \tag{11}$$

Note that this resembles the Langmuir isotherm equation, but there are two important differences. First, the concentrations are expressed in terms of the fraction ions that are component 1, not in terms of the absolute concentrations of component 1. Next note that the coefficient for the second term in the denominator of the right-hand side is $K - 1$ not K, as one would expect if the form of the equation were to be just like the Langmuir isotherm equation. When the exchanging ions have different charges, there is even less similarity with a Langmuir isotherm. In such cases, it is usually not helpful to compare the ion exchange equilibria with the Langmuir isotherm.

Measuring Ion Exchange Resin Parameters

To evaluate the equilibrium parameters, it is usually easier to determine the resin capacity by titration of the hydrogen form of the resin (or hydroxyl form of an anion resin) with a base (or acid for an anion resin), or to saturate the resin with any ion that can be easily analyzed, elute it with a large volume of another ion, and analyze the eluate.

The equilibrium constant for a binary system can be evaluated by equilibrating the resin with different concentrations of the two ions of interest and analyzing. Often it is only necessary to analyze for one component and to determine the final concentration and resin loading for the other component by difference, provided changes in solution concentrations are large enough to determine the differences accurately. If a graphical determination is desired, a log-log plot of $\underline{M}_1/M_1$ versus $\underline{M}_2/M_2$ should give a slope of n_1/n_2, and the position of the line will give the value of K. If the data fall along a straight line with the proper slope, that is evidence that the simple ion exchange mass action equation [Equation (7)] that assumes approximately constant activity coefficients is appropriate. Then K will be the value of $\underline{M}_1/M_1$ when $\underline{M}_2/M_2 = 1$. The fit of the data to a smooth curve with approximately the expected slope provides an indication of the scatter or uncertainty in the data as well as the accuracy of the assumption of constant activity coefficient, that is, a constant value for K.

Note that when the valences of the two ions are the same, one can make a plot much like that usually used to evaluate the Langmuir constants. From the last equation, one can find that a plot of $1/\underline{M}_1$ versus C/nM_1 should give a straight line with

$$\text{slope} = \frac{n}{K\underline{C}} \tag{12}$$

and

$$\text{intercept} = \frac{n(K-1)}{K\underline{C}} \tag{13}$$

This lets one use the analogy to a Langmuir isotherm, but note that the parameters K and $\underline{C}$ appear in both the slope and the intercept. However, it is not difficult to evaluate the parameters since

$$K = 1 + \frac{\text{intercept}}{\text{slope}} \tag{14}$$

When the two exchanging ions have different valences, there is no convenient way to use the Langmuir-like graph.

Multi-component Ion Exchange Equilibrium

Since ion exchange operations require at least two ions (one displacing the other on the resin), multi-component ion exchange refers to situations with three or more cations (or anions). Multi-component ion exchange can be viewed as a series of individual binary exchanges, and Equation (7) can be applied to each pair of ions. If there is a total of m cations (or anions for anion exchange) in a solution equilibrated with a cation exchange material, there will be $m-1$ independent forms of Equation (7) to describe the binary equilibrium, or $m-1$ values of K to be determined. Since there are m concentrations in the resin to be determined from the equilibrium relations, one other independent equation is needed, the total cation balance:

$$\sum A_m = n_m[\underline{\mathrm{M}}_m] = R \tag{15}$$

where R is the total capacity of the resin. Note that penetration of anions into the resin and change in activity coefficients with solution concentration and composition have been ignored. For most dilute solutions, this is a good approximation. If the resin has a particularly strong affinity for one or more of the ions present, the Ks may show more change with composition. If anion penetration is significant (for cation exchange), an equivalent number of cations must be added to the cation balance in the resin.

In principle, multi-component ion exchange can be predicted from a series of $m - 1$ binary measurements, which for many systems gives satisfactory results. As noted, however, there are situations and factors that could cause inaccuracies in such an ideal approach.

There are numerous ways to evaluate multi-component ion exchange equilibria, and readers can choose the approach that is most suitable for their problems. However, this author prefers to see the $m - 1$ binary equilibria constants evaluated, and it is usually easier to evaluate those parameters by using binary exchange measurements. Of course, if those parameters remain constant in the multi-component systems, those are the only parameters needed except for the exchange capacity of the resin. However, it is wise to check this by making at least a few multi-component equilibria measurements to be sure that the multi-components measurements can be predicted reasonably accurately with the parameters determined from binary measurements. It will be most appropriate to make the multi-component measurements with concentrations as close as possible to the concentrations expected in the operating system.

In some respects using this simple extension of the binary exchange to predict multi-component equilibrium is much like using the multi-component Langmuir equilibrium equation where parameters obtained from single component data and some of the factors that could cause problems in the expansion to multi-component situations are similar for both situations. For instance, interactions between the ions in either phase could cause significant error in the ideal treatment just described. Nevertheless, as with multi-component adsorption, considerable equilibrium data are required to describe multi-component systems even using idealized equations, and it is desirable to search for simple approaches such as this to minimize the data needed. More elaborate and accurate relations usually require even more data, so the simpler relations are likely to be used, or at least tried first.

As noted, when idealized equilibrium relations are used, it is best to check the results at conditions as near as practical to those that will be used. To minimize the effects of deviations from the ideal approach, it may even be advisable to make the binary measurements at conditions as near to those expected in the application as practical. Generally this will mean measurements at the same normality as the feed solution. Since ion exchange involves exchange of equal equivalents of ions between the solution and resin, there is no net change in the normality of the solution during batch equilibration or as solutions pass down an ion exchange bed; only the composition of the solution changes. The actual feed composition is one concentration that should be tested, but one should be aware that in multi-component systems, a great variety of compositions could develop at different positions in an adsorption bed. This is as true for ion exchange as it is for adsorption.

Co-Ion Penetration and the Donnan Effect

When ion exchange is used in waste management operations, the concentration of the ion to be removed from solution is usually low. However, the concentration in the resin will not be low. The concentration of metal ions in the resin is set largely by the resin capacity and will be essentially independent of the solution concentration. The total concentration of cations in a cation exchange resin is

$$n_1[\underline{M}_1] + n_2[\underline{M}_2] = R + C_a \tag{16}$$

R is the concentration of ion exchange groups (anion groups for cation resin and cation groups for anion exchange resins) attached to the resin polymer, and C_a is the normality of electrolyte anions that diffuse into the cation resin from the solution (or cations that diffuse into an anion resin). For high capacity polymer-based ion exchange materials, R is normally between 2 and 5 molar. Since that concentration is much higher than the concentrations in most dilute wastewaters treated by ion exchange, one would not expect the penetrating electrolyte concentration C_a to contribute significantly to the total concentration of exchanging ions in the resin.

Actually, anions are excluded from cation resins (and cations are excluded from anion resins) by Donnan exclusion; hence, the concentration of solution anions in a cation resin is much less than the concentration in the external solution. This effect can be explained by considering a solution of a single salt (cations and anions) in equilibrium with a cation resin. Of course, if there is only one salt in the external solution, the resin contains only one cation, that of the salt. Note that at equilibrium the activity of the salt must be the same in the solution and in the resin. Then, neglecting activity coefficients (assuming first that the activity coefficients are approximately the same in both phases),

$$[M_1^{n_1}]^{n_c}[C^{-n_c}]^{n_1} = [\underline{M}_1^{n_1}]^{n_c}[\underline{C}^{-n_c}]^{n_1} \tag{17}$$

where C is the anion and n_c is the (negative) charge of the anion. Writing this in terms of the distribution coefficient for the anion gives

$$\frac{[\underline{C}^{-n_c}]}{[C^{-n_c}]} = \left\{\frac{[M_1^{n_1}]}{[\underline{M}_1^{n_1}]}\right\}^{n_c/n_1} \tag{18}$$

For dilute solutions $[M_1^{n_1}] \ll [\underline{M}_1^{n_1}]$; therefore, $[\underline{C}^{-n_c}] \ll [C^{-n_c}]$. That is, the concentration of salt (anions) in the resin is much lower than the concentration in the already dilute external solution. Thus for dilute systems, one can usually neglect the penetration of anions into cation resins. Likewise, one can neglect the penetration of cations into anion resins when dilute solutions are involved.

Anion Complexes of Metal Cations

As noted previously, the selectivity of the most common ion exchange materials for different ions comes largely from the valence of the ions, and, with a few notable exceptions, separation factors between ions with the same valence will be modest. Although the selectivity between ions with modest separation factors can be used in elution chromatography to achieve essentially any degree of separation desired, the separation that can be achieved using the load-elute operations that are so common for large volume wastewater treatments will be limited. A few cases, however, should be mentioned where extremely selective ligand interactions can be used to make metal separations highly selective. That is, a selected cation can sometimes be removed without essentially removing any of the other cations from the solution.

A ligand is a chemical group, often an anion, that interacts strongly with selected metal ions to form "complexes." An example is the interaction of chloride ions with ferric [iron (III)] ions. In an acid chloride solution, the ferric ions do not all exist as free ions, but there is a series of ferric ion and chloride ion complexes such as

$$Fe^{3+} + Cl^- \rightleftarrows FeCl^{2+} \tag{19}$$

$$FeCl^{2+} + Cl^- \rightleftarrows FeCl_2^+ \tag{20}$$

$$FeCl_2^+ + Cl^- \rightleftarrows FeCl_3 \tag{21}$$

$$FeCl_3 + Cl^- \rightleftarrows FeCl_4^- \tag{22}$$

Ferric iron is not a major source of pollution, but it is used here as an example. Other transition metals can behavior similarly with selected ligands. (In alkaline solutions, the ferric ion is likely to hydrolyze and precipitate as hydroxide.) The fraction of the ferric ions that becomes highly complexed and the extent of complex formation (number of chloride ions associated with a typical ferric ion) increases with increasing concentrations of chloride ions. Note that some of the ferric ions will have sufficient chloride ions attached and that the complexes will have negative charges; that is, they will be anions. These anions can be removed by anion exchange materials, and since ligand complex formation can be highly specific, ferric iron can be removed selectively from almost all of the other metal ions in the solution, or from all metal ions that do not form anion complexes. The selectivity is determined by the selectivity of the ligand, not of the resin. The selectivity of the separation may be many times higher than selectivities using cation exchange.

It may appear that since only a small fraction of the iron may be in anion complexes, a complete removal of ferric iron would not be possible. However, the complex formation reactions are in equilibrium, and removal of the $FeCl_4^-$ from the solution at one point in the ion exchange column causes the equilibrium to shift; so just downstream more $FeCl_4^-$ will be formed and be exchanged with the resin. This process will continue until essentially all of the ferric iron is removed. Of course, the equilibrium loading of $FeCl_4^-$ on the anion resin will depend upon the $FeCl_4^-$ concentration, so it is desirable to have as much of the ferric iron as practical in this form. This means that higher loadings of ferric iron on the resin are possible at higher chloride concentrations. There is likely to be a preference (selectivity) for the larger and (usually) less hydrophilic anions by the resin, so there is likely to be a preference for the ferric chloride complex over chloride ions. However, loading increases with increasing chloride concentration (because of increased concentration of the complex) only up to a point. Once the chloride concentration becomes sufficiently high, the chloride ions begin to compete more effectively for the ion exchange capacity; thus, there is an optimum chloride concentration for maximum ferric iron loading. Similar maxima will be found for other metal ions and other anion ligands.

To be selective for removing individual ions, ligand interactions must not be too common. That is, it may not be practical to find suitable ligands for all metal ions, but their use may be considered, especially when the ligand is already in the wastewater. Some common ligands are Cl^-, F^-, and CN^-, and transition metals (such as iron) are likely to form complexes. Sulfate and nitrate ions form complexes with heavy metals such as uranium and plutonium. Perhaps the most important fact is that the common alkali and alkaline earth metals that are so common in wastewaters are not likely to form complexes with any of the common anions. Also the complexes usually become important only in concentrated solutions, and this may eliminate their importance in many, but not all, environmental applications.

PHYSICAL PROPERTY DATA NEEDED FOR ADSORPTION AND ION EXCHANGE BED DESIGN

Equilibrium Data

As noted throughout this chapter, the most important data required for designing adsorption and ion exchange systems to remove pollutants are equilibrium data; for adsorption, the equilibria data are usually called isotherms. The usual basis for ion exchange equilibria is relatively well understood and has been described earlier. The equilibrium loading capacity of the adsorbent or ion exchange material is the principal factor determining the size of adsorbent bed required. Adsorption equilibria data may be specific to the process stream, temperature, and the adsorbent used. Manufacturers of adsorbents are likely to provide guidance and even useful data. If a commercial adsorbent is to be used, it certainly would be desirable to see what information can be supplied by the manufacturer, especially if the application is believed to be "relatively common" and used many other places. In most cases, it will usually be necessary to test any promising adsorbent with samples from the stream to be treated. This is not specifically to check the reliability of literature or manufacturer's data, but to ensure that there are no unforeseen problems. Unless the stream contains only one solute, it is usually necessary to ensure that no other solute affects the adsorption of the component of interest. Other solutes could compete for adsorbent capacity (multi-component effects); accumulation of even a trace component could eventually "poison" the adsorbent and cause more frequent and possibly expensive adsorbent replacement.

Additional testing can also confirm that the solute is in the same chemical form for which the equilibrium relations are assumed. This can be especially important if the solute is extremely dilute. For instance, traces of a ligand can complex with dilute ions, so removal of trace toxic metal ions may require removal of the metal-ligand complex. Adsorption or ion exchange systems designed to remove the free metal cations may not be effective for complexed ions. Other trace pollutants may be adsorbed on colloid particles. These may be removed in adsorption bed prefilters, not in the bed itself. If the bed uses relatively large adsorbent particles or a fluidized bed of adsorbent without a prefilter capable of removing the colloids, the trace pollutant may not be removed at all. In testing, one may find that a certain portion of the pollutant is easily removed, but the remaining fraction is very difficult to remove, and removal of the last fraction of the component may not follow the predictions of column performance. This could indicate that the pollutant exists in two chemical forms, and the rate of equilibrium adjustment between the two forms is slow. The adsorption bed may remove only one of the forms effectively.

Rate Data

It is obvious that for an adsorber to be effective it must be possible to transfer the solute from the fluid to the particles as quickly as needed to meet the design specifications, usually specifications on the concentration of solute remaining in the fluid leaving the adsorption bed. Before discussing the driving forces and specific methods to predict adsorption bed performance, we discuss the different resistances and coefficients to describe those resistances.

Fluid Film Resistance

For a solute to move from the fluid to an adsorbed state on the solid, it must first diffuse through the region of fluid near the adsorbent particles where the flow velocities are low. This region is usually "modeled" as a stagnant film, but that is only a convenient simplification. The mass transfer rate is described as

$$\text{rate} = k_f a(C - C_i) \tag{23}$$

where k_f is the fluid "film" mass transfer coefficient, a is the adsorbent surface area per unit volume of bed (a quantity easily calculated from the particle size and the fraction of bed volume occupied by adsorbent), C is the concentration of solute in the bulk liquid, and C_i is the concentration of solute in the fluid at the adsorbent surface. Values for k_f can sometimes be obtained for the adsorbent to be used, but since the external fluid mass transfer coefficient is assumed to be largely independent of all properties of the adsorbent except its shape, reasonably good estimates can be obtained from correlations obtained from other similar systems. For spherical or most granular adsorbents, one can usually obtain a reasonable estimate by using correlations developed for spherical particles.

Perhaps the most quoted correlation for mass transfer to spherical particles is that of Rantz and Marshall [35]. This equation does give reasonable estimates, but it was developed from data on individual spheres, not from spheres in packed beds. A later study of data from packed beds by Wakao et al. [36] reviewed data on spheres in packed beds and correlated the data with the equation

$$\mathrm{Sh} = 2 + 1.1\mathrm{Sc}^{1/3}\mathrm{Re}^{0.6} \tag{24}$$

Here Sh is the Sherwood number, $2r_p k_f/d_p$; Sc is the Schmidt number, ν/D; Re is the Reynolds number; and ν is the kinematic viscosity of the fluid (fluid viscosity divided by fluid density).

The reader should not expect this equation, or any correlation, to give very precise estimates. The Wakao et al. correlation was developed from data taken at Reynolds numbers between 3 and 10,000. The scatter in the data used in the correlation sometimes exceeded a factor of 3, so one should not expect it to predict film coefficients much more accurately than that. There are several reasons why the data from investigators can be so different. Probably the principal difference is in the way the particles are packed in beds and in the slight deviations from uniform size in the adsorbents used. Essentially all "uniform" spheres are really narrow "cuts" of particle sizes over a small range. It is also possible that some of the adsorbents in the studies were not exactly spherical. Deviations from the correlations could be more significant if granular shapes deviate significantly from spherical.

Diffusion Resistance Within the Adsorbent Particle

There is no accurate theory nor even an empirical correlation for predicting effective diffusion rates within the pores of adsorbent or ion exchange particles. Obviously the diffusion coefficients will be lower than those in the free fluid because of (1) the smaller void fraction in the particle, (2) the tortuosity factor caused by random directions and "dead end" pores, and (3) wall effects in relatively small pores. The dead end pores and the variation in direction of the pores can be considered separately by including these effects in a special term, usually called the tortuosity factor. However, since diffusion coefficients and tortuosity factors must be determined experimentally, there may be little merit in separating these effects. Tortuosity effects can be lumped into the effective diffusion coefficient.

Pore diameters can be particularly important in determining the effective diffusion coefficient, especially when the pores become very small. Since most adsorbents contain a variety of pore diameters and even individual pores are not uniform in size, it may be most practical to evaluate the effective diffusion coefficient from experimental data on mass transfer rates. For rigid adsorbents such as silica gels, activated carbons, etc., pore diameters can be estimated from adsorption of gases such as nitrogen, but this is not an option for water-swollen organic ion exchange resins. Also note that some adsorbents such as gel-type ion exchange materials may not have pores in the usual sense. Instead, the solute may diffuse through a "mesh" of the polymer.

Information on mass transfer rates in common adsorbents and ion exchange resins with common solutes is available in journals. However, there is a good chance that the information needed will not be available; measurements are likely to be required to obtain effective internal diffusion coefficients for most systems.

Diffusion within ion exchange particles (as well as diffusion in liquid films around ion exchange particles) involves electrostatic as well as concentration gradient driving forces. Since ion exchange involves exchange of ions with the same electric charges (sign), it is obvious that the diffusion of one ion to or from the resin cannot get ahead of the movement of the other exchanging ion without a net electric charge difference developing between the resin particles and the surrounding solution. Such a charge difference would resist further diffusion of the ion with the lesser mobility until the ions diffusing in the opposite directions are moving with equal (net) rates of movement of electric charge. Even within the particle, the diffusion of different ions must proceed under restraints that prevent a continued buildup of an electric charge along the diffusion path. The total force on diffusing charged ions is described by the Nernst-Planck equation:

$$\text{flux} = D\left(\nabla C + \frac{zF}{RT}\nabla\phi\right) \tag{25}$$

where z is the charge on the ion, F is the Faraday constant, R is the gas constant, T is the temperature, and ϕ is the local electric potential.

In practice, the electric forces are so strong that neither ion can diffuse faster than the counterdiffusing ion for a significant time; hence, the local electric field gradient, or the second term on the right side of the equation, can be evaluated by assuming that the counterdiffusing fluxes are the same. Then the counterdiffusion of two ions can be calculated rather easily numerically, but the calculations can be complicated for multi-component systems.

For most engineers applying ion exchange, the important thing to know about counterdiffusion of ions is not the integrated solutions for counterdiffusion of ions, but the fact that ion exchange rates are not determined by the diffusion coefficients of individual ions but by the diffusion coefficients and concentrations of all ions involved. Most analyses of ion exchange bed behavior do not seem to take such complications into account, but the reader is advised to be aware of the complication and its potential effects. If measured data on ion exchange rates are fit to equations for diffusion of the individual ions as one would do for molecules in adsorption, the observed effective counterdiffusion coefficient will lie between the diffusivities of the individual exchanging ions. That means that the rate at which sodium ions exchange with calcium ions is likely to be different from the rate at which sodium ions exchange with hydrogen ions. If ion exchange rate processes are to be described in the same terms as adsorption processes (and one often tries to describe ion exchange breakthrough fronts by using the same methods that are successful for adsorption), then the rate parameters used should be measured with the same ions and approximately the same concentrations as one expects to see in the applications.

Fortunately, the diffusion coefficients of ions in solutions do not differ greatly, so counterdiffusion effects in liquid films are rather easily estimated. However, because of the more restrained space within ion exchange resins and the significant variation of the ability of different ions to bring water molecules of hydration into the resin, there can be more significant differences in ion diffusion coefficients with the resin. The major differences are likely to occur when there are large differences in the size of the hydrated ion. Anion complexes with metal ions may be much larger than the simpler anions that they are replacing and thus have much lower diffusion coefficients.

Diffusion within adsorbents or ion exchange particles can involve all of the solute or ions present or only a portion of them. In most common ion exchange systems, essentially all of the ions are assumed to be free to move within the resin. However, in some adsorption systems, especially in chemical adsorption, the adsorbed components can be bound relatively firmly to the surface of the adsorbent pores and not be free to diffuse further into the adsorbent particle. In such cases, only the "free" molecules within the pores that have not become attached to the pore walls will be free to diffuse further into the adsorbent. This is called "pore" diffusion.

In other cases, adsorbed molecules may diffuse along the pore surfaces without spending significant time in the fluid phase within the adsorbent pores. This is called "surface diffusion" and can be important in many gaseous systems. Mass transfer in adsorbents and ion exchange materials can involve any of these mechanisms or any combination of them. Each has its own characteristics and can affect the adsorption rate. The ways in which each of these mechanisms can affect adsorption and ion exchange behavior are discussed more quantitatively in a later section, which describes predictions of adsorption and ion exchange breakthrough fronts and extraction of rate data from such fronts.

Pressure Drop

Pressure drop is an important consideration because it affects energy costs for operating adsorption and ion exchange systems, but extreme pressure drops can also increase the cost of vessels because thicker bed walls are required, which need stronger and more rigid adsorbents. The last factor is especially important for gel-type ion exchange materials and a few gel-type adsorbents. Useful equations help predict the pressure drop across packed beds of adsorbents or any granular or near spherical particles. A widely used equation was reported by Ergun [37]. The friction factor is predicted to be

$$f = \left[\frac{150(1-e)}{N_{\text{Re}}} + 1.75\right]\frac{1-e}{e^3} \tag{26}$$

where N_{Re} = Reynolds number = vd_p/ν
e = void fraction on the bed
v = superficial fluid velocity in the bed
d_p = adsorbent particle diameter
ν = kinematic fluid viscosity
f = friction factor = $[d_p/\rho e^2 v^2][dP/dZ]$
ρ = fluid density

The pressure drop across an adsorption bed can be calculated from the friction factor:

$$\Delta P = \frac{2fv^2\rho L}{g_c d_p} \tag{27}$$

where g_c is a conversion factor needed when using English units or nonsymmetric units. It is not necessary when using SI units. The Ergun equation was developed from an analogy to flow through tubes. The effects of e and $1-e$ on the pressure drop were predicted from an analogy to an effective cylindrical pore diameter with the same surface area as the void volume. Despite the obvious differences between the flow path through randomly packed beds of spheres and straight circular channels, this equation does give a relatively good prediction of the pressure drop.

Note that the equation consists of two terms. The first term corresponds to a pressure drop that is proportional to the velocity. The friction factor includes the velocity squared, but this term is divided by the velocity in the Reynolds number. This leaves the pressure drop proportional to the fluid flow rate. This term dominates the pressure drop at low velocities and corresponds to viscous (or creeping) flow.

The second term does not include the Reynolds number in the denominator and thus gives a contribution to the pressure drop that is proportional to the velocity squared. This term corresponds to inertial or turbulent flow. It may seem strange that the pressure drop can be predicted by adding these two terms. In normal channel flow the pressure drop is calculated from only one of the terms; the term used depends upon the Reynolds number. This may be where the different shape of the complex path around particles in a randomly packed bed of spheres and the circular channel analog has an effect. In a straight tube, the transition between viscous and turbulent flows can occur over a relatively small range of Reynolds numbers. However, in flow through a packed bed, there are likely to be regions, such as those near the points of contact between particles, where the flow rate will remain viscous (laminar) at average velocities high enough that most other regions of the flow will be turbulent. The flow around the particles is likely to include numerous constrictions and expansions. Constrictions (converging flow) tend to stabilize viscous flow, but expansions (diverging flow) tend to generate eddies and back flow that act much like turbulent flow, even under conditions that otherwise would correspond to viscous flow.

These arguments do not necessarily predict that the simple addition of the two terms in the Ergun equation should predict the pressure drop, but they do suggest that the transition from viscous to turbulent flow would not necessarily be as sharp as it is in straight channels. There will be viscous regions and viscous contributions to the pressure drop even at higher velocities and eddies somewhat like turbulence even at low velocities.

Other equations are used to predict the pressure drop in packed beds such as adsorbers. Perhaps the next most common one is by Leva

[38]. Generally these equations are no easier to use and offer little improvements in accuracy. The scatter in available data is greater than the differences between the correlations, and one should not expect exact predictions of the pressure drop. Deviations from the predictions could well be 50%, or even more. The equations are intended to apply to uniformly sized particles, but all particle collections contain some disparity in size. Thus, some of the differences between your results and the predictions may come from differences in the size distributions in your bed and in those used to develop or check the correlations. Similarly, there could be differences in the sphericity in the two sets of particles. Remember that even when you use perfectly uniform and perfectly spherical particles, the correlations may have been developed from data taken with somewhat nonuniform particles and with some variation in the particle size.

The Ergun equation should only be used for spherical or near spherical adsorbents. Most granular adsorbents can be considered near spherical, but one should use caution if an adsorbent is used which has obviously flat or long particles. Correlations are also available in handbooks for pressure drops through beds with other common packing shapes, such as rings or saddles, but most adsorbents are spherical, granular, or approximately right circular cylinders (pellets). These shapes are usually treated as approximately spherical.

Fluid (Axial) Dispersion

Dispersion within the adsorption bed occurs because of molecular diffusion, velocity gradients, and eddy motion. The importance of uniform packing has already been mentioned. It is also usually desirable to use adsorbent particles with as narrow a distribution in diameters as practical to reduce dispersion.

Axial dispersion incorporates the effects of nonuniform flow distributions, wall effects, and eddy mixing that occur within the bed. Good design is needed to eliminate the effects of poor flow distribution. Effective flow distribution requires careful packing of the bed as well as good design of the inlet distributor. The objective in packing the bed is to avoid variations in the void fraction within the column. Good operations can usually be achieved with some variation in the void fraction in the axial direction, but radial variations in void fraction are always detrimental because that results in axial flow at different radial positions (Figure 7). It is usually helpful to have a significant portion of the total pressure drop occur within the bed or in the outlet distributor rather than in the inlet fluid distribution system. It is equally important to design the fluid

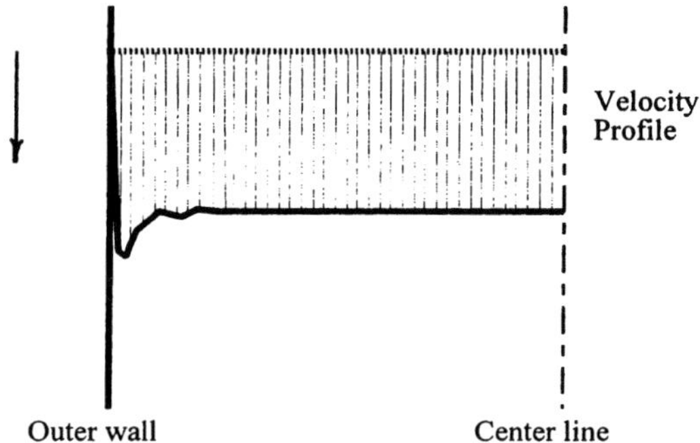

FIGURE 7 Typical velocity profiles in packed beds showing the radial dependence of axial velocities that are most significant near the wall. The wall is shown on the left and the centerline of the bed on the right. The velocity is zero at the wall, but reaches a maximum at a distance less than one particle diameter from the wall. The average velocity fluctuates with decreasing amplitude as one looks further from the wall and becomes statistically approximately constant only a few particle diameters from the wall.

removal system carefully. When the fluid must be transported from one bed to another, it is also important to minimize mixing in the connecting piping.

However, radial variation in the void fraction cannot be eliminated completely, even with careful packing of the bed. Wall effects result from higher void fractions near the wall and may be the principal source of void fraction variations. The higher void fractions occur near the wall because no adsorbent particles can penetrate the wall, so the void fraction at the wall is essentially unity. The average void fraction then declines with distance from the wall over approximately one adsorbent radius. Then the average void fraction may actually be a minimum at approximately one particle radius from the wall, and the local average void fraction may then oscillate somewhat with radial position as it approaches the overall average void fraction of the bed. The resulting axial flow pattern is shown in Figure 7.

The importance of wall effects is determined largely by the ratio of bed diameter to adsorbent particle diameter. A "rule-of-thumb" says that the ratio should be at least 8 or 16 to hold the wall effects to acceptable levels. With higher ratios the "wall region" where the average void fraction differs significantly from that of the bulk bed becomes a smaller fraction of the entire bed. Actually, the lowest acceptable ratio probably depends upon how effectively the other sources of dispersion and mass transfer resistance have been eliminated. Thus, these rules apply to the usual packed beds, and such ratios are probably reasonable for most industrial scale waste and environmental applications. However, in some cases where there is great incentive to reduce dispersion, significantly larger ratios may be needed. One recent study suggests that wall effects could not be ignored until the column diameter was increased to 1 m. The particle diameter was not given, but probably was less than 1 mm (a ratio of at least 1000).

PROCEDURES FOR DESIGNING ADSORPTION COLUMNS

General Considerations

There are several potential goals for an adsorber, so there is no general approach to optimization. This section will identify those factors that affect adsorber capital and operating costs. A typical two-column adsorption operation is shown in Figure 8. Two beds are used so that one can be in operation at all times, and the other bed is being desorbed and prepared for return to service when the first bed becomes loaded. Valves (not shown) are used to divert the feed alternately from one bed to the next after the first bed becomes saturated or after the breathrough curve

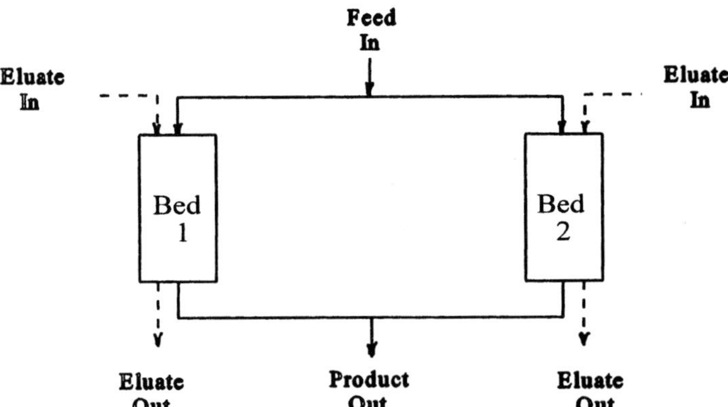

FIGURE 8 Operation of two adsorption beds: one bed is always in operation while the other bed is being regenerated; valving is not shown. Fluid follows the solid lines to the bed that is in service (adsorbing), and regeneration fluid follows the dashed lines to the other bed while it is being regenerated. The two beds alternate being in service and being regenerated.

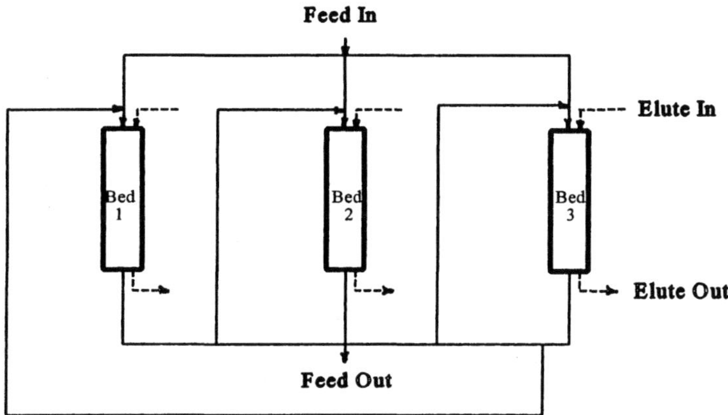

FIGURE 9 Operating three beds so that two sequential beds can be in operation (in service) while the third is being regenerated; valving is not shown. Fluid follows solid lines going into one bed and then into a second bed; regeneration fluid follows dashed lines to the bed being regenerated. Periodically the first bed becomes saturated and is taken off-line to be regenerated; the second bed then becomes the first bed; and the bed that was just regenerated becomes the second bed. This illustrates how a larger fraction (in this case two-thirds) of the beds can be kept in service and operated in series when the regeneration time is much shorter than the adsorption time.

shows too much solute. The first bed is then regenerated while the second is used as the adsorbent bed. This arrangement suggests that desorption (regeneration) is at least as fast as the adsorption step, and that is often the case, especially when the adsorbate in present in the feed at very dilute concentrations and the ratio of the adsorbent loading to the solution concentration (the distribution coefficient) is very large. This is usually the case in environmental problems. Actually, in some cases, it may be possible to have several columns in operation and only one being desorbed if the desorption time is much shorter than the adsorption time. When possible, this may be preferable because it permits greater use of the available bed and adsorbent capacity, and this is most likely to be the case when solutes are being adsorbed from dilute feed solutions.

An alternate multiple bed arrangement is shown in Figure 9. Such systems can become relatively complex and contain considerable piping and valves. In the figure the feed can go to any of the three beds, and the effluent from any bed can go to the next bed serially. Then the loaded bed can be isolated and regenerated. Alternatively, one can arrange the piping to allow the effluent from any bed to go to any other bed, not just to the the next one in a series, but such arrangments can become complex as the number of beds increases, so the simpler arrangement of sequential beds is more likely to be used. In the illustration, two beds are operated in series while another bed is desorbed. When the first bed becomes completely loaded and the front has passed into the second bed, the feed point is changed to the entrance to the second bed, the effluent from the second bed is diverted to the inlet of the freshly desorbed bed, and desorption of the first bed begins. That is, bed 2 replaces bed 1; bed 3 (freshly desorbed) replaces bed 2; and bed 1 replaces bed 3 (and desorption begins on bed 1). This arrangement may be desirable if the front is broad and/or if it is desirable to load the beds as fully as possible so that the pollutant can be recovered at as high a concentration as possible.

Operating Costs

In most problems, the principal operating costs are for the energy and/or reagents required to desorb or regenerate the beds and the pressure drop required to pump the fluid through the bed. With gaseous systems, regeneration usually takes place at higher temperatures and/or at lower pressures. These systems are often called "thermal swing adsorption," "pressure swing adsorption," or "vacuum swing adsorption," depending upon which parameter is changed to desorb the bed. The energy required to heat the regeneration gas (or the bed) is likely to be a major cost [39]. When the beds are regenerated off-site, often by the vendor, the cost of regenerations is the price charged for the regeneration plus the shipping cost and the loss of adsorbent from the regeneration. Pressure drop is more likely to be important in cases where small particles are required, either because high performance is required or because very slow adsorption rates would be involved otherwise. Regeneration of adsorbents used with liquids is more likely to involve the use of reagents (changing the pH or using exchanging ions). However, in some cases such as adsorption of hydrophobic solvents from water, one may drain the bed and use thermal (and/or pressure) swing regeneration. Regeneration of ion exchange beds essentially always requires the use of a reagent.

The condition of the recovered solute can also affect operating costs. It is often desirable to recover the solute at a high concentration. If the solute is to be recovered for use, it is usually desirable to recover it as a high concentration product and avoid costs for further concentration. For products to be marketed, shipping costs per unit mass of a product are usually lower for highly concentrated materials. Thus, both adsorption and desorption conditions could affect costs via the concentration or purity of the desorbed product. Generally, high adsorbent loading will give the highest product concentration. Desorption conditions that leave only very low residual loading on the adsorbent in the bed usually allow removal of the solute to lower levels in subsequent adsorption tests. Reversed flow directions for desorption can assist in achieving high desorption concentrations and leave any residue of contaminant on the portion of the bed near the normal fluid inlet and farther from the normal exit. Spreading of the adsorption and desorption fronts respectively can play an important role in the degree of column loading and the efficiency of desorption, and these will be discussed in a later section. Since most adsorption isotherms used industrially are likely to be "favorable" (a term to be described later that refers to Type I isotherms), it is often desirable to desorb under conditions with very low residual equilibrium loading so that the "unfavorable" shape of the desorption equilibrium isotherm will not have a major effect.

Capital Costs

For many environmental problems, capital costs can be an important part of the overall costs. Capital costs include the costs of the adsorption bed, the adsorbent, and the associated pumps, instrumentation, etc. The costs for piping, pumps, instrumentation, etc., are usually estimated as fractions of the "major equipment" costs such as the cost of the beds themselves. These fractions can be quite high, exceeding the cost of the major equipment by a fewfold. For more details on cost estimates, consult one of the standard texts on design and cost estimation for process systems [40]. When greater accuracy is needed, actual layouts of the piping and specifications for the instrumentation should be prepared and used to make the capital cost estimates.

Some adsorbents are relatively inexpensive and will not be a major part of the capital costs, but other adsorbents made of special materials designed to give very high specificity or very high adsorption rates can be expensive enough to make their contribution to capital costs a major concern. Furthermore, if the adsorbents have a short life, the cost of replacement adsorbent will need to be included with operating costs.

Selecting the Bed Size

Optimization of an adsorber design can involve selection of the proper size bed. The diameter of the bed is usually set by the required fluid flow rate to give an acceptable pressure gradient, and optimizing the bed size usually means selecting the best bed length. For a given volumetric flow rate, the pressure drop is likely to be directly proportional to the bed length and (approximately) inversely proportional to the square of the bed diameter. For high pressure operations, the cost of increased wall thickness needed for larger diameters may also be a consideration. The length of the bed is likely to be the major optimization parameter. Generally the bed will have to be large enough to operate for an acceptable period of time without adsorbate breakthrough exceeding the allowable limits. In environmental applications, this is often the allowable release rate or allowable concentration in the effluent. Increasing the size of the bed increases the capital costs for the bed and the adsorbent and the operating costs through the increased pressure drop. However, longer beds can decrease the operating costs if desorption is required less frequently and if the adsorbate can be recovered at a higher concentration.

The following discussion will focus upon how a given adsorption bed will perform. By exploring the performance of different design options, an optimum design for any particular application can be approached.

Summary of Adsorption/Ion Exchange Bed Design Goals

Before explaining details on the calculations needed to design adsorption columns, it will be helpful to outline what the designer is trying to do and the design procedure that will be suggested. First a procedure will be suggested for predicting the performance of any adsorption bed. The resulting evaluations of bed performances can be repeated to explore different bed (column) sizes, regeneration frequency, regeneration conditions, etc., to select the optimum design for the application.

The suggested procedure for predicting bed performance involves starting with very simple and approximate estimates and then considering more complications. Thus, the more analysis that one employs the more accurate the results, but it is desirable to keep the calculations simple until the approximate size of the bed is known before more complex calculations are begun. The degree of detailed calculations required for an application depends upon the properties of the system and the degree of accuracy needed. Some systems will have very rapid adsorption or exchange rates and favorable isotherms (see the earlier section "Classification of Isotherms"); then very simple calculations can give excellent estimates of the behavior of the systems. Other systems may involve less favorable and even complex isotherms, and more detailed calculations will be needed to describe adsorption bed behavior accurately.

There are considerable differences in the accuracy needed for applications. Careful and accurate design calculations that result in accurate predictions of bed performance provide a basis for optimizing the capital and operating costs to obtain the most desirable system. Of course, costs are required to carry out a careful design, and there is an appropriate optimum total effort. If there is little hazard or risk involved in failure to achieve desired performance and/or if gross overdesign of the column for increased safety factor does not add enough cost to justify time and effort for more careful designs, there may be little need to perform more than simple preliminary calculations and add "safe" additional column capacity. Generally, the need for greater accuracy and reliability increases with the overall cost or importance of the adsorption system. Excessive capacity for a very large adsorption unit could be costly, especially if several such units are to be built. Even with moderate size units, very expensive adsorbents, or high operating costs, perhaps due to expensive desorption/regeneration costs, would require detailed and near optimal design. For environmental systems, more detailed analyses may be needed to give suitable assurance that the bed is sufficiently large to meet the needs and requirements. Adding safety factors is not safe if you do not know how much to add. Greater care is often needed to recover the desorbed "product" (which could be a pollutant) at high concentrations, so the condition of the desorbed product may also be a factor in deciding how much effort and detail to include in the adsorber design.

Computer codes can be used to design adsorption or ion exchange beds, but the available codes are usually not as easily used nor as readily available as the more standard routines commonly used for steady-state continuous process steps, such as distillation, absorption, etc. When suitable computer codes are available and understood by the user, they are recommended. The following sections describe how to design adsorption beds without the use of packaged computer codes, but the material should be helpful to those using computer codes as well as illustrating the calculations made within the codes.

Estimation of Adsorber Performance

Maximum Loading

Estimating adsorber performance means estimating the degree of solute removal as a function of time, which is estimating the breakthrough curve. The crudest estimation of adsorber performance can be obtained simply from the adsorption capacity of the adsorbent in the bed. That is, one can assume that the adsorbent becomes immediately saturated with the inlet (feed) stream and that a front moves down the bed until the bed becomes loaded (saturated with the inlet fluid) (Figure 10). The idealized case in Figure 2 is shown by the solid curve where the advancing front is very sharp, but the more realistic (or common) case is illustrated by the dashed curve. This is a "diffuse" front in which the concentration drops toward zero (where the adsorbent initially contains no solute) more

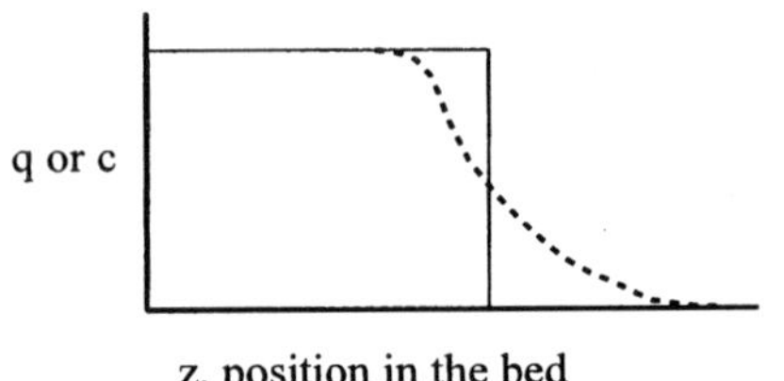

FIGURE 10 Sharp (idealized) concentration profiles (the solid curve) can be used to crudely estimate how long an adsorption bed can operate before requiring regeneration, but such estimates will always be optimistic because a realistic loading front will be diffuse, as shown by the dashed curve. Estimation of the width of the diffuse front is important to optimization of adsorption operations and for ensuring that breakthrough concentrations are maintained at acceptable levels.

slowly. The front can be seen by examining the loading on the adsorbent along the length of the bed after a certain amount of the adsorbate has been added (Figure 10) or by measuring the concentration in the fluid as a function of time until solute begins to escape from the bed (breakthrough) (Figure 11). The only information needed for such a simple estimate is the loading on the adsorbent that would be in equilibrium with the inlet fluid. This value can be obtained from equilibrium curves, usually called "isotherms," or from measurement of the equilibrium loading at one concentration, that of the inlet fluid. With these simplifications, the time that a bed can operate without desorption is estimated to be

$$t_{\max} = \frac{Vq_{\text{equil}}}{UC_{\text{inlet}}} \tag{28}$$

As simple as this approach is, it can be reasonably accurate for several systems, namely those with very favorable isotherms (later shown to be associated with sharp loading fronts) and rapid mass transfer rates. This

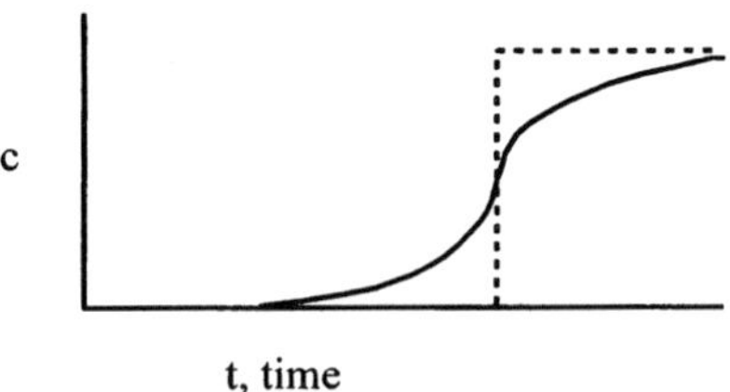

FIGURE 11 The differences between sharp idealized concentration profiles and realistic diffuse profiles are also evident in the concentration breakthrough curves.

will always give an optimistic estimate, which is the maximum column capacity or the longest time before breakthrough. In practice, it is desirable to include some safety factor in the estimate of the time that a given column will break through or for the volume of a bed (column) required to operate for a given time between regenerations. This safety factor is required because the loading front will not be sharp in all cases. Instead, the front will be diffuse as shown by the dashed curve in Figures 10 and 11. If the spread of the front is minor, adding only a little extra bed volume or making the operating time between regenerations only slightly shorter may give satisfactory performance. Then this simple approach may be satisfactory to predict performance. However, for reliable predictions of the bed performance, one should have reliable ways to estimate the behavior of this diffuse region—that is, to estimate how much bed (column) height (or volume) is associated with the diffuse region of the front where the loading on the bed goes from the equilibrium to essentially zero. That is, the "additional" bed height should be estimated by a rational method.

The spread of the diffuse front can be caused by the shape of the equilibrium curve (isotherm), by slow (finite) adsorption rates, and by axial dispersion (a term discussed earlier that accounts for molecular diffusion in the axial direction, eddy mixing in the region between adsorbent particles, and nonuniform flow in the bed). Much of the rest of this section will be devoted to explanations of ways to predict the amount of bed length or volume that needs to be used to account for the width of the diffuse front. The shape of the isotherm can enhance the "spreading" of the diffuse front (an unfavorable isotherm), can contribute to sharpening (reducing the width) the front (a favorable isotherm), or can be neutral and not contribute to spreading or shrinking of front width. Slow adsorption rates and axial dispersion always contribute to spreading the diffuse front. When the isotherm contributes to shrinking the front and adsorption rates and dispersion contribute to spreading the front, a constant front width and a constant shape front can develop when the effects of the two opposing factors become equal.

Initial and Early Breakthrough

There is often strong interest in the maximum removal efficiency of an adsorption bed. That is, when a "fresh" adsorption bed is placed in service, how far will it reduce the concentration of a contaminant? This initial concentration can be a function of several factors. Note first that the initial contaminant concentration in the effluent is actually a part of the breakthrough curve, but it corresponds to the "short time" case, and some of the simplest forms of the solutions for breakthrough curves correspond to "long time" cases. This means that the solution described here to predict breakthrough curves at longer times may not be accurate for the short time periods.

Note next that the initial or early breakthrough curve can be strongly dependent upon the initial state of the adsorbent in the bed. If the bed is not completely desorbed, there will be some contaminant left on the adsorbent or ion exchanger, and this can contribute to the initial breakthrough. Even small traces of contaminant left on the adsorbent can be the dominant cause of an initial breakthrough. This initial contaminant in the adsorbent can be taken into account in some of the approaches described in the following section only if the contaminant is initially distributed uniformly in the bed. If the contaminant is more concentrated near the bottom (outlet) of the bed as it is likely to be if regeneration is done in the bed and in the same direction as the adsorption flow. If the adsorbent is regenerated off-site in another facility, the regenerated adsorbent is likely to be well mixed before it is replaced in the bed, and any residual contaminant concentration is likely to be distributed uniformly in the bed.

If even a trace fraction of the contaminant exists in more than one chemical form, this could contribute to the initial breakthrough, especially if one form of the contaminant does not adsorb. Finally, note that initial breakthrough curves can be extremely difficult to measure because most good adsorption beds reduce the contaminant concentrations initially to levels that are difficult to detect and especially to measure accurately. Although this period may be of interest, it is often difficult to follow accurately when contaminants must be removed to levels that challenge available analytical chemistry methods. The following section will discuss the prediction and evaluation of breakthrough curves over the regions where the concentration of the solute (contaminant) is a significant fraction of the inlet concentration. The equations and other solutions can be extrapolated to any level of solute concentration desired, but the reader should remember that trace contaminant on the initial adsorbent and/or different forms of the solute, rather than the extrapolated curve, could determine the initial and early breakthrough.

DYNAMICS OF PACKED BEDS AND DESIGN OF ADSORBENT OR ION EXCHANGE MATERIALS

To design adsorption systems, one needs to know the adsorption equilibrium just discussed, the rate processes involved in the adsorption process, and the hydrodynamics of the adsorption bed (pressure drop and mixing phenomena in the bed). There are also special mechanical design problems, such as uniform distribution of the fluid over the cross-section of the bed and prevention of channeling or nonuniform flow in the bed, which may be addressed by standard designs from bed manufacturers but must be considered for "in-house" custom designed beds.

As noted earlier, the isotherms (equilibrium conditions) play the most important single role in the performance of an adsorption column. In those cases, the size of the column (bed) and the amount of solute removed by the bed can be estimated crudely from the adsorption isotherm. Adsorption will take place first near the entrance of the bed. The region near the fluid entrance eventually becomes saturated with the solute, that is, in equilibrium with the incoming fluid. At that point, no more adsorption will take place in that region of the bed, and the fluid will then carry solute into deeper regions of the bed, further from the fluid inlet. The region of the column in which the adsorbent is saturated then grows with time, and the boundary of this region (sometimes called the loading front) moves toward the fluid outlet. The behavior of a simple sharp loading front is illustrated in Figure 10. The concentration of the adsorbate in the effluent from the bed (Figure 11) remains near zero until breakthrough occurs; then the effluent concentration rises quickly to the inlet concentration.

In the cases described, the loading front is sharp, and the capacity of the bed can be estimated relatively accurately simply by assuming that little or no solute will leave the bed until sufficient fluid has entered it to carry enough solute to saturate the adsorbent in the bed [that is, by using Equation (28)]. In all cases, the adsorption front will be at least somewhat diffuse, and the concentration of solute in the fluid will not drop from the inlet concentration to zero instantaneously. If the length of the mass transfer zone (loading front) can be estimated, the bed length needed for removal of a specified amount of solute can be estimated simply by estimating the bed length needed if the mass transfer front were short (breakthrough was sharp), as described in the previous section, and adding to that length that of the mass transfer zone. (Alternatively, for a given bed volume, one could stop loading the bed sufficiently early before enough adsorbate enters the bed to saturate it and thus compensate for the length of the mass transfer zone in terms of time.) Several factors can affect the length to the mass transfer zone, and these will be discussed next.

Shape of the Equilibrium Isotherm

The equilibrium isotherm affects the length of the mass transfer zone as well as the maximum adsorption capacity of the bed (that is, the capacity of the bed if the entire adsorbent were loaded sufficiently to be in equilibrium with the inlet fluid). The maximum capacity of the bed is determined by a single point on the isotherm, the concentration in the adsorbent in equilibrium with the inlet fluid. However, the concentration in the fluid within the mass transfer zone varies from zero (if the adsorbent initially contains no solute) to the inlet concentration, and all of the equilibrium isotherm covering that region affects the mass transfer zone. The qualitative behavior of the mass transfer zone can be predicted simply by observing the shape of the isotherm over this region.

Simple isotherms can be described as "favorable," "neutral," or "unfavorable." For adsorption, when solute is being added to the bed, the isotherm can be described as favorable if it is concave downward; that is, if it has a negative curvature. See curve A in Figure 6a, which is a Type I isotherm. Such isotherms are common. Remember that the Langmuir isotherm, the Freundlich isotherm, and many other common isotherm shapes are concave downward and thus are "favorable" for loading (adsorption) processes. Even Type III isotherms have negative curvature at very low concentrations. Such isotherms would be considered favorable if the highest concentration in the bed (the inlet concentration) were low enough that the curvature was negative over the concentration range used in the bed.

These isotherms are favorable because their shape contributes to decreasing the length of the mass transfer zone. In fact, with no dispersing forces, such as slow mass transfer or axial diffusion, any finite length mass transfer zone will shorten with time as it passes through the bed; with sufficiently long beds the length of the mass transfer zone approaches zero. Thus, this type of isotherm is most likely to give sharp loading fronts and short mass transfer zones. Then the crude analysis of bed performance based simply upon the adsorption capacity of the bed is most likely to be realistic, that is, a reasonable estimate.

The role of curvature of the equilibrium isotherm on the length of the mass transfer zone is illustrated by examining the simplest case when there is local equilibrium within the bed, that is, when there is no axial dispersion from the flow patterns and no mass transfer resistance. This means that at all locations in the bed the solid adsorbent and the fluid are in equilibrium. For local equilibrium, it is necessary for any mass transfer phenomena involved in the adsorption to be (infinitely) rapid and for the fluid to pass through the column in plug flow (with no axial dispersion). Of course, these conditions can never be maintained fully in real systems, but with very small adsorbent particles and low flow rates, bed performance can approximate local equilibrium reasonably well for many systems.

A mass balance on the adsorbate flowing through the column with local equilibrium is given by

$$\frac{\partial C}{\partial t} + v\left(\frac{\partial C}{\partial x}\right) + \frac{1-e}{e}\left(\frac{\partial q}{\partial t}\right) = 0 \tag{29}$$

The first term accounts for the accumulation or rate of change of solute concentration in the fluid at a specific location in the bed with time. The second term describes the net flux of the adsorbate from that region by convection, and the last term describes the rate at which adsorbate is removed from the fluid in the region by adsorption. With local equilibrium, the concentration in the solid is always in equilibrium with the concentration in the fluid. The rate of adsorption then results from changes in the fluid concentration. The concentration in the adsorbent is related to the concentration by the equilibrium isotherm. That is, the adsorbent loading, q, is a function only of the concentration, C. Thus, with local equilibrium, the second and third terms in Equation (29) can be combined since

$$\frac{\partial q}{\partial t} = \frac{\partial C}{\partial t}\left(\frac{dq}{dC}\right) \tag{30}$$

Note that since the solid and fluid are in equilibrium, dq/dC is the slope of the equilibrium isotherm. This relation can be used to eliminate $\partial q/\partial t$ in Equation (29). The rate at which the front moves down the column is

$$v_{\text{front}} = \left(\frac{\partial z}{\partial t}\right)_C = -\left(\frac{\partial C}{\partial t}\right)_z \Bigg/ \left(\frac{\partial C}{\partial z}\right)_t$$

$$= v\left(1 + \frac{1-e}{e}\frac{dq}{dC}\right)^{-1} \tag{31}$$

Thus, the rate at which the front moves down the bed is a function of the slope of the equilibrium isotherm. If the isotherm is curved, the slope will vary with concentration. For the simplest (linear) isotherm its slope is constant and independent of concentration. Slopes of favorable isotherms are greatest at low concentrations and least at higher concentrations. This means that the portions of the front with the highest concentrations will move through the column faster than the portions with lower concentrations, and the highest concentrations are on the back portion of the front (Figures 6a, 12). If the front begins with a relatively wide mass transfer zone (Figure 12), the leading part of the zone will have low concentrations and progress down the column at a lower rate than the trailing regions with higher concentrations. The higher concentration regions will eventually catch up with the lower concentration regions, and the length of the mass transfer zone will decrease.

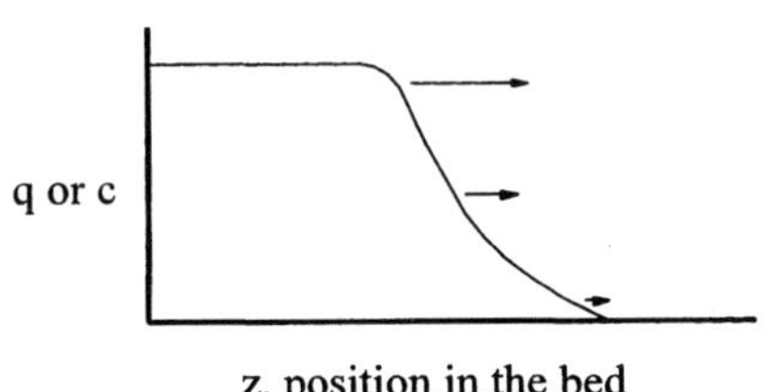

FIGURE 12 Relative movement of different portions of a diffusion concentration front in an adsorption bed operating with a favorable isotherm (idealized case with no mass transfer resistance). The region with higher concentration moves more rapidly than the region with lower concentrations. This causes the diffuse region of the concentration front to decrease in length with time (as the front moves down the bed); in realistic cases mass transfer resistance will prevent the width of the diffuse front from decreasing all the way to zero.

Conversely, an equilibrium curve that is concave upward (positive curvature) as a Type II isotherm would be unfavorable. Fortunately, such curves are not as plentiful, and one is less likely to select and apply adsorption systems with unfavorable isotherms. It is usually undesirable to use a system with a highly unfavorable adsorption isotherm.

Linear isotherms are neutral since they contribute neither to spreading nor contracting of the mass transfer zone. Note that from Equation (31), with a linear isotherm, the rate at which a given concentration portion of the front moves down the column is independent of the concentration. In other words, with local equilibrium all portions of the front move at the same rate and the shape of the front does not change. A sharp front (short mass transfer zone) remains sharp; a diffuse front remains diffuse and maintains the same shape and dimensions (thickness/width). Linear, or approximately linear, isotherms may be common in dilute systems such as those most often encountered when removing trace pollutants from waste streams. Remember that the Langmuir isotherm approaches a linear isotherm at very low concentrations. Thus, the favorable behavior of Langmuir systems is not evident at sufficiently low concentrations. However, highly specific adsorbents with great affinity for the adsorbate may reach high loadings even at trace gas/liquid concentrations and deviate far from linearity.

Just because practical isotherms are likely to be favorable does not mean that one is likely to escape the effects of unfavorable isotherms. If the adsorbate is being desorbed (eluted) from the adsorbent filled bed, a favorable (loading) isotherm becomes an unfavorable desorption (unloading or elution) isotherm (Figure 13). Remember that in the mass transfer

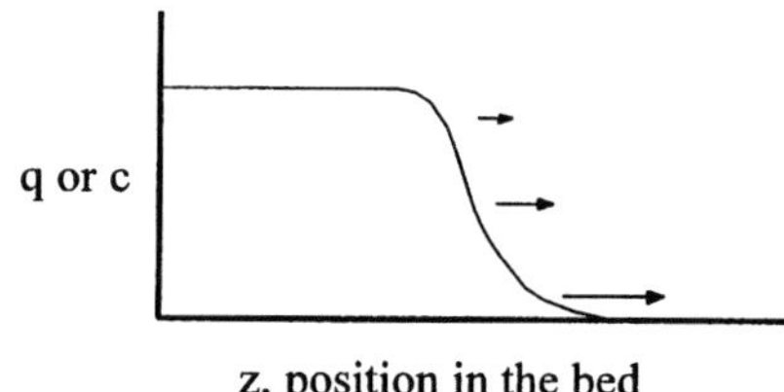

FIGURE 13 Relative movement of different portions of a diffuse concentration front in an adsorption bed operating with an unfavorable isotherm (idealized case with no mass transfer resistance). The region with lower concentrations moves more rapidly than regions with higher concentrations. This is the opposite of cases with favorable isotherms, and the diffuse front widens with time (as the front moves down the bed). Mass transfer resistance in realistic cases adds to spreading of the diffuse front with time.

zone for adsorption, the regions of low concentrations are in the forward portion of the zone. However, in the mass transfer zone for desorption, the low concentration region is in the rear portion of the desorption mass transfer zone (the desorption front). For a concave downward isotherm (negative curvature), the region with low concentrations moves down the column more slowly than the region with high concentrations. For adsorption, this causes the length of the mass transfer zone for loading to decrease with time or position down the column. However, with a desorption front, the slower motion of the region of low concentrations causes the mass transfer zone to increase in length. The length of the mass transfer zone increases steadily with time, or distance down the bed. This, of course, is unfavorable for desorption. Fortunately, elutions usually are carried out under conditions sufficiently different (lower pressures, higher temperatures, or different solution concentrations) that the importance of the unfavorable desorption isotherm can be minimized. Even if the isotherm remains unfavorable for desorption, in many such cases effective desorption is still possible because the equilibrium concentrations under desorption conditions may be very low. Conversely, if the adsorption isotherm were unfavorable (positive curvature as in a Class II isotherm), the isotherm would be favorable for desorption.

In many waste management applications, the main goal in desorption of the solute from the bed is not complete recovery of the solute but preparation of the bed so it can operate effectively in the next adsorption cycle. Of course, solute left on the bed will reduce its capacity for adsorbing additional solute during the next adsorption cycle. As noted in the previous section, residual solute on the adsorbent can play an important role in limiting the ability of the bed to reduce the solute concentration initially leaving the bed. Small amounts of remaining solute will not be a serious problem provided that solute does not leave the column during the next adsorption cycle. That means that the adsorbate remaining on the bed should not be located near the fluid outlet.

When an unfavorable desorption isotherm (a favorable adsorption isotherm) makes complete desorption of solute from the bed difficult or impractical without using excessive volumes of eluate, one can consider desorbing the bed in the reverse direction. This type of operation is always more effective in preparing the bed for the next adsorption cycle, since any remaining solute will be moved toward the end of the bed that is used for the fluid inlet during the adsorption cycle. This reversed direction desorption should be considered when it appears to be impractical to desorb the bed completely. At least this approach does not transfer unremoved solute closer to the fluid outlet, but the unfavorable desorption still reduces elution performance. When the desorption cycle is able to remove essentially all of the solute, the direction of desorption is not so important, and it may be practical to avoid any additional expense required to operate the bed with flow in both directions (during different operating cycles). To operate a bed with flow in either direction, it is usually necessary to provide restraints (screens or porous plates) above and below the adsorbent bed. This is more difficult if the adsorbent or ion exchange material swells during loading or during desorption. In such cases, the upper screen can be spring loaded so that it always maintains sufficient force on the swelling/shrinking resin to keep the bed intact.

Effects of Rate Processes on the Mass Transfer Zone

The preceding discussion assumed that local equilibrium existed throughout the bed. Some adsorption systems can be operated so that this condition is approximated. In most cases, however, it is only practical to operate adsorption beds under conditions where the local equilibrium assumption is not valid. For instance, increasing the fluid flow rate increases the adsorption system's productive capacity, but this could increase the length of the mass transfer zone because the mass transfer phenomena necessary for adsorption may not be able to "keep up" with the higher mass transfer rates need for the high fluid flow rate. Obviously, one would gladly sacrifice small increases in the length of the mass transfer zone if much greater bed throughput could be achieved. Also, increasing the size of the adsorbent particles reduces the pressure drop across the column, but this could also increase the length of the mass transfer zone because the larger adsorbent particles have less surface area per unit volume, longer diffusion paths within the particles, and greater axial dispersion. All of these increase the effective resistance to mass transfer (or decrease the effective mass transfer rates).

Incorporation of the effects of mass transfer resistances into estimations of the length of the mass transfer zone is often needed to select optimum bed operating conditions and the optimum adsorbent and adsorbent particle size. If bed size were based upon the adsorption capacity of the adsorbent alone, almost any bed shape could be used with the same result. Only the volume (quantity) of adsorbent in the column would be important. The bed diameter would be estimated from the total fluid throughput required, as noted in Equation (28). If one had chosen a very short bed, it may not be possible to maintain the steep concentration gradients required for thin beds. That is, the length of the mass transfer zone could be a significant fraction of the total bed length, and the bed could be operated for only short periods between regenerations. However, in more realistic situations, one is more likely to select an optimum velocity through the bed and an optimum bed length. The following discussion will address the question of how to estimate the performance, usually the breakthrough curve, of an adsorption bed operating with a high throughput where mass transfer resistances are important considerations.

Diffusion Resistance in Films and Pores

To be adsorbed, the adsorbate must diffuse through a fluid film surrounding the adsorbent particle, then diffuse into the particle, and finally attach to the surface or adsorption site. Of course, the fluid film surrounding the adsorbent particles is not truly stagnant and does not have a sharp boundary, but it is convenient to approximate the diffusion resistance to the surface of the particle by an imaginary stagnant film surrounding the particles. Remember that the mass transfer flux through the film is traditionally given as

$$\text{flux} = k(C - C_i) \tag{32}$$

where C is the concentration in the (bulk) liquid, and C_i is the concentration in the fluid at the particle surface.

Diffusion within the adsorbent particle can occur by several mechanisms. The solute can diffuse through pores within the particle. The concentration in the pore at the outer radius of the solid (at the apparent surface with the fluid) will be the same as the concentration in the fluid at the interface, C_i. The solute usually adsorbs on the surfaces of the pores very rapidly, so adsorption kinetics usually does not necessarily affect the overall adsorption rates significantly. Since the adsorption rate is very rapid, one can assume that the concentration on the pore surface is in equilibrium with the fluid in the pore. In those cases, the concentrations in the pore and on the solid are related by the equilibrium expression. Diffusion within the pore is likely to follow Fick's law:

$$\text{flux} = -D\nabla C \tag{33}$$

The diffusion process can also be described in terms of the total concentration of the adsorbates in the solid, in the fluid within the pores, and adsorbed on the pore walls:

$$q_t = C + q \tag{34}$$

where q_t is the total concentration of the adsorbate (solute), C is the concentration in the pore, and q is the concentration on the pore. All terms are expressed as moles of the adsorbate per volume of solid and pore, and all terms are functions of position within the solid particle.

For a linear isotherm, q is always proportional to C, so the terms on the right can be combined:

$$q_t = (1 + K_d)C \tag{35}$$

where K_d is the slope of the equilibrium curve, or q/C. Fick diffusion within the particle can then be described in terms of an apparent diffusion coefficient:

$$\text{flux} = \left(\frac{D_s}{1 + K_d}\right)\nabla q_t \tag{36}$$

This equation recognizes that only a portion of the total solute in the particle is available to diffuse in the particle. If the adsorbed material is not stationary but moves by surface diffusion or bulk diffusion in the solid, additional terms need to be included. The diffusion coefficient within the particle pores is not necessarily the same as the diffusion coefficient in the bulk fluid. This coefficient may account for the void fraction of pores in the solid and the orientation of the pore within the particle (a tortuosity factor) as well as wall effects. In principle, these factors can be incorporated into the expression for the mass transfer rate separately, but in many cases it is impractical to separate them. The term in parentheses can be viewed as an effective diffusion coefficient that incorporates all of these factors.

Similar relations to that in Equation (36) can be written for any other isotherms, but the simple proportionality does not hold for nonlinear isotherms, which means that the effective diffusion coefficient is then a function of concentration. When the isotherm is nonlinear, the distribution coefficient, K_d, is not constant but varies with concentration. This makes the expression more difficult to use, but numerical calculations with the effective diffusion coefficient varying with concentration can be made.

As described, "pore diffusion" usually refers to cases where the solute adsorbed on pore walls is not free to diffuse; hence, the adsorption process proceeds into the particle by diffusion of free (unadsorbed) solute in the pores. However, sometimes all of the solute in the adsorbent (or ion exchange material) can diffuse. Remember that ions in ion exchange materials are not attached to any ion exchange site in the resin but are free to move (diffuse) within the resin particle. Then the diffusion flux can be proportional to the gradient of the total solute concentration, and the effective diffusion coefficient in Equation (36) is simply the diffusion coefficient for the solute in the resin.

Surface Diffusion

In other cases, the adsorbed material is not rigidly attached to any point on the surface, but it can also diffuse into the particle along the surface. This is known as surface diffusion. Surface diffusion contributes to the mass transfer rate, but it is not always easy to separate the effects of surface diffusion from other factors that determine the effective overall diffusion coefficient. As noted in the previous section, when the isotherm is linear, surface diffusion and pore diffusion are additive, and there is a constant effective diffusion coefficient within the adsorbent particle. When the isotherm is not linear, the two diffusion effects will not be additive in such a simple way. When surface diffusion makes the major contribution to diffusion within the adsorbent particles, the diffusion process resembles the case where all (in this case, essentially all) solute in the solid particles is free to diffuse.

Solid Diffusion

In other cases, it may not be meaningful to think in terms of pores and surfaces. Many ion exchange materials used to remove metals from dilute aqueous solutions may be more appropriately viewed as gels or swollen polymers containing fixed ionizable sites which can exchange counterions with the surrounding solution. In those cases, the concentration of the adsorbate in the solid can be viewed simply as the concentration in an apparently homogeneous media. That is not much different from using the overall concentration $(q + C)$ in the adsorbent and an effective diffusion coefficient. The important point to remember for most ion exchange processes is that all of the ions in the resins are free to diffuse, and diffusion can be viewed as a bulk diffusion in the solid, not as diffusion within a pore. The exceptions will be the cases where ions are known to be closely associated with specific resin sites. These could include diffusion of hydrogen in weak acid resins and diffusion of certain ions in other highly selective systems.

Approximation of Diffusion Within Particles as an Effective Film Resistance

When all of the solute in the solid particles is free to diffuse (ion exchange, surface diffusion, and solid phase diffusion), the diffusion process can be approximated by an effective film resistance in the solid phase, much like that usually used to describe mass transfer resistance in the fluid phase. That can simplify analyses of adsorption loading fronts. Glueckauf [41] noted that the rate at which a mobile solute diffuses into a spherical particle can be represented approximately by an effective "solid film" coefficient equal to $15D/r^2$, where D is the actual diffusion coefficient in the solid particle and r is the particle radius. Additional discussion of the approximation and how it can be used will be given later in this chapter when calculations of the front shape and thickness are presented. However, at this point, the reader is reminded that it does not apply to cases with significant "pore diffusion" when all of the solute is not free to diffuse but is held on the pore surfaces.

Macro-Reticular Adsorbents

The usual discussion of mass transfer resistance in adsorption or ion exchange includes resistances for the fluid film surrounding the adsorbent and for diffusion within the particle. Diffusion within the particle can occur in the pores or along the surfaces of the pores (i.e., surface diffusion). However, with macro-reticular adsorbents—those with two or more pore sizes—diffusion within the particles can be slightly more complicated. As noted earlier, these particles could be constructed such that they are clusters of smaller adsorbent particles held by a binding material. The openings between the small subparticles may be much larger (perhaps an order of magnitude or more) than the pores within the small subparticles. The diffusion coefficient within the larger pores can be much larger (perhaps by several orders of magnitude) than the coefficient in the small subparticles. The very small diffusion coefficient within the material of the subparticles is usually the reason for building macro-reticular particles; hence, extremely small primary particles can be used.

Diffusion within a macro-reticular particle usually involves diffusion into the particle through the larger pores followed by diffusion into the subparticles. In some cases, the diffusion rate through the larger pores can be rapid enough that essentially all of the internal diffusion resistance will be in the small particles; in other cases, the diffusion paths within the small subparticles will be so short that all diffusion resistance is in the large pores, with higher diffusion coefficients but longer diffusion paths. If the macro-reticular structure is selected because of slow diffusion rates in the smallest particles, it is likely that the optimum small particle diameter and the size of pores in the binding material would be selected to leave some resistance in each. However, commercial adsorbents are usually made for numerous applications, and control of the mass transfer processes could be different for different applications. In either case, the discussions given for one particle diameter can apply to situations where the mass transfer rate control is in either size particle except that the user and interpreter of the results need to know which particle diameter (diffusion path, radius) to use. The shape of the loading curve versus time or the shape (spreading) of a pulse or loading front would not tell which resistance is controlling the mass transfer. It is most important to know where the major diffusion resistance occurs when estimating the effects of changes in diameter of the particle cluster or when the diameters of the small subparticles are changed.

In some cases, the resistances in the larger pores and within the small subparticles make important contributions to the total mass resistance. Obviously, assuming that all of the resistance is in pores of one size will not be adequate for those cases. Gray and Do [42] reported the effects of micropore, macropore, and surface diffusion on adsorbent loading rates and evaluated the results using data from carbon dioxide adsorption on an activated carbon. However, the results were limited to loading of individual particles and did not include descriptions of adsorption bed performance under these conditions.

For long beds, diffusion resistance in macro-pores (between subparticles) and micro-pores (within the subparticles) can be approximated by adding the two resistance terms [49, p. 246]:

$$R_{\text{int}} = \frac{r_p^2}{15D_p} + \frac{r_c^2}{15K_dD_c} \tag{37}$$

where c refers to the subparticle (crystals for zeolites). This approximation is based upon that of diffusion in the solid as an effective film and may not be a good approximation when mass transfer in the smaller particles is by pore diffusion.

Pore and Surface Diffusion with Linear Isotherms

The mathematical solution to column breakthrough curves with plug flow, linear isotherms, and adsorption rates controlled by the effective fluid film surrounding the particle was reported as early as 1930 by Furnas [43] and others. A useful asymptotic solution applicable to very long beds was given later by Klinkenberg [44]

$$\frac{C}{C_0} = \frac{1}{2}\operatorname{erfc}[z_f^{1/2} - t_f^{1/2}] \tag{38}$$

where

$$z_f = \frac{kK_d z}{v}\left[\frac{1-e}{e}\right] \quad \text{and} \quad t_f = k\left(t - \frac{z}{v}\right)$$

The variable t_f is a dimensionless time, and z_f is a dimensionless distance describing the bed length. A "long" bed where this asymptotic solution is applicable is defined in terms of z_f. Note the correction in t_f for the time required for the breakthrough to occur even if no adsorption occurred. This recognizes the time it takes for fluid to move down the bed and reach the fluid outlet.

Useful equations for describing the adsorption loading front with linear isotherms and diffusion into the adsorbent particle have been presented by Rosen [45,46]. The solution for long columns reduces to the simple expression

$$\frac{C}{C_0} = \frac{1}{2}\operatorname{erfc}\frac{w - t_R}{2w^{1/2}} \tag{39}$$

where

$$w = \frac{15D}{R^2}\frac{K_d z}{v}\left[\frac{1-e}{e}\right] \quad \text{and} \quad t_R = \frac{15D}{R^2}\left(t - \frac{z}{v}\right)$$

Again this solution applies only to long columns, i.e., large values of the length parameter w. (Rosen also presents more complex calculations for "short" beds where the concentration profile has not yet approached this asymptotic form.) The solution again takes the form of an error function complement. That is, the asymptotic form of the solution for systems controlled by film resistance is similar to the solution for systems controlled by diffusion within the particles.

Axial Dispersion

The other form of "resistance" that can impede mass transfer is axial (longitudinal) dispersion. Of course, strictly speaking, this is not a mass transfer resistance, but dispersion in the direction of flow spreads the front (or the mass transfer zone) in the same manner as a mass transfer resistance and the effect is similar. Axial dispersion often also accounts for molecular diffusion, eddy mixing in the expanding and contracting flow channels around the adsorbent particles, and nonuniform flow (or channeling) in the bed. However, with carefully designed flow distributors and sufficiently large bed diameters (relative to the adsorbent diameter), the effects of flow distribution can be reduced. Axial molecular diffusion will usually be important only in beds with very small particles and very low flow velocities; such conditions are more likely to appear in laboratory liquid chromatography than in industrial adsorption equipment. Industrial systems are likely to involve larger particles which make axial eddy diffusion and nonuniform flow more important than molecular diffusion.

Strictly speaking, the effects of nonuniform flow and mixing or diffusion cannot be lumped exactly into a single dispersion coefficient and describe all effects of the two phenomena. For instance, nonuniform axial flow distribution will not result in "back diffusion" (against the flow) predicted by a dispersion coefficient, but most behavior of nonuniform flow in the downstream flow direction can be approximated well by an effective dispersion coefficient.

A mathematical solution for the effects of dispersion with a linear isotherm and very rapid mass transfer rates was given by Lapidus and Amundson [47]. The asymptotic solution for very long beds was given by Levenspiel and Bischoff [48]:

$$\frac{C}{C_0} = \frac{1}{2}\mathrm{erfc}\left[\frac{1 - t/t_L}{2(D_a t/vzt_L)^{1/2}}\right] \tag{40}$$

where $t_L = (z/v)\{1 + K_d(1 - e)/e\}$.

Again the asymptotic solution has the same error function complement form. Although asymptotic solutions are only valid for very long beds, these solutions are adequate for describing many, perhaps most, practical environmental adsorption systems with linear isotherms. They are especially likely to be adequate when very high removal efficiency is needed, since that often means that the bed will have to be relatively long. Of course, the time desired between regeneration will also affect the bed length. When short bed solutions are needed, the original references or more detailed textbooks on adsorption should be consulted; the text by Ruthven [49] is recommended because it provides a convenient summary of many available solutions. The solutions will, however, be considerably more difficult to use, so the need for additional mathematical complexity should be considered. It appears to this author that some of the more complex solutions are seldom used.

Multiple Resistances

Before describing the simplest ways to use these solutions for linear isotherms and only one source of mass transfer "resistance" (film resistance, diffusion in the adsorbent particles, or axial dispersion), it will be better to first consider the effects of two or more of these resistances and then the uses of all of these cases can be considered. For pairs of resistances, solutions are available, but they are difficult to use except with linear isotherms, when resistances can be added.

As noted several times in showing the asymptotic solutions for linear isotherms, all of the solutions have a similar form involving an error function complement. The combination of the resistances for linear (isotherm) systems was suggested by Glueckauf [50], who noted that for the asymptotic long bed solutions, diffusion into a spherical particle could be described by the relation

$$k_{\text{eff}} = \frac{15D}{r^2} \tag{41}$$

where k_{eff} is the effective film coefficient due to diffusion resistance within the particle, D is the effective diffusion coefficient in the adsorbent particle, and r is the radius of the adsorbent particle. The reader may note that this term also appeared in the Rosen solution. Cases with linear isotherms and diffusion resistance in the fluid film and within the adsorbent particle can be treated as if there were two films, and the resistance of two such films can be added. This approach can be extended to include the effective resistance caused by axial dispersion, and such extensions are given in most textbooks on adsorption [51] or chromatography. Adding the three resistances gives

$$\frac{1}{K_d k} = \frac{D_a}{v^2}\frac{1-e}{e} + \frac{r}{3k} + \frac{r^2}{15K_d D_p} \tag{42}$$

where D_p is the effective diffusion coefficient in the adsorbent particle (or in the pores). This expression is used in chromatographic analytical chemistry systems which prefer to use systems with linear isotherms and reflects the additive nature of the linear systems.

Because the forms of the asymptotic solutions for bed behavior with linear isotherms are so similar, it is not possible to tell from the shape of the breakthrough curve which mass transfer resistance is the most important contributor to the overall resistance. That is, one can only determine from a single breakthrough curve the total or overall resistance. In principle, one could vary the flow rate through the bed and the particle size and see different diffusion behavior depending upon

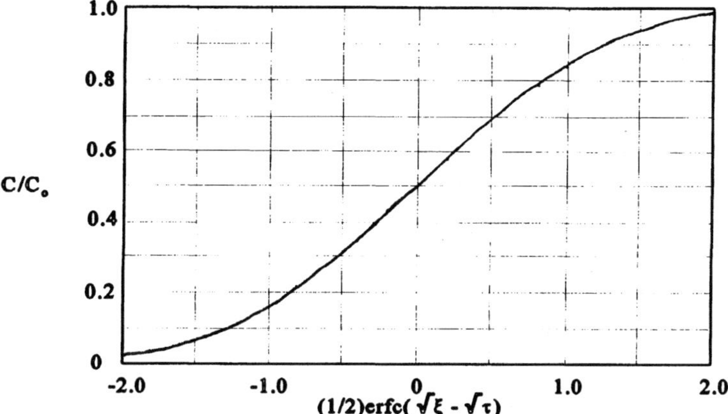

FIGURE 14 Spreading of the diffuse breakthrough front in an adsorption bed operating with a linear isotherm. Without a favorable isotherm, mass transfer resistance causes the front to widen at a regular rate. This plot shows the mass transfer resistance incorporated into dimensionless units of time. It can be used to predict the width of adsorption fronts when the mass transfer resistance is known or to estimate the mass transfer resistance when experimental breakthrough fronts are available.

the contributions from each term. Then the contributions of each term could change differently. Although it is relatively easy to change the flow rate, many investigators will not have significantly different particle sizes of otherwise identical adsorbents available.

These relatively simple error function complement equations are believed to be the most widely used models for describing or analyzing adsorption bed breakthrough behavior. This is due partially to the relatively common occurrence of linear isotherms, especially in dilute systems used in most analytical chemistry chromatographic adsorption separations. However, these equations are also likely to be used (sometimes inappropriately) for nonlinear systems simply because they are so much easier to use. Nonlinear isotherms will be discussed later, and the more appropriate equations for those cases will be discussed then.

It is relatively easy to extract values for the overall resistance from breakthrough curves with linear isotherms and asymptotic (long beds) conditions. One approach is illustrated in Figure 14, which is a plot of the asymptotic breakthrough curve for linear isotherms. To obtain the best values for the resistance, the entire breakthrough curve would have to be "fit" to the error function complement equation with the total resistance used as the parameter to be varied in seeking a minimum for the sum of the squares of the difference between the predicted breakthrough curve and the measured curve.

However, good values for the resistance can be extracted from the data by using only two points on the experimental breakthrough curve. Figure 14 is plotted in terms of dimensionless time. In the following illustration the two points chosen to evaluate the resistance will be where the fraction breakthrough is 0.8 and 0.2. As can be seen in the figure, the difference between the dimensionless time at those two points is

$$\sqrt{t_{f,0.8}} - \sqrt{t_{f,0.2}} = 0.82 - (-0.82) = \sqrt{\frac{k}{v}}(\sqrt{t_{0.8}} - \sqrt{t_{0.2}}) = 1.64 \quad (43)$$

Times with subscript f are dimensionless; the others are real times with dimensions. The numbers in the subscripts denote the fraction breakthrough at that time. The right side of this equation comes from the definition of the dimensionless time in the error function complement equations. To evaluate the resistance, one only has to evaluate the real times when the concentrations in the breakthrough curve are 0.8 and 0.2 times the inlet concentration. Then

$$k = \frac{2.69v}{\sqrt{t_{0.8}} - \sqrt{t_{0.2}}} \quad (44)$$

Almost any two (different) points or concentrations on the breakthrough curve could have been used, but selecting different points would only alter the value for the constant, not the resulting value for mass transfer resistance. However, there are merits in selecting values not far from the ones used here. The points were selected far apart so that the difference in the times could be large, but the values were also selected so that the normalized concentrations were not so close to either 0 or 1, where it is difficult to measure them accurately. Also note that these points lie on both sides of the center of the mass transfer zone. (The center of the mass transfer zone occurs when the concentration is 0.5 times that in the inlet fluid.) It is also worthwhile to measure the time between the 50% and 20% breakthroughs and the time between the 80% and 50% breakthroughs. These values should be approximately the same. If they are not, this is one early indication that the measured breakthrough curve is not following the complementary error function form. Of course, if the data do not follow the form of the equation, one should be wary of evaluating parameters with this equation.

Similar analyses are commonly used in laboratories that use pulses rather than "step" increases in solute concentrations. Then the spreading of the pulse, which represents both a loading and a desorption front, can be used to evaluate the mass transfer resistance. The rate at which the pulse center moves down the bed can be used to calculate the slope of the linear isotherm. Although there are merits in using pulses rather than the larger quantities of solute needed to run tests on breakthrough fronts, this discussion has focused on breakthrough curves in the hope that it will make the extension of the discussion into nonlinear isotherms easier for the reader.

Conversely, if one has found a value for the mass transfer resistance, a predicted breakthrough curve can be evaluated simply from Figure 14 with only a simple calculator. For instance, one may measure the breakthrough curve in a laboratory or pilot scale column and then estimate the breakthrough curve for a variety of potential columns during optimization of a design. First one can estimate the midpoint of the mass transfer curve from the slope of the equilibrium curve (distribution coefficient):

$$t_{0.5} = \frac{(K_d + 1)z}{v} \tag{45}$$

The distribution coefficient can also be extracted from the earlier experimental bed measurement by measuring the time for the 50% breakthrough concentration to occur and rearranging this equation to solve for K_d. However, because it is important to know that the equilibrium isotherm is approximately linear, it is desirable to measure the equilibrium concentration at several concentrations.

To predict the rest of the breakthrough curve, one can choose several fractional breakthrough concentrations between 0 and 1 and measure the difference in the normalized time on Figure 14 that corresponds to that concentration and the dimensionless time that corresponds to a normalized concentration of 0.5. Then the real time that corresponds to that dimensionless time can be calculated:

$$\sqrt{t_x} - \sqrt{t_{0.5}} = \sqrt{\frac{v}{k}}(\sqrt{t_{f,x}} - \sqrt{t_{f,0.5}}) \tag{46}$$

A sufficient number of points can be evaluated to define the predicted breakthrough curve over any range desired. The time corresponding to the center of the mass transfer zone will increase linearly with the length of the bed, and the difference width of the mass transfer zone (as determined at any other fraction breakthrough) will increase as the square root of the bed length. This means that the mass transfer zone constitutes a smaller fraction of larger beds than shorter beds.

A brief comment should be made about the possibility that the breakthrough curve is not in the shape predicted by the error function complement. Using only two points to evaluate the mass transfer resistance does not include any assessment of the agreement of the assumed equation to the data. As noted, it would be best to fit the equation to all of the measurements over the entire range of the mass transfer zone, but one should at least be aware of any obvious deviations from the assumed form. For instance, the breakthrough curve should have symmetry. The difference between the times when the concentration is 0.8 and 0.5 times the inlet concentration should be the same as the difference between the times when the concentrations are 0.5 and 0.2 times the inlet concentration. Symmetry should be seen for any other equal deviation in opposite directions from the midpoint of the mass transfer zone. Major inequalities in these time differences should be one sign that the data are not fitting this equation. That could mean that the isotherm is not linear. More detailed comparisons can be made by obtaining the effective mass transfer parameters from the points on Figure 14 and then using the figure to plot the entire predicted breakthrough curve for comparison with experimental data.

BREAKTHROUGH CURVES FOR NONLINEAR ISOTHERMS

When considering the effects of nonlinear isotherms on the length of the mass transfer zone (often summarized by the slope of the breakthrough curve at the mid-point), the nonlinear effects appear in two ways. First remember that the nonlinearity contributes to spreading or contracting of the mass transfer zone, as described earlier. Favorable isotherms (negative curvature) cause the length of the zone to decrease with time or position down the bed, and unfavorable isotherms (positive curvatures) will contribute to further spreading of the zone. The mass transfer resistances then add to these effects. That is, for unfavorable isotherms, mass transfer resistance and the shape of the isotherm both contribute and cause the zone to expand more rapidly. However, the more common cases involve favorable isotherms, and the effects of isotherm shape and mass transfer resistance act in opposite directions. One factor acts to shorten the zone, and the other acts to lengthen it.

The discussion of breakthrough curves with nonlinear isotherms will be limited largely to favorable isotherms because unfavorable isotherms are expected to be far less common among practical environmental and waste management problems. Although interesting, they give wide fronts that tend to make their use undesirable. This means that one is likely to look for another adsorption system or even another separation method. The major reason to consider unfavorable isotherms is that they are likely to occur in the desorption (regeneration) cycle; remember that systems with favorable adsorption isotherms have unfavorable desorption isotherms. The discussion focuses on the adsorption step because that is usually (but not always) the step that is operated under the more difficult conditions, and in some cases the only step of concern to the reader. Desorption may even be carried out off-site by a contractor. Usually one changes the temperature or other condition for desorption sufficiently that it is rapid and approximately complete.

For sufficiently long beds, the opposing effects of spreading from the mass transfer resistances and contracting from the favorable shape of the isotherm reach a "steady state," and the mass transfer zone (or breakthrough curve) develops a "constant pattern." Then as the front moves further down the bed, neither the length of the zone nor the shape of the concentration profile in the zone changes. Design of columns where the development of these constant patterns needs to be taken into account requires complex equations that are relatively difficult to use, and many or most adsorption beds used in waste and environmental processes are likely to be long enough that the constant pattern solutions will be adequate. However, one should be aware that more accurate designs can be made with short beds using the more complex analyses. These solutions are likely to be available only for the usual case where the initial front is very sharp, that is, when the length of the initial mass transfer zone is very short. Note that in these cases, use of the asymptotic solution will result in a conservative design. If the feed is introduced to the bed as a step function (a sharp front), the length of the mass transfer zone will increase with time and the distance it moves down the bed. Then by predicting the length of the mass transfer zone for a very long bed, one finds that the actual length of the mass transfer zone will be less than that estimated until the constant pattern conditions are reached. That is a conservative estimate since the design would call for a slightly longer bed than that actually required. (Of course, if the feed were introduced as a broad or diffuse front with the mass transfer zone longer than the asymptotic constant pattern front, perhaps because of a large mixed surge volume upstream of the bed, the broad front would become sharper rather than broader as it approached the same constant pattern conditions.)

Langmuir Kinetics (the Thomas Model)

An analytical solution was reported by Thomas [52] for "Langmuir kinetics." The local rate of adsorption was assumed to be

$$\frac{\partial q}{\partial t} = k[C(q_{\text{sat}} - q) - b(C_0 - C)] \tag{47}$$

This is a "kinetic" type rate expression which does not take into account the diffusion processes directly. It treats the rate as a chemical reaction whose rate is governed by mass action and with the concentrations uniform throughout the interior of the adsorbent particles. However, at equilibrium when the rate approaches zero, the equilibrium conditions become those of the Langmuir isotherms because this rate expression is essentially the same as that used in developing the Langmuir isotherm. Although one can question how well this type of rate expression describes what is really happening inside an adsorbent particle, the results have been used with considerable success to describe loading fronts and to predict adsorption bed behavior under a range of conditions.

The Thomas solution is

$$\frac{C}{C_0} = \frac{J(bs, t_T)}{J(bs, t_T) + [1 - J(s, bt_T)]\exp[(b-1)(t_T - s)]} \tag{48}$$

where

$$J(x, y) = 1 - \int_0^x \exp(-y - x) I_0[2(ys)^{1/2}]\, ds \tag{49}$$

$$t_T = kC_0\left(t - \frac{z}{v}\right)$$

$$s = kq_0 z \frac{10e}{ev}$$

$$I_0[z(ys)^{1/2}] = \sum_0^{\infty} \frac{(ys)^n}{n!n!}$$

It is readily apparent that this solution is not very simple. Nevertheless, it is much simpler than the general solutions for most nonlinear isotherm systems. Note that it is even useful in the transient case before a constant pattern is established.

Mass Transfer Resistance Models with Nonlinear Isotherms

Remember that to understand the diffusion processes within adsorbent particles with nonlinear isotherms, it is important to consider whether the adsorbed molecules can diffuse further once they are adsorbed or if they remain essentially stationary after being adsorbed. That is, with nonlinear isotherms, there are significant differences in the behavior of systems controlled by pore diffusion and those controlled by bulk or surface diffusion. In many cases where molecules are adsorbed from a gas, the adsorbed molecules may move further into the particle by surface diffusion, and there are gel-like adsorbents (including some of the most popular ion exchange resins) where the solute remains always relatively free and can diffuse through the solid.

When the adsorbed molecules remain immobile after adsorption, they are usually strongly associated with a specific site or portion of the adsorbent pore surface, and the mass transfer rate is dominated by diffusion of unadsorbed molecules within the adsorbent pores. With linear isotherms, the concentration of adsorbed molecules is proportional to the concentration in the pores; thus, the flux of solute in the pores via surface or solid phase diffusion is proportional to the flux from diffusion of free molecules in the pores. In such cases, all of the mass transfer mechanisms with the adsorbent particles can be lumped into an effective diffusion coefficient which can be expressed in terms of the concentrations within the pores or, more commonly, in terms of the total concentration of solute molecules in the adsorbent. However, this proportionality does not hold for systems with nonlinear isotherms, so for rigor of discussion, these two parallel mechanisms for diffusion need to be treated separately. Of course, there will be cases where only one of the two diffusion mechanisms is important, and there will be other cases where both mechanisms make significant contributions to the overall mass transfer rate. Most solutions consider only limiting cases where a single diffusion mechanism is important in the solid.

As with linear isotherm systems, it is difficult to use column breakthrough data alone to determine the mechanism or the contribution of each mechanism to mass transfer within adsorbent particles. That is, if breakthrough curves are used to estimate the effective diffusion coefficient, one is not likely to tell from the shape of the curve which mechanism or combination of mechanisms is important to the mass transfer rate. This results partly because the shapes of the constant pattern breakthrough fronts predicted for the different mechanisms are not sufficiently different. Other factors that make if difficult to determine the mechanism are additional mass transfer resistances (film resistance and axial dispersion) and unavoidable scatter in the data. Additional data from different experiments may be useful in resolving the different contributions to the mass transfer resistance, but they will not necessarily be conclusive. In most cases, the inability to resolve the different contributions to the mass transfer rates should be a warning to take measurements of breakthrough fronts at conditions near the concentrations expected in the production adsorption beds and to avoid larger extrapolation in concentrations that would have been acceptable for linear systems.

First consider the case where the adsorbed material is not able to diffuse, namely pore diffusion. With an irreversible isotherm, this means

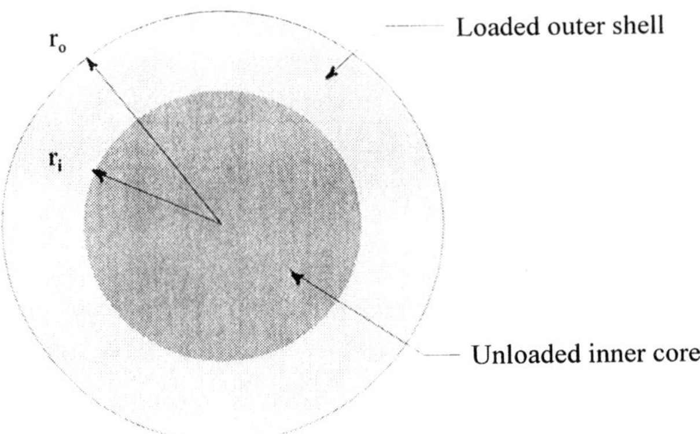

FIGURE 15 Shrinking core concentration profile that can occur within porous adsorbent particles with an "irreversible" isotherm. Adsorption rates are controlled by diffusion through the "loader" shell of the particles.

that, for adsorption, outer regions of the adsorbent will become saturated with solute, while regions near the center of the adsorbent particles will remain completely free of solute. Profiles of adsorbed molecules within the particles will remain a sharp step function with the outer shell saturated with solute and the inner core containing no solute (Figure 15). With time, the outer saturated region will expand inward, and the inner core will shrink. This is called the "shrinking core problem" and is described in texts on heterogeneous kinetics [53] and adsorption [54]. The rate at which the outer (saturated) shell advances toward the center is controlled by diffusion within the pores of the shell. When the concentration of adsorbed molecules is much higher than the concentration in the pores (the usual case), the diffusion can be treated as a quasi-steady-state problem where the movement of the shell boundary and accumulation within the pore fluid can be ignored when calculating the flux through the saturated shell.

Although the solution for loading of a particle with a linear isotherm only approaches saturation as time approaches infinity, the loading of a particle with an irreversible isotherm becomes saturated after a finite time, that is, when the sharp boundary between "loaded" portions of the particle and the "unloaded" core moves to the center of the particle and the diameter of the "core" has shrunk to zero:

$$t_{\text{max}} = \frac{q_{\text{sat}} R^2}{6 D_p C_{\text{surface}}} \tag{50}$$

When the concentration at the particle surface is constant, the total (or average) concentration in an adsorbent particle and time are related as follows:

$$\frac{t}{t_{\max}} = 1 - 3\left(1 - \frac{q}{q_{\text{sat}}}\right)^{2/3} + 2\left(1 - \frac{q}{q_{\text{sat}}}\right) \tag{51}$$

A sketch of concentration profiles within the particles for pore and bulk diffusion is probably sufficient to illustrate that the loading rate which will be proportional to the gradient of the concentration of diffusing molecules (bulk molecules or "free" molecules in the pores) will be related quite differently to the average concentration of solute in the adsorbent particle; therefore, one should not expect mathematical models and design procedures based upon bulk diffusion to be accurate for cases controlled by pore diffusion. However, when the solute adsorbed on the adsorbent pore surfaces is free to diffuse along the pore surfaces (surface diffusion), it usually constitutes the major part of the solute in the adsorbent, and the concentration profiles and loading rates are similar to those predicted by bulk diffusion. The models and equations derived for bulk diffusion then can apply. Of course, the actual mechanism will be diffusion along the pore surfaces, but the equations to describe the phenomena are similar.

With the mass transfer rate controlled by diffusion within the adsorbent particles, for many adsorbents constructed of clusters of smaller particles, there could be a contribution to the mass transfer resistance from diffusion within the larger particles (and, thus, between the smaller particles in the macro-pores) and/or diffusion within the smaller particles (in the micro-pores). If the diffusion resistance is largely in either the macropores or the micro-pores, the rate expressions describing the diffusion process will be similar but would need different values for the diameter and diffusion coefficients. (As pointed out earlier, it is usually not possible to determine which region of the particles controls the mass transfer resistance simply from the shape of breakthrough or loading curves. It is usually necessary to change the diameters of either the larger particles or the smaller particles while holding other parameters constant to determine which resistance is more responsible for the diffusion resistance.)

The solution for loading of adsorbent particles by pore diffusion with intermediate values of the Langmuir constant k is more complex, but it is relatively easy (but not fast) to calculate the profiles numerically. Note that the diffusion equation is

$$\frac{\partial C}{\partial t} = D_p \nabla C \tag{52}$$

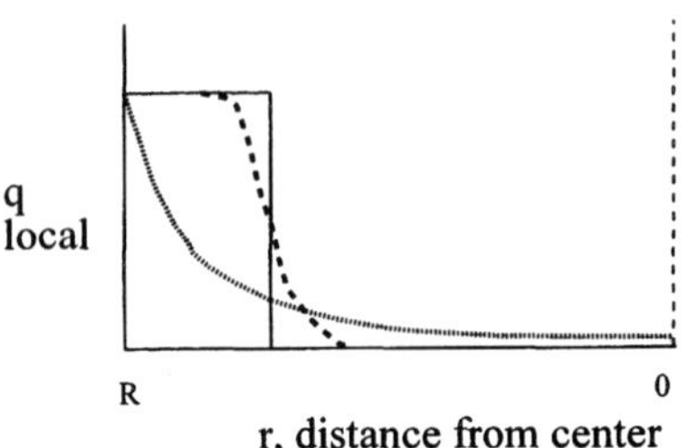

FIGURE 16 Concentration profiles that can occur in an adsorbent particle with a Langmuir isotherm. At high concentrations (when the product of the concentration and the Langmuir constant is high—greater than approximately 10), the Langmuir isotherm is approximated by an irreversible isotherm, and the concentration profile on the solid inside a porous adsorbent follows the shrinking core model in the solid curve. When the concentration is low (the product of the concentration and the Langmuir constant is low), the Langmuir isotherm is approximated by a linear isotherm, and the concentration profile follows the dashed curve for diffusion of mobile components within a particle. For intermediate concentrations, the concentration profile on the solid takes shapes that are immediate between the two extreme cases as illustrated by the dashed line.

which can be rewritten in terms of the total local solute loading in the particle by solving the Langmuir equation for C and substituting the relation into the diffusion equation:

$$C = \frac{q/q_{\text{sat}}}{k(1 - q/q_{\text{sat}})} \tag{53}$$

This means that the effective diffusion coefficient is not constant but depends upon the local concentration of solute in the adsorbent particle. As expected, the resulting profiles are intermediate between the sharp step profile of the irreversible isotherm and the diffuse profile of the linear isotherm. An example is Figure 16. Obviously for small values of k, the profile will be more similar to that for the linear isotherm, and the concentration profile within the particle would resemble the dotted curve in the figure. But as k increases, the profile becomes sharper and begins to resemble the limiting case of the irreversible isotherm. The sharp profile for an irreversible isotherm is shown by the solid curve, and an intermediate value of k is shown by the dashed curve. Although the concentration fronts within the particles are different for pore diffusion and for bulk diffusion, it still can be difficult to tell from the shape or behavior of breakthrough curves alone which diffusion mechanism is most important [55]. If the diffusion mechanism is assumed incorrectly, there is risk in extrapolating the results to other conditions (flow rates, particle diameter, etc.). If an internal diffusion coefficient is determined from experimental adsorption rate data using equations that apply to any transport mechanism, the resulting diffusion coefficients should only be used in equations describing the same mechanism. In other words, it is important to know what basis was used to evaluate rate parameters such as diffusion coefficients.

Analytical Solutions for Breakthrough Curves

Loading fronts for linear isotherms were discussed earlier, so we only need to focus upon the irreversible isotherm. However, remember that the linear isotherm gives a continuously spreading front, and after long times (long beds) the equations take on simpler asymptotic forms. The asymptotic front is symmetrical and has the form of an error function complement, and the rate parameters can be evaluated from two (or more) suitable points on the concentration front. Breakthrough curves for nonlinear isotherms can be quite different, and the irreversible isotherm is one of the "most different" cases.

Irreversible isotherms, like all favorable isotherms, give a constant pattern after long times (long beds), and the front spreading will not increase after that. With favorable isotherms, the shape of the front will not necessarily be as symmetrical as the fronts for linear isotherms. Although solutions exist for breakthrough curves with favorable isotherms for beds that are too short for the constant pattern to develop, this presentation will focus only on the simpler asymptotic (constant pattern) solutions; fortunately for most "deep beds" used in environmental control systems, this is likely to be appropriate. As pointed out later, with all other parameters (flow rate, particle diameter, mass transfer coefficients) the same, the more favorable (the more curvature) the isotherm, the shorter the bed required to approach constant pattern conditions.

The solutions for the loading front with an irreversible isotherm were given by Hall et al. [56] and Cooper [57]. The asymptotic (constant pattern) solution suggested by Hall et al. where the adsorption rate is controlled by the resistance in the fluid film surrounding the adsorbent particles is

$$\frac{C}{C_0} = 1 - 0.07759 \left\{ 2.39 - \frac{ka_p v}{qF} \left[\frac{t - ve/F}{q_{\text{sat}} v / C_0 F} - 1 \right] \right\} \tag{54}$$

where k is the film coefficient, a_p is the external area of the particles per unit volume of bed (a quantity that can be calculated from the solid fraction in the bed and the diameter of the adsorbent particles), v is the volume of the bed, and F is the volumetric flow rate of fluid in the bed.

As described earlier, when adsorption rates are controlled by diffusion within the particles, there are two cases to consider. In one case, diffusion takes place only in the pores of the adsorbent. In that case, the concentration profile moves into the particles as described by the shrinking core equation. In these cases, Hall et al. suggest the approximate asymptotic solution

$$\frac{C}{C_0} = 1 - 0.07759\left\{2.39 - \frac{15D_{\text{pore}}(1-e)v}{r^2}\left[\frac{t - ve/F}{q_{\text{sat}}v/C_0F} - 1\right]\right\} \quad (55)$$

This is much like the fluid film solution except for the term in brackets.

The second limiting case for internal diffusion resistance occurs when the mass transfer rate is controlled by diffusion of the adsorbed material. This is usually described as "solid diffusion," but it could just as well describe surface diffusion. In this case, Hall et al. [56] suggest the constant pattern solution

$$\frac{C}{C_0} = 1 - \frac{6}{\pi^2}\sum\frac{1}{n^2}\left\{-n^2\frac{\pi^2 D_s q_{\text{sat}}v}{r^2C_0F}\left[\frac{t - ve/f}{q_{\text{sat}}v/C_0F} - 1\right] + 0.64\right\} \quad (56)$$

Here, D_s is the effective diffusion coefficient in the solid. Of course, these are the only two limited cases for mass transfer rates controlled by diffusion within the adsorbent particles (and for irreversible isotherms) because both pore diffusion and solid/surface diffusion could contribute to the mass transfer rate. However, remember that one is not likely to be able to determine from the shape of the breakthrough curve alone which asymptotic case applies; there is a better chance to tell the two mechanisms apart from data with two particle sizes.

Continued effort has been made to develop more general and more powerful methods for predicting the breakthrough curves for systems with nonlinear isotherms [58,59]. These approaches can take into account multiple resistances and most expected types of mass transfer within the solid particles. However, these methods are usually numerical and involve moderate to highly complicated codes. They are often available largely to the investigator developing and constantly improving the codes. Although all potential users will not necessarily have access to the computer codes needed, these developments are good indications of what can be done numerically and what is likely to be widely available in the not too distant future, probably incorporated into some of the large process simulation codes.

The following section describes a more limited approach that is simple enough that many useful calculations can be made for nonlinear isotherms with relatively little programming or computations. The approach can even be applied graphically much like the classical graphical methods for distillation or absorption tower design. It is worthwhile to become familiar with the simpler graphical approaches even when more powerful numerical tools are available. One can sometimes gain insight and spot or check for errors in input data by using the simpler methods.

Generalized Graphical or Numerical Method

The author has recently written two papers [60,61] describing more generalized methods for predicting or analyzing breakthrough curves that readers should find helpful for some of the more complicated single component systems with nonlinear isotherms. This method is based upon similar analyses developed for countercurrent operations, such as gas absorption/stripping or liquid extraction, and the reader may want to review the chapters describing those operations while studying this section. The method is limited to those systems with "favorable" nonlinear isotherms and to those systems where the mass transfer resistance comes from the fluid phase, from a solid phase where the internal (solid phase) resistance can be described by the Glueckauf approximation, or from both phases. These are systems in which the mass transfer rate is determined largely by diffusion within the solid itself and/or by surface diffusion, but not where mass transfer is controlled by pore diffusion. These cases of solid diffusion are believed to be relatively common and thus should include many applications. However, the method is suitable for any shape of favorable isotherm, and the calculations can be made numerically or graphically.

Adsorption or ion exchange operations that are applied on commercial applications often have favorable isotherms. Adsorption processes with unfavorable isotherms are likely to have breakthrough fronts that spread so badly that such systems may be unable to compete with alternative removal methods. Examination of the discussion of binary ion exchange will reveal that simple exchange of one ion from a feed solution that initially contains only that ion with a resin loaded completely or largely with another ion will always be favorable, usually highly favorable. For adsorption from very dilute systems, the isotherm may be approximately linear, but relatively simple methods for handling such systems have been available for some time and were described earlier. Isotherms for dilute ion exchange systems are likely to remain favorable.

The more generalized method suggested here is based upon the same equations used in developing the exact solutions described earlier for a few specific isotherms. The only difference is that the equations are presented in a manner where numerical or graphical methods can be

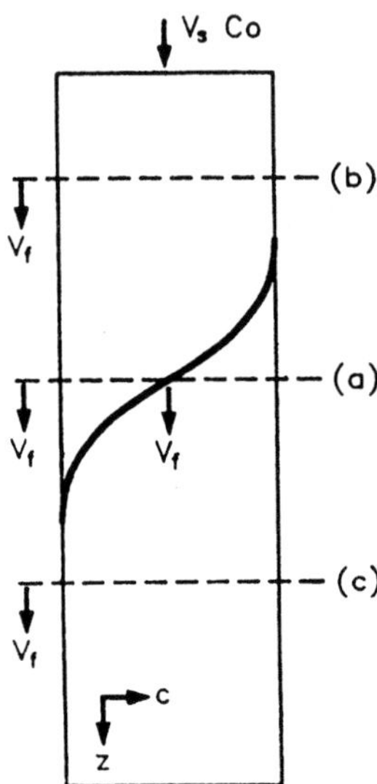

FIGURE 17 After the concentration profile in a bed with favorable isotherm reaches a "constant pattern," the profile will appear to be stationary to an observer moving down the bed with the same velocity as the front.

used relatively easily. The analogy to countercurrent operations is evident when the constant pattern front is viewed by an observer who moves "down" the bed at the same rate as the moving front. To such an observer, the front would appear to be stationary, but the fluid would appear to be moving down the bed at a rate equal to the actual fluid flow rate minus the velocity of the front (that is, minus the velocity of the observer). The solid adsorbent or ion exchange material would appear to be moving upward at a velocity equal to the velocity of the front (Figure 17). Note that this is essentially like a countercurrent system described in the chapters on absorption or liquid-liquid extraction. The only differences are in the end conditions (discussed later), but they are not fundamental differences.

As with the other countercurrent systems, the bulk concentrations in both phases can be described by an operating line obtained from a material balance. The material balance equates the solute lost to the fluid over a portion of the bed to the solute gained (adsorbed) by the adsorbent over the same portion of the bed. For use with this method, the portion of the bed to be covered by the material balance is all of the bed above (upstream of) any specified position within the mass transfer zone—that is, a position within the breakthrough curve. To establish the operating

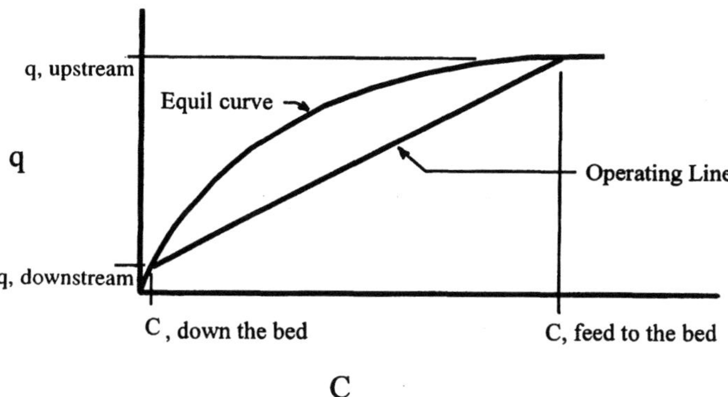

FIGURE 18 The equilibrium isotherm and the apparent operating line seen by an observer traveling down the bed at the same velocity as the adsorption front. Note that the operating line crosses the equilibrium curve at positions far (infinitely far) upstream at the feed concentration and at positions far (infinitely far) downstream at the concentration in equilibrium with the initial adsorbent in the bed.

line across the breakthrough curve, material balances will be written as an equation that applies to every position in the mass transfer zone. Note that for each of these material balances, the concentrations far upstream will be the same. That is, the concentration in the fluid will be the same as the fluid feed concentration, and the concentration in the solid will be the concentration that is in equilibrium with that concentration in the fluid. See Figure 18, which shows the equilibrium curve. The concentration in the solid far upstream will be the point on the equilibrium curve that corresponds to the fluid feed concentration on the horizontal axis. The material balance can be written as

$$(v - v_f)C_{\text{in}} + qv_f = (v - v_f)C + v_f q_{\text{sat}} \tag{57}$$

In this balance, v is the fluid velocity in the bed, v_f is the velocity at which the front moves down the bed, C is the concentration in the fluid at the arbitrary location within the front where the material balance is made, q is the concentration of the solute in the solid particles at the position where the material balance is made, C_{in} is the concentration of the solute in the fluid entering the bed (in the feed fluid), and q_{sat} is the concentration of solute in the solid at the top of the bed after it is equilibrated (saturated) with the inlet (feed) fluid. This equation can be solved for q as a function of C, and the result is shown at the "operating line" in Figure 18. The concentrations C_{in} and q_{sat} are the feed concentration and the concentration of solute in the solid that is in equilibration with the feed concentration. Once C_{in} is specified, q_{sat} can be obtained from the equilibrium curve (isotherm). First one should recognize that the operating line given by Equation (57) is a linear equation and that q is a linear function of C.

The easiest way to plot Equation (57) is by noting the two points where the straight line crosses the equilibrium curve. Since two points define a straight line, they will define the operating line given by Equation (57). Near the fluid inlet, the solid adsorbent is in equilibrium with the inlet (feed) fluid. Since this is also a point on the operating line, the equilibrium curve and the operating line must cross the isotherm (equilibrium curve) at that point. Alternatively, one could let the arbitrary position on the constant pattern front move further and further down the bed until one reaches a position where the concentration of solute in the solid is essentially the same as the original concentration in the solid before the liquid feed was started, that is, at the initial state of the bed. Usually, the initial bed will have little or no solute on the solid, so this point is likely to be the "origin" where both q and C are zero. In Figure 18, the bed is assumed, for the general case, to have a small initial concentration of solute; so the lower intersection of the operating line and the isotherm is not shown exactly at the origin. Once these two points on the operating line are located, the entire line can be drawn by connecting the two points with a straight line.

With both the equilibrium curve and the operating line known for a countercurrent system, the number of transfer units (NTU) can be easily calculated as illustrated in Chapters 3 and 6. The NTU for any portion of the bed can determined by integrating the change in concentration divided by the driving force. If the NTU are to be based upon the transfer resistance in the fluid phase, the integral is

$$\mathrm{NTU} = \int \frac{dC}{C - C_i} \tag{58}$$

where C is the concentration in the bulk fluid, and C_i is the concentration in the fluid at the fluid-solid interface at that position in the bed. If the mass transfer resistance in the solid phase can be neglected, the value for the concentration in the bulk fluid at any position can be found on the operating line, and the concentration at the interface at that same position in the bed would be the concentration that would be in equilibrium with the solid at that position. The concentration at the interface can be obtained by drawing a vertical line from the bulk composition to the operating line and drawing a horizontal line from that intersection to the isotherm (Figure 19a).

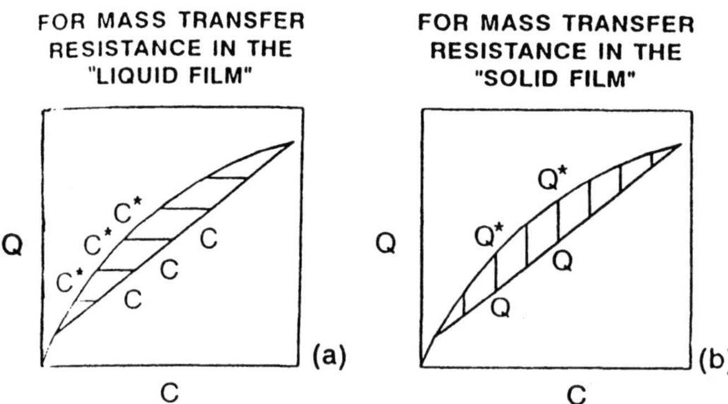

FIGURE 19 (a) Horizontal tie lines connecting the bulk concentrations on the operating line with the interfacial concentration at the adsorbent surface when all of the mass transfer is in the fluid phase. (b) Vertical tie lines connecting the bulk concentration on the operating line to the interfacial concentrations on the adsorbent surface when all of the mass transfer resistance is in the solid phase and when mass transfer within the solid can be approximated by diffusion within the solid.

If the mass transfer resistance is in the solid phase, then

$$\text{NTU} = \int \frac{dq}{q_i - q} \tag{59}$$

The concentration in the solid at the solid-fluid interface is q_i, and the concentration in the bulk of the solid phase is q. Values for q and q_i can be found from the operating line and the isotherm (equilibrium curve), respectively. One can select values for q on the operating line, and corresponding values for q_i lie just above q on the isotherm (Figure 19b).

Note that the NTUs are based on the concentration in the fluid or in the solid. In either case, the NTU is a function of the change in concentration. That is, so many transfer units are needed for the concentration to change from one value to another value, and the number of transfers needed for that change is the integral in Equations (58) or (59) evaluated between those two concentrations. If one integrates over a small change in concentration, the integral will have a relatively small value, but if one integrates over a larger change in concentration, the integral will have a larger value. Thus, one can think of the NTU as a function of the concentration change (either the change in concentration in the fluid, if the NTU is based upon the fluid concentrations, or the change in the concen-

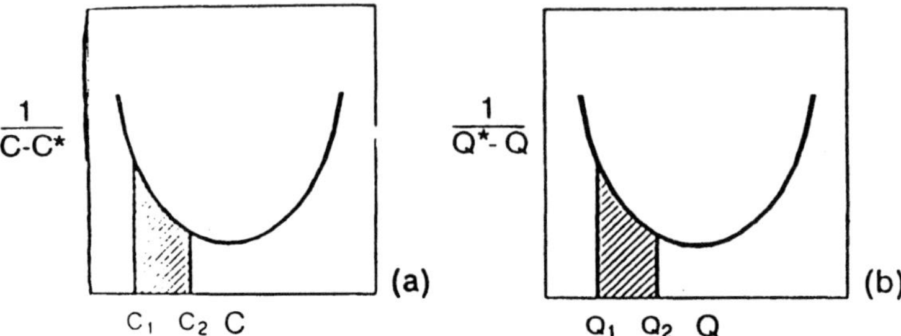

FIGURE 20 Integrating from the 50% breakthrough point to different concentrations above and below it to determine the number of transfer units (NTU) corresponding to different concentrations on the concentration front.

tration in the solid, if the NTU is based upon the concentrations in the solid). In most other countercurrent operations, absorption/stripping, or liquid extraction, the integration is carried out from one end of the bed (column) to the other end, or from the inlet concentration in one phase to the outlet concentration in that phase. For the constant pattern front in an adsorption bed, this cannot be done because at the end points of the integrations the driving force would be zero. That would make the integrand infinity.

Instead, it is possible to begin the integration at some intermediate point in the breakthrough front and integrate in either or both directions (Figure 20). The suggested place to begin the integration is at the mid-point of the breakthrough when the concentration in the fluid is one-half the difference between the inlet and initial outlet concentrations, that is, midway between the lowest and highest values in the front. This is often called the 50% breakthrough point, and, when the breakthrough curve is symmetrical, this point is important because it represents the capacity of the bed when saturated with the feed fluid. Of course, this is an arbitrary place to start the integration, and there could be a valid reason to start with any other concentration on the breakthrough curve. Note that the operating line shows that the mid-point in the concentration in the fluid corresponds also to the mid-point in the concentration in the solid because the bulk concentrations—those given by the operating line—are linearly related. Furthermore, a plot of the breakthrough curve in terms of q/q_{sat} looks the same as the breakthrough front plotted in terms of C/C_0 because the two terms are linearly related by the operating line [Equation (57)].

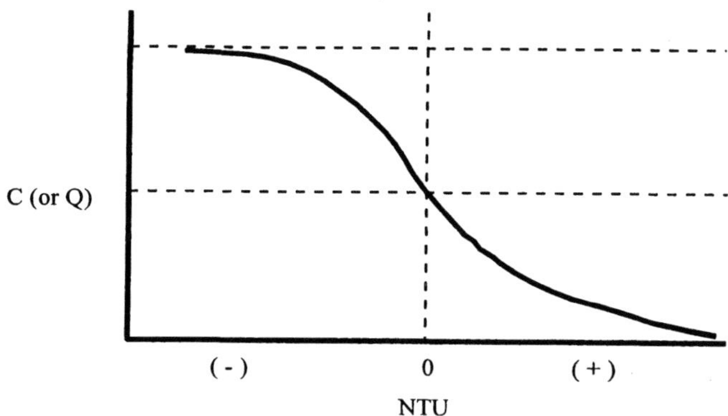

FIGURE 21 Concentration as a function of the NTU corresponding to it. This graph is dimensionless; distances from the 50% breakthrough point are divided by the height of a transfer unit (HTU).

If one integrates in one direction, the changes in concentrations will be positive, but if one integrates in the other direction, the changes in concentration will be negative because dC and dq will be negative. Thus the NTU corresponding to changes in concentration in one direction from the mid-point will be positive, and NTU for changes in the other direction will be negative.

Although the integration gives the NTU as a function of bulk concentration, it is more instructive to plot the bulk concentration as a function of the NTU. In Figure 21 this looks very much like a breakthrough curve—and it is, plotted in terms of dimensionless, instead of actual, distances. The breakthrough curve can be converted into dimensions by multiplying NTU values by the height of the transfer unit (HTU), that is, by $(v-v_f)/k$, where k is the appropriate mass transfer coefficient corresponding to the concentration driving force used to determine the NTU. That is, use k_f if the NTU is based upon concentrations in the fluid phase, and k_s if the concentrations in the solid are used. Of course, one could use overall mass transfer coefficients as well as individual film coefficients. (See Chapter 3.)

Although the breakthrough fronts are described in terms of distances (positions in the bed), they can also be described in terms of the time at which the front moves through a given position in the bed. This is the way breakthrough fronts are usually measured experimentally. To make that conversion, one only has to divide the position (the length dimension described above) by the negative value of the velocity of the front down the bed, $-v_f$. This gives the concentration exiting from the bed as a function of time. Remember that the linear operating line indicates that the relative concentration in the solid, expressed as the fraction of the difference between the initial concentration in the solid and the concentration in equilibrium with the feed fluid, and the relative concentration in the fluid, expressed as the fraction of the difference between the initial concentration that is in equilibrium with the concentration in the initial solid in the bed and the inlet concentration, are proportional; so the plot is obtained in the same manner for calculations based upon concentrations in the fluid and for calculations based on concentrations in the solid.

Remember that the NTU is negative when the integration is carried out in one direction and positive when carried out in the other. The value of the NTU was set at zero at the mid-point of the breakthrough curve. This means that the times are set at zero when the mid-point of the concentration breakthrough curve passes the exit of the bed. The times will be positive as the concentrations are greater than the mid-point (that is, after the mid-point has passed through the bed), and the times will be negative for the lower concentrations that occur before the mid-point passes through the bed. Since this analysis applies to fully established concentration pattern breakthrough curves, the shape of this curve would be the same for any length of a long bed. One could integrate the NTU to values (dimensionless distance) in either direction of the concentrations that correspond to the conditions for either direction (toward the inlet concentration or toward the initial concentration in the bed before fluid was introduced), but the value of NTU (distance or time) would approach $\pm\infty$ as one approached either end condition. That should be expected since the constant pattern applies exactly only to infinitely long beds, and this illustrates why this analysis cannot tell exactly where the breakthrough curve appears; it can only tell its shape.

To determine the position of the breakthrough curve in the bed, it is necessary to know the rate at which the front moves down the bed, and that can be determined from the isotherm. Since the solute is being transferred from the fluid to the solid, the velocity of the front can be estimated simply

$$v_f = v\left[\frac{C_{\text{in}} - C_0}{q_{\text{sat}} - q_0}\right] \tag{60}$$

Then the time at which the mid-point exits the bed is likely to be approximately

$$t_{1/2} = \frac{L}{v_f} \tag{61}$$

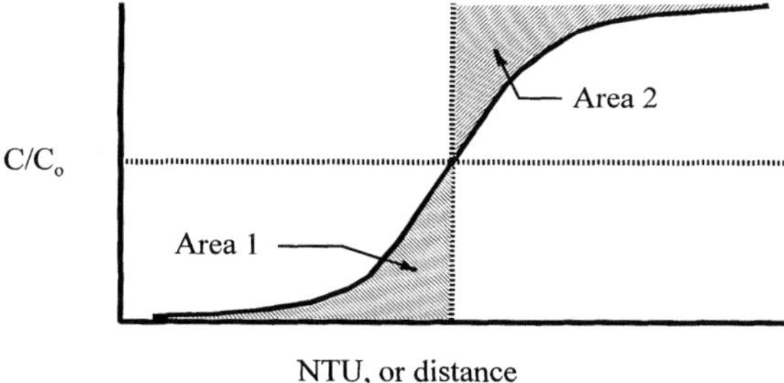

FIGURE 22 Determining if the breakthrough curve is symmetrical. A simple, but not complete, test is to see if the areas in the two regions marked are equal. This is necessary, but not sufficient, for symmetry, but nonsymmetry is usually evident from this test. More exact tests would involve comparing individual positions on the curve equal distances from the 50% breakthrough point.

This relationship is exact if the breakthrough curve is symmetrical, that is, if the shaded areas in Figure 22 are equal. If the breakthrough curve is not symmetrical, one can offset the reference point an appropriate distance from the 50% breakthrough point so that the areas under the curves will be equal.

Mass Transfer Resistance in Both Phases

The discussion has been limited to cases where the resistance to mass transfer resides principally in one phase. For many dilute systems, the resistance is likely to be principally in the fluid phase, but some adsorption or ion exchange systems with very small pores, like zeolites, may have so much solid phase resistance that the fluid phase resistance would dominate only for exceptionally dilute systems. For linear isotherms, the control of the mass transfer resistance would remain in the same phase for all concentration, but for favorable isotherms such as those covered by this approach, the relative contributions to the total mass transfer resistance will change with concentration. High distribution coefficients favor mass transfer in the solid phase because there is a relatively larger driving force available for the solid phase diffusion, and this means that for favorable isotherms the controlling resistance is more likely to be in the fluid phase at low concentration than at higher concentrations. The distribution coefficient changes with concentration for nonlinear isotherms and decreases with increasing concentration for favorable isotherms.

When the mass transfer resistance is essentially entirely in one phase, it is necessary to describe the NTU (and HTU) in terms of the mass transfer coefficient and concentrations in that phase because the concentration driving force in the other phase will be very small, even approaching zero. Alternatively, one could express the mass transfer rates in terms of "overall" coefficients rather than in terms of the coefficients for individual films. This is described in more detail in Chapter 3, but the concept applies here as well. Instead of defining the driving forces as the difference in concentrations across the film (the difference between the concentration in the bulk and the concentration at the interface), one could define the driving force as the difference between the bulk concentration in the reference phase and the concentration that would be in equilibrium with the bulk concentration in the other phase. Using the fluid as the reference phase,

$$\text{rate} = K_f(C - C^*) \tag{62}$$

where C^* is the concentration in the fluid in equilibrium with the bulk solid. Of course, a similar expression could be written in terms of concentrations in the solid phase. The letter K is used to describe the overall mass transfer coefficient to distinguish it from k, which is usually used for mass transfer coefficients for individual films. Then

$$\text{NTU} = \int \frac{dC}{C - C^*} \tag{63}$$

If all of the resistance is in the fluid (or the reference) phase, the overall mass transfer coefficient will be the same as the individual film coefficient because the interface concentration in the other phase is essentially the same as the bulk concentration. However, when there is significant resistance in the other phase, the overall mass transfer coefficient and the individual film coefficients will be different, perhaps significantly different.

Although overall mass transfer coefficients are commonly used in absorption and liquid-liquid extraction, usually because there is insufficient information on individual film coefficients, remember that the overall film coefficient includes an additional approximation. Use of overall mass transfer coefficients assumes that the sum of the two film resistances is proportional to the overall resistance. This is strictly true for linear isotherms, but it is not generally true for nonlinear isotherms. When data are available on the individual film coefficients, it is not necessary to revert to the use of overall mass transfer coefficients in adsorption. One can follow the approaches that are better known in absorption and liquid-liquid extraction for such cases.

The overall mass transfer coefficient can be related to the individual mass transfer coefficients, but the relationship is a function of the distribution coefficient, the ratio of the concentration in the two phases at equilibrium. (See Chapter 3.) Thus, for nonlinear isotherms, such as the favorable isotherms discussed in this section, the relative contribution of the two film mass transfer coefficients changes over the front, and even if the film coefficients are constant, the overall mass transfer coefficient will vary over the breakthrough front. That makes the overall mass transfer coefficient concept less useful for problems where the isotherm is highly nonlinear.

Where resistance is significant in both phases, it is better to account for the actual resistance in both phases if information on the film coefficients is available; this is possible with an approach that is similar to concepts more often used in absorption, stripping, and liquid-liquid extraction applications. This approach [62] utilizes the mass balance on the solute across the two films and recognizes that the inventory of solute in the thin "films" can be neglected. Then the flux of solute across one film is the same as the flux across the other film. This can be expressed as

$$\text{rate} = k_f(C - C_i) = k_s(q_i - q) \tag{64}$$

The equation can be rearranged as

$$\frac{q_i - q}{C - C_i} = \frac{k_f}{k_s} \tag{65}$$

which describes a straight line with slope $-k_f/k_s$. At any point in the bed, the concentrations in the bulk and at the interface of both phases are related by this equation. Thus, to obtain the interfacial concentrations at any point in the bed, one constructs lines from different points on the operating line (remember that the operating line represents the bulk concentrations in each phase at different positions in the bed) to the isotherm (equilibrium curve) with a negative slope equal to the ratio of the individual film mass transfer coefficients (Figure 23). By drawing a series of parallel lines between the operating line and the isotherm with this slope, one can locate the interfacial compositions (on the isotherm) and the bulk concentrations (on the operating line) for all positions in the front. Thus, the NTU can be obtained for different concentrations in either phase in the same manner described previously, except that the interfacial compositions are obtained from this connecting line with a slope of $-k_f/k_s$ rather than the horizontal lines used for cases with mass transfer control all in the fluid phase or the vertical lines used for cases with mass transfer control in the solid phase.

The difference between the bulk concentrations (C and q) and the interfacial concentrations (C_i and q_i) can then be used to evaluate the integral in Equations (58) or (59) and thus the NTU. The procedure is identical to that used for cases with mass transfer control in only one

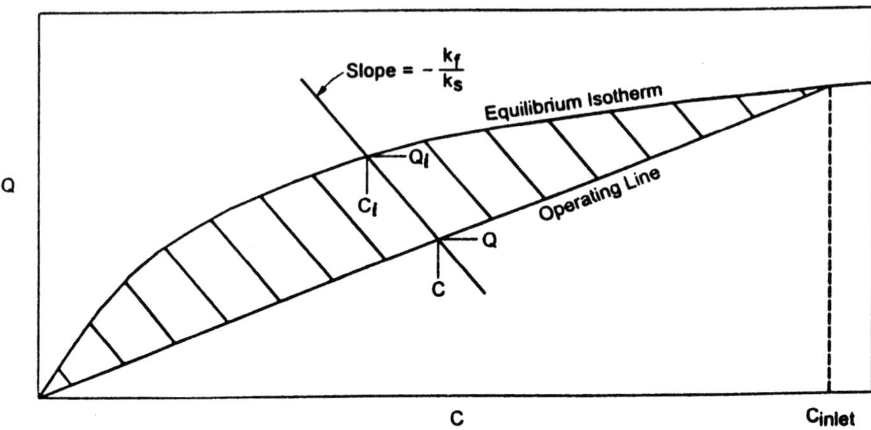

FIGURE 23 Tie lines connecting the bulk concentrations on the operating line to the interfacial concentrations on equilibrium curve when there is significant mass transfer resistance in both phases. The mass transfer resistance is assumed to be described by a solid phase diffusion coefficient and thus is an effective solid film resistance. The slopes of tie lines connecting the concentrations in the bulk and the interface at each point in the bed are the ratios of the mass transfer resistances in the two phases.

phase, except that the interfacial concentrations rather than the concentration in equilibrium with the other phase is used. If all of the mass transfer is in one phase, the interfacial concentration will be the concentration in equilibrium with the bulk of the other phase (the phase with no significant mass transfer resistance), and there will be no difference in the NTU calculated using overall coefficients or individual film coefficients.

Uses of This Approach

This generalized approach to study of the appropriate constant pattern isotherms can be used to predict the breakthrough isotherms when the appropriate mass transfer coefficients and the isotherms are known, or it can be used to determine the mass transfer coefficients from experimental data. Once the coefficients are known, they can be used to predict the behavior of other beds with different lengths, fluid velocities (providing one knows the way mass transfer coefficients change with fluid velocity), or concentrations. Predictions of the breakthrough curves involve first plotting the isotherm, drawing the operating line from the known end conditions of the bed, integrating to determine the NTU as a function of concentration based upon one of the phases, and converting the dimensionless distances to real distance or time.

The technique for determining the mass transfer coefficients from experimental breakthrough curves may be less obvious. The breakthrough curve cannot be used to evaluate both mass transfer coefficients directly. If one should know the ratio of the two film coefficients or know that only one film controls the mass transfer rate, the absolute value of each coefficient could be evaluated in essentially the same way used to evaluate a single coefficient. The first steps needed are the same as those listed above. That is, one uses the isotherm and operating line to integrate and find the NTU as a function of concentration in the breakthrough curve. The NTU curve is independent of the mass transfer coefficient. The shape of the dimensionless breakthrough curve should be the same as the measured breakthrough curve; the dimensionless NTUs are proportional to the actual dimensioned distance of time shown on the experimentally measured curve. One can determine the actual time required for the concentration in the breakthrough curve to change by a given amount, say $C/C_0 = 0.2$ to $C/C_0 = 0.8$ and convert this time into distance in the bed by multiplying the time by V_f. From the plot of the breakthrough curve in dimensionless units, the NTU change for the same change in C/C_0 can be found. The HTU is the ratio of the actual distance to the dimensionless distance, and the mass transfer coefficient is the HTU divided by the fluid velocity, V.

When the ratio of the two mass transfer coefficients is not known, one needs to know one of the coefficients. For example, one could estimate the mass transfer coefficient for the fluid film using standard correlations for mass transfer in packed beds. The calculation of the other coefficient is not straightforward since the slope of the tie lines is not known. This makes the evaluation of the second mass transfer coefficient a trial-and-error problem, and several slopes may have to be investigated to match the width of the front to experimental data to evaluate the second coefficient. If the equilibrium curve is a Langmuir isotherm, one can use the curves already evaluated by the author and described below.

Example Solutions for Ion Exchange and Langmuir Isotherms

This approach is illustrated in the two papers by the author. The first paper illustrates cases with mass transfer in individual films for binary ion exchange. The calculations were made numerically because equations were available for the ion exchange isotherm. When dealing with many simple binary ion exchange problems, one can use the figures already

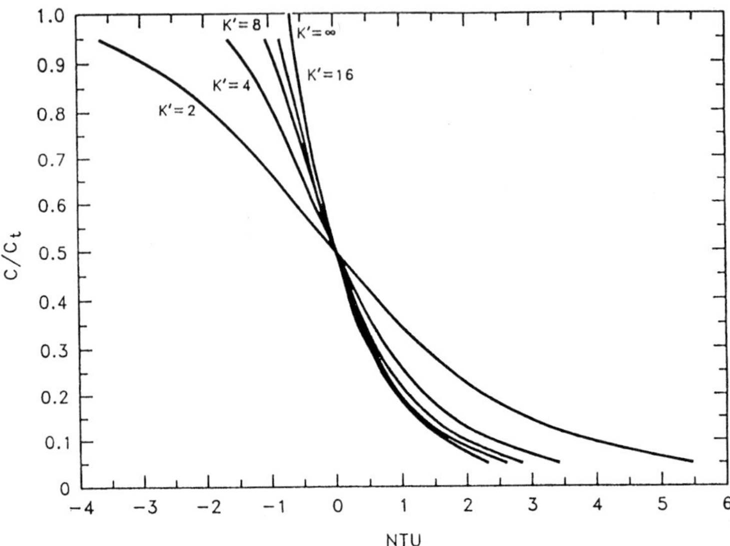

FIGURE 24 Normalized fronts for ion exchange of ions with equal charges and mass transfer resistances in the liquid phase.

calculated and presented. The isotherm equations assumed that the activity coefficients were constant in both phases, so the equilibrium (mass action) coefficients were constant over the breakthrough front. (See the section "Ion Exchange Equilibrium.") For exchange of ions with equal charges, the equilibrium expressions can be solved relatively easily, and the integration of the NTU can be done analytically to give an expression much like the solution of the front for a Langmuir isotherm, but for exchange of ions with different charges the solution for the equilibrium curve (isotherm) is more difficult, and analytical solutions for the NTU are not believed likely to become available.

Dimensionless breakthrough curves were calculated for three cases of ion exchange: exchange of ions with equal charge, exchange of monovalent ions for divalent ions, and exchange of divalent ions for monovalent ions. In each case the bed of ion exchange material is assumed to be loaded completely by one ion, and the solution is assumed to contain only the exchanging ion. For these cases, the results for each case of ion exchange could be shown as a single family of curves with the equilibrium constant as a parameter. When the mass transfer resistance is controlled by the resistance in the liquid (fluid) phase, the ion exchange breakthrough curves for ions with equal charge are shown in Figure 24. When the mass

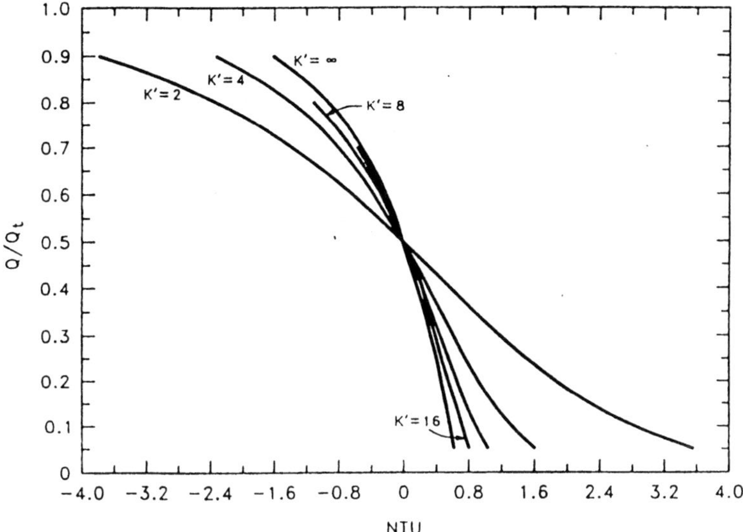

FIGURE 25 Normalized fronts for ion exchange of ions with equal charges and mass transfer resistances in the solid phase. Mass transfer in the solid phase is assumed to be described adequately by diffusion through the solid particle, not in pores.

transfer resistance is controlled by the solid film, the breakthrough curves are shown in Figure 25. As the equilibrium constant for the ion exchange increases, the breakthrough front becomes sharper and approaches that for an irreversible front. As the equilibrium constant decreases, the front in dimensionless distance units (NTU) becomes more diffuse and broader. Of course, the actual width of the breakthrough front is determined by the mass transfer coefficient (that is, by the HTU) as well as the equilibrium constant (the curvature in the isotherm). The shape of the front, however, is determined by the equilibrium constant alone, not by the mass transfer coefficient (as long as the control of the mass transfer process remains in the same film).

Although only a selected number of values of the equilibrium constant are illustrated in these calculations, the shape of breakthrough curves for other values of the equilibrium constant can be estimated by interpolating between these curves or by repeating the calculations for the exact value of the equilibrium constant of interest. The curves shown are believed to be sufficiently close together for extrapolations to be as accurate as most measured breakthrough fronts. Thus, for simple binary ion ex-

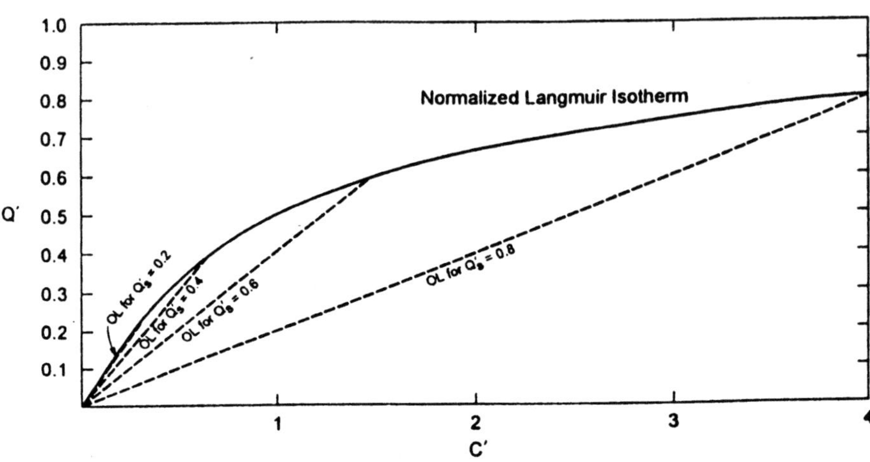

FIGURE 26 A normalized Langmuir isotherm with isotherms that correspond to different maximum loading of the adsorbent. Note that the effects of the curvature in the isotherm become more evident as feed concentrations increase—that is, when the maximum concentration on the solid increases.

change, these curves may be adequate to predict the breakthrough curves or to estimate the mass transfer coefficients for many cases. Remember that, although the curves were calculated for the cases where all of the mass transfer resistance was in one phase, they can be used for other cases if the overall mass transfer coefficients are used.

The second paper on this approach to predicting the shape of breakthrough curves presents dimensionless breakthrough curves when there is significant resistance in both phases, and the results are illustrated for single component Langmuir isotherms. Although the Langmuir isotherm was selected to illustrate the prediction of concentration fronts with resistance in both phases, the choice of the Langmuir isotherm was arbitrary; the approach could have been illustrated for ion exchange equilibrium or for any other isotherm as well.

Note that the Langmuir isotherm can be written in dimensionless units with a single parameter as

$$\frac{q}{q_{\max}} = \frac{kC}{1+kC} = \frac{C'}{1+C'} \tag{66}$$

where $C' = kC$. This simplifies all Langmuir isotherms into a single equation with the fraction of adsorbent loading being a function of only the normalized concentration variable C'. With this normalized form for the

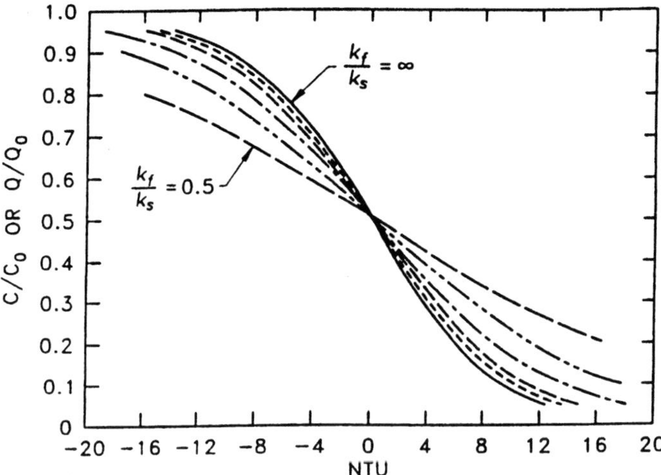

FIGURE 27 Normalized concentration fronts for a Langmuir isotherm based upon concentrations in the solid phase and a feed concentration that corresponds to loading on the solid that is 20% of its maximum capacity. Individual curves correspond to $k_f/k_s = 0.5, 1.0, 2.0, 5.0, 10.0, \infty$.

isotherm, all NTU (or the concentration front) calculations can be made in terms of the dimensionless concentrations for any value of k or q_{max}. This normalized isotherm is shown in Figure 26. Note that several possible operating lines are shown for the case where the initial bed contains no solute. For low feed concentrations, the maximum loading can only be significantly less than the maximum capacity. Since the operating line extends to the point where the adsorbent is in equilibrium with the feed concentration, the upper position of the operating line will depend upon the feed concentration. Thus each operating line in Figure 26 corresponds to a different feed concentration. The breakthrough front calculations (evaluation of NTU) were reported for cases where the initial adsorbent in the bed contained no solute; hence, all operating lines to be considered began at the origin and terminated at different points on the isotherm (Figure 26). Figures 27 through 34 show breakthrough fronts for different values of the fraction of the adsorbent capacity loaded when the adsorbent was saturated with the fluid at the feed concentration. This could be expressed as different values of kC_{feed} or C'_{feed}. Within each figure are different curves; each corresponds to a different ratio of the mass transfer coefficients. This normalization allows a great range of Langmuir isotherms and adsorption conditions to be described by a relatively small

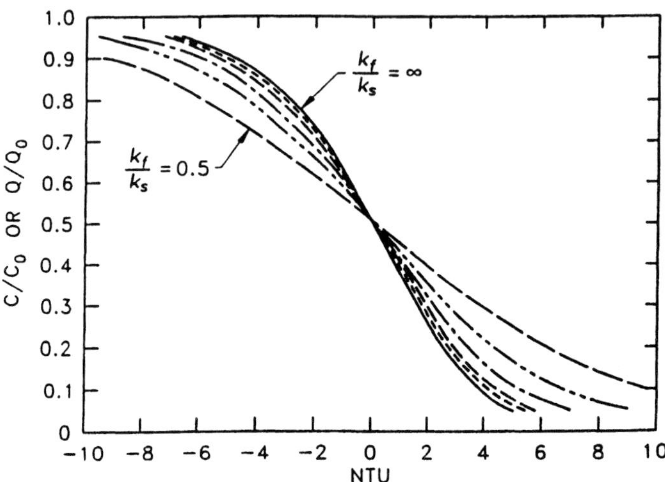

FIGURE 28 Normalized concentration fronts for a Langmuir isotherm based upon concentrations in the solid phase and a feed concentration that corresponds to loading on the solid that is 40% of its maximum capacity. Individual curves correspond to $k_f/k_s = 0.5, 1.0, 2.0, 5.0, 10.0, \infty$.

number of figures. These calculations were performed numerically, but the same results could have been obtained from graphical integrations. Separate sets of figures are needed to describe breakthrough curves based upon the fluid phase and upon the solid phase.

Figures 27 through 29 can be used directly to predict the shape of breakthrough curves when the isotherm and mass transfer coefficients are known. As in the example of single component adsorption processes, the figures can also be used to evaluate the mass transfer coefficients from experimental breakthrough data. The procedures are essentially the same, except it is more likely to be practical to evaluate only one mass transfer coefficient from a breakthrough curve. There are differences in the shape of the different breakthrough curves, which, in principle, would allow one to use the shape to determine the ratio of the mass transfer coefficients and the spreading of the front in real time units to determine the value of the coefficients. However, it is unlikely that one will be able to determine the shape accurately enough to determine the ratio of the coefficients with acceptable accuracy. Instead, one is more likely to want to know the value of one transfer unit and determine the value of the ratio from the width of the front and thus obtain the value for the other mass transfer coefficient. For instance, one may be operating with spherical or granular

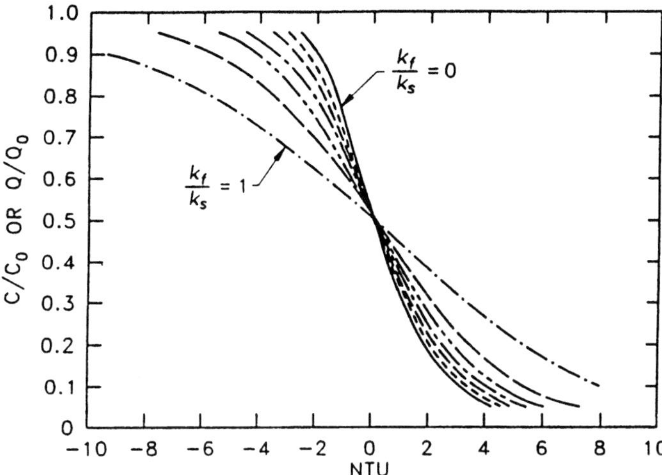

FIGURE 29 Normalized concentration fronts for a Langmuir isotherm based upon concentrations in the solid phase and a feed concentration that corresponds to loading on the solid that is 60% of its maximum capacity. Individual curves correspond to $k_f/k_s = 0.5, 1.0, 2.0, 5.0, 10.0, \infty$.

shaped adsorbent or another adsorbent shape for which relatively good correlations are available for estimating the value of the fluid phase mass transfer coefficient. Then the spreading of the breakthrough front can be used to estimate the value of the mass transfer coefficient in the solid phase.

Extrapolation of Results from a Single Breakthrough Curve to Different Flow Rates

If an adsorption or ion exchange process is tested with one flow rate (velocity through the bed) and the velocity used in the test is greater than the velocity that is expected to be used in the application, it is very easy to use the preceding analyses to predict the shape and width of breakthrough curves at other fluid velocities. Remember that if the same fluid feed concentration and the same adsorbent is used, the operating line and the dimensionless breakthrough curves will be the same for any velocity. That means that the shape, but not the width, of the breakthrough curve will be independent of the fluid velocity (unless the relative contributions of the fluid and solid "film resistances" change). Thus, one only needs to adjust the time (or distance) scale for the new velocity. For each position

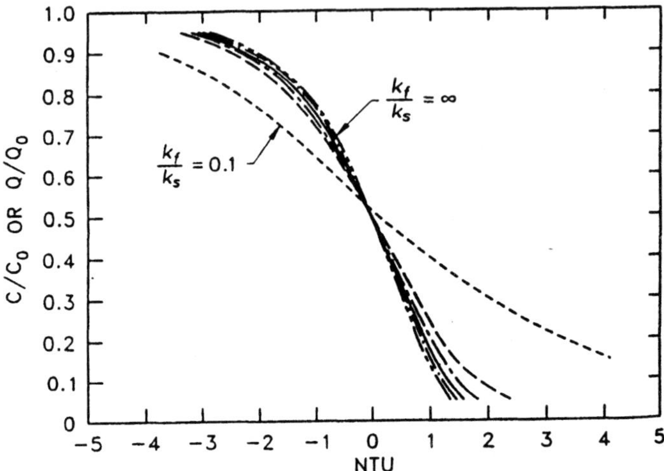

FIGURE 30 Normalized concentration fronts for a Langmuir isotherm based upon concentrations in the solid phase and a feed concentration that corresponds to a loading on the solid that is 80% of its maximum capacity. Individual curves correspond to $k_f/k_s = 0.1, 0.5, 1.0, 2.0, 5.0, 10.0, \infty$.

C/C_0 on the breakthrough curve, there is a dimensionless distance (NTU), a real distance for the velocity used in the first test, and a real distance for the new velocity. These are related as follows:

$$\text{NTU} = \frac{z_t}{[\text{HTU}]_t} = \frac{z}{\text{HTU}}$$

where z is the distance from the mid-point of the breakthrough curve. The subscript t refers to the test conditions, and the lack of a subscript refers to the new (extrapolated) conditions of interest. Then

$$z = z_t \frac{\text{HTU}}{[\text{HTU}]_t}$$

Thus, one can take values for z_t at various values of C/C_0 from the test curve and calculate the value of z that should correspond to the same value of C/C_0 under the extrapolated conditions provided that the ratio of the HTUs is known.

When the reference phase is the fluid phase, the ratio of the HTUs is

$$\frac{\text{HTU}}{[\text{HTU}]_t} = \frac{(v - v_f)/k}{(v_t - v_{ft})/k_t}$$

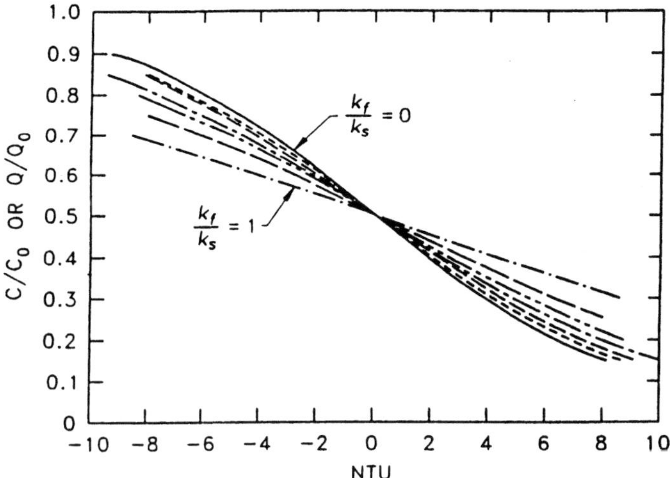

FIGURE 31 Normalized concentration fronts for a Langmuir isotherm based upon concentrations in the fluid phase and a feed concentration that corresponds to a loading on the solid that is 20% of its maximum capacity. Individual curves correspond to $k_f/k_s = 0.0$, 0.05, 0.1, 0.2, 0.3, 0.5, and 1.0.

The mass transfer coefficient, k, would correspond to the fluid phase with this equation. Subscript for the fluid phase were omitted for simplicity. (If the HTUs were based upon the solid phase, one would use the velocity of the front, v_f , instead of $v - v_f$, and the mass transfer coefficients would be for the solid film.)

With the same fluid feed concentration, the velocity of the front will be proportional to the fluid velocity because the adsorbent will become loaded to the same concentration in the solid, and the rate at which the front moves down the bed will be proportional to the feed velocity. Thus the ratio of the velocity terms is simply the ratios of the velocities themselves:

$$\frac{v - v_f}{v_t - v_{ft}} = \frac{v}{v_t}$$

If the mass transfer is controlled by an effective solid film, k_s would not be expected to depend upon the fluid velocity. Then

$$z = \frac{v_f}{v_{ft}} z_t = \frac{v}{v_t} z_t$$

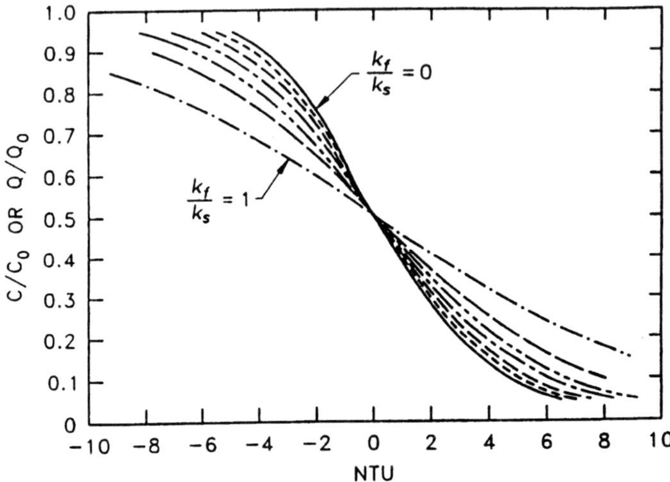

FIGURE 32 Normalized concentration fronts for a Langmuir isotherm based upon concentrations in the fluid phase and a feed concentration that corresponds to a loading on the solid that is 40% of its maximum capacity. Individual curves correspond to $k_f/k_s = 0.0, 0.05, 0.1, 0.2, 0.3, 0.5$, and 1.0.

However, if the mass transfer rate is controlled by the fluid film, k_f will change when the velocity is changed. If k_f depends upon the fluid velocity in the same manner predicted by the Rantz-Marshall equation [63] [Equation (24)], then $k_f \propto v^{0.6}$. Then $z \approx z_t(v/v_t)^{0.4}$.

Thus, much can be learned from a single breakthrough curve, and the results can be used to predict the breakthrough curves for a range of fluid velocities. Note that with information on one breakthrough front, predictions could be made without even knowing much about the isotherm except that the isotherm is favorable and, thus, forms a constant pattern front. However, to extrapolate the results from a breakthrough curve to other feed concentrations, it is necessary to know the equilibrium isotherm, and the shape of the breakthrough front is likely to change at least somewhat. To extrapolate to other feed concentrations, it is necessary to determine the NTU for different concentrations (fractions of breakthrough) for the test and proposed extrapolated conditions. The HTU can be determined from a match of the dimensionless breakthrough curve to the experimental curve. If the same flow rate is used in the test and in the extrapolated condition, the HTU is likely to be essentially the same in both cases. Then the extrapolated breakthrough curve can be obtained by multiplying different values of NTU on the dimensionless breakthrough

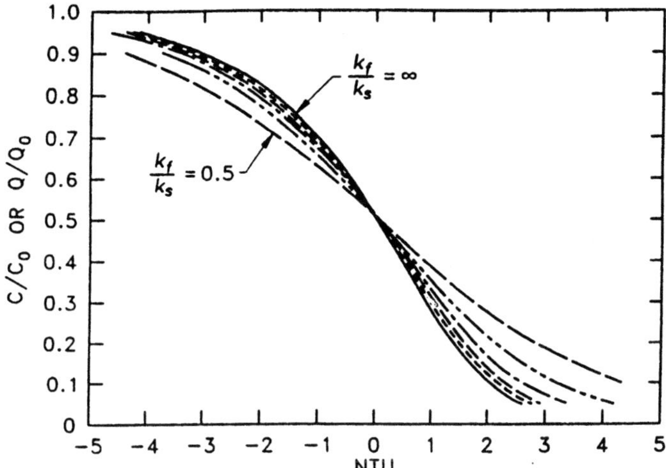

FIGURE 33 Normalized concentration fronts for a Langmuir isotherm based upon concentrations in the fluid phase and a feed concentration that corresponds to a loading on the solid that is 60% of its maximum capacity. Individual curves correspond to $k_f/k_s = 0.0$, 0.05, 0.1, 0.2, 0.3, 0.5, and 1.0.

curve for the extrapolated concentration by the HTU to obtain the new breakthrough curve. If the flow rate is changed between the test conditions and the extrapolated conditions, it is necessary to take the change in velocity into account in the same manner discussed above and make that change to the HTU.

Remember that, although it has been more convenient to discuss the breakthrough curves in terms of distances down the bed at a given time, most breakthrough curves are measured as concentrations at a given position as a function of time. However, for constant pattern conditions, these two ways of seeing breakthrough curves can be easily compared. The time from the mid-point on the breakthrough curve can be related to the distance from the mid-point by the velocity of the front:

$$t = -\frac{z}{v_f}$$

Thus,

$$\frac{t}{t_t} = \frac{z/v_f}{z_t/v_{ft}}$$

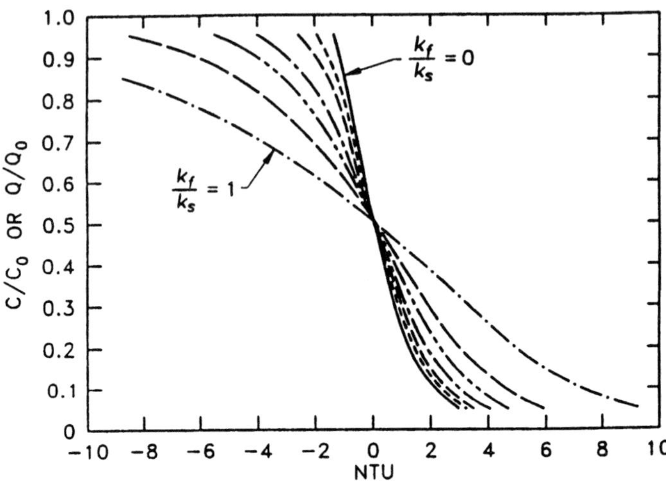

FIGURE 34 Normalized concentration fronts for a Langmuir isotherm based upon concentrations in the fluid phase and a feed concentration that corresponds to a loading on the solid that is 80% of its maximum capacity. Individual curves correspond to $k_f/k_s = 0.0$, 0.05, 0.1, 0.2, 0.3, 0.5, and 1.0.

Alternative Use of Linear and Irreversible Models to Describe Some Nonlinear Fronts

Although the foregoing lengthy discussion recommends the generalized approach for constant pattern fronts, some may prefer to stay with more conventional equations, and there can be important applications that do not have constant pattern fronts. The irreversible isotherm with constant pattern is a particularly simple breakthrough front, and one could use that equation and the graphs in Figures 23, 24, and 27–34 if the isotherm is approximated relatively well by an irreversible isotherm. As just pointed out, even a Langmuir isotherm is approximated well by an irreversible isotherm for large values of kC_{feed}, that is, for values of q_{sat}/q_{max} that approach unity. Likewise, binary ion exchange processes that are highly favorable for the ion in the feed fluid can be approximated by an irreversible isotherm. The principal reason for using the irreversible isotherm is simplicity of the equation. One can estimate when it will be satisfactory to approximate the real isotherm with an irreversible isotherm by inspecting the isotherm and the operating line. For instance, when the mass transfer rate is controlled by resistance in the fluid phase, one can inspect the driving force for mass transfer over the concentration range of interest by drawing horizontal lines connecting different points on the operating line with corresponding points on the isotherm and on the vertical axis, the irreversible isotherm. When there is no significant (relative) difference in the distances between operating line and isotherm and between operating line and vertical axis, there will be little difference between the breakthrough front predicted from the actual isotherm or the irreversible isotherm.

When the bed is too short for a constant pattern front to develop, the general solution for the breakthrough front can be very complicated and difficult to predict. There are no simple general approaches to this problem, and complex numerical calculations may be required. However, there is a limiting case where simple solutions can be used—when the isotherm can be approximated by a linear isotherm over the concentration region of interest. The Langmuir and many other isotherms are approximated by a linear shape at very low concentrations. Although any negative curvature will eventually result in a constant pattern front, it may take a very long bed for the pattern to be developed, and very short beds may be approximated better by assuming that the isotherm is linear. In all such cases, one can note that for isotherms with slight negative curvature over the region of interest, the linear approximation is "conservative" in the sense that the width of the breakthrough curve will be somewhat less than that predicted. On the other hand, the mass transfer coefficients evaluated by using a linear approximation would be somewhat high. In the rarer case where there is a slight positive curvature to the approximately linear isotherm, the "errors" would be in the opposite direction.

BREAKTHROUGH FRONTS FOR MULTI-COMPONENT SYSTEMS

A multi-component system is one with two or more solutes, but for practical uses the term applies only when the presence of one or more components affects the adsorption of other components. As noted earlier, the solutes may affect the adsorption of other components by competing for the adsorption capacity or, possibly, by interacting with each other in the fluid or on the adsorbent. It is this effect of one component on adsorption of another component that makes a system act as a "multi-component." system. If two or more components are not affected by the presence of the other, the dynamic and equilibrium behaviors of the bed will be essentially the same as that predicted by considering the behavior of the bed with each component separately. This could be the case for very dilute systems. In fact, one could define "dilute" as the concentration at which the components do not affect the adsorption of other components. For Langmuir isotherms, this would mean when the concentrations are low enough that the linear approximation is adequate. For several components, this means when

$$\sum k_{Li} C_{i\text{feed}} \ll 1 \tag{67}$$

Note that for a multi-component Langmuir system, "curvature" begins to appear (or have an effect) when the summation becomes significant, not just when $k_{Li}C_{i\text{feed}}$ becomes significant.

As noted when describing ion exchange equilibrium, dilute concentration alone does not ensure that multi-component solutions behave as individual components. It is only when the exchanging ions are all dilute *relative to the total ion concentration* that the different exchanging components "may" act approximately independent of the concentration of the other exchanging ions.

Many environmental adsorption problems involve removal of the last traces of a pollutant, and there is a reasonable likelihood that the system will be dilute and behave approximately as if the adsorption followed a series of linear isotherms. However, remember that the adsorbent properties affect the range where the linear (single component) approximation can be used. That is, the values of k_{Li} as well as the concentration determine if the sum is much less than unity, that is, near zero. If the adsorbent has a particularly high affinity for one or more of the solutes (that is, a sufficiently high value for k_{Li}) other than the solute of interest, then the system may have to be treated as multi-component, even for apparently low solute concentrations in the fluid.

Multi-component adsorption involves the movement of different fronts down the bed. The adsorbent ahead of these fronts will be in the initial state of the bed prior to the beginning of the adsorption operation. The last front (and the adsorbent very near the fluid inlet) will be in equilibrium with the feed fluid. If there are n components, there could be as many as n fronts. Components with the least affinity for the adsorbent will be in the forward portions, those with the greatest affinity for the adsorbent will move more slowly down the bed and be largely in the fronts nearer the fluid inlet. The fronts may be sharp or diffuse, but only sharp (favorable) fronts will remain similar in very long beds.

The simplest and most commonly used analyses of multi-component systems are based upon local equilibrium. Taking mass transfer effects into account as well as multi-component displacement makes it more likely that computer based numerical solutions will be required.

Most analyses of multi-component systems are likely to be based upon relatively simple isotherms, such as the multi-component Langmuir isotherms. As noted when discussing multi-component equilibria, one practical reason for using the Langmuir isotherm is the large amount of data required to evaluate and validate more complex isotherms. We can describe a single component isotherm graphically, even when we have no suitable algebraic equation to describe it. In principle, we could use graphs (perhaps fitting the portion of interest with an equation) to correlate and predict adsorption bed performance for several components. For instance, one could measure isotherms for each component under different concentrations of the other components. Interpolation of single variable plots is straightforward, either numerically or graphically. However, consider the problems of simply plotting isotherms for five or six components. Many graphs would be required, and interpolating values between the data points could be complex. Even if such interpolations are practical, there are seldom sufficient data for systems with many components.

Since multi-component adsorption problems are expected to be less common in environmental operations than in some other applications because of the greater likelihood that concentrations will be low (often sufficiently low for the linear single component approximation to be used), few details on multi-component adsorption bed behavior will be presented. For such detail see Hellferich and Klein [64]. This book contains detailed descriptions of the concepts of multi-component adsorption. A shorter description of the field can be obtained in Ruthven [65]. One improvement that has taken place is the increased use of computers to simulate front development and movement (Katti and Guiochon [66]). Linda Wang and her associates at Purdue University have completed several studies of multi-component adsorption and multi-component ion exchange [67,68].

BREAKTHROUGH FRONTS FOR NONISOTHERMAL SYSTEMS

All discussion of adsorption processes presented here have assumed that the system is isothermal. Of course, any change in bed temperature will change the equilibrium conditions greatly and, perhaps, even the mass transfer parameters slightly. Temperature variations can result from heat added or removed from the bed or from heat that is generated within the bed from the adsorption process itself. Introduction or removal of heat into the bed, either through the outside wall or through internal tubes, causes radial and longitudinal temperature variations, which may be difficult to analyze. Most adsorption beds, however, are simply cylindrical with no direct heating or cooling. Even if the bed is to be regenerated by raising the temperature, it can be accomplished by heating the inlet fluid (gas) rather than the bed. This approach becomes increasingly likely as bed diameter increases.

Temperature gradients resulting from the heat of adsorption/desorption are common and result in longitudinal temperature variations. Temperature effects are more likely to be important in gaseous systems than in liquid systems because of the lower volumetric heat capacity of gases and the greater likelihood that small variations in temperature will cause important changes in the adsorption equilibrium. Temperature effects are also more likely to be important in cases with high solute concentrations. Obviously, the greater the amount of solute adsorbed, the greater the heat generated.

Again, because so many environmental problems deal with low concentrations of solutes, nonisothermal operations will not be covered in detail here. However, the environmental applications are not expected to be limited totally to isothermal conditions. More detailed texts (e.g., Ruthven [69] or Yang [70]) should be consulted. In some ways, temperature can act much like another component as temperature fronts (as well as concentration fronts) move down the column. Temperature fronts can produce column temperatures that are higher or lower than the inlet or exit fluid temperatures.

APPLICATIONS FOR ADSORPTION IN ENVIRONMENTAL AND WASTE PROCESSING

When to Use Adsorption

Although there are no completely accurate general criteria for when adsorption processes are preferred over other separation methods, under some common conditions they are often selected. First, remember that adsorption operations are especially attractive for removing components from dilute solutions. Since many pollutants occur in off-gases and water discharge streams in very low concentrations, adsorption is often an attractive choice for contaminant removal. The vast majority of adsorption operations are transient and carried out in packed beds. With dilute systems, it is often possible to operate a bed for a considerable time between regenerations. Bed size is determined principally by the amount of material to be loaded on the bed. Bed size is a lesser function of the fluid velocity through the bed; the effect on bed size comes from front spreading, which increases with fluid velocity. For systems with highly favorable isotherms, decreasing the solute concentration allows longer operating times between regenerations, higher flow rates, or smaller bed sizes, and the economics for adsorption tend to become increasingly favorable as solute concentration decreases. Note, however, how this generalization assumes a strong affinity for the contaminant and a favorable isotherm.

If the isotherm is linear, bed loading is proportional to the fluid concentration. In such cases, the volume of fluid that can be treated between regenerations becomes largely independent of the fluid concentration. The mid-point of the breakthrough curve will be independent of the concentration for linear isotherms, but the volume treated before the concentration on the breakthrough curve becomes greater than accepted limits (usually much less than the mid-point) will remain dependent upon the feed concentration if the accepted efficicnt concentration remains the same, since the concentration at any point in the breakthrough curve will be proportional to the inlet concentration.

Another advantage of adsorption processes results from the small "stage heights" or "height of transfer units" that can be achieved in packed beds. The analyses of mass transfer in this book have focused upon transfer units, but most readers will remember that these different ways of looking at mass transfer resistance are closely related. (See Chapter 3 for a discussion of stages and comparisons with NTUs.) When large numbers of transfer units are involved (required), the operating line and the equilibrium curve are likely to be close to parallel, and those are the conditions where the numbers of transfer units and stages are likely to be proportional—actually equal under some definitions of transfer units. Small HTU means that numerous transfer units can be obtained in relatively modest scale equipment, especially if small adsorbent particles can be used. For this reason chromatography is so effective in analytical chemistry. The many transfer units allow adsorption processes to remove large fractions of the solute, and adsorption is more likely to become attractive for removing highly toxic pollutants where extremely high removal efficiencies are required.

The combined ability to work well with dilute systems and to remove large fractions of the contaminants makes adsorption especially attractive to back up other methods that may remove the bulk of a contaminant. Although adsorption processes are sometimes practical for removing solutes at high concentrations (such as pressure swing adsorption processes), other separation methods (membrane methods, liquid extraction, gas stripping, etc.) often may be more practical. However, it may not be practical to use those other methods to reduce the contaminant concentrations to acceptable levels for release or reuse of the fluid. Then an adsorption step can be added to remove the last traces of the contaminant before discharge or reuse.

Adsorption beds also have the desirable ability to handle brief surges or changes in solute concentration (and to a lesser extent changes in flow rate) that sometimes cause significant solute to escape through other separation equipment. If a pulse of higher solute concentration enters an adsorption bed at any time except just before the front reaches the outlet of the bed and the bed is to be removed from service and regenerated, the front advances more rapidly but significant contaminant is not necessarily released to the output stream from the bed. A pulse in the flow rate can also advance the front more rapidly and increase its width, but for systems with favorable isotherms the steady-state or constant pattern front shape and width will eventually be reestablished. If a freshly regenerated bed is placed on-line downstream before the bed is approaching its regeneration time (Figure 8), even a disturbance just prior to regeneration would be "contained" and not necessarily result in contaminant release. Thus, adsorption may have additional attraction for applications where the feed rate and/or composition are not steady. To a first approximation, the adsorption system can be designed based upon the "average" composition and flow rate. Only when the deviations from the average become especially severe would one need to account for the composition or flow rate variations.

Adsorption systems normally do not release or add contaminant to the effluent stream. This comment is made principally to contrast adsorption with liquid extraction and absorption systems which always must contend with the small solubility of essentially any solvent in water or the finite vapor pressure of any absorbent in air. If the solvent or absorbent is not toxic, this may not be a critical issue, but most organic solvents are not desirable components to add to effluent water streams. In such cases, adsorption may be used as backup systems to remove traces of the solvent or absorbent, not necessarily to remove traces of the original contaminant.

Examples of Adsorption and Ion Exchange in Waste Management

Adsorption and ion exchange are so common in environmental and waste processing that it is not practical to list or describe all of the present or potential applications of adsorption to waste management, and the following applications represent the author's judgment of interesting and representative cases. There was no effort to make the examples complete;

that would be an impractical task with results that quickly become out of date. These examples include removal of materials from gaseous and liquid wastes. They also include concentration and fixation of toxic materials and recovery of valuable materials. Perhaps the most common adsorbent used in environmental and waste systems is activated carbon, and it thus receives the most extensive discussion.

Removal of Pollutants from Gas Streams

Organic vapors are common pollutants in air emissions. They may be hydrocarbons such as motor fuels, chlorinated compounds, or organic solvents. If the vapor molecules are nonpolar or have hydrophobic groups, the vapors are likely to be adsorbed on activated carbon or other hydrophobic adsorbents [71]. Activated carbon is most useful when the concentration of organic solvents is low; otherwise replacement or regeneration of the carbon can be expensive [72]. (For high concentrations incineration or catalytic oxidation may be a more practical approach for removing organic vapors if suitable facilities for incineration or oxidation are available.) If the vapor is an acid or other polar material, a polar adsorbent such as zeolite, silica gel, or alumina may be preferred. In such cases, the effects of water vapor in the gas stream can be important. Water will compete with other polar molecules for the adsorbent surface and may reduce the effective capacity for the pollutant. However, the adsorbed water is also a polar material, and the pollutant could be adsorbed on or in the bound water. Adsorbed water can enhance adsorption of polar materials on some adsorbents [73], but it can hinder adsorption of hydrophobic organic compounds [74]. Water can also adsorb on activated carbon and reduce the adsorption of other components, including organic vapors, but water usually interferes less with adsorption on activated carbon than with adsorption on more hydrophilic materials such as zeolites or silica gel.

Most physically adsorbed gases can be regenerated by raising the temperature, and the adsorption capacities are usually low at temperatures much above 100°C for highly volatile materials. However, there are important cases where one would like to remove pollutants from hot gases. This need arises in fuel processing where sulfur compounds need to be removed from "syngas" (mostly mixtures of CO, H_2, and N_2) prior to forming the gases into more useful hydrocarbon products, but this needs to be done at the process temperature so that the sensible heat of the gas will not be lost. This is also an example of a pollution control problem where the separation does not have to take place directly on an emission stream; it could occur on a process stream prior to emission.

The separation of hydrogen sulfide from syngas or natural gas prevents environmental pollution, because if left with the syngas or natural gas, the sulfur compounds would eventually reach the environment, probably as sulfur oxides.

When high temperature adsorption is necessary, a chemisorption process is more likely to be effective. For treating high temperature syngas, mixed oxides of iron and zinc have been studied and shown to be effective for removing H_2S and COS from coal gasification syngas [75], and other mixed oxides have also been studied [76–79]. A sintered mixture of zinc oxide and titanium oxide (or a zinc titanate) is often used, but the zinc appears to be the active agent that reacts with the hydrogen sulfide. Other oxides may be added as binding agents to hold the active particles together. To regenerate chemisorbed beds, it is often better to use chemical regeneration reagents rather than simply higher temperatures or lower pressures. In the high temperature process using mixed oxides, small concentrations of oxygen are effective for regeneration. The chemisorbed sulfur compounds are oxidized, and the sulfur is desorbed as SO_2 rather than H_2S, which was originally adsorbed. The regeneration is exothermic, and the temperature must not get too high. Zinc metal sublimation can occur if the temperature exceeds approximately 600 to 750°C. The temperature also should not be too low. In addition to causing slow regeneration rates, low temperatures can result in the formation of zinc sulfate, which has a larger molecular volume than the oxide or sulfide and can put strain on the particles and cause incomplete regeneration of the adsorbent. The maximum temperature for zinc sulfate formation is also in the same temperature range as that for zinc metal volatilization, so the regeneration is likely to be carried out over a narrow temperature range at approximately 750°C.

The adsorption of sulfur oxides and nitrogen oxides at low temperatures with organic polymer beads containing tertiary amine groups was discussed earlier. Selected adsorbents such as alumina and alumina with copper or calcium oxides incorporated in the pores can be used to remove SO_2 at higher temperatures [80,81]. One new laboratory alumina produced by a sol-gel process and incorporating copper oxide could adsorb sulfur dioxide at temperatures up to 500°C and be regenerated at temperatures above 500°C [82].

Adsorption can involve almost irreversible chemical reactions, for example the reaction of acid gases such as sulfur oxides with limestone or other basic oxides like those used in electric power plants. (Most limestone scrubbers use slurries of limestone, but "dry" systems with other adsorbents are also used or proposed [83].) With essentially irreversible chemical reactions during adsorption, there is no clear distinction between adsorption (separation) equipment and chemical "reactors." However, it is still convenient to refer to such systems as separation processes, rather than as reactors, when the objective of the operation is to separate or remove a component of the mixture because one then recognizes that the competing processes are likely to be adsorption or other separation processes.

One important and interesting application of adsorption removes a radioactive gas and does not necessarily require regeneration. Radon is a serious problem in uranium mill tailings, and is especially important in the residue from uranium ore treatment facilities that once treated high assay ores containing high concentrations of radium as well as uranium. Many homeowners have found radon to be a serious health hazard in their own homes, and its source is likely to be the small concentrations of uranium or thorium that exist naturally in the soil and rock under many neighborhoods. Radon is a noble gas and has no strong chemical affinities. However, it is highly hydrophobic and collects on activated carbon [84]. Radon adsorption can be modest and affected by the water content of the air (and thus the water on the carbon), but the adsorption can be sufficient to remove radon from vents or off-gas streams from uranium ore residue. Because radon isotopes have short half-lives, the radon can decay as fast as it is adsorbed, so the carbon bed may not saturate and not require a regeneration step.

Adsorption of Pollutants from Liquid Streams

Since the only liquid streams that can normally be discharged to the environment are aqueous (water), the examples given focus on removal of toxic materials from water. However, other pollution control problems could be solved best by removing the pollutant or the pollutant forming material from an internal process stream before it reaches an effluent or product stream. Sulfur removal from syngas was a gas phase example of such a case. Removal of sulfur and nitrogen compounds from crude oil prior to refining the "crude" into fuels and other products is another example.

Activated carbon is a relatively inexpensive but highly effective adsorbent for removing a wide range of hydrophobic materials. As noted, it can be used to remove some hydrophilic components from either gases or water as long as they have a hydrophobic group. Applications for activated carbon are much older than the current phase of interest in environmental and waste problems. Activated carbon has been used for much of this century to treat some drinking water, but more recently its uses to remove toxic organic compounds from drinking water have become more evident. The use of activated carbon on water supplies is sometimes considered one of the possible ternary treatment methods from municipal water supplies, perhaps the most common ternary treatment method. It is likely to become more common as we identify more potentially toxic organic compounds in raw water feed to water treatment plants and/or as the raw waters become more contaminated. Activated carbon can also be used in homes to treat drinking water, but almost all U.S. municipal water supplies are relatively free of toxic materials.

Activated carbon is a standard method for removing dissolved oils in water supplies or in discharge waters. It is effective in removing many hydrophobic halogenated solvents such as trichloroethylene (TCE), carbon tetrachloride, polychlorobiphenyls (PCBs), and similar compounds. A great many insecticides can also be adsorbed on activated carbon. Although the effectiveness of activated carbon does depend upon the particular compound of interest, it is likely to have some effectiveness for essentially any compound with a hydrophobic part to its molecule. Even when the organic compound is polar and has a moderately high solubility in water, it can be adsorbed by activated carbon if some part of the molecule is hydrophobic [85]. Even compounds that ionize can be adsorbed on carbon if they have sufficient hydrophobic groups, but equivalent numbers of ions with the opposite charge will, of course, have to be adsorbed or become associated with the carbon to maintain electrical neutrality. In such cases, the hydrophobic group is expected to be adsorbed on the surface, and the polar or ionized groups would be oriented away from the carbon surface. If these groups are ionized and then have identical charges, the groups may repel each other and lower the adsorption capacity. A recent study [86] showed that adsorption of ionized organic compounds is greatest in the pH range where the compounds are not ionized. There is a less dramatic improvement in adsorption in the ionizable regions when additional salt (electrolyte) is present. This suggests that at higher ionic strengths (higher salt concentrations), the thinner electrical double-layer thickness reduces the repulsive effects from the ionizable groups on the adsorbed compounds. Activated carbon is not likely to be extremely selective among the different hydrophobic compounds until the loadings become higher than those usually seen in environmental problems. At some point, of course, the selectivity of the carbon for the more hydrophobic components and the effects of other adsorbed materials can become important.

One very large volume wastewater is the "producer water" from flushing petroleum from underground formations. The volume of water in these operations is usually much larger than the volume of oil removed, especially for "stripper wells" that attempt to obtain more oil from the formation than can be recovered practically by pumping alone. Producer water obviously becomes saturated with the components of the petroleum and usually contains considerable additional petroleum in the form of fine droplets. The most important components in the water are those that are most toxic and/or have the highest solubilities. Generally, the components of most concern are benzene, toluene, ethylbenzene, and xylene (BTEX). Activated carbon can remove those compounds [87]. BTEX can be removed from activated carbon by heating (usually using steam) or by use of volatile solvents such as methanol or acetone. However, the regeneration may be incomplete and fail to return the carbon to its initial adsorption capacity. The reason for this loss of capacity is believed to result from accumulation of petroleum droplets on the adsorbent. Since petroleum wets the hydrophobic surface of the adsorbent, it can coat the surface and fill the pores of the adsorbent. Since some components of the petroleum are not as volatile or as easily removed by a solvent regeneration, it can be difficult to remove all of the petroleum from the adsorbent particles. A recent study investigated methods for "protecting" adsorbent beds from the accumulation of petroleum, and coalescence devices upstream of the adsorption bed appeared to be the most effective approach [88]. Once the adsorption bed is protected sufficiently that several regenerations can be achieved before the adsorbent is discarded, it becomes practical to consider more effective, but more costly, adsorbents. Polymer adsorbents such as Amersorb 572 produced by the Rohm and Haas Company have higher capacities for BTEX and can be attractive if they can be regenerated and reused many times.

Some treatments of municipal or industrial water supplies use powdered activated carbon, which is simply slurried with the water and then filtered to remove the carbon and the adsorbed contaminants (or the odor forming components). The filtration could be carried out along with precipitation or metal hydroxides that remove metal ions by adsorption or ion exchange on the precipitate surfaces. Such batch contact of an adsorbent with a solution can be mechanically simple and involve little capital cost, especially if filtration equipment is already being used. This type of batch contact is also not sensitive to the presence of other solids that would plug conventional packed beds, and the adsorption treatment can be applied to unclarified feed water. However, it does not give exceptionally high removal efficiencies or maximum solute loading on the adsorbent, and it does not usually allow regeneration and reuse of the carbon. Such operations can only be adequate when the adsorbent is highly selective, when the required removal efficiencies are not too high, and/or when the adsorbent is relatively inexpensive. Since the carbon becomes a solid waste, this approach usually does not minimize solid waste volume.

Powdered activated carbon can also be used with other water treatment operations such as bioreactors, which may be used for destroying some toxic components [89]. The role of an adsorbent in an activated sludge reactor is complex because it can involve removal of undestroyed contaminants and retention of contaminants in the bioreactor for destruction. One important role of the activated carbon is to hold some contaminants in the activated sludge for longer residence times because the solids have longer residence times in the digester. An important result of this increased residence time is the ability of the carbon to dampen the release of adsorbed contaminants that enter the digesters in sudden pulses [90]. We simply point out that in-reactor adsorbents appear to be helpful in many cases.

For high removal efficiencies and for maximum loading of the adsorbent, it is usually preferable to use a packed bed of adsorbent, and activated carbon comes in a variety of granular forms with particle sizes suitable for efficient packed bed operations. As with any packed bed operation, it is necessary to clarify the feed to remove most of the solids in the water. Carbon, like a number of adsorbents or like granular filter beds, can remove the particles as well as adsorb the pollutants. Removal of particles is usually not considered a merit of packed beds because it can eventually plug the bed. It is usually far better to remove the particles in an upstream filtration step. The estimation of required adsorption bed sizes was discussed earlier.

Activated carbon can also adsorb some metals from aqueous solutions without changing the carbon surface, say by adding significant surfactant molecules. These metals are usually noble (e.g., mercury, gold or, to a lesser extent, cadmium and lead), whole molecules may have hydrophobic properties. Other metals may be adsorbed if they become associated with hydrophobic ligands. Some or all of the "adsorption" may involve chemical reactions, even reduction of the ions to metal molecules and adsorption of the molecules, or they may involve the presence of ligands, added intentionally or present originally. Regardless of the mechanism of the adsorption, this can be an inexpensive way to remove some toxic heavy metals from waste or environmental waters. Adsorption of metals on activated carbon can be selective if highly selective ligands are involved that make only specific metals hydrophobic; otherwise there may be more selective adsorbents for some of these metals. Reed and Nonavinakere [91] investigated removal of cadmium and nickel from waste streams, and Reed and Matsumoto [92] modeled cadmium removal. Reed [93] studied the removal of lead from activated carbon. Carbon molecular sieves have been reported to be useful for removing some toxic metal ions [94]. The more toxic metals often do not interact strongly with water and thus are not highly hydrophilic. An example of other metals that can become associated with organic ligands and then adsorbed by carbon was given by Chang and Ku [95], who investigated the removal of copper on activated carbon after the copper ions had been chelated by EDTA.

Carbon molecular sieves are usually more expensive than activated carbon, but they can be especially effective for removing small and moderate-sized molecules such as acetone, benzene, chloroform, carbon tetrachloride, dichloroethane, and HCN from water [96]. Although the actual adsorbent may be more costly, the ability to obtain an adsorbent with higher performance because of equilibrium and fibrous shape may sometimes offset the additional cost of adsorbent.

Polymeric adsorbents can also be selective for organic compounds, and the nature of the pore surfaces of polymers may be better controlled with a specially prepared polymer. Styrene polymers can be effective adsorbents. One study [97] on adsorption of chlorinated volatile organic compounds (trichloroethylene) from groundwater found Dowex Optipore to be the best of approximately 50 adsorbents tested. This adsorbent could be easily regenerated at 90°C and showed no change in adsorption performance over three cycles. The cost of polymeric adsorbents, however, is usually greater than the cost for activated carbon, up to $60/kg versus approximately $5/kg for activated carbon. Macro-porous polymers have been filled with solvents that enhance the removal of other organic components from water [98]. The macro-porous polymer is hydrophobic and preferably wet by the highly insoluble solvent that is responsible for most of the organic removal from the wastewater. Obviously, this approach could be mentioned in the chapter on liquid-liquid extraction (Chapter 6) because the principal mechanism for removal of the organic contaminant is probably much like dissolution of the solvent into the solid, but it is mentioned here instead because the operation behaves much more like adsorption using packed beds of solid adsorbents. (The reader will note that there are several cases mentioned in this book where the best chapter to present a topic can be questionable.)

The behavior of surfactants should be mentioned. Many surfactants will adsorb on activated carbon, and, as one would suspect, the hydrophobic part of the surfactant molecule is the part most likely to become attached to the carbon surface. This can result in ionic groups being selectively oriented on the adsorbed surfactant and extending from the carbon surface. These ionic groups are then hydrophilic, and adsorption of sufficient surfactant on activated carbon can make the carbon appear to be hydrophilic and then adsorb (or ion exchange) metal ions. The "adsorption" of metal ions on activated carbon can follow the mechanism just described, particularly if the surfactant is introduced prior to exposure of the carbon to the metal ions, or the surfactant molecules can first become associated with the metal ions and surround them so that the carbon appears to be adsorbing a hydrophobic cluster of molecules. From adsorption measurements alone, it can be difficult to determine details of the carbon surface or exactly how the metal ions and surfactant molecules are attached to the surface, but in many cases it may be sufficient to determine the affinity and capacity of carbon surfaces for selected metals as a function of metal ion and surfactant concentrations. Similar behavior can be seen with other normally hydrophobic surfaces.

Surfactants have been reported to be useful for removing adsorbed organic materials from activated carbon [99]. If the carbon is used to adsorb organic materials from water, regeneration of the carbon with surfactants significantly decreases the adsorption capacity of the carbon for future cycles. This could result from a change of the carbon surface properties because of the retention of surfactant, probably making the carbon less hydrophobic. Of course, to be practical, the regeneration must produce a concentrated solution/dispersion of the contaminant for disposal, destruction, or recovery.

Another important group of adsorbents with important applications in environmental and waste problems are metal oxides, particularly hydrous metal oxides. Silica is probably the most important (often hydrated) metal oxide likely to be used in environmental and water treatment because of its properties and relatively low cost. Although the most common form of silica is sand crystals, it is not likely to be an effective adsorbent, because its surface area (area per unit volume) is not very high, or to have significant hydration. Similarly, mineral forms of other metal oxides often have too little surface to be attractive as adsorbents. However, if these materials are formed as gels precipitated from solution, highly porous material with very large surface areas can be formed. If the porous solids are not heated to high temperatures, they may retain much or most of their porosity and OH groups.

Inorganic materials can be hydrophobic and adsorb organic materials such as volatile organic compounds. Dealuminized type Y zeolites (sometimes called DAY zeolites) and dealuminized ZSM-5 zeolites have been shown good isotherms for removal of trichloroethylene [100]. The adsorption capacities of these hydrophobic zeolites can be higher than those of other hydrophobic inorganic adsorbents, such as silicalite, because of the greater pore surface area. Inorganic adsorbents have potential advantages over activated carbon because they are not burned during regeneration; stronger oxidation conditions can be used to regenerate inorganic materials.

The properties of metal oxide surfaces depend strongly upon the thermal (and possibly chemical) treatment of the material. Fresh gel that has been lightly dried may have large quantities of OH (hydroxide) groups on the surface, making the material essentially an ion exchange material. These oxides are amphoteric and can act as anion exchange materials in acid solutions by exchanging OH groups with anions in aqueous solutions. In alkaline solution, the oxide can act as a cation exchanger by exchanging protons for metal ions in aqueous solutions. The pH at which the material switches from an anion exchanger to a cation exchanger depends upon the metal oxide, but for the most common materials (silica, titania, zirconia), it is usually near neutral (not far from 7). In slightly acid or alkaline solutions, these materials will be hydrophilic and often relatively weak ion exchange materials. (Of course, silica is not stable in solutions with very high pH values.)

Thermal treatment can make these metal oxides much stronger physically, but that also decreases the surface area and the hydrophilic properties of the surface. As the oxides are "fired" at higher and higher temperatures, the smallest pores begin to disappear and many of the hydroxide groups on the surface are lost, but a physically strong adsorbent material that is suitable for many industrial application is likely to need some heat treatment. Silica gels that retain sufficient hydrophilic properties can be used as drying agents, and packets of silica gel are often placed in consumer products to help them remain dry during shipment. If the oxides are fired to sufficiently high temperatures, they may lose almost all their hydrogen (hydroxide), and the surfaces may even become hydrophobic. They can also eventually lose nearly all of their internal surface area and approach solid spheres, but for some metal oxides this requires a very high temperature, especially for materials such as zirconia. Kim et al. [101] prepared fresh hydrated titanium and ferric iron hydroxide mixtures and studied removal of cobalt, an important radioactive activation product in nuclear reactors.

Clays and zeolites are useful as both adsorbents and ion exchange materials [102]. Zeolite adsorption properties result from their high surface areas. Zeolite surfaces are usually highly hydrophilic and particularly useful for removing water, not a common need in waste or environmental operations. Their ion exchange capacity results from the mobility of "free" metal ions to neutralize negative charges that result from the aluminosilicate basic structure. These ions are usually located near the entrance of the zeolite cavities and thus affect the size of the openings to the cavities that make up most of the adsorbing surface area of the zeolite. Zeolites can also be used in aqueous solutions as ion exchange materials, and they have been shown to have significant selectivity for cesium and strontium, common components in wastewaters from some nuclear facilities [103]. More recently, titanosilicates have received considerable attention for some of the same applications and have shown significantly improved selectivities [104,105].

The selectivity of inorganic materials such as clays for metal ions can be altered by changing the spacing of layers in the clay structure or the openings for metal ions in other inorganic particles. The spacing between clay layers can be expanded to allow larger ions to enter the space between layers, and "pillar" type structures can be used to hold the layers for the metal ions to enter. Other inorganic materials, such as cyanoferrates and titanates, have been developed with openings that are highly specific for retaining important radioactive contaminants such as strontium and cesium.

When metal ions are to be removed from wastewaters, one is likely to look for ion exchange materials, and the organic ion exchange resins are more frequently used than inorganic ion exchange materials. As noted in the earlier description of these materials, the standard or more common ion exchange materials may not show strong specificity for a particular metal ion. Because many toxic metals have valences greater than 1, a standard cation exchange material will show selectivity for removing divalent ions over monovalent ions from dilute solutions. There are, however, small but growing numbers of relatively specific ion exchange materials that are selective for single metal ions or for a few metal ions.

Wastes that contain metal oxides can also be used as inexpensive adsorbents of ion exchange materials for removing metal ions from solutions. Although these may not be the most effective adsorbents or ion exchange materials, they can be relatively inexpensive, and if they are to be sent to a solid waste disposal anyway, there may be merit in using them to help hold selected metal ions in the disposal site and minimize leaching of toxic metals from the site. Chromium(VI) has been shown to be "adsorbed" (or ion exchanged) on some fly ashes [106]. Such behavior can be important in removal of toxic metal ions and in predicting how toxic metal ions will be transported in landfills that contain other wastes such as fly ash.

In many cases, selective ion exchange materials (or adsorbents) can be developed for metals which have highly insoluble salts with some anions. For instance, mercury and some other heavy metals have highly insoluble sulfides, and organic resins with thio (sulfur) groups have been found to be highly selective for mercury and other metals with insoluble sulfides. These resins can be relatively inexpensive to produce, compared to the cost of some other selective resins. Even if the ion exchange part of the resin is not participating in the selectivity for mercury and other heavy metals, it may still be useful to have the ion exchange groups available to help "swell" the resin so that the heavy metals can reach the interior parts of the resin more easily.

Other resins have been reported to be selective for specific ions. Amberlite IRA 958, a strong base anion resin, was reported to remove cyanide complexes (ferro- and ferricyanides) selectively [107]. Quantitative values of the selectivity were not given, but the principal competing anion appeared to be sulfate.

At least one new ion exchange resin incorporates two groups into the resin to enhance its selectivity for specific ions [108]. The most active functional group on this resin is probably diphosphonic acid, but there are also more traditional sulfonic acid groups in this resin. The resin has a high selectivity for several multivalent ions that are environmentally important, such as zinc, lead, manganese, cadmium, uranium, and nickel. These also happen to be ions that interact strongly with the phosphonic acid groups. The role of the sulfonic acid groups appears to be in improving the access of the phosphonic acid groups to the metal ions, probably by swelling the resin so that there is more water penetration and better mobility for the metal ions in the resin. This resin is an adaptation in an ion exchange form of a solvent extraction method developed earlier for removal of components from solutions of nuclear waste [109].

Metal oxides (or hydroxides) are also potentially important adsorbents. Some precipitation operations that use iron or aluminum oxide precipitation to remove metal ions may also involve adsorption of metals on the oxides. It may be necessary for the surface to be sufficiently hydrated that the metals can exchange for hydrogen in the OH groups on the surface. In such cases, the adsorption will be highly pH dependent [110]. Such materials are being studied for removing radioactive contaminants released from the nuclear accident in the former Soviet Union [111]. Other adsorbents being developed for other uses could also be important for such applications [112,113]. Nickel can be removed from slightly alkaline solutions (pH from 8 to 10) containing waste silica particles made in the refining of pottery clay [114]. Several inorganic solids are relatively specific for removal of selected ions. Some ferrocynates, such as cobalt potassium cyanoferrates and copper cyanoferrates, are highly specific for removal of cesium ions, and cesium is an important contaminant from some nuclear reactors and other nuclear facilities. In fact, it is one of the most significant contaminants in fallout from the nuclear accident in the Ukraine. Besides using pellets of the compounds themselves, one can also incorporate complex anions of materials such as copper cyanoferrates onto anion resins and have a relatively porous and convenient size ion exchange particle for use in columns [114].

Among the most selective resins for individual metal ions are those that incorporate specific groups which bind to specific metal ions. Crown ethers, cryptans, and other groups that can "chelate" an ion by binding with several of its coordination points are especially effective. Such groups can use geometrical properties such as ring size on crown ether or cryptan groups to enhance the selectivity for metal ions that fit most easily in such cavities and whose coordination features most closely match those of the metal ions. These groups can be bound to silica or other hydrated oxide particles as well as to organic resins. Although a number of highly selective adsorption/ion exchange particles have been developed by companies such as IBC, Inc. of Provo, Utah, the materials are often very expensive and not yet suitable for many large-scale environmental and waste treatment systems. Although the cost of such selective material is likely to decline as the demand for them increases and larger-scale production begins, synthesis of some groups, such as crown ethers, is probably difficult enough to keep the cost for making materials using those groups from becoming very cheap unless significantly less costly synthesis methods are found. Some increase in selectivity for certain metal ions can be achieved by using less costly ligand in the resins and creating a geometric arrangement of the ligands or active sites so that several ligands can interact with each metal ion [115]. This can be accomplished by forming the metal complexes with the ligands and incorporating the complexes into the polymer. Such a procedure should leave the ligands arranged in such a way that they can most easily coordinate with specific metal ions with the geometric arrangement that the specific metal ions prefer. The "normal" or more common procedure for creating ion exchange resins involves first forming the polymer and then adding the active sites. Thus the more common procedure places the active sites at more random positions within the polymer.

Since crown ether and cryptan compounds can also be used in liquid-liquid extraction to remove toxic metals very specifically, we consider some factors that could favor the use of extraction or adsorption. The liquid-liquid extraction process always offer two potential advantages over adsorption operations which use selective binding groups attached to particles. One is the ability to select different diluents to hold the selective extractant. Different diluents can have properties, such as dielectric constants, that considerably increase the extractant selectivity. The second advantage is the ability to operate countercurrent systems more easily and employ scrubs and washes to improve the separation. However, there are also three important potential disadvantages to these extractants in solvent extraction systems. The potential toxic effects of adding extractant or diluent to the water stream has been mentioned earlier. The second disadvantage is the low, possibly almost zero, loss of extractant. There are always small entrainment and solubility losses of solvent (and extractant) used in any extraction system. When the extractant is very costly, such losses could be most important. The third disadvantage is the ability of packed bed operations to incorporate many transfer units into relatively small pieces of equipment.

Natural materials (those used with little or no pretreatment to enhance their adsorption properties) used as adsorbents include peat and cellulosic materials, such as wood. These relatively inexpensive adsorbents are suitable for dye removal from wastewater streams [116,117]. When there is no need to recover or concentrate the contaminant, such as when the contaminant and the adsorbent can be destroyed by combustion or another mechanism, such low cost adsorbents may be preferred.

Organic biomass and microorganisms can play important roles in adsorption of many metals, especially heavy metals, which are more likely to be toxic. Whole organisms like seaweed have been shown capable of adsorbing metals such as cadmium [118]. In this study, the "adsorption" was fit to a Langmuir isotherm, but, as noted, metal uptake by organisms may be by ion exchange as well. Perhaps the most common form of heavy metal adsorption by biomass occurs in essentially all activated sludge bioreactors that are used so commonly for wastewater treatment to remove biodegradable organic materials. Bioreactors are technically alternatives to separation processes for removing organic contaminants, but other separation processes (metal removal) occur within bioreactors. Significant fractions of heavy metals can become attached to organic mass in either the anaerobic or aerobic regions of activated sludge bioreactors [119–122], and a variety of metals can be removed [123]. Metals can be removed by sterilized sludge biomass as well as living biomass [124]. In one set of experiments with sterilized municipal sludge, the sludge preferred Cd to Zn to Ni. The "adsorption" could be described very well by a Langmuir isotherm, but the strong effect of pH on the "adsorption" suggests an ion exchange mechanism.

Living and dead cells of specific microorganisms have been seriously proposed for selective removal of uranium from a number of wastewaters [125]. Macaskie and co-workers have evaluated several organisms for uptake of several metal ions [126,127].

In some cases, microorganisms can be incorporated into pellet, which can be used for adsorption operations much like other commercial adsorbents [128]. Usually it is most convenient to incorporate the biomass into gel particles. Biomass from most common bacteria is relatively inexpensive to create, and if it can be economically incorporated into gel particles suitable for metal adsorption/ion exchange operations, they could be an economical source of adsorbents. Biopolymers can also be modified chemically to introduce additional ligands.

Metals can become attached to microorganisms and/or biomass in several ways. For instance, many organisms and biomass contain organic phosphate groups which could contribute to heavy metal uptake. Biomass often contains significant ion exchange capacity which is most evident in the uptake and exchange of light metals [129]. There is no more than moderate selectivity for the lighter metals, and the ion exchange effects are relatively evident. However, heavy metals which often have highly insoluble phosphates may be held more tightly to the biomass. In some cases, there could even be a redox reaction which results in an insoluble form of the metal somewhere within the individual cells. Heavy metals can be removed from biomass with acid treatment, and this may be some indication of ion exchange mechanisms for some of the metal attachment.

Lignin is a major by-product of the paper industry that has often been viewed more as a major waste than as a product. Ionizable groups have been added to Kraft lignin, and the product was especially effective in removing lead and copper from wastewaters [130]. The basic materials, especially the lignin, are very inexpensive, and the modified lignin-based adsorbent can be prepared easily and should be significantly less expensive than most other adsorbents with similar selectivities.

The great importance of adsorption to waste and effluent processing results principally because adsorption is so practical for treating very dilute fluids, gases or liquids, and that is the reason that this chapter includes more details than the chapters on most other separation methods. Adsorption has the advantage over absorption or solvent extraction because it is not necessary to keep the relative contribution of each phase approximately equal—that is, have flow ratios that are of the same order of magnitude. This permits the exploitation of very high distribution coefficients. Adsorption has the advantage over distillation and stripping operations for low concentration because it can remove the trace contaminant from the bulk liquid, even when the volatility of the trace contaminant is lower than the volatility of water.

The principal disadvantage of adsorption often results from the fact that it is not a continuous operation like distillation or absorption. The importance of this disadvantage declines as the concentration of the contaminant decreases and the distribution coefficient increases because regeneration is required less often. Ion exchange is generally the preferred approach for removing inorganic ions from dilute solutions. Solvent extraction is able to compete with ion exchange when the concentrations are higher but usually becomes increasingly less competitive as the concentration of the contaminant is decreased.

One should always consider adsorption methods when the concentrations are sufficiently low. Of course, the availability of a suitable adsorbent and the capabilities of competing methods, especially gas stripping or distillation for highly volatile contaminants, should also be considered. When the concentration of the contaminant is relatively high, it often is more economical to use other methods only to remove the bulk of the contaminant and reduce the concentrations to the levels where adsorption can be used for removing the last traces of the contaminant and meeting the required effluent concentrations. In such cases, adsorption or ion exchange should be considered "polishing" steps to remove the last traces of the contaminant. The cost of adding relatively small adsorption or ion exchange equipment to remove the last traces of a contaminant often justifies the introduction of an additional processing step.

SUMMARY

Adsorption operations are frequently used to remove small concentrations of pollutant from gaseous and liquid streams. These are often emission streams. A variety of natural materials and manufactured materials can be used in adsorbents. Adsorbent beds usually operate in a non-steady-state manner with the bed alternately loaded with solute during the adsorption step and then release their solute during a regeneration (or desorption) step. A continuous fluid processing rate can be maintained by using one or more beds and desorbing one bed while the other is being used.

Once the proper adsorbent is selected, the diameter of the adsorbent bed needed can be estimated by calculating the superficial flow velocity that will give a reasonable pressure gradient within the bed. The next step is to determine a suitable operating time for a given bed depth (or the bed depth needed to adsorb for a given time). A crude approximation of the time when solute will "break through" the bed and appear in the exit stream can be obtained from the capacity of the bed for the solute at the concentration of the feed fluid. This will always overestimate the time for breakthrough because it does not account for spreading of the adsorption "front" as it passes through the bed. However, for highly "favorable" isotherms and rapid mass transfer coefficients, the crude estimate will be reasonable, and only a modest fractional reduction in the operating time (or a modest increase in the bed length) will be necessary to account for spreading of the loading front and reduced operating time before breakthrough occurs.

However, when mass transfer is not rapid or the isotherm is not favorable, the correction can be significant, and sufficient bed length must be allowed for spreading of the front. This requires estimation of the mass transfer resistances (film resistance, diffusion within the adsorbent particles, and axial dispersion in the bed). Mathematical expressions for long beds and linear isotherms are simple and easy to apply. One can estimate the appropriate rate parameters from breakthrough curves and apply them to other operating conditions.

Mathematical expressions for breakthrough in beds with nonlinear isotherms are more complex. For many cases, this chapter has recommended that the shape of the breakthrough be estimated by approximating the equilibrium with a linear or irreversible isotherm (a limiting case of a favorable nonlinear isotherm). Since either of these will give only an approximation, the choice usually should be the one that appears to approximate the actual isotherm more closely. Although it is difficult to estimate quantitatively the errors introduced by approximating the isotherms in this manner, the directions of the errors from the approximation can be determined.

Multi-component adsorption must be considered when one or more of the solutes occupy a significant part of the adsorbent capacity or otherwise affect adsorption behavior of the other solutes. This considerably complicates estimation of the breakthrough fronts. In fact several fronts can be progressing down the bed. Nonisothermal effects are more likely to be important in gaseous systems with high heats of adsorption and high solute concentrations. Temperature fronts then may progress down the bed and complicate the analyses. These conditions are less likely to be important in environmental problems because dilute solutes are more likely to be involved. The reader is referred to more advanced and specialized books and papers for estimating breakthrough in such systems.

REFERENCES

1. Seibert, K. and M. Burns. "Adsorption Separations in Stably Fluidized Beds." Preprints for the Topical Conference on Recent Developments and Future Opportunities in Separations Technology. AIChE Annual Meeting. Miami, FL. Nov. 12-17, 1995. P. 105.
2. Chetty, A. and M. Burns. *Biotech. Bioeng. 38*, 963 (1991).
3. Keller, G. *Chem Eng. Prog. 91* (10) 56 (1995).
4. Saska, M. et al. *Sep. Sci. Technol. 27*, 1711 (1992).
5. Nakhla, G. F. et al. *Environ. Technol. 13*, 181 (1991).
6. Abuzald, N. and G. F. Nakhla. *J. Environ. Sci. Technol. 28*, 216 (1994).
7. Abuzald, N. et al. *J. Water Res. 29*, 653 (1995).

8. Abuzaid, N. and G. F. Nakhla. *Environ. Prog. 15*, 128 (1996).
9. U.S. EPA. "Cleanup of Releases from Petroleum USTs: Selected Technologies, *EPA/530/UST-88/001* (1988).
10. Crittenden, J. C., D. W. Hand, H. Arora, and B. W. Lykins, Jr. *J. AWWA*, 74–82 (Jan. 1987).
11. Hutchins, R. A. *Chem. Eng.* Feb. (1980).
12. Keller, G., *Chem. Engr. Prog. 91* (No. 10), 56 (1995)
13. Van Vlient, B. V. and W. J. Weber, Jr. *J. Water Pollution Cont. Fed. 53*, 1585 (1981).
14. Klein, J. et al. *Novel Adsorbents and Their Environmental Applications*. AIChE Symposium Series 309, Vol. 91, p. 72 (1995).
15. Diaf, A. and E. J. Bechman. *Novel Adsorbents and Their Environmental Applications*. AIChE Symposium Series 309, Vol. 91, p. 49 (1995).
16. Diaf, A. et al. *J. Appl. Polymer Sci. 53*, 857 (1994).
17. Juang, R-S. et al. *Sep. Sci. Technol. 31*, 1915 (1996).
18. Carturla, F. et al. *J. Colloid Interface Sci. 124*, 528 (1988).
19. Ruthven, D. M. *Chemical Eng. Prog.* (1988).
20. Inui, T. and S. B. Pu. *Sep. Technol. 5*, 229 (1995).
21. Keller, G. *Chem. Engr. Prog.* 56, Oct. (1995).
22. Svec, F. and J. M. J. Frechet. *Science, 273*, 205 (1996).
23. Robinson, S., W. Arnold, and C. Byers. "Multi-component Ion Exchange Equilibria in Chabazite Zeolite." ACS Symposium Series 468. *Emerging Technologies in Hazardous Waste Management II*. W. Tedder and F. Pohland, Eds. ACS, Washington, DC, p. 133 (1991).
24. Amphlett, C. B. *Inorganic Ion Exchangers*. Elsevier, Amsterdam (1964).
25. Qureshi, M. and K. G. Varshney. *Inorganic Ion Exchangers in Chemical Analysis*. CRC Press, Boston (1990).
26. Clearfield, A. *Chem. Rev. 88*, 125 (1988).
27. Sebesta, F. et al. "Evaluation of Polyacrylonitrile (PAN) as a Binding Polymer for Absorbers Used to Treat Liquid Radioactive Wastes." U.S. DOE Report No. *SAND95-2729* (1995).
28. Sebesta, F., J. John, and A. Motl. "Inorganic-Organic Composite Absorbers with Polyacrylonitrile Binding Matrix." Proceedings of the SPECTRUM '94 Conference. Atlanta, GA. Aug. 14-19, 1994. Vol. 2. P. 856.
29. Sebesta, F. "Development of PAN-Based Absorbers for Treating Waste Problems at U.S. DOE Facilities. Proceedings of the ICEM95 Conference (Am. Soc. of Mechanical Engineers). Berlin, Germany. Aug. 1995. Vol. 1. P. 41.
30. Watson, J. S. "Improvements in Macro-reticular Particles for Ion Exchange and Gel-Permeation Chromatography." In *Recent Developments in Ion Exchange 2*, P. A. Williams and M. J. Hudson, eds. Elsevier Applied Science, London. p. 277 (1990).
31. Watson, J. S., *Sep. Sci. Technol., 28*, 519 (1993).
32. Carta, G. et al. "Characterization of a Novel 'Gigaporous' Sorbent for Chromatography." Paper presented at the AIChE Annual Meeting, Los Angeles, CA. Nov. 17-22, 1991.

33. Pimenov, A. V., et al. *Sep. Sci. Technol. 30*, 3183 (1995).
34. S. Brunauer, L. S. Deming, W. E. Deming, and E. J. Teller. *J. Am. Chem. Soc., 62*, 1723 (1940).
35. Rantz, W. E. and W. R. Marshall. *Chem. Engr. Prog. 48*, 173 (1952).
36. Wakao, N., et al. *Chem. Eng. Sci., 33*, 1375 (1973).
37. S. Ergun, *Chemical Engr. Progress, 48*, 89 (1952).
38. M. Leva. *Chemical Engr. 56*, 115 (1949).
39. G. Carta. "Analytical Solutions for Periodic Mass Transfer in Fixed-Bed Sorption Separation Processes." *Fundamentals of Adsorption*. L. I. Liapis, Ed. Engineering Foundation. 145–154 (1987).
40. Peters, M. and K. Timmerhaus. *Chemical Process Design*. McGraw-Hill.
41. Glueckauf, E. *Trans. Faraday Soc. 51*, 1540 (1955).
42. P. G. Gray and D. D. Do, "Dynamics of Carbon Dioxide Sorption on Activated-Carbon Particles," *AIChE Journal, 37*, 1027–1034 (1991).
43. C. C. Furnas. *Trans. AIChE, 24*, 142 (1930).
44. A. Klinkenberg. *I&E Chem. 46*, 2285 (1954).
45. J. B. Rosen. *J. Chem. Phys., 20*, 389 (1952).
46. J. B. Rosen. *I&E Chem. 46*, 1590 (1954).
47. L. Lapidus and N. R. Amundson. *J. Phys. Chem., 56*, 984 (1952).
48. O. Levenspiel and K. B. Bischoff in *Advances in Chem. Engr. 4*, 95 (1963).
49. D. M. Ruthven. *Principles of Adsorption and Adsorption Processes*. Chapter 8. Wiley Interscience (1984).
50. E. Glueckauf. *Trans. Faraday Soc. 51*, 1540 (1955).
51. D. M. Ruthven, *Principles of Adsorption and Adsorption Processes*. p. 243, Wiley Interscience (1984).
52. H. C. Thomas, *J. Am. Chem. Soc., 66*, 1664 (1944).
53. O. Levenspiel, *Chemical Reaction Engineering*.
54. D. M. Ruthven, *Principles of Adsorption and Adsorption Processes*. p. 243, Wiley Interscience (1984).
55. Huang, Run-Tun, T.-L. Chen, and H.-S. Weng, *Sep. Sci. Technol. 29*, 2019 (1994).
56. K. R. Hall, L. C. Eagleton, A. Acrivos, and T. Vermeulen. "Pore- and Solid-Diffusion Kinetics in Fixed-Bed Adsorption Under Constant-Pattern Conditions." *I&E Chem, Fundamentals, 5*, 213–223(1966).
57. Cooper, R. *I&E Chem. Fundamentals, 4*, 308 (1965).
58. Ma, Z., R. D. Whitley, and N.-H. Wang. *AIChE J. 42*, 1244 (1996).
59. Yoshida, H., M. Yoshikawa, and T. Kotaoka. *AIChE J. 40*, 2034 (1994).
60. Watson, J. S., *Sep. Sci. Technol., 30*, 1351 (1995).
61. Watson, J. S. Submitted to *Sep. Sci. Technol.*
62. Watson, J. S. Submitted to *Sep. Sci. Technol.*
63. Rantz, W. E. and W. R. Marshall *Chem. Engr. Prog. 48*, 173 (1052).
64. F. Hellferich and G. Klein. *Multi-component Chromatography*. Marcel Dekker, New York (1970).
65. D. Ruthven. *Principles of Adsorption Processes*. John Wiley, New York (1984), Chap. 9.

66. Katti, A. M. and G. Guiochon. *AIChE J.*, 1722 (1990).
67. Ma, Z., R. Whitley, and N-H. L. Wang. *AIChE J.* 42, 1244 (1996).
68. Ernest, M., R. Whitley, Z. Ma, and N.-H. L. Wang. *Ind. Engr. Chem. Res.*, 36, 212 (1997).
69. D. Ruthven. *Principles of Adsorption Processes*. John Wiley, New York (1984).
70. Yang, R. T. *Gas Separation by Adsorption Processes*. Buttersworth. Boston, 1987.
71. K. E. Noll, Y. P. Hang, E. Kalili, and J. N. Sarlis. "Adsorption Characteristics of Activated Carbon, XAD4 Resin, and Molecular Sieves for the Removal of Hydrogenous Organic Solvents." *Fundamentals of Adsorption*. A. I. Liapis, Ed. Engineering Foundation. 441–450 (1987).
72. U.S. EPA. "Cleanup of Releases from Petroleum USTs: Selected Technologies." *EPA/530/UST-88/001* (1988).
73. H. L. Fleming. "Adsorption with Aluminas in Systems with Competing Water." *Fundamentals of Adsorption*. A. I. Liapis, Ed. Engineering Foundation. 221–234 (1987).
74. Wilson, D. J, et al. *Sep. Sci. Technol. 29*, 2073 (1994).
75. R. E. Ayala and B. M Kim. "Modeling and Analysis of Moving Bed Hot-Gas Desulfurization Processes." *Environ. Progress. 8*, 19–25 (1989).
76. J. M. Gamgwal, J. M. Strogner, and S. M. Harkins. "Testing of Novel Sorbents for H_2S Removal from Coal Gas." *Environ. Progress, 8*, 26–34 (1989).
77. Lew, K., A. F. Sarofim, and M. F. Stephanopoulos, *AIChE J. 38*, 1161 (1992).
78. Lew, K., A. F. Sarofim, and M. F. Stephanopoulos, *I&EC Research, 31*, 1890 (1992).
79. Lew, K., A. F. Sarofim, and M. F. Stephanopoulos, *Chem. Engr. Sci. 47*, 1421 (1992).
80. Yeh, J. T. et al. *Environ. Progr. 6*, 44 (1987).
81. Deng. S. G. and Y. S. Lin. *I&EC Research, 41*, 559 (1995).
82. Deng, S. G. and Y. S. Lin. *Novel Adsorbents and Their Environmental Applications*, AIChE Symposium Series No. 309, Volume 91, p. 32 (1995).
83. Khare, G. P., et al. *Environ. Progress, 14*, 146 (1995).
84. Tassan, N. M., T. K. Ghosh, A. L. Hines, and S. K. Loyalka. *Sep. Sci. Technol. 30*, 565 (1995).
85. E. Costa, G. Callija, L. Marijuan, and L. Cabra. "Kinetics of Adsorption of Phenol and P-Nitrophenol on Activated Carbon." *Fundamentals of Adsorption*. A. I. Liapis, Ed. Engineering Foundation, 195–198 (1987).
86. D. O. Cooney and J. Wijaya. "Effects of pH and Added Salts on the Adsorption of Ionizable Organic Specie onto Activated Carbon," *Fundamentals of Adsorption*. A. I. Liapis, Ed. Engineering Foundation. Pp. 185–194 (1987).
87. Heilshorn, E. D. *Chem. Engr.*, 40 (Feb. 1992).
88. Gallup, D. L., E. G. Isacoff, and D. N. Smith. *Environ. Progress, 15*, 197 (1996).
89. Leipzig, N. A., and M. R. Hocenbury. "Powered Activated Carbon/Activated Sludge Treatment of Chemical Production Wastewaters." *Proc. of the 34th*

Industrial Waste Conference. May 8, 9, 10, 1979. Purdue University, pp. 195–205. (1980).

90. Singh, A., S. E. Shelby, and D. K. Perkins. "Use of PACT Treatment to Control In-Plant Elevated Chlorinated Organic Discharges at a Plastics Manufacturing Facility." *Proc. Of the 51st Industrial Waste Conference*. May 6-8, 1996 Purdue University, p. 525 (1996).
91. Reed, B. E. and S. K. Nonavinakere, *Sep. Sci. Technol. 27* (1992).
92. Reed, B. E. and M. R. Matsumoto, *Sep. Sci. Technol. 28*, 2179 (1993).
93. Reed, B. E. *Sep. Sci. Technol. 30*, 101 (1995).
94. Sakoda, K, K. Kawazoe, and M. Suzuki, *Water Res. 21*, 712 (1987).
95. Chang, C. and Y. Ku, *Sep. Sci. Technol. 30*, 899 (1995).
96. Cal, M. P., S. M. Larson, and M. J. Rood. *Environ. Prog., 13*, 26 (1994).
97. Mackenzie, P. D. and M. M. Grade. "Polymeric Sorbents for CVOC Abatement." Paper presented at the 1996 Annual Meeting of the AIChE. Chicago, Nov. 12, 1996.
98. Van der Meer, A. B. and P. E. Brooks, Jr. *Environ. Progress. 15*, 204 (1996).
99. Bhummasobhana, A., J. Scamehorn, et al. "Use of Surfactant-Enhanced Carbon Regeneration for Wastewater Applications." Preprints for the Topical Conference on Recent Developments and Future Opportunities in Separations Technology., AIChE Annual Meeting, Miami, FL. Nov. 12–17, 1995. P. 300.
100. Chandak, M. and Y. S. Lin. "Sorption and Diffusion of Volatile Organic Compounds on Hydrophobic Zeolites." Paper presented at the 1996 Annual Meeting of the AIChE, Chicago, IL. Nov. 12, 1996.
101. Kim, K-R, K-J Lee, and J-H Bae. *Sep. Sci. Technol. 30*, 963 (1995).
102. Olguin, et al. *Sep. Sci. Technol. 29*, 2161 (1994).
103. Kim, B. T., et al. *Sep. Sci. Technol. 30*, 3165 (1995).
104. Zheng, Z., R. Anthony, et al. "Ion Exchange of Cesium by Crystalline Silico-Titanates." Preprints for the Topical Conference on Recent Developments and Future Opportunities in Separations Technology. AIChE Annual Meeting. Miami, FL Nov. 12-17, 1995. P. 288.
105. Botrun, A., L. Bortun, and A. Clearfield. "Synthesis and Investigation of the Ion Exchange Properties of Alkali Metal Titanosilicates." Preprints for the Topical Conference on Recent Developments and Future Opportunities in Separations Technology. AIChE Annual Meeting. Nov. 12-17, 1995. P. 47.
106. Dasmahapata, G. P. et al. *Sep. Sci. Technol. 31*, 2843 (1996).
107. Bessent, R. A., P. A. Luther, and C. W. Eklund. "Removal of Cyanides from Coke Plant Wastewaters by Selective Ion Exchange - Results of Pilot Plant Testing Program." *Proc. of the 34th Industrial Waste Conference*. May 8, 9, 10, 1979. Purdue University. Pp. 47–62 (1980).
108. Totura, G. *Environ. Progr. 15*, 208 (1996).
109. Horwitz, E. P., et al. *Solvent Ext. and Ion Exchange. 11*, 943 (1993).
110. Lou, C.-S. and S.-D. Huang. *Sep. Sci. Technol. 28*, 1253 (1993).
111. Puziy, A., G. Bengtsson, and H. Hansen. "Evaluation of Novel Strontium Adsorbents for Selective and Safe Reduction of Foodstuff Radioactivity." Preprints

for the Topical Conference on Recent Developments and Future Opportunities in Separations Technology. AIChE Annual Meeting. Nov. 12-17, 1995. P. 71.

112. Zheng, Z., R. Anthony, et al. "Ion Exchange of Cesium by Crystalline Silicotitanate." Preprints for the Topical Conference on Recent Developments and Future Opportunities in Separations Technology. AIChE Annual Meeting. Miami, FL. Nov. 12-17, 1995. P. 288.
113. Kato, T. et al. "Adsorption of Nickel Ion with Industrial Waste Quartz Particles." Preprint for the Topical Conference on Recent Developments and Future Opportunities in Separations Technology. AIChE Annual Meeting. Miami, FL. Nov. 12-17, 1995. P. 94.
114. Singh, I. J. and B. M Misra. *Sep. Sci. Technol. 31*, 1695 (1996).
115. Zeng., X. And G. M. Murray. *Sep. Sci. Technol. 31*, 2403 (1996).
116. Al Mansi, N. M. and M. M. K. Fouad. *AMSE, 45(2)*, 41 (1994).
117. Al Mansi, N. M. *Sep. Sci. Technol., 31*, 1989 (1996).
118. Da Costa, A. C. A., and F. P. DeFranca *Sep. Sci. Technol. 31*, 2371 (1996).
119. Gould. M. S., and E. J. Genetelli. "Heavy Metal Distribution in Anaerobic Sludge Digestion." *Proc. 30th Industrial Waste Conference*. Purdue University (1995).
120. Gould, M. S. and E. J. Genetelli. *Water Research, 12*, 505 (1978).
121. Hayes, T. D. and T. L. Theis. *J. Water Pollution Control Federation, 50*, 61 (1978).
122. Hayes, T. D., W. J. Jewell, and R. M. Kabrick. "Heavy Metal Removal from Sludges Using Combined Biological/Chemical Treatment." *Proc. 34th Industrial Waste Conference*. May 8, 9, 10, 1978. Purdue University (1980).
123. Patterson, J. W. and S.-S. Hao. "Heavy Metal Interactions in the Anaerobic Digestion System." *Proc. 34th Industrial Waste Conference*. May 8, 9, 10, 1978. Purdue University, pp. 544–555 (1980).
124. Solari, P., et al. *Sep. Sci. Technol. 31*, 1075 (1996).
125. Shumate, S. E. et al. *Biotechnol. Bioeng. Symp. No. 10*. 27 (1980).
126. Macaskie, L. E. *Biotechnol. Letters, 6*, 71 (1984).
127. Macaskie, L. E., and A. C. R. Dean. *Biotechnol. Letters, 7*, 627 (1985).
128. Watson, et al. "Improvements in Macro-reticular Particles for Ion Exchange and Gel-Permeation Chromatography." In *Recent Developments in Ion Exchange 2*, P. A. Williams and M. J. Hudson, eds. Elsevier Applied Science, London. p. 277 (1990).
129. Watson, J. S., C. D. Scott, and B. D. Faison. "Evaluation of Cell-Biopolymer Sorbent for Uptake of Strontium from Dilute Solutions." *Emerging Technologies in Hazardous Waste Management*. ed. by D. W. Tedder, and F. G. Pohland, ACS Symposium Series No. 422. pp. 173–186 (1990).
130. Szlag, D. et al. "Modified Lignin for the Removal of Lead and Copper from Aqueous Process Streams." Paper presented at the 1996 Annual Meeting of the AIChE. Chicago, IL. Nov. 12, 1996.

3
Absorption and Stripping

Absorption involves removal of one or more components from a gas to a liquid; conversely, stripping is the transfer of one or more components from a liquid to a gas. Both are important in environmental operations. Absorption is used to remove pollutants from gas streams, including effluent streams, and gas stripping is used to remove volatile components from wastewater and groundwater. The scale of absorption and gas stripping operations can vary from small laboratory flasks filled with caustic or lime solutions, through which exhaust gases with acid vapors from small experiments are "bubbled," to very large industrial facilities such as the flue gas scrubbers built for coal-fired electric power plants.

Because of the similar spelling, "absorption" is frequently confused with "adsorption," which is the transfer of a component from a gas (or liquid) stream to a solid. Technically, adsorption means transfer of a solute to a solid surface, and absorption can mean dissolving a solute inside a liquid or solid. In some cases, it may not be evident if the solute is collected on the surface of a solid or within the solid. Many (or most) solid "adsorbents" are highly porous, so the solute can be adsorbed on internal surfaces. Other solids could be gel-like materials that absorb organic compounds. There can be confusion on absorption in solids, and at least one reference given in Chapter 2 used the word "absorption." However, in this book, any transfer of a solute to a solid is covered in Chapter 2 without determining whether the solute is retained in the solid or on the solid surface. Absorption and adsorption methods can be used to remove components from gas streams, so there could even be other reasons for confusing the methods. Absorption operations are usually carried out in packed towers or spray towers. Adsorption operations are usually carried out in packed beds of adsorbents. For more detailed discussion of adsorption operations, see Chapter 2.

Stripping operations can be carried out in towers (columns) or gas-liquid mixers like most absorption operations, but some environmentally important "single stage" stripping operations are carried out in underground regions (in situ) where volatile components are dissolved in groundwater. In situ stripping operations are often viewed as less expensive than ex situ operations, especially if the groundwater from ex situ treatment in towers cannot be returned to the ground. This situation can result when the exiting or "uncontaminated" groundwater does not meet all acceptance standards for injection, perhaps because the groundwater does not have an acceptable pH, and the preference of in situ operations results at least partially from current U.S. regulations on reinjecting groundwater. Since more and more complex separation operations could be carried out above ground than underground, there are reasons to question the logic for distinguishing reinjection of groundwater after removal of a contaminant from similar in situ operations, but in situ stripping of volatile organic compounds (VOCs) is generally accepted even if the treated groundwater (and possibly like the surrounding uncontaminated groundwater) has properties such as a low pH that would prevent reinjection of water under current regulations.

Absorption and stripping operations can be used together to remove the components from one gas stream and to concentrate the component in another gas stream. If the equilibrium conditions in the stripper and the absorber are different, it is possible to recover the removed components at higher concentrations, perhaps where the recovered components can be reused. One method for altering the equilibrium conditions is to use different temperatures in the absorber and the stripper. Lower temperatures essentially always favor absorption of gases, and higher temperatures favor stripping. Thus, strippers are likely to be operated at higher temperatures than absorbers. However, the equilibrium conditions also can be altered chemically. For instance, acid gases are often readily absorbed in caustic solutions, but by lowering the pH to neutral of even slightly acid conditions, they may be readily stripped from the liquid.

Although adsorption and stripping are not necessarily used together, it is still worthwhile discussing them together because the equipment, analyses, and design methods used for the two operations are similar. In fact, the analyses are also similar to those used in liquid-liquid extraction (Chapter 6), and Chapter 6, which describes those operations, will refer to this chapter for analyses. Although there are often reasons for using different types of equipment for the two operations, there are almost as many variations in the equipment used for each operation as there are variations between the two operations. The equipment for the two operations will be discussed, and then some common design considerations for the two operations will be mentioned; finally a few specific examples of environmental applications for these operations will be discussed.

SOME COMMON APPLICATIONS OF ABSORPTION AND STRIPPING

Absorption is a standard operation for removing selected components from gas streams. In environmental applications, absorption is, of course, likely to be chosen to remove toxic components from discharge gas streams. One of the large-scale absorption operations removes SO_x and NO_x from exhaust gas from industrial and utility power plants. Other important absorption operations remove H_2S from natural gas and petroleum and coal process gases. (These may not be viewed as "environmental" separations because they do not deal directly with effluent streams, but they do remove components that cause environmental problems. Instead the removal of the contaminants occurs before they get to the effluent streams.) Both of these examples are "acid gases" that are usually absorbed in basic (alkaline) solutions or solvents. Similarly, basic gases such as ammonia can often be removed effectively by acid solutions or solvents. In other cases, organic components of gas streams can be removed effectively by organic solvents. Remember, as noted, absorption may be an attractive alternative for removing many toxic and valuable components from internal as well as effluent gas streams, like the removal of H_2S from natural gas.

Perhaps the most obvious use of gas stripping in environmental applications is in the removal of VOCs from wastewater or groundwater. Gas stripping, especially, can be carried out in very simple batch operations (often called gas sparging) as well as in more efficient towers with countercurrent flow of the two phases. If the VOC in the water is sufficiently volatile, that is, has a high Henry's law constant, one stage of stripping may be sufficient for removing the contaminant sufficiently. Values for the Henry's law constant for several common contaminants in water are given in Table 1. High volatilities also give high concentration in the gas phase from stripping operations, and with high volatilities the VOC can be removed to a very low concentration in a liquid by batch stripping, continuous sparging with the gas, without using an excessive volume of gas.

Batch stripping can be carried out in situ, with the aquifer serving as the reservoir, or in batch tanks. However, when there is a need to minimize the amount of gas to be used or when it is desirable to recover

TABLE 1 Selected Henry's Law Constants for Common Contaminants at 25°C

Compound	Henry's law constant [moles per m^3 (gas)/moles per m^3 (liquid)]	Reference
Bromoform	0.018	29
Chloroform	0.150	30
1,1,1-Trichloroethane	0.703	30
Trichloroethylene	0.392	30
1,4-Dichlorobenzene	0.137	31
Tetrachloroethylene	0.723	31
1,1,2,2-Tetrachloroethane	0.011	31
Toluene	0.277	31

the VOC at as high a concentration as possible, it is better to consider a countercurrent system such as a packed tower. Countercurrent operations make better use of the gas.

When gas stripping is used to remove VOCs from water, the gas from the stripper containing the VOC also will be covered by regulations, and discharge of the remaining trace of VOC to the atmosphere may not be permitted. At present, discharge of small quantities of simple hydrocarbons may be allowed, but discharge of most solvents, essentially chlorinated solvents, is severely restricted. In such cases, essentially all the VOC must be removed from the gas before it can be discharged. If the VOC concentration in the strip gas is sufficiently high, it might be possible to recover the bulk of the contaminant by condensation and remove the last traces with another operation, perhaps with a carbon adsorption bed. In either case, there is usually significant moisture in the gas from a stripper (it is essentially saturated with water at the stripper temperature), and it is often desirable to cool the gas (air is the usual stripping gas) and remove as much of the water as practical to minimize its effects on of VOC adsorption on the carbon. Membranes have also been used to remove VOCs from air coming from gas stripping operations. The degree of removal required is determined by the release requirements. However, it may be more practical to "reuse" the air for further stripping the VOC; reused air will not be released and, thus, will not necessarily have to meet release requirements. Reuse of the air reduces the requirements for pollutant removal, reduces pollutant release by reducing the rate of air release to only that rate at which the air leaks to or from the system.

Activated carbon can remove many VOCs from water as well as from air, and the reader may wonder why it could be better to remove the VOC by gas stripping and transferring it to a carbon bed rather than using the carbon bed to remove the VOC directly from the water. One reason is the higher adsorption loadings of VOCs that often can be obtained on an activated carbon when VOC is removed from air rather than water, provided the moisture content in the air is sufficiently low. Another reason is the ease of doing the stripping in situ; the advantages of acceptable in situ operations have already been discussed.

It may be desirable to chill the air from a stripper to reduce the moisture content before a carbon bed is used to remove the last VOCs. The small volume of condensed water can be recycled to the liquid feed going to the stripper so that no contaminant need be released by this path. Also, in some cases, the concentration and volatility of the contaminants may allow most of the contaminants to be collected in condensers as "free" organic liquid. The condenser could remove two liquid phases, namely the VOC and the condensed water. Separation of the two phases produces a highly concentrated VOC liquid and a contaminated water that can be recycled to the stripper. The carbon adsorption beds would only have to remove the last traces of contaminant that were not condensed.

DESIGN OF ABSORPTION AND STRIPPING EQUIPMENT

Absorption and stripping operations are usually carried out in multistage towers filled with packing material, but some applications were just noted that require only a single stage or a few stages that can be achieved in spray towers, a sparged vessel, or other simpler device. Process design of a countercurrent multistage tower usually involves using a tower with suitable packing to minimize axial mixing of the gas and liquid phases, establishing sufficient gas-liquid interfacial area for good mass transfer (acceptably short effective "stage heights" or "transfer unit" heights), providing sufficient cross-sectional area (diameter) to handle the required gas rate, and providing sufficient height in the tower to achieve the required number of effective stages. The process engineer is likely to select the type of packing or tower internals used and the diameter and height of the tower, but not its structural design. This book stresses process design, not mechanical or structural design aspects.

If the tower diameter and length are to be specified with actual dimensions, rather than more basic terms such as stages or transfer unites, the designer must know the type of packing to be used and the performance of the packing materials in terms of the height of a stage or transfer unit under the expected operating conditions in the tower. The height of a stage (or a transfer unit) is a measure of mass transfer performance described later. It depends upon the interfacial mass transfer performance of the packing material and the flow rates used. This information is usually sufficient to select the dimensions of a tower that could be supplied to a manufacturer or designed by a structural engineer.

Tower Diameter

Tower diameter is determined by the allowable liquid and gas flow velocities through the packing material used in the tower. Maximum flow rates, called "flooding" rates, are the rates at which the packing can no longer effectively separate the two phases and/or maintain good mass transfer performance. Under flooding conditions, a portion of one phase (liquid or gas) can appear in the effluent of the other phase. Such back mixing greatly degrades the performance of the tower, and the effective number of stages (or transfer units) decreases considerably. In absorption or stripping towers, the most evident effect of excessive velocity may be entrainment of liquid mist into the gas phase. For gas-liquid systems (such as absorption and stripping), the gas phase flow rate is more likely to limit the tower capacity, and the tower diameter is then likely to be set by the required gas rate and the acceptable gas velocity through the packing. A simple rule-of-thumb is that the pressure drop for the gas should not exceed an inch or two of water per foot of tower. Generally, higher pressure gradients can be allowed for smaller packing materials, and they may have to be maintained even lower for larger packing materials. However, even if lower pressure gradients can be allowed, larger packing materials still allow much greater gas velocities because the flow velocities to cause the same pressure gradient are much greater.

At very low liquid flow rates, there may not be sufficient liquid in the tower to wet the packing material effectively, and the mass transfer performance cannot be as good as one would expect with the packing fully covered with liquid. Thus, there can be an optimum region of flow rates over which the tower will be most effective. Fortunately, this region is relatively broad, and a tower does not have to operate over a narrow range of flow rates. A tower designed to operate moderately close to its maximum rates, but safely below the flooding rates, can usually be effective at significantly lower flow rates.

The flow capacity of a packing material is specified in terms of allowed volumetric flow rates per unit area of tower cross-section; so the cross-sectional area of the tower is estimated by simply dividing the required volumetric flow rate by the capacity of the packing material, the flow rate per cross-sectional area that can be used with the packing material. Of course, for a circular-cross-section tower, the diameter is

$$\text{diameter} = 2\sqrt{\frac{G}{\pi v}} \tag{1}$$

where G is the volumetric flow rate, and v is the allowable gas velocity in the tower. The allowable flow rates (flooding rates) are quite different for different packing materials.

Tower Length

Tower length is proportional to the number of stage or transfer units required to make the desired separation. Stages or transfer units can be viewed as dimensionless distances or lengths. They are determined from the desired concentrations of the effluent streams. The height equivalent to a stage or the height of a transfer unit is the factor that converts the dimensionless distances into dimensional distances. These factors are determined by the mass transfer capabilities of the packing material and the operating conditions, i.e., flow rates. The calculation of the required number of stages or transfer units and the heights of stages or transfer units will be discussed in more detail later in this chapter. First consider the type of absorption and stripping equipment used most often.

The numbers of stages and transfer units are similar dimensionless units of tower length that express the "amount" of separation that takes place. These terms are interchangeable, and in most situations either term can be used. However, there are some differences, as is discussed later.

The first task in determining tower height is to find the number of stages or transfer units required, and the second task is to determine the height of a stage or transfer unit, which is the dimensional unit that converts number of stages (or transfer units) to tower dimensions. The terms in the height of a transfer unit can be defined more clearly in terms of the factors that determine its value and units than can the height of a stage. The height of a transfer unit is proportional to the fluid velocity and inversely proportional to the mass transfer coefficient for adsorption (or stripping). However, either the height of a transfer unit or the height of a stage can be evaluated experimentally by measuring the separation (number of stages or number of transfer units) achieved in a given height of tower with specified flow rates.

Equipment for Absorption and Stripping

Because of the wide range of applications and sizes of streams used in absorption and stripping, several types of equipment are used. Perhaps one basis for grouping equipment types is based upon whether the operation requires or needs countercurrent multiple stages. Multiple stages provide opportunities for removing the component from one phase to the other (gas to liquid in absorption) with less of the removing phase (the liquid in absorption or the gas in gas stripping). A simple material balance will show (see the section "Process Design for Absorption and Stripping") that this also recovers the material at a higher concentration than can be achieved with single stage operations.

Single Stage Equipment

Equipment using a single absorption stage (or approximately a single stage) can be very simple. It can consist of little more than a tube inserted into a tank or drum to bubble gas through a liquid to absorb a component from the gas or to strip a component from the liquid. A slightly more complex case would be a spray of water or other absorbent into the air to strip a component to the water or to absorb a component from the air.

Single stage systems are more likely to be used when the equilibrium strongly favors the desired absorption or stripping, when high removal efficiency is not necessary, or when physical problems make it difficult to operate with more complex multiple stage devices. When the equilibrium strongly favors absorption, the system can be considered essentially irreversible. In those cases, there is little or no advantage in the countercurrent operations used in multiple stage operations. Then a single stage is sufficient and is capable of removing large fractions of the solute. Absorption of acid gases into caustic solutions are examples of systems that usually favor absorption so strongly that they behave almost like irreversible systems. Nevertheless, even in these cases it is still necessary to provide enough interfacial area for sufficient mass transfer to occur between the two phases; therefore, the single stage bubble or spray systems need to generate enough bubbles or drops and sufficient phase residence time in the equipment to approach sufficiently close to equilibrium and/or achieve the desired separation.

In some applications high fractional removal is not necessary, but high fractional removal is usually desired. One example is batchwise stripping of a component from a liquid in a tank. Although it may be necessary to remove a large fraction of the material eventually, in a batch operation, relatively slow removal rates can be tolerated if the slower processing

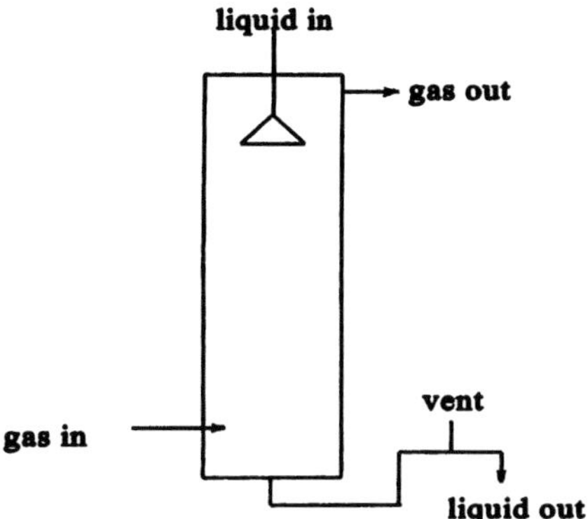

FIGURE 1 Absorption or gas stripping tower.

rates are acceptable. The slower processing rate may be the alternative to higher capital costs that could be required for high efficiency equipment.

One relatively common physical problem that can hinder the use of multiple stage equipment (or most complicated equipment) more than single stage equipment is formation of solids. Solids that attach to surfaces can plug or otherwise hinder the operation of many pieces of countercurrent towers. Although different packing materials have different capabilities for handling small concentrations of solids, they all have limitations on the concentrations of solids that can be present. In some applications with high solids concentrations, it may be more advantageous to use very simple equipment that can tolerate the solids rather than attempt to operate higher performance equipment that will be strongly affected by them. The absorption of sulfur oxides in lime solutions is an example where solids are formed, and it is difficult to use towers with small packing or with more complex internals. Fortunately, in that application, equilibrium highly favors absorption (equilibrium approaches irreversibility), and a single stage of absorption is adequate—actually considerably better than real equipment can usually achieve. (However, the size and complexity of some sulfur oxide absorbers, especially those units used in the electric power industry, make it difficult to describe them as "simple."

A common piece of equipment used for absorption and stripping that can handle moderate concentrations of solids is a spray tower (Figure 1), which in many cases approximates a single stage. However, it could do better or worse than a single stage. The liquid enters the top of the tower through one or more spray heads, which may look and operate much like a shower head in a bathroom. Although the spray heads may appear to be simple devices, it is usually worthwhile to seek high performance devices and thus use spray heads that form small and approximately uniform droplets. The open spray tower can usually handle significant quantities of solids that could be formed in the tower, but most spray heads will not handle liquids with significant concentrations of solids. In such a case, weirs and other liquid distributors can be used. The gas enters the spray tower at the bottom and is prevented from leaving the tower with the liquid by a liquid seal at the bottom of the tower. The seal is achieved by maintaining a gas-liquid interface a short distance above the bottom of the tower. In Figure 1, the liquid interface is maintained at a desired position by a "jack leg," which is a vented exit pipe that rises approximately the desired level for the liquid interface.

Multistage Absorption/Stripping Equipment

Spray towers and bubble columns can have more than one stage; usually the stage height is relatively large, and other devices are more appropriate if many stages are needed. Although it is possible to use spray towers to achieve several stages, it is often better to add packing to the column to provide greater interfacial area, more uniform flow distribution, and/or reduction in axial mixing (which usually has effects similar to nonuniform flow distribution).

The most common types of multistage absorption and stripping equipment consist of towers with internal packing or structures. A multistage tower may be similar to a spray tower except that it is filled with a packing material or a series of trays. The internal packing materials or trays used in absorption and stripping towers are likely to be similar to those used in distillation towers. The need for distributing the liquid flow over the cross-section of the tower and to maintain the liquid seal at the bottom of the tower is similar to the spray tower. The distribution of liquid flow in tray towers is usually incorporated into the design of the trays since the liquid must be redistributed at each tray.

Many packing materials are used in absorption and stripping towers, but they can be classified as random packing and structured packing. Random packing is a loose stacking of rings, saddles, and numerous other shapes, randomly oriented, that can be dumped into the tower. This author is always fascinated by the ingenious shapes that have been developed to give high interfacial area (usually related to high packing surface area), low pressure losses, and ease of packing fabrication.

Structural packing is constructed of multiliths that fit into specific size towers, and again considerable ingenuity has gone into the design on different structured packing for absorption and stripping towers during the past decade or two.

Tray towers are often more expensive than packed towers, but they can offer high efficiency performance, usually with low to moderate pressure loss. Most tray towers now appear to be relatively simple sieve trays with holes or slots for gas to bubble through liquid on the tray. Flaps of other movable covers over the holes or slots allow the use of a wide (usually lower) range of flow rates.

Distribution of the Liquid Flow over the Tower Cross-Section

The performance of a packed absorption or stripping tower (or a distillation tower) can be affected by poor distribution of liquid over the tower cross-section. Hence, a properly designed and effective distributor is required to introduce the liquid approximately uniformly over all portions of the tower cross-section. For small towers, the distributor could be a well-designed spray head, but for towers with large diameters it is probably necessary to design more complex distributors. Several manufacturers have different distributor designs, and there is a great deal of similarity in the general form of distributors for absorption towers, stripping towers, distillations, and adsorption beds. As the tower diameter increases, the distributor is likely to involve an array of pipes or troughs that spread the liquid evenly to different portions of the tower cross-section and introduce it to the tower through separate sprays or overflow points.

There is evidence that even a well-distributed liquid flow can develop a poor liquid distribution after passing through a certain length of some packing materials, especially some "random" packing materials discussed later. Such behavior could be due to an innate property of the packing material, a slight nonuniformity in the way the packing was placed in the tower, or even a slight deviation of the tower from a vertical orientation. Near vertical orientation of absorption towers (or any other towers that handle countercurrent flow of gases and liquids) is very important since there will be a tendency for the liquid to migrate toward the "lower" side of the tower and for gas to flow largely through the "upper" side of the tower. If the displacement of the top of the tower from the bottom of the tower is greater than its diameter and if the liquid always flowed vertically, it would all be on one side of the tower by the time it reached the bottom. Although the results may not be as dramatic when the tower deviates far less from vertical, there can still be some undesirable degradation in performance because poor liquid flow distribution results. Even with less apparent deviations from vertical, the liquid can flow preferentially on one side of the tower (the lower side) and gas can flow preferentially on the other (upper) side.

For very tall towers, some have found it useful to install different distributions at periodic distances down the tower. In these regions, the liquid flow is collected and redistributed across the tower cross-section.

The effects of nonideal liquid flow distributions in packed towers have been discussed by Killat and Rey [1] and Kouri and Sohlo [2]. The effects of poor liquid distribution can be complicated, and the importance of a particular percent deviation from the mean flow can depend upon how the deviations from the mean flow occur. In many cases, such as slight deviations of local liquid velocities up to 25% from a slight tilt in a tower, the tower performance did not degrade greatly [3]. The effects of poor liquid flow distribution are probably most important when the interfacial area and mass transfer rates are high, that is, when the height of a transfer unit with plug flow would be very short. Zuiderweg and co-workers [4] examined different "types" of deviations from uniform flow. These deviations followed patterns that could be generated from imperfect design or fabrication of the liquid distributor or the tower itself. Random deviations in the liquid velocity were usually not as important as systematic deviations, such as those caused by failure to distribute the liquid over the entire tower cross-section (leaving lower velocities near the outside of the tower), a tilted distributor, or a warped distributor.

Random Packing Materials

Random packing consists of individual pieces that are dumped or placed randomly in the tower. The pieces are often rings or saddle-shaped materials. Good discussions of the more common types of random packing material are provided in standard chemical engineering handbooks [5] and in textbooks on mass transfer [6]. The simplest rings are known as Raschig rings, which are short lengths of tubing with the length approximately equal to the outer diameter of the ring. Raschig rings are some of the oldest forms of random packing. More advanced metal rings are more common today, and they usually have slots cut into the sides and portions of the metal bent inward to provide additional surface area. There are also molded plastic or ceramic rings with roughened surfaces and additional surfaces "inside" the ring. There are some variations in slotted rings from different manufacturers, and they are usually sold under different trade names. Rings are manufactured of metal, ceramic, and plastic. The choice of material to use may depend upon the fluids to be used. Plastic rings are often effective and relatively inexpensive, but they have limited strength, are useful only at low to moderate temperatures (not

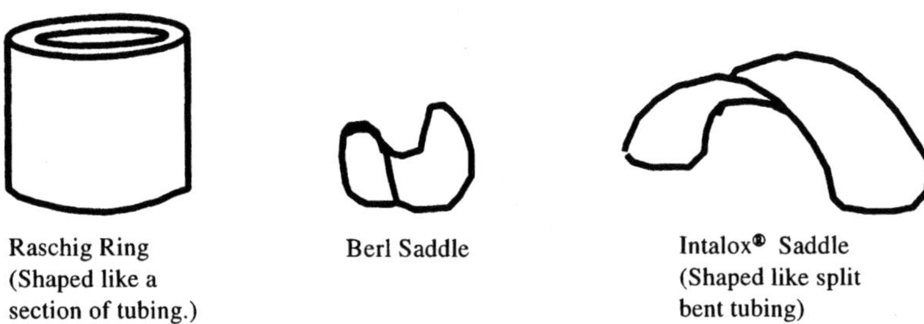

FIGURE 2 Common types of random packing materials.

likely to be a problem for most absorption operations), and are subject to attack by some organic absorbents. The choice of material will be affected by the absorbent used, and it may be desirable for the absorbent to wet the packing to obtained high interfacial areas if low to moderate absorbent flow rates per unit cross-section are desired. Selection of the packing material may also have to consider the tower temperatures or corrosive conditions. Ring-shaped packing materials are available in sizes from fractions of an inch to several inches in diameter (Figure 2).

Saddle-shaped packing comes in similar materials and in the same wide range of sizes. Although all saddle packing materials have the same general shape, there are differences in the method of manufacturing and the exact shape of packing from various manufacturers may be significantly different. These differences can significantly affect cost and performance. Saddle-shaped packing often gives high mass transfer rates and higher interfacial areas than rings, but there can be considerable variations among the available materials. The performances of ring- and saddle-shaped packing materials overlap considerably. Like ring packing, saddle packing also comes in modified forms with holes, slots, or ridges that enhance interfacial areas.

Other random packing materials have been made from wire coils and other shapes. Wire packing is usually expensive, but gives very low pressure drop, which is always desirable but absolutely necessary for systems that must operate under reduced pressure (vacuum). However, very low pressures are less likely to be needed in waste and environmental systems than in other chemical processing industries.

Selecting the Size of Random Packing

The packing size used depends upon the application and the scale of operation. Larger packing sizes will give a lower pressure drop and higher allowable flow rates, but less surface area (per unit volume of the tower) and greater mixing length (greater axial dispersion). As expected, larger packing materials are used in high throughput applications, but they are likely to give less effective mass transfer and thus greater stage heights or greater HTUs. The maximum packing size that can be used will also be set by the tower diameter. In general, one should avoid using packing materials with diameters of more than approximately one-sixteenth the tower diameter. Larger packing materials cause significant nonuniform flow near the tower wall. (Note that there are similar limitations on packing size in adsorption beds because of nonuniform flow near the walls.)

One study [7] suggests that the ratio of tower diameter to packing size be at least 30 for saddle packing and at least 15 for ring packing (Raschig rings). These ratios reflect the importance of different packing densities near the wall and near the center of the tower. The restrictions of the wall and the inability of packing to penetrate the wall mean that the packing density will be less near the wall than it is further inside the tower. Since the packing is added randomly, this difference is statistical, but unavoidable. The average packing density will vary in a complicated matter with distance from the wall, but the pattern of this variation scales approximately with the diameter of the packing material. (The curvature of the wall, however, also has a secondary effect.) This means that the importance of the wall region with a different packing density to the overall tower cross-section decreases as the ratio of the tower diameter to the packing diameter increases. This is the basis for the minimum ratios reported. Those limits are given only as guidelines which reflect the most common operating conditions and tower performance. Even greater ratios would decrease the effects of different flow distributions at the wall even further. The importance of decreasing the negative contribution of wall effects depends upon the effects of other parameters. If a system operates with slow mass transfer rates or large stage heights for any reason, the additional contribution of wall effects may not be important, even if the ratio of the tower diameter to the packing diameter is less than 8. However, in other cases, the wall effect could be an important contributor to stage (or transfer unit) height if other efforts are made to reduce stage heights, even when the ratio of the tower diameter to the packing diameter is much higher. The effects of tower diameter to packing size may not become apparent until the flow distribution, mass transfer rates, and other aspects of the tower design approach optimal conditions.

Details on the performance of a packing material should be sought from the manufacturer. Although the manufacturer may not be able to provide information from tests for the application of interest, there will usually be information on parameters (or information that can be reduced to the parameters) that can be used to estimate the performance in your application, such as pressure drop, maximum (flooding) rates, and mass transfer performance. The maximum flow rates are those above which the effective mass transfer falls, usually because of entrainment of liquid in the gas, often called "flooding." There is likely to be some information on mass transfer rates, but not necessarily mass transfer data on the system of interest. The information may include correlations for mass transfer coefficients, stage heights, or simply data from some relatively standard system. Absorption of carbon dioxide in sodium hydroxide solutions and absorption of oxygen into water are common systems used to test the mass transfer performance of packing materials. The use of these mass transfer parameters to design or predict performance of absorbers or strippers will be discussed later.

Pressure Drop through Random Packing

Since the towers are usually operated with the gas phase as the continuous phase, the most significant pressure drop is for the gas phase. The liquid pressure drop is essentially the static head of liquid since the liquid must be pumped to the top of the tower. Although the resistance to liquid flow is important, it is usually expressed in terms of the liquid holdup in the tower; the importance of resistance to liquid flow can be in the mass transfer performance. The pressure drop for gas flowing through an empty tower (no liquid present) can be estimated from standard equations for packed beds, such as the Ergun equation [8]:

$$\Delta P = K_1 G + K_2 G^2 \tag{2}$$

This equation is usually expressed in terms related to spherical particles. Then the constants can be expressed in terms of the particle diameter and the void fraction in the tower (the fraction of the tower volume not occupied by the solid part of the packing). However, for rings and saddles, these parameters are not so easily defined, so empirical factors have to be placed before each term. In Equation (2) those factors are grouped with the void fraction and diameter dependence; thus, the constants depend upon the packing shape used.

Equation (2) is most useful for showing how the pressure drop depends upon the gas flow rate, assuming that the gas is the continuous phase. Note that there are two terms on the right side: the first represents the pressure drop under viscous flow (flow at low Reynolds numbers), and the second represents the pressure drop under inertial flow (flow at high Reynolds numbers); the respective terms dominate the equation at low flow rates and at high flow rates. Estimates for the constants for Raschig rings, Pall rings, and saddles manufactured by the Norton Corporation are given by Strigle [9]. At low liquid rates, the pressure drop for the gas phase is approximately the same as that for an empty tower, but the pressure drop does increase slowly with the liquid rate. Examples of increasing pressure drop are given for a few packing materials by Strigle, and most manufacturers have such information for their packing materials, probably for operations using air and water. As the liquid rate increases further, the rise in the pressure drop with liquid rate becomes more rapid; this reflects the increasing liquid holdup in the tower. A significant portion of the void volume (the volume of flow channels) becomes occupied by liquid, and some flow paths can even become blocked by liquid. Eventually, there will be appreciable entrainment of one phase in the other, sometimes called flooding, and those conditions represent the maximum allowable fluid flow rates. Usually in absorption and gas stripping operations, the entrainment will be liquid in the gas.

The pressure drop in the gas phase may increase when the liquid flow rate increases, primarily from the increased liquid holdup in the tower (less void space for gas flow) as noted, but to a lesser extent from the velocity and "roughness" of the liquid as it flows over the packing. The liquid holdup is important to the mass transfer rate, and the pressure drop is important to the gas phase. At very low liquid flow rates, the liquid may not cover the packing material completely. Since the mass transfer rates (per unit volume of the tower) are approximately proportional to the interfacial area (per unit volume of the tower), it is important to have all of the packing "wet" by the liquid and as much additional interfacial area as practical. At low liquid flow rates, the liquid holdup is approximately proportional to the liquid flow rate, but as the flow rate increases, holdup increases more rapidly.

Mass Transfer in Packed Towers

As noted in a following section, the mass transfer resistance can reside in the gas phase, the liquid phase, or both phases. The importance of the resistance in each phase depends upon the equilibrium conditions, flow rates, and resulting agitation in each phase. The mass transfer resistance in individual phases is usually expressed as mass transfer coefficients, which are inverses of the resistances; that is, they are equivalent to conductances. For the gas phase,

$$k_G = \frac{\text{rate per volume of tower}}{a(y - y_i)} \tag{3}$$

The subscript G refers to the gas phase, and the subscript i refers to the condition at the gas-liquid interface. The concentration is expressed in terms of mole fraction, y, but other concentration units could be used as long as the mass transfer coefficient, k, is understood to include those units. The interfacial area per unit volume of the tower is usually called a.

The mass transfer rate can also be expressed in terms of an overall mass transfer coefficient based upon concentrations in the bulk phases and not involving concentrations at the gas-liquid interface. The overall mass transfer expression based upon concentrations in the gas phase can be written as

$$K_G = \frac{\text{rate per volume of tower}}{a(y - y^*)} \tag{4}$$

Note that a capital K is used to describe the overall mass transfer coefficient, and the difference between the expression for the overall coefficient and the single film mass transfer is the use of y^* instead of y_i in the driving force. The y^* term is the concentration in a gas phase that would be in equilibrium with the concentration in the bulk liquid. The overall mass transfer expression is not a general expansion of the single film mass transfer expression. Of course, the entire concept of all mass transfer resistance residing in stagnant films of gas and liquid on opposite sides of the interface is a major approximation that can have serious limitations. However, even if the two film approximation for mass transfer were sufficiently accurate, the use of the overall expression for mass transfer involves additional approximations. The important additional approximation is that the slope of the equilibrium curve does not change significantly over the concentration ranges that occur in the system. The approximation is always good when all of the resistance is in the film of the reference phase [the gas phase for mass transfer expressions such as Equations (3) and (4) based upon concentrations in the gas phase]. The approximation becomes less satisfactory as the resistance in the other phase becomes important and the equilibrium curve has significant curvature over the concentration range studied.

A similar expression can be written for the mass transfer coefficient in the liquid phase. In that case, the subscript L represents liquid, and the concentrations are for the liquid phase, usually written x when expressed as mole fraction, but other concentration units also could be used:

$$k_L = \frac{\text{rate per volume of tower}}{a(x - x_i)} \tag{5}$$

Again, a similar expression could be written for an overall mass transfer coefficient based upon concentrations in the liquid phase. Other units of concentration such as moles per liter (molarity) or even mass fractions can be used for the concentration difference driving force, but the mass transfer coefficient must contain the same units.

The overall coefficients are related to the individual coefficients by the expressions

$$\frac{1}{K_G a} = \frac{1}{k_G} + \frac{m}{k_L} \tag{6}$$

and

$$\frac{1}{K_L a} = \frac{1}{m k_G} + \frac{1}{k_L} \tag{7}$$

Here m is the slope of the equilibrium curve at the concentrations occurring at a position in the tower. For many dilute systems, the equilibrium curve is approximately constant, and m is approximately constant—Henry's law constant. Each term on the right-hand sides of these equations represents the resistances from each of the two phases. Note that the relative contribution of each phase to the overall mass transfer resistance is determined by the slope m and by the values of the individual film mass transfer coefficients. Remember that if there is significant curvature in the equilibrium over the concentrations that exist in the tower, m will not be a constant throughout the tower. Then, even if k_G and k_L are constant throughout the tower, K_G and K_L change because of changes in m. The error in using overall mass transfer coefficients then depends upon the variation in m over the tower and the relative importance of the terms in Equations (6) or (7) that contain m, that is, the relative importance of the mass transfer resistance in the nonreference phase.

Many environmental absorption and stripping operations involve dilute solutions in the liquid phase, and the equilibrium curve is linear over the region of interest and described well by the Henry's law constant. (See Table 1, p. 170.)

In most measurements of mass transfer coefficients, one measures the product of the coefficient times the interfacial area per unit volume of the tower, not the coefficients alone. It is, however, possible to determine the interfacial area independently in packed towers and to correlate it as a function of fluid properties, packing material, and operating conditions, largely flow rates. For adsorption operations, the interfacial area is often approximated crudely by the surface area of the packing material provided the liquid flow rate is sufficient to cover all of the packing. As liquid and/or gas flow rates increase, there will be some increase in interfacial area, probably because of ripples or other surface disturbances at the interface. Note that because the interfacial area is so often lumped with the mass transfer coefficient and determined from mass transfer measurements (not the only way to measure interfacial area), other factors may appear in the results that may be confusing at first. For instance, higher flow rates can result in entrainment of liquid from spraying or entrainment of gas from foam formation. It is doubtful that these phenomena decrease interfacial area, or even the actual mass transfer rates. However, these phenomena do decrease the mass transfer performance and will appear to decrease the product of the interfacial area and the mass transfer coefficients if the effects of axial mixing are lumped into the "effective" mass transfer coefficient unless the coefficients are measured by taking into account the back mixing that results from liquid entrainment. Seldom are such phenomena taken into account, and the coefficients are evaluated by ignoring dispersion or back mixing.

The mass transfer coefficients, along with the fluid velocities, determine the height of a transfer unit or (approximately in most cases) the height of the tower that is equivalent to a stage. More details on calculations of transfer units and stages are given later.

Structured Packing Materials

Structured packing materials are gaining applications in the process industries because they often give better mass transfer performance and usually at lower pressure losses. The most likely disadvantage of structured packing is the cost of the packing material. The objective of high mass transfer performance is to reach the same performance in smaller volumes, that is, in smaller towers, and structured packing materials are often good choices to meet these objectives. If the same performance can be obtained with less expensive packing material, but in a larger tower, there is an opportunity for optimizing the cost of the tower against the cost of better packing. There are also other potential advantages of structured packing. As mentioned earlier, they may have less flow resistance, that is, a lower pressure drop. Some structured packing material may also maintain good interfacial area even with low liquid flow rates [10].

Structured packing is usually made of sheets of woven metal, plastics, or ceramics. The earliest forms were probably woven metal structures used in distillation systems, such as vacuum distillation equipment where the pressure drop had to be minimized. Woven metal packing is still available, but structures made from metal and plastic sheets are less costly and appear to have accounted for the greatest increase in the use of structured packing. Since plastic can be cast with numerous holes, some

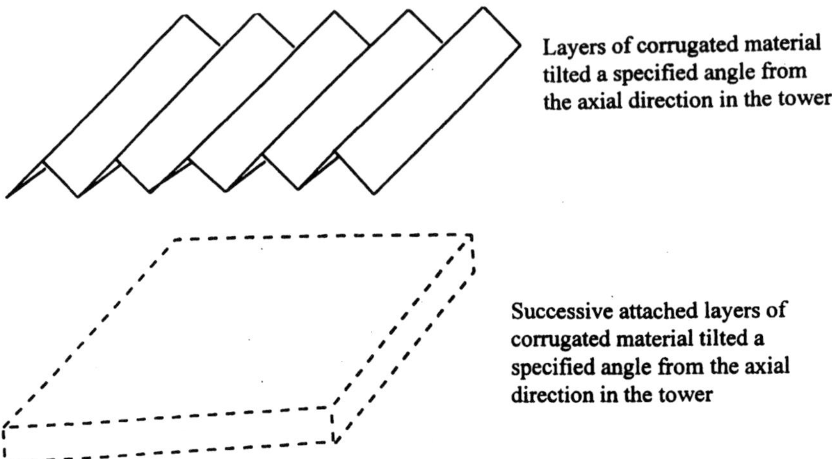

FIGURE 3 Example of structured packing material.

structured packing materials made of plastics look somewhat similar to packing made from woven metal. Furthermore, the metal used in structured packing materials can be made with numerous holes and thus be somewhat similar to the woven metal materials.

The sheets or metal or stiff woven cloth is usually folded to form channels. Structured packing comes in sizes cut or formed to fit specific tower diameters, so the packing must be purchased for the tower diameter to be used. A typical structured packing material is constructed as a series of sections that can be inserted and fit snugly into a tower with a given diameter. This packing material, like many structured packing materials, is fabricated from corrugated metal, but the angles and shapes of the corrugations are selected to optimize tower performance. The corrugated sheets are attached so that the directions of the corrugations cross at a specific angle selected by the manufacturer. Furthermore, note that the corrugations are oriented at a specified angle from the axial direction in the tower. To add randomness to the packing, each segment, which can be several inches thick, is rotated relative to the direction of corrugations in the segment below. The angle of rotation between segments can vary among manufacturers/installers, but it is not likely to be a simple integral fraction of 360°. This prevents exact repetition of mesh orientation anywhere within the tower.

As noted, structured packing made of sheets can also have holes punched so fluids can flow through as well as along the sheets. In some respects, this helps the sheet packing to take on some of the properties of structured packing made from woven materials and provides alternative flow paths and at least some decrease in pressure losses. Adding roughness to the surface of the sheets can also improve mass transfer performance. It is relatively easy for manufacturers of plastic structured packing to mold roughness into the plastic materials from which the packing can be fabricated.

The improvements in fluid capacity (pressure drop and flooding rates) and mass transfer performance over those obtained with the more common random packing can be significant. The pressure drop through several sizes of structured packing from the Koch Engineering Company and Glitch Incorporated was studied by Bravo [11], who correlated the pressure drop in terms of a friction factor, f:

$$\Delta P = \frac{0.193 f(\rho G) v_e^2}{D_p g_c} \left[\frac{1}{1 - h_T} \right]^5 \tag{8}$$

Here D_p is the hydraulic diameter of the packing; v_e is the effective gas velocity, which is the actual superficial velocity divided by the sine of the angle of the corrugations; and h_T is the liquid holdup, which is a function of the Froude number, Fr:

$$h_T = C \mathrm{Fr}^{0.5} \tag{9}$$

where

$$\mathrm{Fr} = \left(\frac{L}{\rho L} \right)^2 \frac{1}{D_p g_c} \tag{10}$$

and L is the liquid flow rate. This equation applies to the group of structured packing materials tested. The constant C depends upon the structured packing material used, with values usually between 3 and 5 for packing materials with equivalent diameters between 0.4 and 1.4 inches.

Since structured packing materials are usually constructed to fit snugly inside a tower of a given diameter, there are no wall effects similar to that discussed for random packing such as rings or saddles, but there are obvious limitations on the size of the corrugations used in smaller diameter towers. Standard packing with large size corrugations is more appropriate for towers with larger diameters. Although rules for selecting structural packing are not as well established as those for random packing, one should normally not consider structured packing with corrugations more than one-eighth of the tower diameter. Because manufacturers fabricate structured packing for specific tower diameters, the manufacturer's advice on the proper packing to use in a given diameter tower is usually incorporated in the equipment catalogue. That is, one is not likely to find a structured packing material available for use in towers with diameters too small for the size of the corrugations in the packing material. Manufacturers provide some guidance on installation of the packing or provide the installation. There may be a more limited selection of structured packing materials for very small diameter towers, and one may be more likely to use random packing for very small towers because the selection of structured packing can be limited. Furthermore, some of the smallest diameter (a few inches) towers may be more suitable for mesh packing if structured packing and low pressure drop are desired.

Structured packing has been designed for a variety of applications, not just for waste and environmental applications, but replacement of older random packing with new structured packing has been reported to improve the performance of VOC air stripping operations [12]. Of course, the degree of improvement depends upon the performance of the original packing material as well as the merits of the new material, and performance can mean different things in different applications (removal efficiency, pressure drop, throughput/flooding rates, etc.). It would be far too general to say that structured packing is (always) superior to random packing, since one should not compare a less appropriate structured packing with an optimum random packing. However, it is probably fair to say that structured packing materials should be considered for most new absorption or stripping operations.

Materials Used in Tower Packing

We have said that the most important factors in the choice of materials to use in the packing may be temperature, corrosion resistance, wettability, and cost. One purpose of the packing material is to increase the interfacial area between the liquid (water) and the gas (usually air). Although the column can be operated with either phase (liquid or gas) continuous, the pressure drop is usually less if the gas phase is continuous, and systems with the liquid phase continuous will usually not be considered. This means that the liquid phase is dispersed into discontinuous drops or segments flowing down the packing material.

If the liquid rate is relatively low and the gas phase is to be the continuous phase, the use of a packing material that is wet by the liquid is more likely to give a larger interfacial area, and thus better mass transfer performance, but that may not be the case when the liquid flow rate per unit area is larger. When the liquid wets the packing, the interfacial area may be approximately the area of the packing material itself. However, once the liquid flow rate is increased significantly, the interfacial area is difficult to predict, and it is better to seek information on the packing material to be used. (It may not always be desirable for the packing material to be wet by the liquid.) Larger liquid flow rates may reduce the interfacial area as liquid fills in crevices in the packing material, such as the points where packing particles touch, or they can increase the interfacial area as ripples appear at the gas-liquid interface or as additional droplets are formed. As mentioned, the interfacial area often is not reported separately or even measured separately. Instead, the product of the interfacial area per unit volume and the mass transfer coefficient are evaluated and reported together.

PROCESS DESIGN FOR ABSORPTION AND STRIPPING

Recall that a countercurrent absorption tower (also called a column) involves introduction of the gas into the bottom of the tower and removal of the gas from the top of the tower. This section describes how to determine the diameter and height of towers needed to make a given separation. The calculations of tower height will first be made in dimensionless units, such as the number of stages or the number of transfer units needed for the separations. Then the dimensionless numbers will be converted to actual tower heights (lengths) needed. The two phases pass through the tower in opposite directions, that is, countercurrently. Most towers will operate with the gas phase continuous, and the liquid will be sprayed into the top of the tower, usually onto the top of the tower packing. A tower without packing is more likely to be used when solids are present in the liquid feed or when solids are produced during the absorption or stripping process, such as when the gas solute being removed forms an insoluble compound in the liquid absorbent or when only a single stage (or few stages) of separation is required.

A small pool of liquid can be maintained at the bottom of the tower to "seal" the liquid outlet and force the gas to go up the tower and exit at the top. It is desirable also to introduce the gas through a distributor so that gas will travel up the tower at a uniform rate at every radial position in the tower. Uniformly distributed flow of both phases is important to column performance. Poor distribution of the liquid or gas flow can degrade the tower performance considerably. A good discussion of both the effects of poor distribution and the design of distribution systems is given by Kleman and Bonilla [13].

Process design refers to designation of the tower concentrations, flow rate, and dimensions. It does not involve design details of the column construction, such as the fabrication of vessels or selection of inlet sprayers or distributors. The tower diameter is determined by the flow capacity of the packing—that is, usually by the allowable gas velocity in the tower. It is often desirable to operate a tower moderately close to the maximum gas rate to obtain optimal mass transfer performance. Of course, the gas rate must be safely below the maximum allowable rate. Generally, it is reasonable to operate a tower with gas rates from 70% to 80% of the maximum rate.

If only the gas rate is specified initially, a proper liquid rate may not be specified until the required mass transfer performance needed for the tower is determined. For absorption processes, an optimization will be needed to set the liquid rate, because increasing the liquid rate may reduce the number of separation stages (or transfer units) required but will produce a more dilute liquid product. The following design procedures are described in numerous textbooks on mass transfer or separations used in chemical engineering curricula. Although the description is complete for most purposes, the reader may want to refer to other texts for a different description or a different point of view as well as for more detail.

Absorption

It is often convenient to think of absorption processes in terms of equilibrium stages, and, in many cases, it is adequate to design absorption towers in terms of the effective number of equilibrium stages. In many separation processes, the tower can be divided into compartments separated by trays or divided into mixer-settler units. The stages (or approximate stages) then are clearly identified. However, most absorption towers are filled with random or structural packing, and the number of effective stages is not obvious from a visual examination of the packing. In those cases, the concept of a certain length of the tower being the equivalent to a stage is a satisfactory basis for designing absorption towers.

In a few cases important to environmental problems, the equivalent number of stages may not be considered sufficiently realistic for the designer to use the staged approach reliably because the effective height of a tower that is equivalent to a stage may be a function of variables such as concentration and, thus, be a function of the position in the tower. In such cases, the height of a tower would not be proportional to the number of stages required. It is then usually better to use another form of dimensionless tower height, the transfer unit. However, this discussion will begin with the ideal stage concept. The discussion of stages will be followed by a discussion of the transfer unit (or rate) approach, and the two approaches will be compared. In certain common cases, the height of a tower that is more likely to be approximately proportional to the number of transfer units may not be estimated sufficiently accurately from

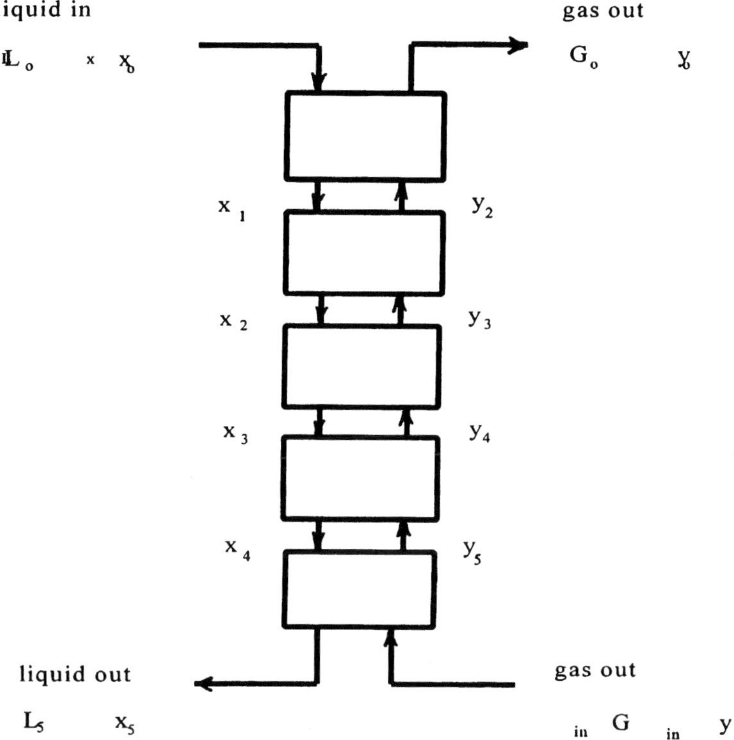

FIGURE 4 Interstage flow in a staged tower.

the number of stages needed. Thus, the differences between stages and transfer units will need to be explained.

In the staged approach, a tower is viewed as a collection of stages, and in each stage the gas and liquid phases are brought together and equilibrated until the two phases are in thermodynamic equilibrium. Then the two phases are separated, and each phase flows separately in opposite directions. The gas phase moves upward to the next higher stage, and the liquid moves downward to the next lower stage (Figure 4). A tower constructed of stagelike units is a sieve tray tower (Figure 5). The transfer unit approach views the tower as two phases moving in opposite directions and in plug flow with mass transfer resistance between the two phases in a continuous manner without individual stages. This transfer is always trying to bring the two phases closer to equilibrium. However, the rate of approach toward equilibrium is limited by the resistance to mass

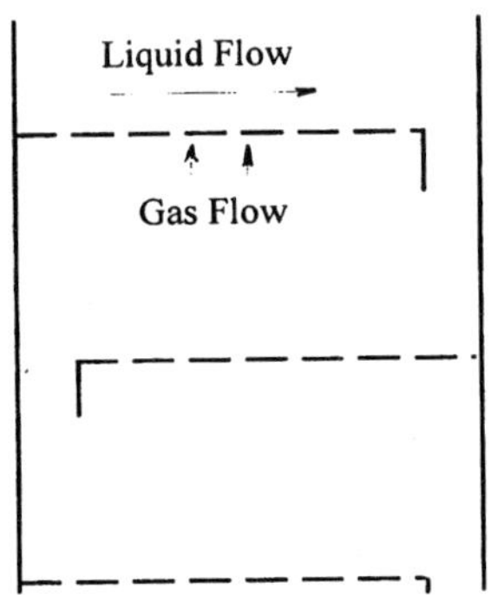

FIGURE 5 Sieve tray tower.

transfer, and the concentrations in the bulk phases never actually reach equilibrium because the movement of the two phases by convection always brings new gas and liquid phases into contact. Although there are significant assumptions involved in the rate- (or transfer unit) based approach, it is generally considered more realistic than the stage approach for packed towers. Usually results differ little in the two approaches, but the instances when they do should be noted.

Equilibrium

When a liquid absorbent and a gas are brought into contact and allowed to equilibrate, solute begins to transfer from the gas to the liquid (in stripping, the direction of transfer is in the opposite direction), and the concentration of solute decreases in the gas and increases in the liquid. Generally, the transfer will not completely deplete the gas of solute. Instead the rate of transfer will decrease and eventually approach zero as the phases approach "thermodynamic equilibrium," and the remaining phases, both containing concentrations of solute, are in equilibrium. This occurs when the chemical potentials of solute in the gas and the liquid absorbent become equal. One can view the equilibrium loading of solute as the capacity of the absorbent for solute, but that capacity depends upon the concentration of solute in the gas.

The "equilibrium" relation between the gas and the liquid absorbent is expressed generally in graphical form, with the concentration of solute in the gas plotted as a function of the equilibrium concentration in the liquid. Experimental measurements of equilibrium concentrations can be plotted with the concentration of the gas on the vertical axis and the concentration of the liquid on the horizontal axis. Although there are no limits to the shapes of equilibrium curves that may be observed, it is not common for equilibrium curves for dilute systems to have much curvature. Many dilute systems can be approximated well by linear equilibrium curves. A linear absorption equilibrium curve follows Henry's law. If the concentrations in both phases are given in mole fractions, Henry's law is

$$y = Hx \tag{11}$$

Here, y and x are mole fractions of solute in the gas and liquid phases, respectively, and H is Henry's law constant. Although many absorption equilibrium curves can be approximated as a straight line (or Henry's law) over a limited concentration range, this is certainly not a general case. However, many equilibrium curves are approximately linear in the dilute regions, and many environmental applications are dilute. (It is even common in environmental papers to assume a linear equilibrium curve, hopefully not without checking the entire concentration range of interest to ensure that the curve is linear.) (See Table 1, p. 170.)

Henry's law can be viewed as a generalization or extension of the "ideal" solution, which follows Raoult's law. In the ideal solution, the vapor pressure of the solute is proportional to the mole fraction, and the constant of proportionality is the vapor pressure of the solute. (Actually, for an ideal solution, the pressure of the absorbent liquid itself is also equal to its vapor pressure multiplied by its mole fraction, but for this discussion the vapor pressure of the liquid absorbent will be assumed to be so low that its presence in the gas can be ignored.) Then

$$p = xP_v \tag{12}$$

or

$$y = \frac{xP_v}{P_T} \tag{13}$$

where P_v is the vapor pressure of the solute, and P_T is the total pressure of the gas phase. Raoult's law can be considered to be a special case of Henry's law, where Henry's law constant is P_v/P_T.

As the mole fraction, x, of solute in the liquid approaches unity, the pressure over the liquid should approach P_v. Thus, the equilibrium curve for an ideal solution could remain linear over the entire concentration range (from $x = 0$ to $x = 1$). Note, however, that if a solute that is completely soluble in the liquid and follows Henry's law, but not Raoult's law, at low solute concentrations, the equilibrium curve will have to eventually deviate from linear somewhere at high concentrations to allow the total pressure to approach the vapor pressure of the solute as x approaches unity. Thus, systems that follow Henry's law (but not Raoult's law) may do so only under a limited range of concentrations.

The thermodynamics of gas and (especially) of liquid mixtures is a large subject that will not be discussed in further detail here. The reader should, however, be aware that considerable effort has gone into predictions and correlations of phase equilibrium data, and any of the standard texts will give additional details. For the remainder of this discussion, the equilibrium will be considered in terms of graphs or equations that can be considered or used as measured empirical curves.

Countercurrent Staged Absorption Towers

In a countercurrent absorption tower, the gas phase enters the tower at one end (the bottom) and leaves from the other end (the top). The liquid enters from the opposite end (the top) and exits from the other end (the bottom) (Figure 4). The tower has several "stages." It is assumed that the performance of a tower can be expressed in terms of an equivalent number of stages. Despite the empirical aspect of applying this concept to packed towers, it remains a useful concept for understanding and extrapolating tower performance. The following pages describe analyses that are presented with more detail in chemical engineering textbooks on absorption and similar mass transfer operations [14]. Also note that such calculations are now done easily with widely available computer codes capable of operating on modest sine computer systems, including the more advanced personal computers. However, the basis for these methods is presented because it is important to have at least a minimal grasp of the calculations are being made and what they mean. This discussion focuses on simpler cases because they can be understood easily and are relatively common in environmental and waste problems.

Consider the top stage in Figure 5. Two streams enter and leave the stage. One is the liquid fed to the top of the tower and, thus, to the top stage. The other is the gas stream from the stage below. Two streams also leave the top stage: the gas stream and the liquid stream going from the top stage to the stage just below it.

As the equilibrium stage was defined, the two streams leaving the stage are in equilibrium. This means that the solute concentrations in the gas and liquid are related by the equilibrium curves just described. Thus, if one knows the solute concentration in a stream leaving a stage, it is only necessary to look at the x position on the equilibrium curve to determine the composition of the other stream leaving that stage.

The most common specification for an absorber design gives the composition of solute in the gas phase leaving the tower, that is, the

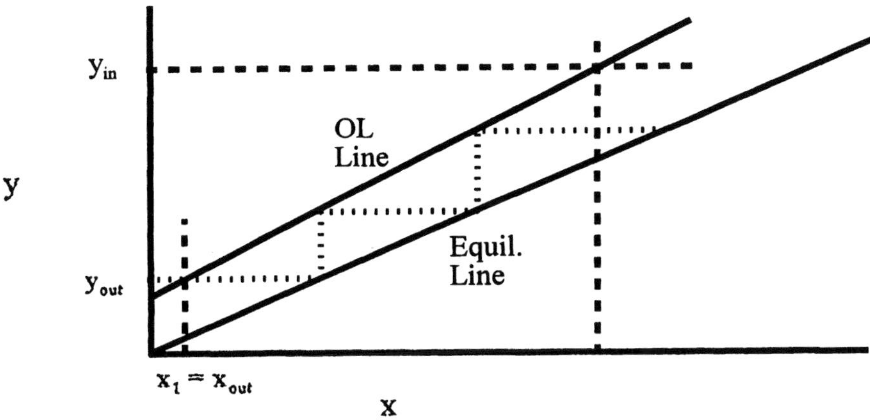

FIGURE 6 McCabe-Thiele plot for absorption with equilibrium curve, operating line, and inlet and outlet liquid and gas concentrations.

composition leaving the top stage. This is likely to be specified because the purpose of the absorber is to reduce the solute concentration to some required concentration in the exiting gas stream. Since the liquid leaving the top stage is in equilibrium with this specified gas composition, the solute concentration in the liquid leaving the top stage can be obtained from the equilibrium curve. In Figure 6 the lower curve is the equilibrium curve that relates the composition of the solute in the gas phase (y) to the concentration of that solute in the liquid phase (x). The composition in the exiting gas is labeled y_{out}, and the liquid leaving the top stage is labeled x_1 or x_{out}. The subscript 1 denotes streams leaving the top of the tower, or the first stage. (Note that y_1 is also the composition of gas leaving stage 1.) The compositions y_1 and x_1 represent a point on the equilibrium curve because the gas and liquid phases leaving each stage are in equilibrium. This is equally true for all stages; the compositions of the gas and liquid streams will be points on the equilibrium curve, but to know where they lie on the equilibrium curve, we must look at material balances to determine another curve that we will call the operating line.

The composition of the gas stream entering the top stage can be labeled y_2 because it describes the gas stream leaving the second stage, the stage immediately below stage 1. If the composition and the flow rates of three of the four streams entering or leaving stage 1 are known, the composition of the fourth stream can be obtained from a solute balance around stage 1. At steady state, there is no accumulation of solute in the stages, and the rate at which solute is transferred to the stage is equal to the rate at which solute is transferred from the stage. Two streams enter and two streams leave the stage. In equation form,

$$L_{in}x_{in} + G_2y_2 = L_1x_1 + G_1y_1 \tag{14}$$

For the moment, consider only dilute systems where the solute concentration is so low that transfer of solute from the gas to the liquid does not significantly affect the total moles of gas or liquid in either phase. Then the flow rate of liquid leaving the stage (or any stage in the tower) is essentially the same as the flow rate entering that stage. That is, the liquid flow rate is essentially the same throughout the tower and is approximately the same as the liquid feed rate to the tower. Similarly, the gas rate entering and leaving all stages in the tower is approximately the same as the gas feed rate to the tower. Then the subscripts on the flow rates can be dropped:

$$Lx_{in} + Gy_2 = Lx_1 + Gy_1 \tag{15}$$

where L and G are the liquid and gas (molar) flow rates to and from every stage in the tower. This equation can be solved for y_2:

$$y_2 = y_1 + \frac{L}{G}(x_1 - x_{in}) \tag{16}$$

Equation (16) gives the composition of the gas entering stage 1 and leaving stage 2 from the composition of the liquid stream from stage 1 and the liquid entering the top of the tower.

At this point one has the same information for stage 2 as originally available for stage 1. That is, the compositions of the liquid entering stage 2 and of the gas leaving stage 2 are known. Following the same pattern used for stage 1, the composition of the liquid leaving stage 2 can be obtained from the equilibrium curve because the liquid stream leaving the stage must be in equilibrium with the gas phase leaving that stage. Then since compositions of three of the four streams entering or leaving stage 2 are known, a material balance around it can be used to calculate the composition of the fourth stream—the composition of the gas stream leaving stage 3 and entering stage 2.

Rather than writing the material balance around only stage 2, it is just as easy to write the balance around stages 1 and 2. For dilute systems where the liquid and gas flow rates are the same for all stages, the composition of the gas stream from stage 3 is

$$y_3 = y_1 + \frac{L}{G}x_2 - \frac{L}{G}x_{in} \tag{17}$$

The solute balance can be written generally for any stage:

$$y_{n+1} = y_1 + \frac{L}{G}x_n - \frac{L}{G}x_{\text{in}} \tag{18}$$

Equation (18) relates the gas composition from stage $n + 1$ to the liquid composition from the stage above it (stage n) and is called the "operating line." The equation describes a line with slope L/G that passes through a point on the x-y graph to x_{in}, y_1 (the conditions at the top of the tower). Thus, knowledge of the conditions at the top of the tower, usually given in the problem specifications, and the ratio of the liquid to gas flow rates is sufficient to construct the operating line.

This procedure for calculating the composition of gas and liquid streams can be used for each stage of the tower. It involves alternately using the equilibrium curve, which relates compositions of the two streams leaving each stage, and the operating line, which relates the composition of gas streams entering the stages to the composition of liquid streams leaving them. With the assumptions of constant liquid and gas flow, the procedure is called the McCabe-Thiele method and is described in most chemical engineering texts dealing with mass transfer operations. The method is also applied to other countercurrent mass transfer processes, such as solvent extraction and fractional distillation, and is mentioned again in similar terms in chapters covering those methods.

It is often most convenient to carry out absorption calculations by the McCabe-Thiele method or by other methods with packaged chemical process simulation computer programs, but when the equilibrium curve is available in graphical form, the calculations can also be done graphically. By alternately using the equilibrium and operating lines, one is effectively taking "steps" between them, as shown in Figure 6. Although it is convenient to draw these steps, the calculational procedure simply involves alternately using the operating line and equilibrium curves to obtain the next composition as calculations proceed down the tower. Packaged computer codes for absorption calculations allow the engineer to quickly explore a wide range of operating conditions and seek optimal systems.

The merit of this procedure is that one can quickly determine how many stages are needed to achieve the desired tower performance. The calculational procedure should be continued until the composition of the gas feed composition is reached. Of course, this probably will not occur with an integer number of stages. It may be obvious that the calculations could have begun at the bottom of the tower. In that case, the operating line would still be derived from a material balance on the solute, but it would be solved to give the composition of the liquid stream leaving the stage above as a function of the gas stream leaving the stage below. It would be the same equation, but with the dependent and independent variables interchanged.

Minimum Liquid Rate

A convenient result of seeing the stage to stage calculations in steps (Figure 6) is the observation that the composition changes more quickly when the operating line and the equilibrium are farther apart. This is evident in Figure 6 even though the curves do not separate greatly. Note also that the concentration in both the gas and liquid phases can only increase with positions further *down* the tower. In no case can the concentration of solute in the gas leaving the top of the tower be lower than the concentration that would be in equilibrium with the inlet liquid, and the concentration in the liquid leaving the bottom of the tower cannot be higher than the concentration that would be in equilibrium with the gas entering the bottom of the tower.

These considerations provide one way to visualize the minimum liquid rate needed to reduce the solute concentration in the gas from its initial (inlet) concentration to the desired or specified lower concentration, and in most cases they will be adequate for determining the minimum liquid rate. In algebraic form,

$$L_{\text{min}}(x^* - x_{\text{in}}) = G(y_{\text{in}} - y_1) \tag{19}$$

This equation defines a minimum ratio of L_{min}/G and thus a minimum slope for the operating line. This operating line intersects the equilibrium line at the conditions at the bottom of the tower, that is, at the upper right portion of Figure 6. Such an operating line is illustrated in Figure 7a. The intersection of the equilibrium curve and operating line is sometimes termed a "pinch," the behavior that one observes when stage by stage calculations down the tower result in smaller and smaller steps that converge at the point of intersection of the operating and equilibrium curves. In Figure 7a the sizes of the steps (that is, concentration changes that correspond to each stage) that represent individual stages decrease as the distance between the equilibrium curve and the operating line decreases, and the calculations approach the pinch point (intersection). An infinite number of stages would be required to reach the intersection. Note also that McCabe-Thiele stage by stage calculations can never cross an intersection of the operating line and the equilibrium curve. With a linear equilibrium curve (Henry's law) or with an equilibrium curve with positive curvature (concave upward), the minimum slope (or the minimum liquid rate) will always be determined from a pinch at the bottom

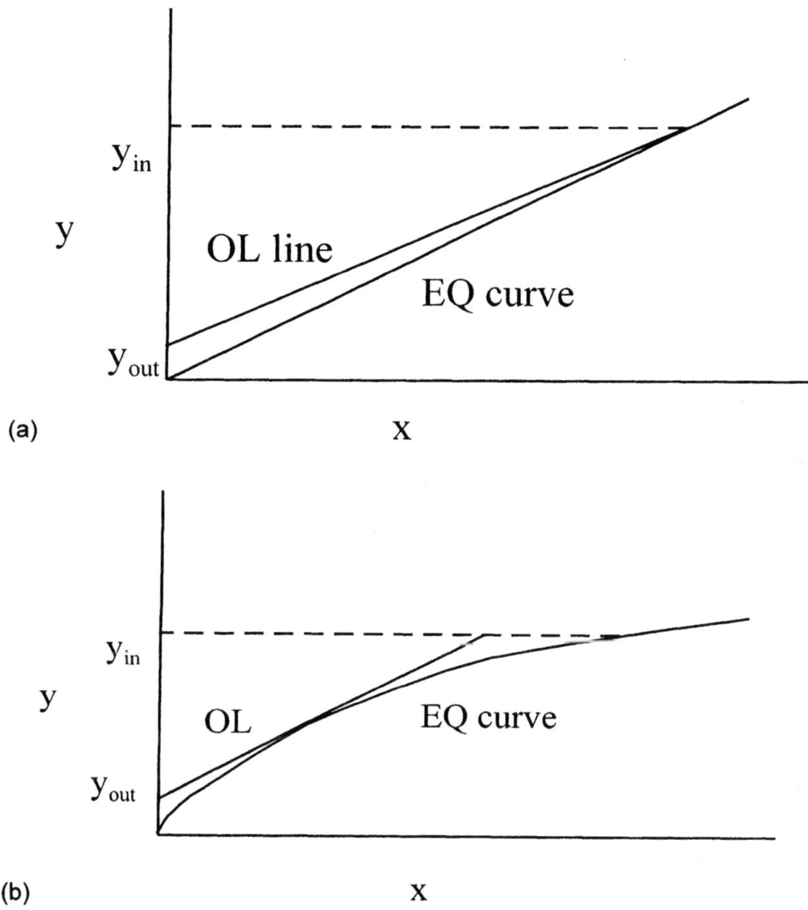

FIGURE 7 Determining the minimum absorbent (liquid) rate. (a) Example with pinch point at bottom of tower (common case). (b) Example with pinch point between bottom and top of tower.

of the tower (upper right end of the graph in Figure 7a). This is the location of the pinch for determining the minimum liquid rate for most absorption systems.

However, one should be cautious when the equilibrium curve (or a portion of the curve) is concave downward, especially sharply concave downward. It is then possible for the operating line and the equilibrium curve to intersect at higher liquid rates than the rate that gives an intersection at the top of the tower. Since a pinch occurs any time the operating line touches (or intersects) the equilibrium curve, it is necessary to use an even higher liquid rate. That is, the minimum liquid rate is the rate above which the operating line neither crosses nor is tangent to the equilibrium curve at gas compositions between the inlet and outlet concentrations. In Figure 7b the operating line intersects the equilibrium curve before it reaches the conditions at the bottom of the tower, and the slope of the operating line that gives this intersection is greater than the line that intersects the equilibrium line at the bottom of the tower. This resulted because the equilibrium curve has considerable negative curvature. Although any shape equilibrium curve is possible, this much negative curvature is not likely among systems common in environmental problems. Nevertheless, by drawing the equilibrium curve and the specified conditions at the top of the tower, it is easy to draw a trial operating line from those conditions to the equilibrium curve at the inlet gas composition. If that line intersects the equilibrium curve at any other point before the inlet gas composition is reached, there will be a higher minimum liquid rate. To determine that rate, a line can be drawn from the conditions at the top of the tower with the minimum slope that just clears the equilibrium curve (or is a tangent to it), and the slope of that line will correspond to the minimum L/V and thus the minimum liquid rate (Figure 7b). We are interested only in that portion of the equilibrium curve between the gas inlet and outlet compositions because these are the only compositions that occur in the tower; an intersection of the operating line and the equilibrium curve or any unusual behavior of the equilibrium curve in other concentration regions is not important to the immediate problem.

With that minimum liquid rate, an infinite number of stages (that is, an infinitely tall tower) would be required. As larger liquid rates (greater than the minimum) are used, the slope of the operating line increases, and fewer and fewer stages are required. However, higher slopes of the operating line (higher L/G) mean higher liquid rates in each stage (or in any volume of the tower) and lower concentrations of solute in the liquid. The larger volume of liquid used and the lower concentration of solute in the liquid can mean that a greater effort (and, perhaps, cost) may be required to recover the solute from the liquid. This illustrates that optimization is needed between the liquid flow rate and the number of stages (or height) of the tower.

Algebraic Expressions for the Number of Stages

When both the equilibrium curve and the operating line are linear, the number of stages can be determined analytically [15]:

$$N_p = \log\left\{\frac{y_{\text{in}} - mx_{\text{in}}}{y_{\text{out}} - mx_{\text{in}}}\left[1 - \frac{1}{A}\right] + \frac{1}{A}\right\} \Big/ \log A \tag{20}$$

$$A = \frac{L}{mG} \tag{21}$$

where A is the "absorption factor" and m is the slope of the equilibrium line. It may be more convenient to evaluate the equation rather than using the graphical method when designing absorbers for systems with linear equilibrium curves.

Nondilute Systems

All discussions so far were limited to dilute systems, that is, to systems where the solute is so dilute that transfer of the solute from the gas to the liquid phase does not significantly affect the flow rate of either phase. Many (probably most) environmental problems will be dilute. When the solute concentration is a significant fraction of the total gas or liquid phase, only a small change has to be made in the operating line to handle the problem more exactly. As long as the gas and liquid have little mutual solubility in each other (and the solubilities are thus not affected significantly by the present of solute), the flow of solute free gas and solute free liquid remains constant over the entire length of the tower, although the total flow of solute and gas or liquid may vary with position in the tower. Denoting the solute free liquid and gas flow rates as $\underline{L}$ and $\underline{G}$ respectively, the material balance from any position to the top of the tower can be written as

$$x_{\text{in}}L_{\text{in}} + y_{n+1}G_{n+1} = x_nL_n + y_1G_1 \tag{22}$$

or

$$x_{\text{in}}\left[\frac{\underline{L}}{1 - x_{\text{in}}}\right] + y_{n+1}\left[\frac{\underline{G}}{1 - y_{n+1}}\right] = x_n\left[\frac{\underline{L}}{1 - x_n}\right] + y_1\left[\frac{\underline{G}}{1 - y_1}\right] \tag{23}$$

since

$$\underline{L} = L_{\text{in}}(1 - x_{\text{in}}) = L_n(1 - x_n) \tag{24}$$

and

$$\underline{G} = G_{n+1}(1 - y_{n+1}) = G_1(1 - y_1) \tag{25}$$

Of course, if x_{in}, x_n, y_{n+1}, and y_1 are all near zero (that is, much less than unity), L_{in} and L_n will be close to $\underline{L}$, and G_{n+1} and G_1 will be approximately the same as $\underline{G}$. Even with the high solute concentrations, the solute balances shown can be plotted as a straight line if the concentration variables used are

$$Y = \frac{y}{1-y} \quad \text{and} \quad X = \frac{x}{1-x} \tag{26}$$

Then the solute balance (the operating line) becomes

$$X_{\text{in}}\underline{L} + Y_{n+1}\underline{G} = X_n\underline{L} + Y_1\underline{G} \tag{27}$$

Solving for Y_{n+1} gives

$$Y_{n+1} = Y_1 + \frac{\underline{L}}{\underline{G}}(X_n - X_{\text{in}}) \tag{28}$$

This equation is a straight line and can be used in the McCabe-Thiele method in the same way that the x-y plot was. It is usually more convenient to work with straight operating lines, because in considering a range of absorbent flow rates one may need to plot several operating lines to determine the minimum liquid rates and to explore the number of stages required for different liquid rates. It is convenient to use straight lines to determine the minimum liquid rate that avoids a pinch when the operating line touches or crosses the equilibrium curve.

However, a single equilibrium curve can be used for all of all calculations when the liquid flow rate (operating line) is being explored. If computer codes are used, the difference between dilute and concentrated systems may not be apparent to the user. If the tower design is to be optimized, it will be necessary to compare the effects of variations in the tower diameter, length, and flow rates on capital and operating costs. Note that if the equilibrium curve were linear in the x-y plots, it would be curved in the X-Y plots. The equilibrium curve is usually plotted only once for a given system, but one may want to try different absorbent (liquid) flow rates and thus try several operating lines. That is why it is usually more convenient to plot several straight lines rather than several curves.

Nonisothermal Absorption

When the solute concentration and the heat of absorption are sufficiently high, temperature changes in the tower can be significant and need to be considered during selection of the equilibrium conditions that correspond to each stage. However, nonisothermal conditions are not covered here in detail because they are not believed to be sufficiently common in environmental absorption problems. However, if significant temperature changes are expected in a tower, one could use a computer program to analyze the problem. Of course, additional data are required for nonisothermal systems. The equilibrium relations must be known at different temperatures, and the heat of absorption and the heat capacities of both phases must be known over the concentration range of interest.

Other Problem Specifications

The preceding discussion focused upon the most common way an absorption problem is stated. That is, the compositions of the liquid and gas feed streams are specified, and the desired removal of solute from the gas is given. Then the problem is to calculate the height of the tower required to make the separation. The McCabe-Thiele method permits one to calculate concentrations down (or up) the tower and thus determine how many stages (or tower height) will be needed. That is the problem specification described earlier. Although this is the most common way to specify the problem, it is not the only way.

For instance, one could specify the feed conditions (flow rates and compositions of liquid and gas streams fed to the tower) and the number of stages and calculate the concentrations in the effluent streams. This type of specification is used to examine the potential use of an existing absorption tower. The calculations are not straightforward, but are likely to be trial and error. The same principles and equations are used, but the effluent concentrations (or slope of the operating line) are assumed as a trial. One could assume the effluent concentrations and then calculate number of stages required to reach the conditions at the other end of the tower. Note that assuming the concentration in one effluent stream specifies the concentration in the other effluent stream, since the composition of that stream can be calculated from an overall material balance on the solute. If the calculations do not give the correct number of stages, a new value for the effluent concentration must be assumed. The procedure is repeated until the correct number of stages is found. Alternatives to the manual trial-and-error calculations include computer simulations which can do the calculations automatically once they are set up. This is another case where the advantages of automated computer calculations are most evident. If the equilibrium curve and the operating line are linear, one can use Equation (20) to calculate the conditions for the required separation.

Another possibility is to specify the feed concentrations, the gas effluent concentrations, and the number of stages and to determine the tower capacity to make the desired separation. Then the allowable flow rates would be set by the tower flooding characteristics and its packing material. The calculations can give the liquid to gas flow ratio that will provide the desired separation in the available number of stages. In this case, a trial-and-error calculation using different slopes of the operating line may be needed to determine which slope gives the separation in the available number of stages. (Be aware that changing the flow ratio or flow rates can affect the height of a stage, so accurate calculations with these specifications may not be simple.)

Although such trial-and-error calculations are not particularly difficult, they may be arduous enough to make one appreciate computer programs. The calculations become increasingly arduous as the number of stages increases because even small changes in operating line slope can significantly change the required number of stages.

Effective Stage Height

The merits of calculating the number of "stages" needed for an absorption separation depends upon the use of that result to predict the required tower height. Since many absorption towers are now expected to be packed towers, not discrete staged units, the use of the stage concept assumes that a specified (or measured) height of the tower will be equivalent to a stage. This is usually referred to as the "stage height" or the "height equivalent to a theoretical stage." Stage heights are measured experimentally, and the measurements must be made under conditions similar to those to be used in the application. This obviously means that the liquid and gas flow rates should be the same since they determine the interfacial area, the shear at the interface, and, thus, the mass transfer rate. It may be less obvious that systems with similar equilibrium relations should be used, but the importance of the equilibrium relations will become evident in the next section when design methods are discussed which are based upon mass transfer rates and not upon stages. If one has data for a given absorption separation in one tower and wants to estimate the effects of using the same system but with different tower height, the stage approach is likely to be accurate as long as the slopes of the equilibrium and the operating line are not greatly different. However, we will defer discussion of how differences in the equilibrium properties and differences in the operating conditions can affect the accuracy of stage-based treatment until after the next section where the alternative rate-based approaches for calculating required tower heights are discussed.

Rate-Based Tower Design

Although the stage concept is useful for many applications, it does not always accurately describe all packed towers. The more realistic description of packed towers is based upon the rate at which the solute is transferred from the gas to the liquid absorbent. This approach is sometimes called "continuous," as opposed to the discrete staged approach. It may also be

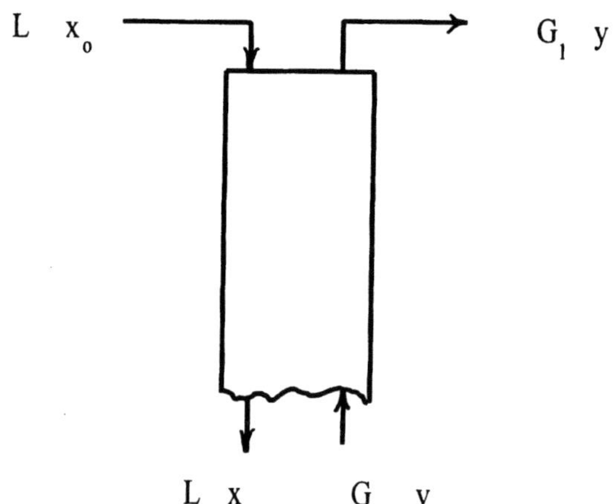

FIGURE 8 Material balances around the top of a tower without stages.

called the transfer unit approach because the results are usually expressed in terms of "transfer units" rather than stages.

The essence of rate-based analysis is illustrated in Figure 8. This figure shows the upper portion of the tower with the compositions (y and x) at a random position in the tower. At that point, or at any point in the tower, solute is being transferred from the gas phase to the liquid absorbent. The rate at which the solute is transferred from the gas to the liquid can be expressed in any of many ways, usually with the form

$$\frac{\text{rate}}{\text{volume}} = k_g a(p_g - p_i) = k_y(y - y_i) \tag{29}$$

This relation describes the rate in terms of a concentration driving force. In this case, the driving force is the difference between the partial pressure (or concentration) of the solute in the gas and the partial pressure of solute in the gas at the liquid interface. This can be described as a "two film" approach because the rate equation describes the diffusion resistance as if it occurred in two stagnant films, one film on each side of the gas-liquid interface. This is a significant simplification of the mass transfer near the gas-liquid interface, but it has proven quite useful. The rate constant, k_g, relates the mass transfer rate to the driving force. The a term is the interfacial area per unit volume of the tower. It has units of length^{-1}. As noted, several driving forces can be used in this expression, but k_g must have units to match the units of the driving force. That is, if the rate constant to be used is based upon one driving force, that driving force should be used in the rate expression. Equation (29) shows rate constants for two driving forces, solute partial pressure and solute mole fraction in the gas. Note also that the partial pressure can be presented in any of several different units, such as atmospheres, newtons per square meter, pounds force per square inch, etc. The mass transfer coefficient used with each of these different sets of units would then have different values.

At steady state, the rate at which the solute is transferred from the gas to the liquid in a differential element of the tower is equal to the rate at which solute enters the differential element by convection. For dilute systems using the driving force expressed in units of partial pressure, this can be written as

$$G\left(\frac{dp_g}{dz}\right) = -k_g a(p_g - p_i) \tag{30}$$

This expression can be integrated from the top of the tower to the bottom to give

$$\int_{p_{\text{out}}}^{p_{\text{in}}} \frac{dp_g}{p_g - p_i} = \frac{k_g a}{G} \int_0^L dz = -\left(\frac{k_g a}{v}\right) L \tag{31}$$

The right-hand integral can be integrated easily as shown. The left-hand integral can be integrated analytically only in a few cases. For the moment, the integral on the left can be simply given a name, the number of transfer units (NTU). Solving for L gives

$$L = \frac{G}{k_g a} \int_{p_{\text{out}}}^{p_{\text{in}}} \frac{dp_g}{p_g - p_i} = [\text{HTU}][\text{NTU}] \tag{32}$$

where

$$\text{HTU} = \text{height of a transfer unit} = \frac{G}{k_g a} \tag{33}$$

and

$$\text{NTU} = \text{number of transfer units} = \int_{p_{\text{out}}}^{p_{\text{in}}} \frac{dp_g}{p_g - p_i} \tag{34}$$

The tower height required for a separation is proportional to the integral defined as the number of transfer units, and the constant of proportionality is the height of a transfer unit. The number of transfer units is used in much the same way as the number of theoretical stages discussed in the previous section.

The meaning of this expression is illustrated in Figure 9. This figure shows the case where the mass transfer from the gas to the liquid is con-

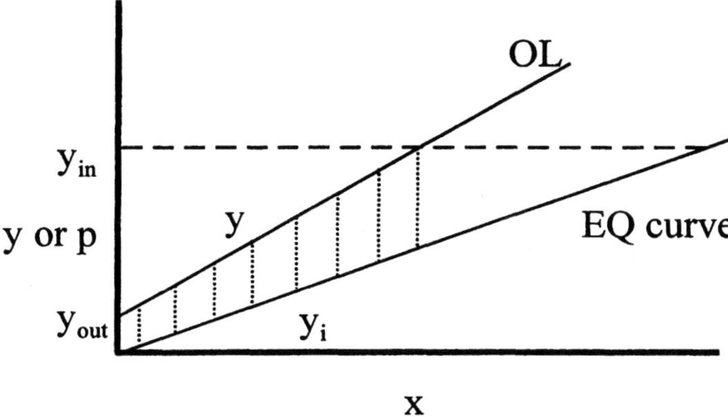

FIGURE 9 Tie lines, bulk concentrations, and interfacial concentrations when mass transfer from the gas to the liquid is controlled by mass transfer resistance in the gas phase.

trolled by mass transfer resistance through the gas to the interface. This refers to the condition where mass transfer resistance within the liquid absorbent is very much lower than the mass transfer resistance through the gas film. Then the concentration drop across the liquid film would be very small, and the difference between the concentration in the bulk liquid and the concentration in the liquid at the interface would be small or could be neglected. The concentration or pressure at the interface at any point in the tower is located just below the concentration at that point on the operating line. This is shown in Figure 9 by vertical "tie lines" that connect in the bulk phase concentrations (on the operating line) to the interface concentrations (on the equilibrium curve). Equation (34) can be integrated graphically, even if no algebraic expression for p_i (or y_i) is available (Figure 9).

Note that the operating line was described in the stage approach as the curve (or line) that related the composition of the liquid leaving a stage in an absorption tower to the composition of the gas coming from the stage below. When analyzing towers with continuous variations in concentration, the same operating line can be viewed as the curve (or line) that relates the composition of solute in the bulk fluid in each phase. By making a material balance around any position in the tower and the top (or bottom) of the tower, the same relation for the operating line is reached:

$$y = \frac{L}{G}x + y_{\text{out}} - \frac{L_{\text{in}}}{G}x_{\text{in}} \tag{35}$$

The only difference is the lack of any subscripts on the liquid or gas compositions within the tower. Subscripts are used for compositions from different stages in Equation (18).

Values of p (or y) and p_i (or y_i) are read from the operating line and the equilibrium curve, respectively, for different positions in the tower, that is, for different values of p (or y). The composition in the liquid at the interface is the same as the bulk liquid for this case where there is no significant mass transfer resistance in the liquid phase; this will not be true when there is significant resistance in the liquid phase, as will be discussed later. Remember that the concentration in the bulk of each phase is given by the operating line. The concentration in the gas at the interface will be in equilibrium with the liquid phase at the interface, and in this case with the bulk liquid also. That means that the concentration in the gas at the interface can be found from the equilibrium curve directly above the concentration in the bulk liquid. Then for that each position in the tower, the concentration in the bulk liquid will be a position on the operating line, and the concentration in the gas at the interface will be on the equilibrium curve directly below the bulk composition. These two compositions are then used to calculate values for $1/(y-y_i)$ for each value of y. When this expression is plotted as a function of y (Figure 10), the area under the curve between y_{in} and y_{out} is the number of transfer units required for the separation.

A similar expression could have been developed for the liquid phase when the mass transfer resistance in the gas film can be neglected. Then, the number of transfer units and the height of the transfer unit would be based upon resistances and concentration in the liquid phase, and both would have different values. The composition in the bulk liquid would be found on the operating line, and the composition in the liquid at interface would be found on the equilibrium curve at the same gas composition. In other words, the concentration in the liquid at the interface would be found by looking horizontally (at the same value of y) toward the equilibrium curve.

When there is significant mass transfer resistance in the gas and liquid phases, the interfacial compositions cannot be measured experimentally, but they can be calculated if the mass transfer coefficients for both phases are known. Since the interfacial films are so thin, accumulation of solute in the films can be neglected. Then the flux of solute through one film is essentially the same as the flux through the other film.

$$k_g a(p - p_i) = k_l a(c_i - c) = k_y(x_i - x) \tag{36}$$

The mass transfer through the liquid film is illustrated with two different units (moles per liter and mole fraction), concentration, and different

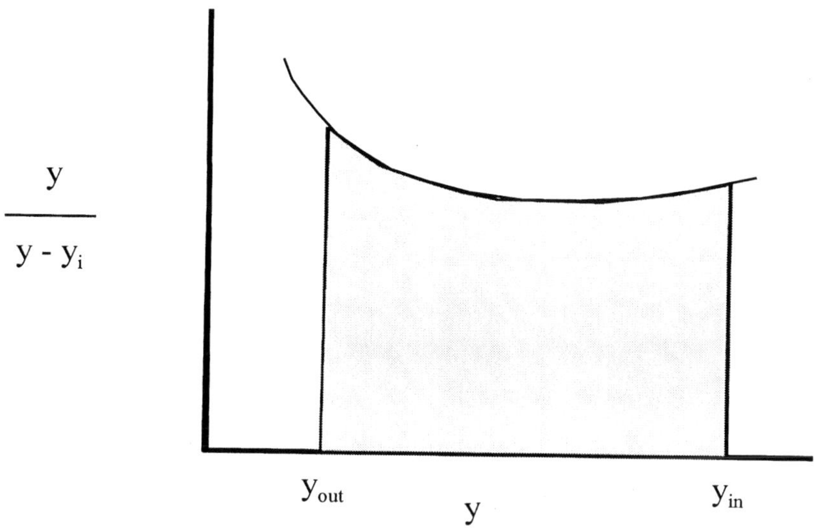

FIGURE 10 Integration to determine number of transfer units (NTU) based upon compositions in the gas phase.

mass transfer coefficients must be used with different units in the driving force. Similarly, different units of concentration could be used in the gas phase. Since the two phases are in equilibrium at the interface, p_i and c_i are related by the equilibrium condition

$$p_i = f(c_i) \tag{37}$$

This function is simply the equilibrium curve discussed earlier. There are two unknowns (p_i and c_i) and two equations [(36) and (37)] to solve to determine the interface concentrations. For example, when the equilibrium is linear,

$$p_i = K_h c_i \tag{38}$$

Then p_i can be eliminated easily from Equation (35), and the equation solved for c_i:

$$c_i = \frac{k_g ap + k_l ac}{k_l a + k_g a K_H} \tag{39}$$

When the equilibrium relation is more complicated, perhaps only available in graphical form, direct solutions for the interfacial concentrations

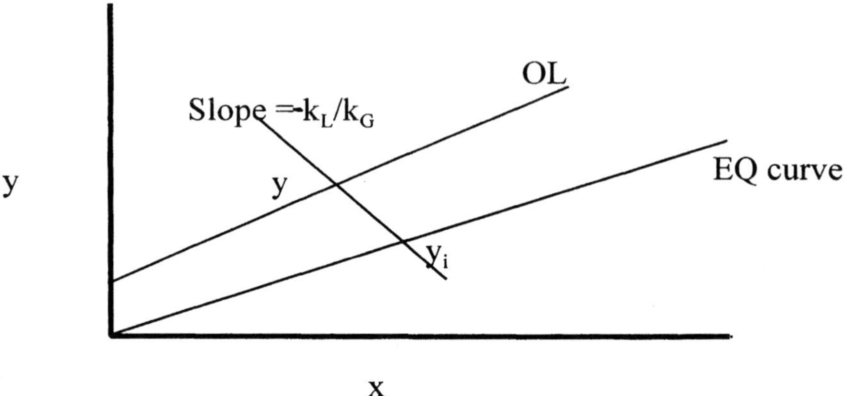

FIGURE 11 Determining the interfacial concentrations with mass transfer resistance in both phases.

may not be so easy, but they can be calculated graphically if the equilibrium curve is available in graphical form.

The graphical solution to the interfacial concentrations depends upon the same equations used above, which equate the flux of solute through each film, Equation (36). The equilibrium curve and operating line are plotted just as they were in the staged analysis (Figure 11). Remember that the operating line gives the concentrations in the bulk streams, and the equilibrium curve gives the concentration at the interface. The problem is to determine which position on the equilibrium curve corresponds to each position on the operating line (i.e., each composition on the operating line).

The concentrations at the interface can be related (connected) to the bulk concentrations given on the operating line by applying Equation (35), which equated the solute flux through the two film resistances. Dividing both sides of Equation (35) by $k_g a(c_i - c)$ gives

$$\frac{p - p_i}{c_i - c} = \frac{k_l a}{k_g a} = \frac{k_l}{k_g} \tag{40}$$

This describes a straight line on a p versus x plot, and the slope of the line is $-k_l a/k_g a$ (Figure 11). Furthermore, the line passes through the bulk compositions c and p and the interface compositions c_i and p_i. Thus, this line connects the bulk composition at any point in the tower with the interface composition at that point. The collection of these lines for different bulk compositions is sometimes called "tie lines" because they tie compositions in the bulk phases to compositions at the interface at the same position in the tower. If the mass transfer coefficients in both phases are known, the compositions at the interface at any point can be determined graphically by drawing a line from the bulk composition (from the operating line) with a slope of $-k_l a/k_g a$ to the equilibrium curve (Figure 11). To determine the interfacial compositions at many (all) points in the tower, one can draw numerous lines with slope $-k_l a/k_g a$. Thus one can evaluate the bulk and interfacial concentrations in either phase at numerous positions along the tower and use those values to evaluate the integral for the NTU over the height of the tower.

Since this graphical approach allows one to determine the interfacial composition, it also gives the driving force at any composition that can be used in the integration for determining the NTU. With tabulated values for the driving force determined graphically, the integration can be done graphically or numerically.

Note that the slopes of the tie lines which connect the bulk concentrations on the operating line with the interfacial concentrations on the equilibrium curve become very steep when $k_l a$ is much larger than $k_g a$. In that case, the driving force is larger in the gas phase; there is a greater difference in the composition in the bulk gas and in the gas at the interface than in the composition in the bulk liquid and in the liquid at the interface. The concentration difference in the liquid phase will be small and possibly even difficult to determine accurately. Conversely when $k_l a$ is much smaller than $k_g a$, the tie lines approach the horizonal, and it will be difficult to determine the driving force in the gas phase accurately.

When the mass transfer resistance is solely in the gas phase, the tie lines are vertical, and the interfacial compositions are read by simply drawing vertical lines from different positions on the operating line. Then the tie lines appear as in Figure 9. On the other hand, when all of the resistance to mass transfer is in the liquid phase, the tie lines are horizontal, and the interfacial compositions are read by drawing horizontal lines from different positions on the operating line. There is no general rule about which phase will contribute the most mass transfer resistance, but one will find a strong trend. In gas absorption systems, the mass transfer resistance is more likely to be largely in the gas phase than in the liquid phase. Although there is no absolute reason why this should always be, there are good reasons to expect this to be so. Absorbents for gases are likely to have a high (or at least moderately high) affinity for the solute, otherwise one would probably select a different absorbent or even a different separation method. Thus, the concentration in the absorbent is likely to be relatively higher than the concentration in the gas phase interface. This usually means that the mass transfer resistance will be largely in the gas phase unless the diffusion coefficients in the liquid are extremely low. Occasionally one may have no alternatives and be forced to use a relatively poor absorbent; then the resistance in the liquid phase is more likely to be a major contributor to the overall mass transfer resistance. In such cases, one may want to consider alternative separation methods such as adsorption.

Similarly, the larger part of the mass transfer resistance is likely to be in the liquid phase when gas stripping is used. This often results because gas stripping is more likely to be used when the solute is relatively volatile. Otherwise, large gas rates would be required. Again there may be cases where the solute is not sufficiently volatile, and the resistance in the gas phase may be a significant portion of the overall resistance. Again, in such cases, it may be wise to consider alternative removal methods such as adsorption.

It is always better to base the analysis upon the phase whose "film" contains more resistance to mass transfer. Although in reality there will always be a finite resistance to mass transfer in both phases, the resistance in one phase can be so much greater than that in the other phase that it is not practical to measure the resistance in the other phase. The effects of having most of the resistance in the other phase on the integral that we call the "number of transfer units" is seen by observing the definition of that integral. Note that the difference in the interface and bulk concentrations in that phase will be very small. In the limiting cases, this difference will be too small to measure accurately. Since the difference is in the denominator within the integral, very small differences in the concentrations will result in very large values for the integral, or the number of transfer units. There is no difficulty from the large value of the integral alone, but with the precision at which the integral can be determined. Although the integral for the NTU is larger for the phase with the least resistance, the height of a transfer unit is corresponding smaller; therefore, the overall tower height could be calculated based upon either phase. The difficulty in basing the number of transfer units on the phase with the least mass transfer resistance results because it can be more difficult to determine the integral (NTU) accurately.

Since there are small errors or uncertainties in all experimental equilibrium curves, the uncertainties could be a significant part of the difference between the equilibrium curve and the operating line. There could be similar uncertainties in the flow rates and the exact slope and location of the operating line. This makes the NTU uncertain but very large. Remember that the very large value for the mass transfer coefficient in one phase will result in a very low value for the height of a transfer unit because the mass transfer coefficient (HTU) appears in the denominator of the expression for the HTU. The total height of the tower, which is (HTU)(NTU), would still be the same as the value obtained by multiplying these terms of the other phase, but the uncertainty in the large value for the NTU makes it more difficult to evaluate using concentrations in the phase with low mass transfer resistance, that is, in the phase with high mass transfer coefficients.

It was easy to discuss in the preceding section how control of the mass transfer rate in each phase depends upon the mass transfer coefficients in each phase. These single phase mass transfer coefficients may not differ greatly from system to system. The other factor that determines the relative contribution of resistance in each phase to the overall mass transfer resistance (and the factor that is more likely to be changed when the solute and/or absorbent are changed) is the equilibrium distribution of the solute at the interface, that is, the equilibrium relationship. Although the effects of the equilibrium curve on the relative contributions of the different phases are very important, the reason is not always obvious. If the solute strongly favors one phase, and the liquid phase absorbents are usually chosen for their affinity for the solute, the higher concentration at the interface provides a higher driving force for absorption. With the higher driving force in one phase, the mass transfer rate may be controlled by the other phase (usually the gas phase for absorption) because the driving force is less. Remember that the driving force in the liquid phase for gas absorption cannot be greater than the solute concentration in the bulk gas because the absorption equilibrium cannot reduce the concentration in the gas at the interface to a value below zero. On the other hand, the driving force in the liquid phase can be much higher if the equilibrium condition places the concentration in the liquid at the interface many-fold higher.

The effect of the equilibrium relation on the relative driving force in the two phases is illustrated in Figure 12. The mass transfer coefficient and the driving force in both phases must be considered when determining which phase (if either) controls the mass transfer rate. If the liquid absorbent has a very high affinity for the solute, the concentration at the liquid side of the interface could be high, perhaps several times higher than that in gas at the interface and even higher than the concentration in the bulk gas. The potential driving force across the liquid film can be increased further by using a solute for which the liquid has a still higher affinity, but the driving force in the gas phase can be affected little by the change of the absorbent liquid since no absorbent can reduce the interfacial concentration at the interface to a value below zero. The dashed line in Figure 12 shows a concentration profile across the two films when a moderately good absorbent is used. The solid curve shows the concentra-

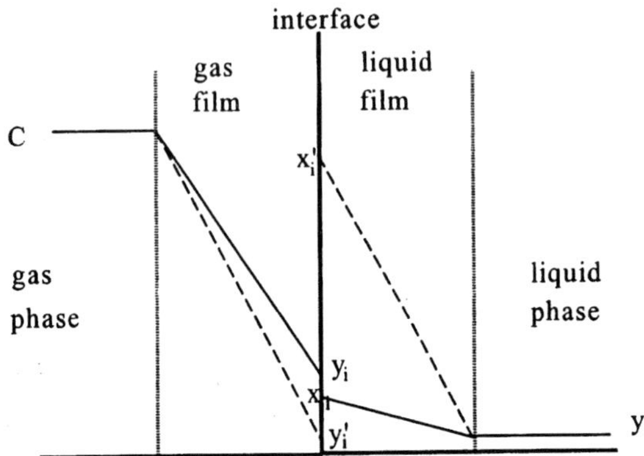

FIGURE 12 Concentration drop across gas and liquid films: importance of equilibrium relationships.

tion profile when a better absorbent is used, and the ratio of the driving forces is changed much more in the liquid film.

Perhaps the best way to understand this effect is to note that the scale on the gas (y or p) axis need not be the same as the scale on the liquid (x or c) axis in Figure 11. Changing the equilibrium relations could be viewed as raising or lowering the equilibrium curve while leaving the operating curve stationary. Changing the scale on the y (or p) axis to reflect the different equilibrium curve means that the actual slope of the tie lines would not look the same because the scales are different. This is another way to illustrate how changes in the equilibrium relation can affect the relative contribution of the resistances in the two phases. It is the precision at which the driving force can be measured that determines the merits of using a particular phase in determining the NTU, and that is affected by the scale of the graph as well as the relative values of the mass transfer coefficients.

Dimensions to Be Used in Transfer Units

Since the driving force and the mass transfer coefficients can be expressed in several units, there are several ways to express the NTU and HTU. Although the NTU has no dimensions, units are implied in the definition, and values for NTU evaluated for one set of units cannot be used with HTU in any other set of units. The HTU has units of length, and proper units must be used. However, the mass transfer coefficient included in the HTU must correspond to the driving force used in the NTU. For instance, the concentrations, and the driving force, in the gas phase can be expressed in terms of mole fraction, partial pressure (any units of pressure such as atmosphere, psig, etc.), or concentration (moles per unit volume). Since there are numerous ways to express driving forces and mass transfer coefficients, there are numerous ways to express NTUs and HTUs. The units in the NTU are important, even if the quantity has no net dimensions.

Estimating Mass Transfer Coefficients

The mass transfer coefficients can be evaluated from experimental data on adsorption or stripping in specific systems by procedures that are essentially the inverse of those described in this chapter for designing adsorption/stripping towers. Most designers of adsorption towers are more concerned with using the resulting mass transfer coefficients than with measuring them. However, it is often beneficial to understand how parameters are obtained. Although it would be desirable to have data for mass transfer coefficients for the same system (packing and fluids), one is more likely to rely upon correlations developed from data on a variety of systems. Flow through towers filled with random (dumped) packing materials or trays is very complex, and there are no exact descriptions of the flow. This alone would make it necessary to rely for the mass transfer coefficients upon experimental data and correlations that are largely empirical.

The most widely accepted correlation for absorption/stripping in random packed columns is that of Onda et al. [16]. The correlation for the mass transfer coefficients in the liquid phase is

$$k_l = 0.005\left(\frac{L}{a\mu_l}\right)^{2/3}\left(\frac{\mu_L}{\rho_l D_l}\right)^{-1/2}(ad_p)^{0.4}\left(\frac{\rho_L}{\mu_l g}\right)^{-1/3} \tag{41}$$

and for the gas phase,

$$k_g = 5.23aD_g\left(\frac{G}{a\mu_g}\right)^{0.7}\left(\frac{\mu_G}{\rho_g D_g}\right)^{1/3}(ad_p)^{-2} \tag{42}$$

The area term, a, in the Reynolds and Peclet number terms is the wetted area per unit volume of tower, and the area term in the ad_p terms is the area per unit volume of packing itself. These terms will differ slightly if the packing is wetted strongly by the liquid.

Although this correlation is widely used, it has received considerable study and is not always as accurate as desired. There can be several reasons for the inaccuracy, but the principal reason may reflect a common problem with many correlations of this type. Generalized correlations are likely to be developed in universities or research laboratories with limited access to large equipment and are thus based largely upon data from smaller equipment. Hence, the correlation is based upon relatively small packing material, but some larger industrial applications may need to use larger packing material to obtain the throughput needed. The use of these correlations can involve extrapolation in at least this one parameter, the packing diameter. It is often desirable to examine the range of parameters used in developing correlations if it is felt that the conditions of the proposed application may be outside the range of common applications.

Djebbar and Narbaitz [17] reviewed the range of parameters used in developing the Onda correlation and assembled a new data base covering a wider range of conditions. They found that the Onda correlation often overestimated the mass transfer coefficients, especially when larger diameter packing materials were used. Overestimating the mass transfer coefficients can result in underdesign of adsorption and stripping towers. Djebbar and Narbaitz suggested the following modification to the Onda correlation:

$$(k_l a)' = \frac{k_l a}{0.834 + 0.293 \ln(d_p)} \tag{43}$$

The mass transfer coefficient without the prime is that calculated by the Onda correlation, and the coefficient with the prime is the corrected value. Note that the corrected coefficients deviate increasingly from the Onda correlation as the packing diameter increases, and the corrected coefficient is lower than the Onda correlation for larger packing materials. Note that this is not a dimensionless correction, and the diameter must be expressed in centimeters.

Djebbar and Narbaitz also discuss the problems reported for overestimating the performance of very long stripping columns [18]. There are several potential reasons for the degradation in the apparent mass transfer coefficient observed in very long towers. Perhaps the most reasonable explanation is poor distribution of the liquid. It may be necessary to provide new liquid distributors periodically along the length of very long (high) towers. However, the reason for degradation in mass transfer coefficients with increasing tower height is still in question. Although it is not possible at this time to explain the phenomenon accurately or to estimate the performance of very high towers accurately, the reader can be warned of the risk in using traditional mass transfer coefficients obtained from the Onda correlation since they may not be accurate for very high towers.

Overall Mass Transfer Coefficients

It is not always practical to determine mass transfer coefficients for both phases; thus, the mass transfer rate is often expressed in terms of a simplified overall mass transfer coefficient, which can be expressed in terms of the liquid or gas phase. For example, for the gas phase the mass transfer rate can be expressed as

$$\text{flux} = K_g a(p - p^*) = K_y a(y - y^*) \tag{44}$$

Here p^* is the pressure of solute that would be in equilibrium with the bulk liquid, and y^* is the mole fraction of gas in equilibrium with the bulk liquid. Thus, in the overall coefficient, the driving force is based on a concentration in equilibrium with the bulk of the other phase, not on the interfacial concentrations. However, if all resistance is in the gas phase (the reference phase in this example), the interfacial concentration (in units of pressure or mole fraction) will be in equilibrium with the liquid, and the overall mass transfer coefficient will be the same as the gas film coefficient; therefore, using overall mass transfer coefficients is the same as assuming that all of the resistance is in one phase.

The height and number of transfer units can be expressed in terms of the overall mass transfer coefficients as well as the individual film coefficients. For instance,

$$L = [\text{HTU}]_{og}[\text{NTU}]_g = \frac{v}{K_g a}\int_{p_{\text{out}}}^{p_{\text{in}}} \frac{dp}{p - p^*} \tag{45}$$

$$= [\text{HTU}]_{ox}[\text{NTU}]_{ox} = \frac{v}{K_x}\int_{x_{\text{out}}}^{x_{\text{in}}} \frac{dx}{x^* - x} \tag{46}$$

where K_x is the overall mass transfer coefficient based upon the liquid phase mole fraction, and x^* is the mole fraction of solute in liquid that is in equilibrium with the bulk gas phase.

The overall mass transfer coefficients can be related to the individual film coefficients when the equilibrium curve is linear. By solving for the interfacial concentration(s) and noting that the flux is the same whether expressed as an overall rate expression or as an individual film expression, one can show that

$$\frac{1}{K_x} = \frac{1}{k_x} + \frac{1}{mk_y} \tag{47}$$

Here m is the slope of the y versus x equilibrium curve. Similar expressions can be derived for linear equilibrium curves in any set of units. In this case it is particularly easy to see the relative contributions of each film to the mass transfer resistance because each term on the right-hand side of

Equation (47) represents the resistance in one film. The role of the equilibrium relation (expressed here as the slope of the equilibrium curve, m) in determining the relative contribution of each film to the overall mass transfer resistance is also evident: increasing m decreases the relative contribution of the gas phase film resistance, whereas decreasing m increases the relative importance of the gas phase film.

Mass Transfer Coefficients in Concentrated Systems

Less has been said here about concentrated systems than in most textbooks on mass transfer and separation processes in general. This reflects the higher probability that environmental problems involve removal of dilute contaminants, but that is not always the case. As noted in the discussion of staged systems, since the flow of solute-free liquid and solute-free gas is constant over the length of the tower, the operating line is linear when expressed in terms of moles of solute per mole of solute-free gas or liquid, X and Y. (Similarly one could have a linear operating line by working in terms of mass or mass of solute per mass of solute-free liquid and gas.) The rate expressions can also be written in terms of X and Y. For instance,

$$\text{rate} = K_X a(X^* - X) = K_Y a(Y - Y^*) \tag{48}$$

In the more general case, the NTU_{og} can be obtained graphically. When the gas is the reference phase, Y^* is located directly below Y, X on the equilibrium curve on the McCabe-Thiele plot because Y^* is in equilibrium with the bulk liquid composition X. Remember that the bulk compositions are shown on the operating line, so the overall driving force for mass transfer at any point in the tower is simply the vertical difference between the operating line and the equilibrium line. Then NTU_{og} can be determined by graphical integration. Similarly, if the overall NTU_{og} were to be determined based upon the liquid phase composition, the overall driving force at any position in the tower (or at any concentration in the tower) would be the horizontal distance between the operating line and the equilibrium curve.

However, these concentration units are not customary, and using this driving force is not consistent with Fick diffusion. One must be careful because there could be more variation in these mass transfer coefficients with concentration. However, convection to or from the interface can be more important than small variations in the mass transfer coefficients. Very high fluxes of solute to (from) the interface can result in significant net flow (the solute diffusion rate) toward (from) the interface, and this equation has to be modified, especially for the gas phase where the fluid densities are low enough that velocities are more likely to be significant. If the rate expressions based on solute-free compositions describe the mass transfer rates satisfactorily (the mass transfer coefficients do not vary significantly with composition or position in the tower), the NTU and HTU can be defined and used satisfactorily in these units.

Related expressions for HTU and NTU can also be derived from these rate expressions for the simpler case of linear isotherms. The expression for NTU_{og} with a linear equilibrium curve is [19]

$$NTU_g = \log\left\{\frac{y_{in} - mx_{in}}{y_{out} - mx_{in}}\left[1 - \frac{1}{A}\right] + \frac{1}{A}\right\} \Big/ \left(1 - \frac{1}{A}\right) \tag{49}$$

where A is the absorption factor, L/mG. Equation (49) is similar to Equation (20) for the number of stages for systems with linear isotherms. A similar expression can be written for NTU_x, NTU_{oY}, or NTU_{oX}. For concentrated systems, however, the gas and liquid velocities will not be constant if the number of moles (or mass) of solute transferred between phases is a significant fraction of the flow rate of that phase. This means that the velocity cannot be kept outside of the integration for NTU.

Comparison of Stages and Transfer Units

The reader should recognize that the use of stages is quite similar to the use of transfer units since they are the results of two conceptual approaches to the same problem. The concept of a stage is usually well understood because it can be constructed in a vessel where the two phases are mixed vigorously enough that the phases approach very close to equilibrium. Of course, in a countercurrent tower with random or structured packing there is no region where the two phases are mixed in such a manner, so the height equivalent to a theoretical stage is simply the height of the tower over which the separation that takes place is the same as that from a single stage. For a given set of operating conditions, this is equivalent to specifying a portion of the tower over which the change in the composition of one phase is the same as the change of composition that occurs in a single stage with those operating conditions (flow rates, compositions, and equilibrium).

More people probably have difficult visualizing transfer units because there is no simple equivalent "model" for a transfer unit that corresponds to the well-mixed tank that we can use to describe a theoretical stage. However, since there are no "mixed tanks" in packed towers, it probably should be just as difficult to see how the theoretical stage concept can be applied to packed towers. The number of transfer units is best described simply as integrals as in Equation (34), and the height of a transfer unit is calculated directly from the fluid velocity in the reference phase and the mass transfer coefficient. Physically, one can think of the number of transfer units as simply one way to express the amount of separation that occurs in the tower. The important point is that the NTU will be proportional to the tower height. The height of a transfer unit is the height of the tower divided by the number of the transfer units that occur over the tower.

For those who prefer a more physical description of a transfer unit, it may be helpful to think of a single transfer unit in the following manner. The integral in Equation (34) contains the change in concentration in the "variable" (dC, dy, or dp) and the driving force for the mass transfer in the denominator. For a portion of a tower to have one transfer unit, the value of the integral has to be unity. That means the change in concentration in the reference phase (change in the numerator) has to be equal to the average value of the driving force during that change. Then one can think of a single transfer unit as the height of the tower over which the change in the concentration in the reference phase is equal to the average driving force over that height. However, this author still prefers to think of both transfer units and stages as dimensionless lengths (or tower heights) without relying upon physical descriptions. Multiplying the dimensionless quantities by the appropriate factor (height of a transfer unit or height of a stage) converts the dimensionless lengths into real dimensions of length.

In the special case where the equilibrium curve and operating line are parallel, the driving force is constant over the entire tower. Then the concentration changes in each transfer unit in the tower (or at any position in the tower) are all the same. The concentration change over each stage is also the same for any region of the tower (Figure 13). Since the driving force (the distance from the operating line to the equilibrium curve in either the vertical or horizontal direction, depending upon which phase controls the mass transfer resistance) is constant, the denominator in the integral for the NTU is a constant. Each transfer unit in the tower corresponds to a change in concentration that is equal to this driving force. This means that the change in concentration for each NTU is the same as the change in concentration for each stage, and then the number of stages required for a given separation is the same as the number of transfer units required.

What happens when the operating line and equilibrium curve are not parallel? The number of transfer units does not equal the number of stages. Consider the extreme case where the solute strongly favors the liquid; the equilibrium curve then lies essentially on the horizontal axis ($y = 0$). The resistance would have to be in the gas phase for gas absorption, and the tie lines would be vertical lines to the x-axis.

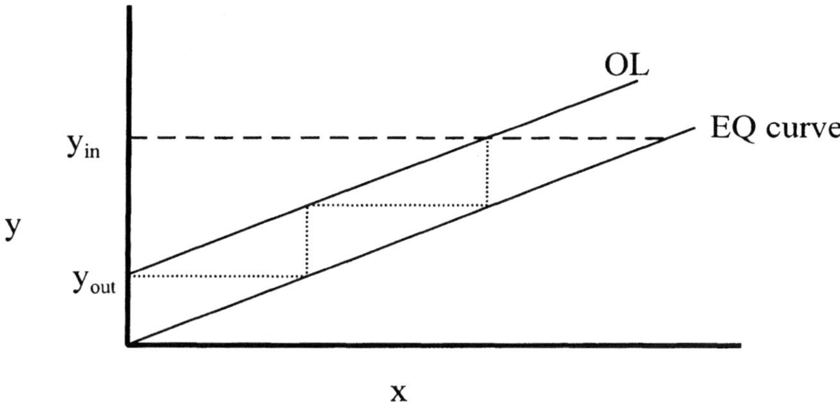

FIGURE 13 Case when operating line and equilibrium curve are parallel, when the number of stages and the NTU can be equal.

Integrating Equation (34) would be simple since y_i is always zero, and the NTU becomes simply $\int dy/y$ or $\log_e(y_{in}/y_{out})$. An infinite number of transfer units (or an infinitely tall tower) is required to reduce the concentration of solute in the gas to zero. However, one equilibrium stage would take the concentration in the gas to zero. In this example the number of stages (one) and the number of transfer units (infinite) are especially different, and this is discussed in more detail in the next section. It should be apparent that one should not attempt to use a stage analysis to design towers with irreversible isotherms. Actually one should consider using transfer unit analysis for any system where the slopes of the equilibrium curve and the operating line differ greatly.

Although the number of transfer units and the number of stages are not necessarily the same under other conditions (other than the straight and parallel equilibrium curve and operating line), both approaches are commonly used. Which is better? Although there is probably some debate about this point, the author generally prefers the NTU approach because it is a more realistic (but not exact) representation of the mass transfer processes taking place in packed towers. However, the stage approach is still frequently used for two reasons. The first is that in most situations in absorption and similar countercurrent operations, the two approaches give similar results, because most towers are operated with the equilibrium curve and the operating line approximately parallel. Of course, if the equilibrium curve is not linear, a line is parallel to a curve only at a single point, or at a few points, not over the entire length of the line.

In optimizing a tower design, the total cost is infinite when the minimum liquid rate (minimum gas rate) is used for adsorption (for stripping) because the tower height would be infinite. On the other hand, tower cost is infinite if the liquid rate is infinite for absorption (gas rate is infinite for stripping). This requires a tower with an infinite diameter—very costly! In optimizing the cost between these extremes, the operating line that spreads the concentration changes approximately evenly over the tower stages is likely to provide the lowest capital and operating costs. This operating line is approximately parallel to the equilibrium curve.

The second reason why the stage approach is used frequently is that it is simpler. This is not a major advantage for most simple single component absorption or stripping systems or when computer design procedures are used, but it can be very important for systems with more than one component absorbing or being stripped at concentrations high enough that the concentration of one component affects the equilibrium curve for the other component.

If the stage approach is not adequate, then definitely try to use the transfer unit approach. Generally, use this approach when the operating line cannot be approximately parallel to the equilibrium curve or when it is not optimal to have them made approximately parallel. Odd-shaped equilibrium curves with high curvature may be one set of circumstances, but, because dilute systems are used in so many environmental and waste treatment operations, this situation may not be common. Perhaps a common case found in environmental systems is that mentioned earlier when absorption is essentially irreversible. Therefore, the equilibrium concentration in the gas is essentially zero, and one stage of adsorption corresponds to essentially complete absorption, and that would require an infinitely tall tower. An infinite gas rate or a zero liquid rate would be needed to make the operating line parallel to the equilibrium curve. In this extreme case, a stage approach will not even yield useful information; a transfer unit approach must be used. Although the irreversible isotherm is a special case, it is approximated sufficiently often in environmental situations that it deserves some further discussion.

Irreversible Isotherms

To achieve extremely high removal efficiencies, it is common in environmental problems to select absorbents that absorb the solute strongly. This may involve chemical reactions between the solute and some component in the absorbent liquid. If it is desirable to recover the solute (perhaps by stripping at a higher temperature), it may not be desirable to have such a strong absorbent since it will make recovery by stripping very difficult or, in the limit, impossible unless the strong affinity of the absorbent for the solute can be changed by other methods, such as oxidation, reduction, or pH changes. However, many pollutants have little or no value, and removal (not recovery) is the principal goal. Then an "irreversible" absorbent may be attractive if the consumption of the absorbent reagent is acceptable.

If the pollutant is an acid (SO_x, NO_x, etc.) or base, the absorption equilibrium can often be made approximately irreversible by using a solution of base or acid which would neutralize the volatile acid (or base) pollutant into a nonvolatile salt. In other cases, a component that reacts with the pollutant to form an insoluble compound can force the equilibrium curve to become essentially irreversible. However, when solid precipitates are formed (such as the neutralization of SO_x by lime), this can interfere with the operation of most packed towers and tray towers.

An irreversible equilibrium absorption curve lies along the horizontal x-axis. Hence, the "equilibrium" concentration of solute in the gas for any liquid composition is essentially zero. Not only does this make high removal efficiencies easier to obtain (fewer NTUs are required), but the NTU analysis is simplified. The simplification is evident from the driving force for mass transfer. The concentration of solute in the gas in equilibrium with the liquid (of any composition) is zero. Then the overall driving force is simply y (or p). For the dilute case, the integration is simply

$$\mathrm{NTU}_y = \int_{y_{\mathrm{out}}}^{y_{\mathrm{in}}} \frac{dy}{y} = \log_e\left(\frac{y_{\mathrm{in}}}{y_{\mathrm{out}}}\right) \tag{50}$$

The solute concentration decreases logarithmically with NTU, or tower height, as one observes for a "first order" process. This means that each section of the tower removes the same fraction of the remaining solute.

With an irreversible equilibrium curve, one should not attempt to design the tower on the basis of stages. With the equilibrium concentration in the gas always zero, a stage would always remove all of the solute since the gas stream leaving a stage is defined as in equilibrium with the liquid leaving the stage. This is also evident from the McCabe-Thiele plot where a single "step" for a stage will take the gas composition to zero. Such an analysis provides little insight and design information; for quantitative analysis of an absorption tower performance with irreversible equilibrium, the transfer unit approach should be used. When the isotherm is irreversible, the axial concentration gradients in the tower can be very high if the mass transfer coefficients are high; the concentration driving forces can be relatively high. High axial concentration gradients can make axial diffusion or dispersion more important; so one may find that the performance may not be quite as good as expected based upon the analysis just described, which does not take axial dispersion into account. Axial dispersion describes molecular dispersion in the axial direction, eddy mixing in the axial direction, and nonuniform flow distribution in the tower. Although the molecular diffusion coefficient is known or measured separately, eddy mixing and molecular diffusion are usually measured together and their effects not separated.

This is an extreme example of a common behavior. If the slopes of the operating line and the equilibrium curve are approximately equal, there usually is little merit in using the NTU approach over the stage approach. However, as the ratio of the two slopes becomes further removed from unity, the two approaches give increasingly different results, and the NTU approach is recommended. For the irreversible system, the ratio of the two slopes is infinity.

Gas Stripping

Stripping operations are frequently used to remove or recover the solute (often a pollutant) from the liquid absorbent. Gas stripping can be used to remove a contaminant, usually a volatile organic compound (VOC), from a liquid process or effluent stream. Gas stripping equipment is much like absorption equipment. Packed towers are particularly effective, but staged, or even open, spray towers (and possibly bubble towers) can be used. Some stripping operations with groundwater may even be carried out in situ (that is, in the ground). In situ operations may, in principle, be simplified stripping operations, but since the concentrations and flow patterns can be complex, reliable design of such systems is often difficult and may require important assumptions about the rock structure, the contaminant distribution, and the hydraulics in situ.

Stripping towers usually look much like adsorption towers, and the mathematical description of stripping operations can be the same as in Figures 5 and 9. Actually, the material balance equations around the upper portion (or the lower portion) of the stripping tower can be the same as the material balance shown for absorption, but the conditions of the inlet streams are different. In stripping, the inlet gas stream has zero or very low concentrations of the solute, but the inlet liquid stream entering the top of the tower has a relatively high concentration. This is the opposite for the conditions for gas absorption. Although the equation for the operating line is the same as that used for absorption, the inlet conditions (concentrations) place the operating line below the equilibrium line rather than above it (Figure 14). Note that for gas stripping, the equilibrium curve lies above the operating line. The higher concentrations in gas stripping towers occur at the top, that is, at the liquid inlet. This

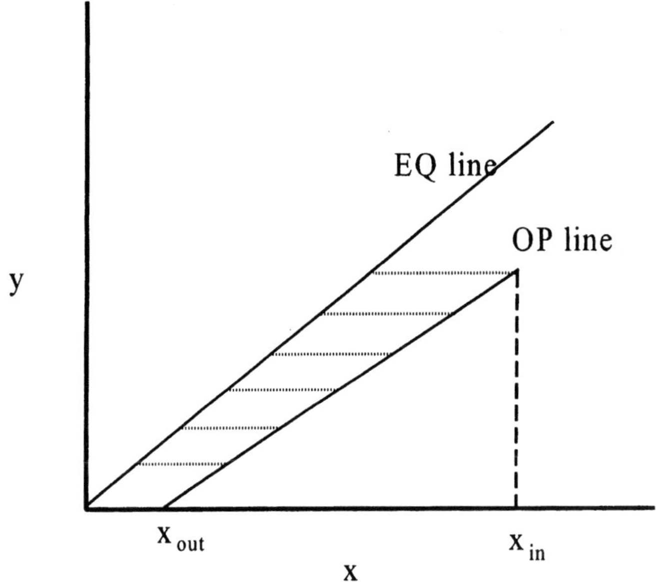

FIGURE 14 Operating line and equilibrium curve for gas stripping and tie lines for mass transfer resistance in the liquid phase.

means that the "top of the tower" is shown in the upper right portion of Figure 14. Horizontal tie lines are also shown for the case where mass transfer control is in the liquid phase. Although in principle either phase can control the mass transfer rate, the liquid film is more likely to be the more important resistance for gas stripping and the gas film to be the more important resistance in gas absorption.

The conditions at the bottom of the tower, where the concentrations in both the liquid and the gas are lowest, are shown on the left side of Figure 14 near the bottom. Conversely, the conditions at the top of the tower, where the concentrations in both phases are the highest, are shown the right side near the top. This situation is opposite that for absorption. In absorption, the lowest concentrations are at the top of the tower, and those conditions are shown on the lower left portion of Figures 6 and 8. The conditions at the bottom of the absorption tower are shown in the upper right portion of Figures 6 and 8. In McCabe-Thiele diagrams, the operating lines lie below the equilibrium curve in gas stripping systems, but the operating lines are above the equilibrium curve in absorption operations.

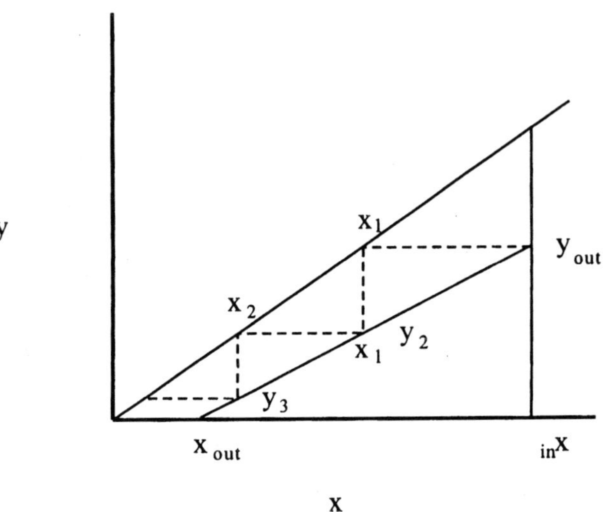

FIGURE 15 McCabe-Thiele diagram for gas stripping.

The number of stages or the number of transfer units is calculated for a stripping operation from Figure 15 in the same manner that they were calculated for absorption in Figure 6, but there is a reversal in the position of the equilibrium curve and the operating line. Although the similarity of the stage-by-stage calculations may be obvious, an example calculation is given in Figure 15. A similar graphical representation of driving force calculations was given in Figure 14. The driving force again is divided into film resistances on both sides of the interface. Calculations for stages, transfer units (NTU) based upon resistances in individual phases, or overall transfer units are calculated in the same manner described for absorption, except the tie lines extend to the equilibrium curve above, rather than below, the operating line. Figure 14 illustrates dilute systems where the operating line is straight in y versus x plots; for concentrated systems it may be desirable to use a plot of Y versus X for concentrated systems to keep the operating line straight and thus easy to plot (and replot for different flow rates). However, there seem to be fewer gas stripping cases than gas absorption cases where the use of concentrated systems with changing phase flow rates need to be taken into account.

Just as there was a minimum liquid absorbent rate required to achieve a given degree of absorption, there is a minimum stripping gas rate required to achieve a given degree of solute removal from the liquid. The number of stages or transfer units required for the separation will decrease as the gas rate is increased above this minimum rate. Higher gas rates reduce the slope of the operating line and thus move the operating line farther below the equilibrium curve, and, as noted for gas absorption, greater distances between the operating line and the equilibrium curve result in larger changes in the concentration in each stage and few stages required for a specified removal rate.

Steam Stripping

In some cases, it is practical to use steam as the stripping agent when a contaminant is being removed from water. Obviously, this raises the temperature of the water to the boiling point at the pressure of the steam stripper; so it is most practical to use when heat is needed to achieve sufficient volatility of the contaminant to have acceptable stripping rates. Steam stripping operations are more likely to be selected for large throughput systems than for small systems. Steam stripping resembles distillation in many ways, and a more detailed discussion is given in the chapter on distillation. If the volatility of the contaminant is not significantly greater than the volatility of water, a large quantity of water will be removed, and it may be necessary to add a reflux system and multiple stages as discussed in the chapter on distillation. Such operations are usually called fractional distillation with steam injection.

Cross-Flow Stripping

Cross-flow operations have been used in gas stripping when large reductions in the contaminant are needed in a few stages and when it is not necessary to restrict the use of strip gas or maximize the concentration of contaminant in the strip air [20]. There are cases where a contaminant such as a nontoxic hydrocarbon is removed with air, and the air can be discharged into the atmosphere without removing the hydrocarbon. Cross-flow operations could be used instead of countercurrent operations with many separation methods in addition to gas stripping. Cross-flow operations can remove larger fractions of the solute in a specified number of stages, but at the expense of larger gas flow rates required for the same solute removal and lower average concentrations of the solute in the strip gas. Because the stripping reagent, usually air, is relatively inexpensive, cross-flow is more likely to be attractive for gas stripping than for gas absorption, liquid-liquid extraction, or other separation methods that use more costly removal agents. Because the concentration of solute in the gas is lower for cross-flow, it is important to consider what treatment of the gas stream is needed after it is used in the stripping operation. When

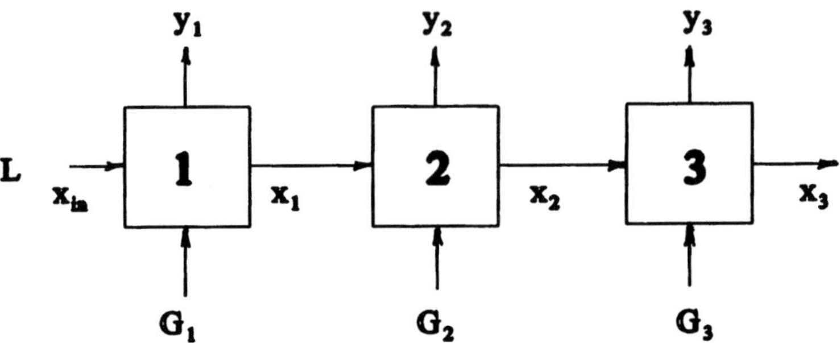

FIGURE 16 Cross-flow operations.

the solute is a nontoxic hydrocarbon or other contaminant that can be discharged directly to the atmosphere, cross-flow operations may be very attractive; when the solute has to be recovered from the gas with high efficiency, it may be less attractive.

Cross-flow operations are illustrated in Figure 16 with three stages. Fresh gas is introduced into each stage. The removal achieved with cross-flow stripping can be very high since the liquid is repeatedly contacted with fresh gas, but the concentration of solute that can be reached in the gas phase is not as high as one could achieve with countercurrent operations. As noted, such operations are more likely to be attractive when the solute can be discharged to the atmosphere or when the contaminant can be easily removed from the air stream at low concentrations. For example, the solute contaminant might be destroyed chemically in the air stream, or a contaminant could discharge into the air but not be acceptable in the aqueous stream.

The equivalent cross-flow in an absorption operation involves contact of the gas with fresh liquid in each stage. Although cross-flow operations give high removal efficiencies from the gas phase, they are not likely to be considered for absorption operations because the volume of absorbent liquid needed is large, and recovery of the contaminant and regeneration of the absorbent are more costly. Air for gas stripping is inexpensive, but absorbents are likely to be costly.

Single Stage Stripping

Before describing multistage cross-flow operations it will be helpful to look at single stage operations; multistage operations, countercurrent or cross-flow, consist of multiple applications of single stages. Although there are obvious merits in performance for using multistage systems, single stage stripping is still practiced. In such cases, it may simply be impractical to construct a multistage system. Some examples of interest are one-time stripping of a pollutant from a tank or from an aquifer by pumping air through the water. The single stage operation requires more gas (usually air) than a multistage operation, and the concentration of the solute in the gas leaving the stripper is lower than that from a multistage system. The higher flow rate and the lower concentration make it more difficult to recover the solute from the gas in another operation, such as condensation or adsorption. For single stage batch stripping to be attractive, the solute usually must be easily released or destroyed. Nevertheless, a one-time or infrequent stripping of a solute from a liquid in a tank may be practical by simple batch stripping. If a pollutant is to be stripped from an aquifer in situ, it may not be possible to use a multistage system.

As simple as batch stripping operations seem, the mathematics may be little simpler than multistage stripping, principally because batch stripping is often not a steady-state operation. The concentration in the tank starts at one value and decreases with time as the stripping continues. The rate of solute removal from a stage is given by a material balance between the solute in the stage and the concentration in the gas leaving the stage:

$$V_l\left(\frac{dx}{dt}\right) = -Q_g(y_g - y_{g0}) \tag{51}$$

where V_l is the moles of liquid in the stage (tank), x is the mole fraction of solute in the liquid, y_g is the mole fraction of solute in the gas leaving the stage, y_{g0} is the mole fraction of solute in the gas entering the stage (usually near zero), and Q_g is the molar gas flow rate through the stage.

If the stripping vessel is a perfect stage, the gas leaving the vessel will be in equilibrium with the liquid remaining in the vessel. Then y_g is a function of x. The time required for the concentration of the solute to reach a concentration x in the vessel is obtained by integrating Equation (51):

$$t = \int_0^t dt = -\frac{V_l}{Q_g}\int_0^x \frac{dx}{y_g - y_{g0}} \tag{52}$$

In the general case, the integration can be made graphically from equilibrium data such as Figure 17. For each value of x between the initial value and the desired lower final value, the value of y_g can be obtained from the graph. Then a plot of $1/(y_g - y_{g0})$ versus x can be integrated to give the desired integral and the time, t. In many cases, there will be no solute in the initial gas, so y_{g0} is zero.

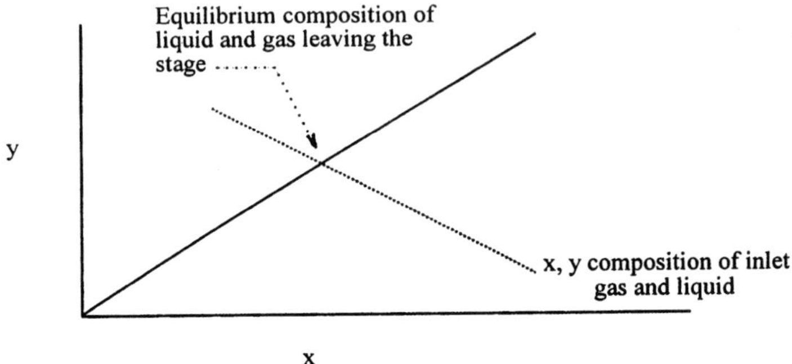

FIGURE 17 Liquid and gas compositions during single stage batch gas stripping.

For the special case where the equilibrium curve is linear and passes through the origin (Henry's law solutions), the integral can be evaluated analytically:

$$y_g = Hx \tag{53}$$

Then the integral becomes

$$t = \frac{V_l}{Q_g} \int \frac{dx}{Hx - y_{g0}} \tag{54}$$

$$= \frac{V_l}{Q_g} \left[\frac{1}{H}\right] \log_e \left[\frac{Hx_{\text{initial}} - y_{g0}}{Hx_{\text{final}} - y_{g0}}\right] \tag{55}$$

This gives a logarithmic decrease in the concentration of solute in the stripping vessel. Since so many systems obey Henry's law in dilute systems, the relation is often adopted, but care should be used in assuming the exponential decrease in concentration without knowing the equilibrium relation. Even if Henry's law cannot be assumed throughout the process, it may be possible to model the last part of the process with the linear equilibrium curve (Henry's law).

For steady-state single stage operations, liquid is fed continuously to the vessel. In this case, the material balance is

$$L(x_{\text{in}} - x) = Q_g(y - y_{g0})$$

This equation can be solved for x if x is the specified parameter:

$$x = x_{\text{in}} + \frac{Q_g}{L}(y - y_{g0})$$

If the stripping vessel operates as a perfect stage, the effluent gas will be in equilibrium with the effluent liquid, so y will be in equilibrium with

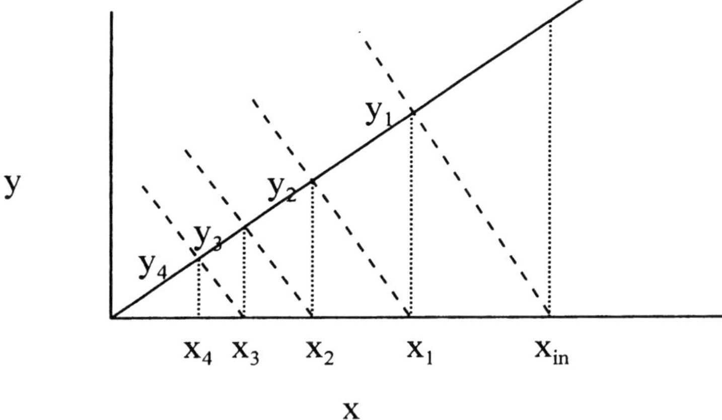

FIGURE 18 Liquid and gas compositions in multistage cross-flow gas stripping operations.

x. If one specifies the liquid and gas flow rates, the values of x and y can be determined from this equation. Note that x and y line on a line with slope Q/L that goes through the point (x_{in}, y_{g0}) with slope $-L/Q$. Since x and y also must lie on the equilibrium curve, the liquid and gas effluent concentrations must lie at the intersection of that line with the equilibrium curve. The operating line for a single stage is shown in Figure 17.

To make calculations for multiple stages of cross-current gas stripping, the concentration of the liquid feed to the second stage will be the effluent liquid effluent from the first stage. Then a line of slope $-L/Q$ can be drawn from the point (x_1, y_{g0}) to the operating line. Here the liquid feed to the second stage is designated x_1 to note that it comes from the first stage. This composition was given no subscript when discussing a single stage, but does need a subscript when different stages are to be used. The same procedure can be followed for any number of stages (Figure 18). In the figure the inlet gas contains no solute, so the dashed lines proceed from the x-axis. If the inlet gas contained some solute, the dashed lines would begin at a point on the dotted lines above the x-axis that correspond to the concentration of solute in the feed gas. Although the figure and the discussion have assumed that the concentration in the gas and the gas flow rate are the same for all stages, those assumptions are not necessary. One could use different values for Q or for y_{g0} for each stage. That would simply mean that the slopes of the dashed lines would be different if the gas rates, Q, were different for different stages. If different concentration of solute were present in the feed gas from different stages, the dashed lines would begin at different positions on the dotted lines above the x-axis.

APPLICATIONS OF ABSORPTION AND STRIPPING

When to Consider Absorption

The choices of absorption over adsorption or other separation methods for removing pollutants from gaseous (almost always air) exhaust streams depend on several factors. Obviously, one must identify a suitable absorbent before absorption can be considered. A few general guides can assist in determining which approach is more likely to be satisfactory for given applications.

First, there is the question of the concentration of the pollutant and the degree of removal required. Usually, absorption methods are more likely to be practical for removing moderately high concentrations of a component from a gas stream. This is a matter of cost since many absorbents would also remove components even at moderate to low concentrations. The practical issue arises because at sufficiently low concentrations an adsorbent bed may become more cost effective. Because absorbents are liquids operating in continuous systems, regeneration of the absorbent (often by gas stripping) can also be done continuously using fluids and usually at low to moderate cost. Regeneration of solid adsorbents, however, is a non-steady-state operation and is likely to be more costly. There is no definite concentration where one method will be more practical. The choice depends upon the capacity of the adsorbent, the ease of regeneration, the properties of the absorbent, and the ease of recovering the pollutant from the liquid absorbents. When extensive treatment is needed for operations such as regeneration, there are usually practical reasons to prefer treating fluids (liquid absorbents) rather than solids.

Because the packed beds usually used in adsorption operations provide so much surface area and a good approximation of plug flow, adsorption is often capable of extremely high removal efficiencies. This means that packed adsorption columns are capable of being designed with large numbers of equilibrium stages or transfer units. Absorption towers, on the other hand, are likely to have larger stage heights, and extremely high numbers of absorption or stripping stages may be costly to build and operate.

When it is necessary to remove a pollutant at high concentrations and reduce the concentration with high efficiency, it may be wise to consider using both absorption and adsorption operations in series with the absorber removing most of the pollutant and reducing the concentration until adsorption becomes more practical for removing the remainder of the pollutant. This may be a particularly attractive option when the absorbent is not water and it has significant vapor pressure. In such cases it is likely to be necessary to remove traces of the absorbent as well as traces of the solute from the exhaust stream before releasing it.

Some facilities prefer to use single operations to treat a stream and may be reluctant to consider using two separation steps for a single problem. Some of this reluctance may result simply from the anticipated additional cost of two operation, but there can be additional reasons. Some separation processes, including gas absorption, could add traces of another component (the absorbent) to the gas stream and complicate the second separation step. Of course, condensers or other devices can be used to remove most or essentially all of such components, but that can be viewed as a third treatment step. All or most of these arguments have merit, and the designer/planner has to take such questions into account when evaluating the merits of using two separation steps operating only in the concentration range in which they are most effective.

When to Consider Stripping

In environmental applications, stripping may be used to remove organic compounds from water discharge streams when the organic compounds have sufficient volatility, and the volatility, usually expressed as Henry's law constant, is often the first parameter of importance in evaluating the merits of gas stripping applications. The organic compounds could be hydrocarbon compounds which enter surface waters or groundwaters after oil spills (not all components from petroleum will be sufficiently volatile for most stripping operations) or an organic solvent such as trichloroethylene that has been washed from a factory floor or leached into wastewater or groundwater from a spill, an old landfill, or other waste disposal site. These environmental problems are common. Stripping operations have become an important component of several "pump-and-treat" operations to remediate groundwater from volatile organic contaminants. (The wisdom of many pump-and-treat operations is being questioned when the sources of the VOCs are "pools" of highly insoluble solvents that can continue to saturate the surrounding groundwater for years, hundreds of years, or perhaps even longer. This problem does not reflect on a limitation in gas stripping, but it reflects the difficulty in getting the bulk of the volatile contaminant to the gas stripping equipment because of its extremely low solubility in water.)

Strippers are also often associated with absorbers. At a higher temperature, stripping is one way to regenerate an absorbent and recover the pollutant for concentration or reuse. An absorbent that is highly effective at a relatively low temperature, usually near ambient, may be easily stripped at a somewhat higher temperature. Other methods are also used to recover pollutants from absorbents, but stripping is certainly one of the most common methods.

In a few environmental applications, the gases "stripped" from groundwater or wastewater may be discharged to the atmosphere, but in most cases an additional operation will be needed to "recover" the toxic components of the stripped gases. The "recovery" operation must be considered in evaluating potential applications of gas stripping. Adsorption of the stripped contaminant on carbon beds is commonly used to recover VOCs after they are stripped from groundwater or wastewater. In some cases, it may be possible to recover the bulk of the contaminant by condensation and use adsorption to recover the last traces of the contaminant. The ability to use condensation depends upon the concentration of the contaminant in the strip gas and the volatility (vapor pressure) of the contaminant. If the concentration of the contaminant in the gas from the stripping operation is greater than the vapor pressure of the contaminant at a reasonable condensation temperature, it will be possible to collect a part (perhaps most) of the contaminant in a condenser. The concentration in the gas leaving the condenser should approach the vapor pressure of the contaminant at the temperature of the condenser. If it is necessary to further reduce the concentration of the contaminant, adsorption or another secondary removal method may be required.

Membrane processes have also been used to recover contaminants after gas stripping. It is also possible to destroy the toxic components by sending the strip gas to a biological treatment facility or to an incinerator to destroy the toxic materials. (However, it has been difficult to obtain approval for incinerators that handle any significant quantities of toxic materials in the United States.)

In some applications, it may not be necessary to remove the contaminant to discharge concentrations. For instance, if condensation or a membrane system removed a large fraction of the contaminant but does not reduce the concentration to one suitable for discharge, it may be possible to avoid further treatment by reusing the gas and recycling it to the stripping operation. Not only does recycling of the strip gas relieve the possible difficulties of removing contaminant to allowable discharge levels, it can also reduce the actual amount of contaminant discharged to the atmosphere. By recycling the strip gas, there is no inherent discharge of contaminant to the atmosphere, but that only means that the release of contaminant can be reduced to that carried in the accidental small leakage of strip gas that can occur throughout the system. Of course, such leaks should be minimized, but the magnitude of the release is less when the leaks occur between the condenser/membrane or other gas treatment and the recycle to the gas stripping equipment. Leaks that occur between the stripping operation and the gas treatment operation would carry a higher concentration of the contaminant.

Example Applications of Absorption

Absorption can be carried out on a grand scale or in small devices associated with experiments. The removal of sulfur oxides from flue gas produced in large coal-fired electric power generation plants is one example of adsorption carried out on a particularly large scale. An electric utility burning high sulfur coal must reduce the sulfur emissions from those facilities, and absorption processes have been used in such applications. The absorbent is usually an aqueous solution or suspension of alkali or alkaline earth compounds (carbonates or hydroxides). The sulfur oxides are acid, and the alkaline absorbent liquid neutralizes the acid and forms sulfate or sulfite compounds with very low volatility. Such neutralization reactions may be essentially irreversible and thus give very favorable equilibrium conditions. When the neutralizing compound is dissolved in the water and the products of the neutralizing reactions are all soluble in the water, the resulting process is relatively straightforward absorption. However, when all or most of the neutralizing compound is insoluble, the process becomes an interesting combination of absorption and adsorption. The soluble gas, in this case sulfur oxides, is "absorbed" into the water, but it may then react with a solid such as suspended particles of calcium carbonate, and that last step is more like an adsorption (chemisorption) step.

Nitrogen oxides can also be removed in alkaline solutions, but the solubility of nitric oxide is so low that the rates of absorption are not practical. Including a reductant such as Fe(II) chelated with EDTA increases the NO_x somewhat, but the removal of SO_x is much greater [21]. The iron can be oxidized to Fe(III) by other components without removing NO_x. Shi et al. have suggested that a different chelating agent (2,3-dimercapto-1-propanesolfanate) is more effective in protecting the ferrous iron from oxidation by air, and better than 50% NO_x removal was achieved in 130 cm long laboratory columns [22]. Oxidation of NO by chromate to nitrate can improve NO removal [23]. The Cr(III) can be oxidized back to chromate and reused.

Absorbent fluids have been proposed, tested, and installed at electric power generation stations. These are essentially all solutions of alkali or alkaline earth compounds. A suspension of limestone, calcium carbonate, is a low cost material to use, but such systems often are not able to make efficient use of the calcium carbonate and leave some of it unreacted because the calcium sulfate and sulfite formed on the particle surfaces block the surface and prevent additional sulfur oxides from penetrating and reacting with the remaining calcium carbonate. These processes also produce large volumes of waste sludge, which presents serious disposal problems, particularly when the sludge contains large fractions of calcium sulfite (rather than calcium sulfate). The long crystals of sulfite make dewatering the sludge very difficult. The sludge may remain a thick paste-like material for a long time and make disposal particularly difficult. The choice of other absorbents may be based upon a desire for better use of the neutralizing reagent, a reduction in the waste volume, or a reduction in difficulties of waste disposal. Some absorbent solutions/suspensions allow better utilization of reagents, and some systems even permit recovery of the sulfur oxides as sulfuric acid and reuse of the neutralizing reagent.

Most acid gases, like sulfur oxides, can be removed from air by absorption in alkaline liquids (as in flue gas desulfurization systems), and basic gases, such as ammonium, can be absorbed in acid solutions. In the large systems described the absorber could be a large spray or packed tower. In small laboratory systems simple small "bubble columns" can be used to remove small quantities of potential pollutants from small gas streams.

Other large absorption units are used to recover sulfur from petrochemical, natural gas, or experimental coal processing systems. In these facilities, the sulfur is often in the lower redox state as hydrogen sulfide. This is another sulfur bearing acid gas, and a basic adsorbent liquid is needed. Solutions of one or more amines are usually used. The amine solutions can reduce the sulfur (hydrogen sulfide) concentration to low and acceptable levels, and they can be regenerated to recover the hydrogen sulfide and reuse the absorbent solution. The hydrogen sulfide can be recovered in a concentrated form for conversion to elemental sulfur or oxidation to sulfuric acid.

Organic compounds are usually most easily absorbed in organic solvents, but are not always good choices for environmental and waste problems. For air streams with low concentrations of an organic contaminant, adsorption may be more effective. However, if the concentration is sufficiently high, absorption with an organic absorbent may be a better choice for removal of the bulk of the contaminant. The alternatives to absorption are likely to include membrane processes. It will then be preferable to select an absorbent with a relatively low vapor pressure to minimize loss of absorbent and contamination of the treated gas with the absorbent. Since most organic absorbents will have strict concentration limits in effluent gas, it is likely that a carbon adsorption bed will be needed somewhere downstream of the absorber. If absorption is used only to remove the bulk of the contaminant, adsorption (probably carbon beds) could be used to remove the last traces of the contaminant. As noted, the carbon beds are likely to also remove the trace of absorbent that could be volatilized during the absorption process.

The solute can usually be separated from the organic absorbent by gas stripping at a higher temperature, by distillation, or a membrane process like pervaporation, depending upon the volatility of the solute and the absorbent.

Examples of Stripping Applications

Gas stripping usually competes with distillation and membrane processes at high concentrations and with adsorption when the solute is at low concentrations. The key factor for determining if gas stripping should be considered is solute volatility. If it is too close to that of water, too much water must be evaporated to remove large fractions of the solute. Therefore, distillation, or possibly membrane, processes are more likely to be attractive. Furthermore, low volatilities would require very high gas rates to remove large quantities of solute from the solution, and one should consider one of the other options. It is usually possible to enhance the volatility of the solute by raising the temperature, but that raises the volatility of water; hence, a distillation or a membrane process like pervaporation should be considered if the temperature is to be raised significantly (where water volatility becomes significant).

Another important criterion in the selection of gas stripping is the desired final concentration of solute in the solution. When the concentration must be lowered to sufficiently low levels, adsorption becomes an increasingly attractive option because many stripping stages or large air rates would be required. Again, one should not feel constrained to using a single method for removing a solute, especially when the solute concentration must be reduced from relatively high to extremely low. Remember that different separation methods are often more economical over different concentration ranges, and it may be better to utilize different separation methods only over the concentration ranges in which they are the most economical. Flowsheets appear neater when only a single

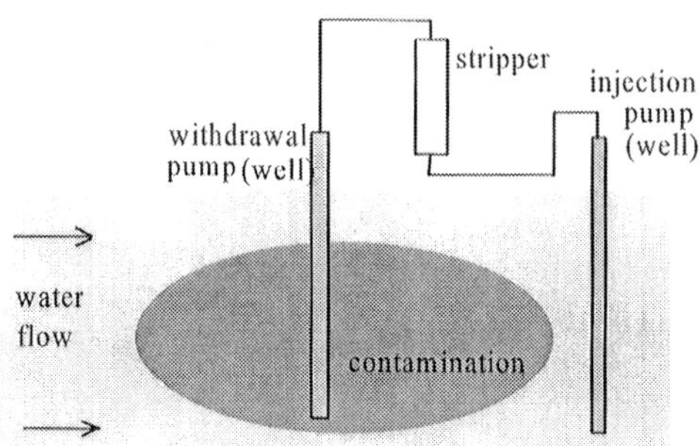

FIGURE 19 Pump-and-treat operation to treat groundwater.

separation step is used, but use of only a single method over every large concentration range may include operations in concentration ranges that are far from optimum for that separation process. Remember that gas stripping could be used to remove the bulk of a contaminant, but not to reduce its concentration to levels sufficiently low to allow the water to be discharged or reused. An adsorption process such as an activated carbon bed may be more economical for removing the last traces of the contaminant. Although there is sometimes a desire for finding a single separation method for a problem because it appears to be "less complex" to use a single method, the two step options can result in significantly smaller equipment and costs which can easily justify the modest increase in "complexity" of the overall system.

Ex Situ Stripping of Pollutants from Groundwater

In attempting to strip volatile solvents from groundwater aquifers, one can drill into the aquifer and pump water from it for treatment. This is called an ex situ stripping operation. It can take place in multistage towers on the surface (above ground) [24]. Such operations are often called pump and treat. If sufficient fractions of the pollutant are removed, the water can be reinjected into the ground through an injection well. The removal of groundwater and its reinjection elsewhere alter the water table near the removal and injection points. If the pumping rates are moderately high relative to the underground flow, the underground flow pattern can be changed significantly. Thus when properly designed, such systems can prevent migration of contaminated groundwater (Figure 19). Of course, the above ground (ex situ) treatment method could use an operation other than stripping. Adsorption on carbon beds is also a relatively common practice, especially for contaminants with relatively low volatilities, but gas strippers are likely choices for highly to moderately volatile contaminants.

Some contaminated plumes are relatively stagnant, and removal wells within the plume will eventually treat the entire contamination problem. In other cases, the plume may be from a source in fast moving groundwater, and it may not be possible to recover all of the contaminants that have escaped from the source. However, the removal well, or a "curtain" of removal wells, may be able to "cut off" the plume and prevent any more of the contaminants from escaping beyond the removal wells as long as the pump-and-treat operations is continued.

The required water treatment rate will depend upon the size or cross-section (width and depth) of the contaminated plume and the groundwater flow rate in the plume. For large plumes, it may be necessary to use several extraction (removal) wells. Underground barriers can be used to direct groundwater plumes toward the extraction wells and reduce the number of extraction wells needed. Wastewater from all the wells can be treated in a single stripping unit or in several units. The cross-section of the contaminated plume is usually estimated from exploratory (characterization) wells drilled in the region suspected to be contaminated. Strong suction from the extraction wells can converge the contaminated plume toward the extraction wells, at least to some extent. Characterization of underground water flow can be difficult, especially when the rock and soil structure are complex and nonuniform. The underground water flow rate can be estimated by using dye tracers, but the contaminant itself can be the most important tracer and reveal much about the underground water flow. Uncertainties in contaminant distribution and groundwater flow can be the limiting uncertainties, or the weakest parameters, in the design of groundwater remediation systems. Design of the actual gas stripping unit using the procedures described in this chapter may be among the more accurate and reliable parts of the overall design. Errors in characterization of the underground structure are likely to be more serious.

If the treated groundwater is not returned to the aquifer, the water table around the removal well will become depressed, and water will flow toward the well from other regions of the aquifer. If the "return" well is located far from the removal well, the groundwater table will be raised in the region around the return (injection) well. Some time will be required for an effective steady-state water table profile to develop. The effective steady-state water table is simply a profile affected by changes in the pumping rate from the removal well. The pumping rate required to "catch" a plume of contaminant depends upon the water flow rate in the plume, that is, in the underground structure. The flow rate depends upon the hydraulic gradient and the porosity in the aquifer. The flow rate through tight clay can be very slow, but some highly porous sandy aquifers may require much higher pumping rates. The treatment rate becomes even more complex when there are large voids such as caves in the aquifer, which are usually known as "krast" systems.

The water table also can vary greatly with time, and the portions of the aquifer near the surface will certainly be affected by seasonal or occasional heavy rainfall changing the groundwater flow. In some regions where the rainfall is highly seasonal, the water table should be significantly different during the seasons. The decision to design the ex situ treatment for the average effective water table or for some higher groundwater flow experienced during periods of heavy rainfall may be determined by the regulatory bodies involved, and such decisions are often based upon local or regional considerations. Regulators may require that the capacity of a treatment system to be based upon the maximum expected flow rate or some fraction of the maximum expected flow.

Many wastewaters and polluted groundwater contain several volatile components. Thus, in principle, these are multicomponent systems that could involve complex equilibria. However, if the systems are sufficiently dilute, usually because of the low solubilities of the volatile components, the volatility of each contaminant may be affected little by the presence of the other components. That is, Henry's law constants (if the equilibria are linear) may not be affected by the low concentrations of the other components. Such cases are common, and the system can be designed as a single component system considering individual contaminants. Thus if one can determine which component requires the highest gas rate and the tallest tower to strip the component to acceptable concentrations, the tower can be designed to meet the needs in removal of that component. Then all of the other components will be removed with higher efficiencies than necessary. It is often obvious which component is the most difficult to remove to acceptable levels—the one that requires the highest removal efficiency or the one with the lowest volatility. If the critical contaminant is not obvious, one can always design stripping systems for each of the suspected critical components and select the tower with the highest liquid rate and the most stages (or transfer units).

Very seldom is the volume of a polluted aquifer or the effective volume of the aquifer being treated known accurately. The range (or volume) of the aquifer affected by a pump-and-treat operation may be a function of the rate at which groundwater is pumped from the aquifer for treatment. This is only one reason why predictions of the time required to reach a certain level of pollutant removal from a relatively stagnant contaminated plume may not be accurate. Once the pump-and-treat operations have begun, any decrease in concentration in the water removed from the ground can be monitored.

There are serious problems with some pump-and-treat operations, and the performance of pump-and-treat operations can deviate far from simplified models. In some quarters, pump-and-treat has developed a negative image. Perhaps the worst case for pump-and-treat operations occurs when there is a high concentration source of pollutant and low, but environmentally significant, concentrations in the water being removed for processing. The key to cleaning up a plume of contaminant in a timely manner is to get the contaminant to the treatment system. If the extraction wells can only bring the contaminant to the treatment system in low concentrations, removal of large quantities of contaminant from the plume may require treatment of unreasonably large volumes of water. An example of such a situation occurs when a polluting solvent is present in droplets or pools of highly insoluble liquid. If the solubility of the solvent is low, the concentration of solvent in the water will be low, at or below saturation. However, some contaminants such as halogenated solvents can be toxic even at low concentrations. That is, a small amount of such contaminants can contaminate extremely large volumes of water. In such cases, the total solvent content in the aquifer may decrease very slowly and require tens or hundreds of years to remove all of the solvent if the treatment must rely upon pumping only the dissolved material from the plume. In other cases, the pollutant may be adsorbed on a component of the solids in the soil, and the concentration of pollutant in the water phase may be only a small fraction of the total pollutant present. Again an impractically long time may be required for an effective cleanup of the aquifer and surrounding soil. If the rate of mass transfer from the "pool" of insoluble or adsorbed pollutant to the groundwater in the removal well is sufficiently slow, the concentration of pollutant in the treated water may decrease initially with treatment time. However, after the pump-and-treat operation is stopped, the concentration may then slowly return to the original value with little evidence that the treatment has done any good. Although a large volume of water has been treated, there could still be sufficient solvent left in the ground to saturate much of the aquifer flow.

These situations have occurred frequently enough that pump-and-treat operations have been criticized. Before beginning a pump-and-treat operation, it is certainly advisable to make significant efforts to ensure that the treatment can be completed in an acceptable time. Otherwise, other approaches should be considered which act more directly on the main sources of the contamination (the concentrated regions of insoluble liquids) and try to avoid dealing only with more dilute groundwater. Pump-and-treat operations can never remove a sparsely soluble contaminant more quickly that the pumped groundwater can carry the contaminant to the treatment facility.

Another less serious problem with air-stripping of VOCs from many groundwater streams is the formation of iron oxide precipitates when there is significant soluble ferrous ion in the groundwater. That is common in many communities, especially in the eastern United States. Salts of ferrous iron are usually soluble at or near neutral pH, but ferric iron quickly hydrolyzes and forms insoluble hydroxides, the familiar "rust" colored solids found so often in plumbing systems in some areas of the country. Air stripping equipment can also be an oxidizing reactor that converts the ferrous iron to ferric iron. Usually the iron oxide/hydroxide precipitates will not degrade the performance of a gas stripping operation until/unless they accumulate in sufficient quantities to plug the stripper or part of the stripper system. The deposition of ferric hydroxide/oxide precipitate on tower packing can become evident after only a few hours of operation. In extreme cases, tower packing can become clogged after prolonged operation.

In Situ Stripping of Contaminants from the Vados Zone or from Groundwater

When it is undesirable to remove the groundwater from the soil or when the contaminant is in the vados zone rather than the groundwater, the removal is more likely to be carried out in situ, that is in the ground. Stripping of contaminants in the vados zone (the region above the water table) can be considered desorption (instead of gas stripping) if the contaminants are adsorbed on solids in the soil, evaporation if the contaminant is free liquid (NAPL), or gas stripping if the contaminant is dissolved in water perched in the vados zone. In many cases, contaminants may be in several or all of these forms or the exact condition of the contaminant may not be known. However, containments in any of these forms can often be removed, usually at a reasonable rate, by flowing gas through the contaminated regions. Although it would be appropriate to discuss removal of contaminants in other sections, depending upon the condition of the contaminant, the subject will be mentioned here to keep the somewhat similar separation operations together. One may not always know with certainty how much of the contaminant is in each form.

Removal of volatile contaminants from the vados zone will be considered first. This has become a very common operation for remediating sites where gasoline or other volatile materials have been spilled. The "stripping" air can be injected into the soil by selectively spaced injection wells and removed from other "removal" wells (Figure 20). For shallow

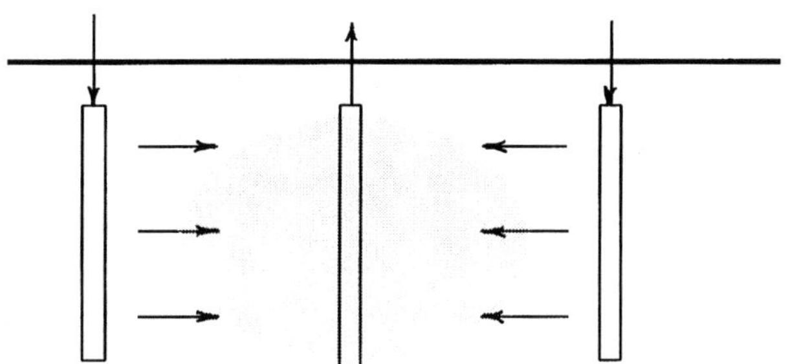

FIGURE 20 In situ gas stripping operation with gas injection.

systems, it may not be necessary to install injection wells. By "pumping" air from the vados zone, a vacuum is produced which draws air form the soil surface through the contaminated region (Figure 21). This is likely to be the preferred mode of operation for spills of volatile contaminants that occur at or near the ground surface and have not migrated far into the ground. For spills over large areas or cases where contamination has spread over large regions, it will be necessary to install several removal wells. This is often called "soil venting." Contaminants in the vados zone may be dissolved in perched water trapped between soil particles (wet soil), or the contaminant could be liquids perched in the soil, that, is soil wet by the VOC contaminant. Obviously the removal of volatile organic

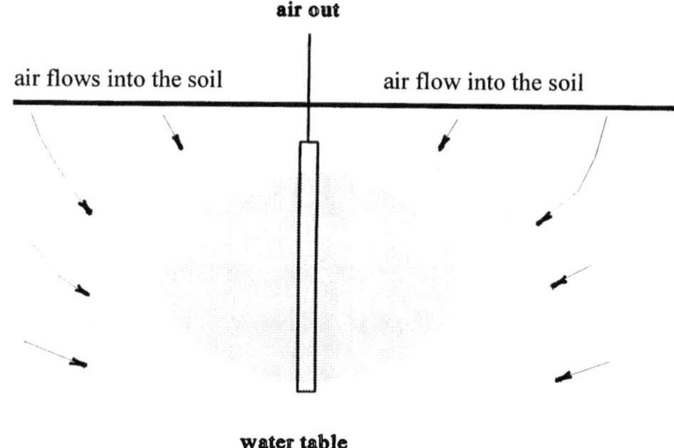

FIGURE 21 In situ gas stripping of shallow contamination without gas injection: soil venting.

liquids by airflow through the soil is evaporation, but the operation is discussed here with gas stripping because the operations are so similar and the user may not even know if the contaminant is largely free or dissolved.

For some contaminants in the vados zone or below the water table, it will be necessary to recover the contaminants and not release them to the atmosphere. Adsorption, membrane systems, and even condensation can be considered for treatment of the air pumped from the ground. The selection of a treatment method depends upon the concentration and properties of the contaminant. This problem is mentioned in other sections that describe those separation methods. Some relatively low hazard hydrocarbons removed from soils in this way may not require treatment of the exhaust air, but chlorinated solvents and many other volatile contaminants will probably have to be recovered and not allowed to escape to the atmosphere.

For many situations, the rate of contaminant removal is not easily predicted, principally because it is not common for the concentration distribution of contaminants or the flow patterns in the soils to be known accurately. It is particularly difficult to estimate even the amount of a contaminant originally in soils when the contaminant is not uniformly distributed in the soil. When a significant portion of the contaminant exists as droplets of nonaqueous liquid (NAPL), core samples of the soil taken to assess the contamination may not capture a representative sample, and may not even detect any pools of NAPL at all. There is never enough time or money to drill a sufficient number of soil core samples to be sure that the distribution of contamination is known reliably. If sufficient wells were drilled to determine the distribution of contamination in some complex sites, the drilling could remove a significant fraction of the soil. In situ operations always have to rely upon incomplete information.

Similarly, it is difficult to assess that contamination after treatment if that contaminant is not distributed uniformly. That is, it is difficult to determine when the treatment can be ended. In some cases, estimates of the extent of remaining contamination and the decision to terminate removal efforts may be based largely upon the concentration of the contaminant in the exhaust gas coming from the stripping operation. The concentration of the contaminant in the gas from the removal well may remain relatively constant for a long time or slowly decline. As long as the airflow through the vados zone becomes saturated with the contaminant, the removal rate may remain relatively constant for a while. Of course, the concentration in the gas declines and approaches zero as the level of soil contamination approaches zero. With time, the contaminant located in the more porous regions which receive the highest gas flow may become depleted first; fortunately, for some recent spills, those may also be the regions with the most contamination. When a spill occurred long ago, there will be more time for the contaminant to diffuse or be transported to other regions of the soil. Note that regions where porous passages allow the contaminant to move within the soil may not be the regions most accessible to air which enters the soil from different positions or directions, so one can never be sure that there are not significant regions of contamination that were not reached with the airflow, especially in highly heterogeneous soils with complex airflow patterns.

The injection of air into the vados zone can do more than "strip" volatile components from the soil; it can also stimulate microbial action that helps accelerate contaminant removal. In some cases biodegradation can make a significant contribution to the removal rate [25]. The air, of course, stimulates aerobic processes, and the importance of biodegradation depends upon the susceptibility of the contaminant to aerobic biodegradation and the availability of sufficient nutrients in the soil to maintain the biological activity. Since most hydrocarbons are more easily degraded biologically than many chlorinated solvents, this is less likely to be a significant factor in removal of solvents like trichloroethelyene, carbon tetrachloride, etc. Although it is possible to add microorganisms and nutrients to help with the biodegradation of many organic contaminants, this addition is not always necessary. Natural organisms can degrade many contaminants. If the contamination has been present for a long time (e.g., at an industrial site), organisms capable of metabolizing the contaminant may have become established in the soil. Otherwise, such organisms may be obtained from other sites that have been contaminated for longer periods with the same contaminants. The presence of significant biodegradation can sometimes be detected by unusual concentrations of carbon dioxide in the stripping gas, by the presence of degradation products in the soil or gas, or by the change in contamination composition because of more rapid removal of the more easily degraded components in the soil.

When the contaminant is in the groundwater (that is, below the water table), it is usually necessary to introduce air into the water, much like in conventional ex situ gas stripping. If it is necessary to recover the contaminant, it will again be necessary to use removal wells as well as gas sparge wells. The removal well may be located above the water table and remove the gas from the vados zone above, but near, the injection point. Such operations have been described or modeled in a series of papers by David Wilson and his co-workers and students [26, 28]. A review of in situ gas stripping and mention of several other models was presented by Fam [27]. Models of in situ stripping have to become increasingly complicated if it is necessary to account for perched liquid VOCs, VOCs dissolved in perched water and groundwater, VOCs floating on the top of the water table, dense VOCs perched on bedrock, and VOCs adsorbed on one or more solid components of the soil. However, the complication that is probably most difficult to model is the heterogeneity of the soils and the distribution of the contamination.

Flow through highly heterogeneous soils can be very difficult to measure and model. Small regions such as cracks or fissures in the rock or packed soil can appear to be minor variations in properties, but they can divert significant fractions of the airflow past parts of the contaminated region. In a similar manner, rock formations can act as barriers and divert the flow in unexpected ways. In some mountainous (or formerly mountainous) regions, the soil can include highly fractured rock formations that are extremely difficult to characterize. Even in regions where the soils are usually assumed to be relatively uniform, there can be man-made intrusions that make the soils heterogeneous. Most excavations for ditches, trenches, buried pipes, etc., do not pack the soil as tightly as the original soil that could have had thousands to millions of years to settle. The regions around those man-made constructions could be paths for high gas flow.

Again, it is not always easy to know the distribution of the contaminant in the soil, even in the soil under the water table. For instance, most organic hydrocarbons and solvents that must be removed from soils and groundwater are only very sparingly soluble in water. If the contaminant is a liquid that is largely insoluble in water, the bulk of the contaminant may be "floating" on the top of the water table if it is less dense than water (LNAPL) or perched on bed rock at the bottom of the aquifer if it is more dense than water (DNAPL). Although the water will be contaminated, probably with the concentration of the contaminant approaching saturation, the bulk of the contaminant may not actually be in the water. The top of the aquifer is not a clearly defined surface since the water table may fluctuate considerably with rainfall, so LNAPL contaminants can become perched above and below the water table (at any one time). Similarly DNAPLs may become perched in the soil within the aquifer or on low permeability bedrock that defines the lower limits of the aquifer.

In many ways the resulting uncertainties in the distribution of a contaminant can raise questions about the need of treatment at all. For instance, with NAPLs spread unevenly within a soil, it would be possible (and often reasonably probable) for a single vertical probe by drilling or "punching" to miss the contamination completely and falsely assume that there was no contamination. On the other hand, striking a single small region of a NAPL could leave the appearance of a highly contaminated soil. The importance of the difficulty in assessing heterogeneous contamination is compounded when one tries to determine when a treatment has proceeded sufficiently for it to cease. If the decision to end treatment and declare the soil clean is based upon a few (or even several) cores of the soil, one could be misled by the statistical finding or not finding of concentrated regions of NAPLs. It is possible to treat a region of soil, remove considerable and measurable quantities of contaminant, and then find posttreatment core samples that suggest higher contaminant levels than were found originally.

Most of the problems discussed earlier about the importance of inhomogeneity in soils to the operation of stripping/venting or volatile contaminants from the vados zone apply to removal from the groundwater within the aquifer. Of particular importance are the regions not reached effectively by the stripping gas. Wilson and his students [28] have explored the effects of nonuniform gas flow on gas stripping rates. There is often considerable uncertainty in the distribution of the contaminant and in the flow patterns in the soil. Although there are some nonintrusive methods for detecting some contaminants and estimating the soil and rock structure, none of the available methods are likely to be as accurate or reliable as desired or needed. In many ways, in situ contamination and soil/rock structures obey an environmental principle something like the Heisenberg principal in physics. One cannot measure the distribution of contamination or the flow properties of a soil completely without significantly altering the soil. However, there is more hope for finding better ways to "image" contamination distributions and flow patterns in soils than there is for avoiding the effects of the Heisenberg principle.

Again, the introduction of air into the aquifer can stimulate biological activities that enhance the removal of some contaminants by biodegradation just as it can stimulate biodegradation in the vados zone. However, there can also be chemical changes caused by introducing air into an aquifer. Many soils, particularly in the eastern United States, contain significant concentrations of ferrous iron, the usual soluble form. Addition of air can oxidize the iron to less soluble ferric iron in the aquifer just as air can from ferric iron precipitants in ex situ equipment. The most important effect from this oxidation can be a decrease in aquifer permeability, especially in the region immediately surrounding the injection wells. This is not necessarily permanent damage to the aquifer since the same microorganisms that oxidize the contaminant, or even different organisms, may eventually reduce ferric iron to ferrous iron and restore the aquifer approximately to its original condition. If the iron hydroxide precipitation occurs only in the contaminated region, any reduction in the flow through that region could reduce the rate at which the contamination is spread.

The rate of contaminant removal from either the vados zone or the aquifer depends upon the vapor pressure of the contaminant or the solubility of the contaminant in the groundwater. Thus, raising the temperature in the underground structure even by a moderate amount can increase the removal rate considerably in some cases. There have been significant efforts in recent years to heat the underground regions to promote VOC removal with in situ operations. The methods have included the use of steam, electrical resistance heating, and even microwave heating (for contaminants near the ground surface). Obviously, heating will complicate the gas stripping operation, and the value of the enhanced removal rate must be compared with the heating cost.

There appears to be a strong preference for in situ operations over ex situ operations. This is probably most apparent for gas stripping operations, but there is similar interest in in situ adsorbents. In situ operations are preferred because they are less costly, or at least are perceived as less costly. Some of their cost advantages occur because there is less need for constructing facilities such as the gas stripping vessel. However, that cost advantage can be offset, at least partially, if there is a failure to achieve effective stripping efficiencies. That is, countercurrent stripping equipment can achieve higher concentrations in the strip gas. That is not a significant advantage if the gas from in situ operations is able to become essentially saturated with the contaminant because of the presence of "neat" liquid VOC in the gas flow path. An additional potential advantage of in situ gas stripping that may actually be more important than saving process vessels is the ability to remove VOCs from the vados zone as well as from the groundwater. That can be important because the major difficulty with ex situ pump-and-treat operations may be the continuous recontamination of the groundwater from liquid VOCs in the groundwater or vados region.

Two other potential advantages of in situ operations may be somewhat more artificial. In many cases, it is difficult to return groundwater after a pump-and-treat operation back to the water table, but if the same treatment is done without removing the water from the aquifer, the question of returning the water never arises. For instance, in some regions of the country, groundwater is normally quite acid with pH values that are significantly lower than the allowable pH for waters returned to aquifers. Then treatment of the groundwater to remove a contaminant such as a VOC may have to be accompanied by a second treatment to raise the pH if the operation were carried out ex situ, but there may be no such requirements for in situ operations. This is characteristic of current U.S. regulations, and it is likely that any regulation will eventually be changed if it dictates a treatment approach that does not enhance the environment. Other related potential advantages of in situ treatment may arise if there is any other natural contaminant in the groundwater that may not have been caused by the industrial or other facility that caused the VOC contamination. For instance, groundwaters in a region that has mineral deposits may have marginal concentrations of minerals that may not be permitted in discharge water, and removal of the groundwater for removal of the VOCs could also impose the requirement for an additional treatment system to remove the other mineral(s). Normally, if the water meets drinking water standards, there may be no problem, but if the facility uses the metal or mineral of interest, it may be difficult to prove that the mineral contaminant did not result from the facility's operations. It is difficult to define "baseline" levels of minerals and be sure that nothing was added to the groundwater; hence, it may be easier to use an in situ treatment.

SUMMARY

Gas absorption is an attractive approach to removal of water soluble gases, usually acidic or basic gases, from exhaust streams, and gas absorption operations are carried out effectively in towers, usually filled with a suitable packing material. Open towers, without packing, give less favorable mass transfer performance and are not likely to be used except where equilibrium conditions are highly favorable to absorption and/or the presence of solid make it difficult to use packing in the tower. Although the standard shapes of random packing (rings and saddles) and trays are still in common use, there is a growing interest in "structured" packing materials that currently are more expensive than the cheaper random packing materials but can give significantly better mass transfer performance.

Absorption is generally most effective for systems with moderately high solute concentrations. When the concentration of the solute becomes too low or when extremely high removal efficiencies are necessary, it may be more practical to use an adsorption method. If the initial concentration is moderately high, but extremely high removal efficiencies are required, one could consider using absorption only for removing the bulk of the solute and using adsorption to remove the last traces of the solute.

Gas stripping can be used to remove solutes from a liquid absorbent, usually by stripping at a higher temperature, or for removing volatile components from a liquid stream. Gas stripping, usually at ambient temperatures, has become a standard method for removing volatile organic compounds (VOCs) from groundwater and wastewaters. Gas stripping becomes increasingly attractive as the volatility of the solute increases, and the volatilities can be increased by raising the temperatures. It becomes less attractive as the concentration or the volatility of the contaminant declines. When solute volatility is in the same range as water volatility (or the solvent's), one may have to consider distillation or membrane processes instead of gas stripping. Gas stripping, like gas absorption, is not usually practical at extremely low concentrations or when extremely high removal is required (that is, when extremely large numbers of stages or transfer units are required). In such cases, adsorption may prove to be either a better removal process or a desirable polishing step.

Absorption and gas stripping operations are usually carried out in packed towers, but simpler single stage systems are also used, especially when only a single stage of removal is adequate. In situ operations are particularly attractive for gas stripping of VOCs from groundwater. In situ operations can affect undissolved VOCs in the aquifer or in the vados zone above the aquifer and can be more effective than ex situ approaches that deal only with the VOCs dissolved in the groundwater. There can also be regulatory advantages for using in situ operations because of the potential presence of other contaminants (such as hydrogen ions, pH) that could make reinjection of the treated water into the aquifer difficult.

REFERENCES

1. Killat, G. R. and T. D. Rey. *Chem. Eng. Prog.* May, p. 69 (1996).
2. Kouri, R. J. and J. J. Sohlo "Liquid and Gas Flow Patterns in Random and Structured Packings." *I. Chem. E. Symp. Ser. 104*, p. B193 (1987).
3. Kunesh, J., *Canad. J. Chem. Eng. 65*, 907 (1987).
4. Zuiderweg, F. J. et al. *Trans. I. Chem. E. 71*, (Part A) 38 (1993).
5. Perry, R. H., and C. H. Chilton, *Chemical Engineer's Handbook, 5th ed.* McGraw-Hill Book Company, New York (1984).
6. Trebal, R. *Mass Transfer Operations, 2nd ed.* McGraw-Hill Book Company, New York (1968).
7. Eckert, J. S. *Chem Eng. Prog. 54*, 57–59 (1961).
8. Ergun, S. *Chem Eng. Prog. 20*, 1996 (1928).
9. Strigle, Ralph F. Jr. *Packed Tower Design and Applications: Random and Structured Packings*. Gulf Publishing. Houston (1994), p. 2.
10. Bravo, J. L., et al. *Hydrocarbon Process 64*, 91–95 (Jan. 1985).
11. Bravo, J. L., et al. *Hydrocarbon Process 65*, 45 (Mar. 1986).
12. Nelson, A. D., R. J. Schmitt, and D. Dickeson. *Environ. Prog. 16*, 43 (1997).
13. Kleman, L. and J. A. Bonilla. *Chem. Eng. Prog.* p. 27 (July 1995).
14. Treybal, R. *Mass Transfer Operations, 2nd ed.* McGraw-Hill Book Company. New York (1968).

15. Treybal, R. E. *Mass Transfer Operations*. McGraw-Hill Book Company. New York (1968).
16. Onda, K., H. Takeuchi, and Y. Okumoto. *J. Chem. Eng. Jpn. 1*, 56 (1968).
17. Djebbar, Y. and R. Narbaitz. *Environ. Prog. 14*, 137 (1995).
18. Roberts, P. V., et al. *Environ. Sci. And Technol. 86*, 30 (1985).
19. Treybal, R. L. *Mass transfer Operations, 2nd ed.* McGraw-Hill Book Company, New York (1968).
20. Gavaskar, et. al. *Environ. Prog. 14*, 33 (1995).
21. Littlejohn, D. and S. G. Chang. *I&EC Research, 29*, 10 (1990).
22. Shi, Y., D. Littlejohn, P. B. Kettler, and S-G. Chang. *Environ. Prog. 15*, 153 (1996).
23. Bart, H-J. and K. Burtscher. *Sep. Purification, 11*, 37 (1997).
24. Okeniewski, B., *CEP, 88*(2), 89 (1992).
25. Chao, J. S., D. C. DiGiulio, and T. T. Wilson. *Environ. Prog. 16*, 35 (1997).
26. Wilson, D. J. *Sep. Sci. Technol. 27*, 1675 (1992).
27. Fam, S., "Critical State of the Art Review of Vapor Extraction." *Proceedings of the 59th Industrial Waste Conference*. Purdue University, May 1995. P. 7 (1995).
28. Wilson, D. J., C. Gomez-Lahoz, and J. M. Rodriguez-Maroto. *Sep. Sci. Technol. 29*, 2387 (1994).
29. Munz, C. and P. V. Roberts. *Water Res. 23*, 589 (1989).
30. Gosset, J. M. *Env. Sci. Technol. 21*, 202 (1987).
31. Ashworth, R. A. et al. *J. Haz. Materials 18*, 25 (1988).

4
Membrane Processes

Membrane separation processes involve the relative transport of molecules across a film or other barrier (the membrane). The film is usually a solid, but it can also be a liquid. (Liquid membranes are closely related to some liquid-liquid extraction systems, and liquid membranes will be mentioned in Chapter 6 but discussed in more detail in this chapter.) The membrane can be a homogeneous material or a composite material with a very thin active layer supported by a stronger but more porous material. In the following, separation of insoluble particles will be classified as filtration processes, and membrane processes discussed in this chapter will be those that separate small or moderate sized molecules. There really is no such clear distinction between filtration and molecular separations with membranes because many "molecules" such as proteins and polymers are large enough to be "filtered." Pressure driven membrane separations of smaller molecules from liquid solvents—separations based upon molecular properties—are usually called reverse osmosis, but separation of macromolecules is usually called "ultrafiltration" because the maromolecules behave much like particles. Many reverse osmosis processes behave as if they involve movement of "dissolved" components within the membranes. Thus reverse osmosis can depend upon chemical characteristics of the molecules rather upon just the size of the molecules.

This chapter is organized into two main parts. The first part discusses membrane processes in general, the structure and performance of membranes, and membrane processes that utilize solid membranes and pressure or concentration differences to force selective components through the membrane. The second part discusses selected membrane processes with different structures (liquid membranes), different driving forces (electrodialysis), or with phase change as well as membrane properties affecting the separation (pervaporation). This division of the chapter results from the difficulty in making a broad and general presentation that covers all of the membrane methods without appearing so abstract that the reader would find it difficult to remain focused on the application of membranes to environmental and waste problems.

The use of membranes for separations is growing rapidly and is believed to have involved approximately $2 billion of sales annually [1] a few years ago; the sales are probably significantly higher today. Membranes have become commonplace in the food, biomedical, and pharmaceutical industries. Membrane technology is now spreading rapidly in process, waste treatment, and environmental areas where applications are likely to experience particularly strong growth during the coming years. In waste and environmental applications, membranes can be used to remove toxic organic gases from effluent air, to remove (and often recover) dissolved organic compounds from water, to purify water for reuse or discharge, for removing dissolved salts (metal ions) from solutions, for recovering selected metal ions or organic compounds from aqueous solutions for recycle of either the component being removed or the solution, for removing selected components from effluent gases, for recovery of spent acids (and/or caustic), and probably for numerous other applications that will be identified in the future. These different applications utilize different membrane operations.

To gain even a general understanding of the importance of membrane separations, to estimate where membrane separations should be considered, and to understand the basis of design and evaluation of membrane separations, it will be necessary to consider some of the different types of membrane operations. Membrane operations differ in the driving force used to force material through the membrane, the fluid phases being treated, the membrane materials used, and the shape or geometry of the membranes and the membrane cells that are employed.

Membrane separation processes can be carried out by using any of three driving forces: pressure differences, concentration differences, and/or electric potential differences. This chapter deals principally with processes driven by pressure and concentration differences. Electrically driven systems (electrodialysis) will be described but covered in less detail. The rate of growth in electrodialysis does not seem likely to be as high as the rate of growth for some of the other membrane processes that will be covered in more detail. The pressure and concentration driven processes appear to have received the most attention in recent years and probably have the greatest potential for expanded applications to more environmental and waste problems. Pressure and concentration driven processes will be discussed separately, but in many cases some aspects of both processes are similar and will be discussed together. The membrane structure and several aspects of membrane cell design are similar and will be discussed together. For gas separations there is no inherent difference between concentration driven separations and pressure driven separations because the pressure and concentration of all components are approximately proportional to each other. Thus the principal difference between pressure and concentration driven systems occurs for separation of liquid systems. In liquid systems a considerable fraction of the membrane discussion concerns pressure driven reverse osmosis because of the current and probable continuing importance of reverse osmosis in waste and environmental separations.

The selection and design of membrane systems involves the usual questions of material selection, type of equipment to use, and equipment size. The materials selection involves finding the best membrane material as well as the best materials to use for constructing the equipment. The unique materials selection for membrane processes, of course, is the membrane material itself. The membrane should be compatible with all of the fluids involved and not degraded too rapidly by use. It should also have high selectivity for separation and a high permeation rate. A unique materials problem with membranes is fouling, which can result in a decrease in the permeability or selectivity of the membrane with use or exposure to the fluid being processed. Fouling, discussed in more detail later, can result from degradation of the membrane material and/or from collection or formation of solids on the membrane surface.

Membrane processes can be staged so that a complete separation does not always have to be made in one pass or a few passes through a membrane. However, staging of large numbers of membranes is usually not as simple as staging in other "equilibrium" based multistage systems such as adsorption, extraction, or distillation; so it is usually desirable to achieve the desired separation in one pass or in a few passes through a membrane. (Isotope separation by gaseous diffusion is an example where numerous membrane stages were applied successfully, but that process produced a valuable product, which justified the additional difficulty of using so many stages.)

Countercurrent flow of fluids across both sides of membranes can resemble countercurrent flow in absorption or solvent extraction processes, but the analogy is not always exactly the same because permeation of the solute(s) though the membrane is usually highly irreversible. This means that the phases on each side of the membranes are not in equilibrium. In such systems, there may be less or no advantage to countercurrent flow in the downstream (low pressure) side of the membrane. In some membrane systems, there may be no true countercurrent flow of fluid in the chamber downstream of the membrane. The entire flow in the downstream (lower pressure) chamber of pressure driven processes such as reverse osmosis may come from fluid that permeates the membrane.

The size of membrane equipment is determined largely by the permeation rate that can be achieved. For process design, one may conclude that a certain number of square meters of membrane will be required for the separation. This area could be packaged into several arrangements, which will be discussed as different types of membrane separations are introduced. However, there can be mass transfer resistance in the fluids as well as in the membrane itself, and the hydrodynamics of different membrane equipment can also affect film mass transfer resistance and have an important secondary effect on the membrane area needed.

Many developments and new applications in membrane separations are aimed at finding less costly ways to replace expensive and energy intensive separations such as distillation. In other cases, the development may seek separation methods that operate at ambient or moderately low temperatures that can be used to separate heat sensitive molecules that would be degraded significantly by alternative methods, such as distillation, that require higher temperatures. This latter case is one reason why membranes have been adopted so readily by the food and pharmaceutical industries. However, heat sensitivity of components is less likely to be a major consideration for environmental applications.

MEMBRANE PERMEABILITY

For applying membranes, it is often safer to think in terms of permeabilities that do not imply a particular mechanism of transport. The permeability of a membrane can be described as

$$P = \frac{\text{Flux}}{p_u - p_d} \qquad \text{or} \qquad P = \frac{\text{Flux}}{c_u - c_d}$$

where P is the permeability, p_u is the partial pressure of the permeating component on the upstream side of the membrane, and p_d is the partial pressure of the permeating component on the downstream side of the membrane. Alternatively, permeability can be expressed in terms of the concentration of the permeating component on the upstream side of the membrane, c_u, and the concentration on the downstream side of the membrane, c_d. The permeability can be defined for both or all components in the system, and the selectivities of the membrane can be expressed by the ratios of the permeabilities of the different components.

Permeability is a composite of several phenomena and can be a function of the concentrations in the liquid or gas being treated. Nevertheless, it often varies only slightly with concentration, pressure, or flux; thus it is a practical concept to use. In a few cases it is clearly preferable to define the permeability in other terms, one such being the permeation of hydrogen through metals such as palladium. In this case, the solute which diffuses through the metal membrane is atomic hydrogen dissolved in the metal. Since the concentration of atomic hydrogen in the metal is proportional to the square root of the hydrogen partial pressure rather than the hydrogen pressure, it is preferable to use the difference in the square root of pressure on both sides of the membrane as the driving force rather than the difference in the pressure. Such cases are less likely to arise in environmental processes. One would like to think that dilute gas and liquid streams, such as those that are more likely to arise in environmental separation systems, will usually give constant values of the diffusion coefficient(s) in the membrane and the "solubility" of components in the membrane. This is a reasonable hope, but membranes that have high affinities for the component of interest can have nonconstant distribution coefficients for such components, even for dilute (or low pressure) systems because the component may not be "dilute" within the membrane.

The effect of temperature on membrane permeability deserves a short discussion. Normally the permeability to gases increases with increasing temperature, and it is often desirable to operate a membrane process at temperatures as high as the membrane can tolerate. Higher temperatures, on the other hand, may decrease membrane selectivity, and other factors could cause the permeability behavior to be different than normally expected. If the gas "dissolves" in the membrane, higher temperatures could reduce the "solubility" of the gas sufficiently to reduce the advantages of higher diffusion coefficients at higher temperatures. The permeation rate can even decrease with increased temperatures.

Many solid organic membranes go through a "glass transition" temperature, T_g. At temperatures above T_g, the permeability coefficient may be essentially independent of solute partial pressure, and the flux will be proportional to the difference in the partial pressure on the opposite sides of the membrane. In this region, the permeability, P, will behave much like adsorption equilibrium and diffusion coefficients and vary with temperature as

$$P = P_0 \exp\left(\frac{-E_P}{RT}\right)$$

where P_0 and E_P are constant parameters, R is the gas constant, and T is the absolute temperature. However, at temperatures below T_g, the permeability is likely to be a stronger function of solute pressure and may vary less regularly with temperature. In some regions it may even decrease with increases in temperature. This can occur if (as noted) the "solubility" of the solute in the membrane decreases with increasing temperature faster than the diffusion coefficient increases. With any membrane where permeation depends upon diffusion of solute dissolved in the solid parts of the membrane or solute adsorbed on the membrane, the temperature dependence of adsorption as well as diffusion can affect the permeability. Thus, abrupt changes in the adsorption phenomena with temperature will affect permeation, perhaps greatly.

SOLID MEMBRANES

When one thinks of membranes, solid films of materials usually come to mind, and these are certainly the most common membranes. Thus the major portion of this chapter is devoted to solid membranes. Two shorter sections discuss liquid membranes and electrodialysis. Electrodialysis also utilizes solid membranes, but the driving forces are electrostatic potential rather than concentration or pressure.

Pressure Driven Membrane Processes

Filtration processes can be viewed as pressure driven membrane processes, and there are certainly similarities between filtration of very fine particles and membrane separation of molecules. Nevertheless, in this book, filtration of particles is treated separately from separation of molecules with membranes.

In separating a low concentration or trace component from a fluid, it is useful to consider two groups of situations. In one case, the membrane will be more permeable to the major component, that is, to the liquid or gas. In most environmental problems, the streams will be water or air. In the second situation, the membrane will be more permeable to the low concentration or trace component. One is likely to prefer the second situation because then it is only necessary to pump (or recover) the small amount of the trace component through the membrane. That would usually result in less energy consumption than forcing all of the bulk material through the membrane. Important cases where the membrane selectively permeates the minor component do exist, and, where possible, it is usually preferable to remove the minor component from the major component. That is usually more desirable in other separation systems as well since less material has to be transported through the membrane or into the other phase in separations methods based upon two phases.

However, there are also important applications where the membrane is more permeable to the major component (the solvent) and where there are no practical membrane systems available that are selective to permeation by the minor component. This is the case for reverse osmosis, one of the more common membrane applications important to waste and environmental problems and the membrane method that will receive the most significant coverage in this chapter.

Reverse Osmosis

Separation of water from solutes by forcing the water through a membrane that rejects the solute is called "reverse osmosis." It resembles filtration of small particles, but the mechanism for solute rejection can involve phenomena other than the size of the molecule and the size of the membrane pore openings. The term "reverse osmosis" comes from the fact that the process appears to be the reverse of the older (but still important) osmosis processes. In osmosis, a solvent diffuses across a membrane from a solution with a lower solute concentration into a solution with a high solute concentration. The net effect of osmosis is to dilute the solution with the higher concentration and thereby concentrate the solution with the lower concentration. The driving force for osmosis is the difference in the osmotic pressure of the solvent, and osmosis processes cease when the concentrations, and thus the osmotic pressures, on both sides of the membrane become equal, so there is no more driving force.

Osmosis can be viewed as a concentration driven process; the driving force is the concentration of the solvent across the membrane. However, the concentration difference also involves a pressure difference, and osmosis could also be considered a pressure driven process. An osmotic pressure exists across an osmosis membrane because of the concentration difference, and if the membrane is permeable to solvent (water), solvent will permeate by osmosis from the dilute solution into the more concentrated solution. The pressure driving force is the difference in the "osmotic pressure" in the two solutions. This is just one more example where pressure and concentration driving forces are, as mentioned earlier, not always distinctly different bases for grouping membrane processes.

In reverse osmosis, pressure is applied across the membrane that exceeds the osmotic pressure difference and in the direction opposite to the osmotic pressure difference. This results in solvent flow from the high concentration (and high applied pressure) side of the membrane to the low concentration side. Since the solvent flows in a direction opposite to that in normal osmosis, the process is usually called reverse osmosis. In reverse osmosis, the driving force is the applied pressure on the solvent, and the applied pressure must exceed the osmotic pressure difference across the membrane since the osmotic pressure will oppose the reverse osmosis transport/permeation.

Interest in large-scale reverse osmosis operations accelerated significantly with the development of high flux (high for that time) cellulose acetate membranes by Loeb and Sourirajan [2]. The function of reverse osmosis membranes in water desalination is essentially the same as the function of reverse osmosis in the treatment of wastewater, rejection of electrolyte (and usually most other dissolved components) and permeating the water. In the following decades, there have been significant advances in reverse osmosis membranes, and this has included the development of different membrane materials. This development of new membrane material is continuing; the new polyamide membranes are of particular interest.

Several membrane materials, such as cellulose acetate, which initiated the modern interest in reverse osmosis membranes, are capable of high rejection factors, often 90% or much more. This means that at least moderately high purity water can be produced, purities that are suitable for many uses, and concentrations of contaminants are often low enough for discharge. However, as more and more water passes through the membrane, the concentration of solute increases on the high pressure side of the membrane. This increases the osmotic pressure, and higher applied pressures are required to overcome the osmotic pressure. The higher concentrations also are likely to decreases the purity of the permeate water since the concentration of solute in the permeate often increases as the concentration of the solute in the high pressure side of the membrane is increased.

Reverse osmosis functions to concentrate the solute, not to remove the solute completely from all of the water. Although most separation processes are not capable of completely separating components, except with infinite numbers of stages, the limitation of reverse osmosis in this respect is a little different. Membranes that permeate water and reject the solute function more like evaporation than adsorption. (Note that evaporation could be one separation method that competes with reverse osmosis.) One can evaporate a solution to dryness and remove all or essentially all of the water that is not chemically bond to the solute. However, reverse osmosis requires that at least some liquid remain with the solute so that a fluid can be pumped through a reverse osmosis cell. Although this makes it inherently impossible for reverse osmosis systems to produce a liquid free reject stream, there are other limits to the practical osmotic pressure and the degree of rejection that always set far lower limits on the degree of concentration that can be allowed in the reject stream.

When a membrane does not completely reject a solute, or does not reject it adequately, one can consider two stages of reverse osmosis, or one can use a polishing step with another method such as ion exchange to remove the last traces of the solute.

As noted, membrane separations, however, also are not likely to be able to remove all of the water and leave a pure solute, usually a solid. It is obvious that with all membrane processes it will be necessary to have sufficient water remain to "flow" through the membrane cell (the membrane filled equipment). However, with reverse osmosis, the limitation is even more restrictive. With nonvolatile solutes such as electrolytes, the osmotic pressure builds up as the concentration of solute increases, and the pressure required to force more water through the membrane becomes impractically high once the concentration becomes too high. That is, the osmotic pressure difference across the membrane becomes too high for further concentration of the solute to be practical. For most applications, further concentration of the solute usually becomes impractical when the concentration reaches a value of only a few molar. This also means that reverse osmosis processes will only be practical for dilute solutions, but remember that many environmental streams are very dilute; so this limitation may not prevent the use of reverse osmosis in many environmental and waste applications.

Those who have read Chapter 2 will note that adsorption is also usually most practical for dilute systems. This means that reverse osmosis will compete with adsorption and ion exchange for some applications, but reverse osmosis is more likely to be the choice for those applications where an aqueous stream is too dilute for most other competing separation methods, but also too concentrated for optimum use of adsorption or ion exchange. Since reverse osmosis is usually involved in removing ionic materials from aqueous streams, the competition at extremely low concentrations is likely to come from ion exchange. Reverse osmosis is not usually highly selective among the solutes, so all or most of the electrolytes are usually rejected (removed from the permeating water). When the contaminant is only a trace fraction of the total electrolyte and a selective ion exchange material is available that can concentrate the contaminant many-fold by not removing the other bulk electrolyte, ion exchange will have a significant advantage over reverse osmosis. When all or most of the electrolyte must be removed and the concentrations are above trace levels, reverse osmosis is likely to be attractive.

Reverse osmosis is important in waste processing and environmental control for removal of pollutants from a discharge stream, groundwater

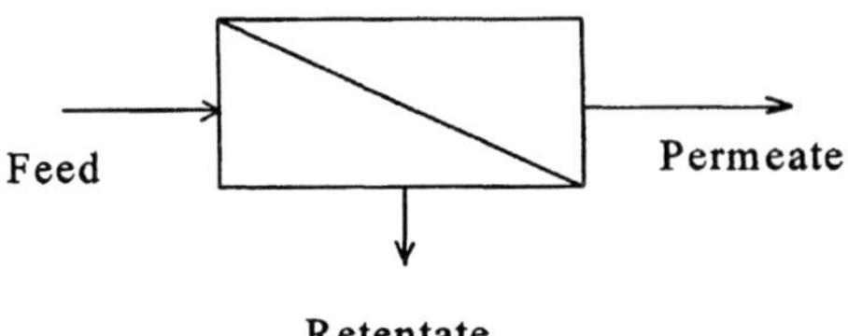

FIGURE 1 Common symbol for membrane units in flowsheet diagrams. The diagonal line represents the membrane, and the arrows suggest a feed stream divided into one stream going through the membrane (permeate) and one not (retentate).

or surface water. The products are usually a purified water stream, hopefully suitable for release to the environment or for reuse, and a more concentrated solution. Although one may prefer that a reverse osmosis membrane selectively reject only the pollutants of interest, many (or most) solutes are rejected; thus, the polluting solutes are likely to be concentrated along with the nonpolluting solutes. Although reverse osmosis usually does not concentrate the solutes selectively (individually), there is usually some selectivity. It is common for membranes to reject divalent cations more effectively than monovalent cations. However, since the most common reverse osmosis application—desalination—is trying to remove alkali metals effectively, the reverse osmosis systems will be designed for removing the monovalent cations and removal of divalent cations is simply a little more effective.

Since reverse osmosis is essentially a method for separating the solvent from all of the solutes, not a method for separating the different solutes, it is also likely to compete with evaporation, which (for electrolytes and other nonvolatile solutes) produces the same results, but perhaps with a greater consumption of thermal energy. It is the significant reduction in energy consumption that has allowed reverse osmosis to penetrate markets (applications) that otherwise would probably utilize multi-effect evaporation.

A pressure driven membrane process like reverse osmosis can be viewed as a simple division of a solution into two streams: the permeate stream with little solute and the reject stream with a high concentration of solute. The symbolic representation of a membrane step is shown in Figure 1. The one stream entering the membrane cell is divided into two streams and is used in schematic process diagrams to denote a variety of membrane processes. The slash across the "box" symbolically represents the membrane. Note that one stream appears to pass through the membrane, and the other stream is rejected or does not pass through the membrane. Of course, this is only a symbolic representation, and the real internal structures of membrane cells, which will be discussed later, are much different. In reverse osmosis, one stream is depleted in solute (the permeate stream), and the other stream is enriched in solute (the reject stream). Simple material balances on the solute and the solvent (or the overall flow) relate one of the three streams to the composition of the other two streams:

$$F_F = F_P + F_R \tag{1}$$

$$F_F c_F = F_P c_P + F_R c_R \tag{2}$$

There can, of course, be material balances on each solute when more than one solute must be taken into account. The applied pressure difference, the membrane properties, and the solution composition determine the permeate flow rate per unit area of membrane. The size of the membrane is determined by the total permeate flow rate, and thus the size of the reject stream, desired. The designer usually has control over the relative size of the permeate and reject streams by controlling the size of the membrane cell (membrane area), the applied pressure difference, and/or the feed rate.

However, both the rejection of solute and the permeation rate can be functions of the solution composition and the applied pressure (or the permeation flux). Since the concentration will change with time in a batch system (or, for a more common continuous system, over the length of the membrane cell), the material balance given above can be applied either to local (or instantaneous for a batch system) conditions or to the overall performance of the cell, that is, to the total permeate and reject streams. The overall performance is an average of what is happening over the entire surface of the membrane (or over a specific period of time for a batch system). To understand the membrane process, it is necessary to look at the performance locally at each position on the membrane surface.

The rejection of solutes by reverse osmosis membranes can be defined from the concentration of solute in the permeate stream, the stream passing through the membrane:

$$R = 1 - \frac{c_P}{c_R} \tag{3}$$

The concentration in the reject stream (at the high pressure side of the membrane) is c_R, and the concentration in the permeate at this point in the membrane is c_P.

The rejection of membranes can be a function of the permeation rate (thus the pressure drop across the membrane) and the solution composition. There are various mechanisms or models proposed for the rejection, and the dependence of rejection upon permeation rate and solution concentration can be different for each model. The agreement of such models with experimental data provides the principal justification for using such models to design new membranes or to extrapolate designs to new ranges of operating conditions. However, most models cannot take into account all of the possible phenomena that take place in membrane transport, and care should be taken when using models to extrapolate the behavior of membrane systems far beyond the available experimental data.

Theoretical Models for Membrane Performance (Reverse Osmosis)

Although there will be no attempt to cover all membrane models in this book, two of the most important classes of models will be discussed. The first, and probably the most obvious, model of a membrane is a "pore model." This is a direct analogy to a filter. A membrane with small pores will certainly reject molecules that are larger than the pores. However, such membranes may also reject (but not completely) molecules that are smaller than the pores, for several reasons. In some cases, molecules of the component that is not rejected (the one that permeates better through the membrane) may be adsorbed on the pore surfaces and block the pore or hinder permeation by the nonadsorbed specie (usually the solute in reverse osmosis). The adsorbed specie can then move across the membrane (down the pores) by flow of adsorbed molecules and by diffusion of adsorbed molecules as well as by flow down the pores. Of course, a membrane also can have a variety of pore sizes, so some pores may be small enough to reject molecules that could pass through other pores.

The second class of models is based upon dissolution and diffusion of species within the membrane. In these models, the membrane does not even have to have pores, and when pores are introduced into such models, they may be introduced as imperfections that prevent the membrane from giving the maximum rejection possible. In such models, rejection results because the solubilities of the two components in the membrane are different, often greatly different, or the diffusion coefficients of the different components in the membrane are different. If a reverse osmosis membrane can "dissolve" only water and not solutes, the rejection of such a membrane would then be essentially 100%, provided there were no imperfections such as pores or cracks that would let solute "leak" through the membrane by other mechanisms. If the solute had a smaller, but finite, solubility in the membrane, the rejection would be high, but less than 100%. The solubility of solvent and solute in a membrane could follow any equilibrium relation, and the equilibrium relations would determine the maximum rejection that could be obtained at any concentration in the feed solution if there were no holes, cracks, or other imperfections in the membrane.

Although it may be convenient to think of dissolving the solvent (water) in the membrane, one should not carry the image of dissolution too far. Extremely small pores that are selectively wet by the solvent can give essentially the same effect. Note that "nanoporous" membranes, which are certainly viewed as porous materials, usually show significant rejection of salts. Thus, reverse osmosis membranes can be thought of as an extension of nanofiltration to even smaller pore sizes, but, historically, reverse osmosis membranes became popular before nanofiltration.

Rejection of a solute by a reverse osmosis membrane can involve different mechanisms for transport of the permeating and rejected components through the membrane. For instance, transport of the permeating component could result principally from diffusion of disolved components through the membrane, while the flux of rejected component through the membrane could result principally from flow through pores or membrane imperfections. (Remember that no membrane can be absolutely perfect, and when high rejection is involved, the presence of even small imperfections can be an important factor in membrane performance.) If the dissolution of solvent in the membrane were only a simplification of adsorption of solvent on the membrane pores that block the permeation of solute, the presence of some solute flux could result from a distribution of pore sizes; larger pores may be too large for the adsorbed solvent to block them completely. Whatever the exact mechanism(s) for solvent and solute transport, this simple model, which views solvent transport as a dissolution-diffusion phenomena and solute transport as flow through pores and other imperfections, can explain much of the behavior of some membranes in rejecting solutes over a range of concentrations and pressure differences.

From the user's point of view, it is more important to know the rejection as a function of concentration and operating conditions; it may be less important to understand the mechanism and the theoretical relation of rejection to these parameters. However, those developing new membranes need to understand the mechanisms of rejection and the theories or models based upon these mechanisms to guide them in selecting new approaches to membrane preparation. Furthermore, a reasonable model of the mechanism of permeation of solute and solvent that follows experimental behavior over a significant range can add confidence in the interpolation or even extrapolation to other conditions.

Osmotic Pressure

As noted earlier, in reverse osmosis, pressure must be applied to overcome the osmotic pressure across the membrane as well as the flow resistance through the membrane. When two solutions with different concentrations are separated by a permeable membrane, the osmotic pressure difference between the solutions will cause the solute to diffuse through the membrane from the side with a lower concentration of solute to the side with a higher concentration of solute if sufficient "reverse pressure" is not applied, and this is contrary to the goal of reverse osmosis. Osmosis, of course, dilutes the high concentration solution. If the process were allowed to continue, osmosis (diffusion of solvent to the high concentration solution) would continue until the two concentrations were equal, that is, until the osmotic pressure becomes the same on both sides of the membrane. To prevent osmosis, a counterpressure must be applied that is equivalent to or greater than the osmotic pressure. To achieve useful reverse osmosis rates, the applied pressure must be significantly greater than the opposing osmotic pressure. The principles of osmosis and the osmotic pressure are explained in more detail in most textbooks on physical chemistry; the following discussion is a brief summary of the basic principles.

The osmotic pressure results from differences in the chemical activity of the solvent in the different solutions. The presence of solute lowers the concentration and the activity of the solvent. This is usually expressed thermodynamically as

$$p_{\text{osmotic}} = \frac{RT}{v_s} \ln a_s \tag{4}$$

where the temperature (in kelvins) is T, R is the gas constant, v_s is the partial molar volume of the solvent, and a_s is the chemical activity of the solvent.

Most solutions are sufficiently dilute (the solvent concentration is often many times greater than the solute concentration) that the osmotic pressure can be approximated by the van't Hoff equation,

$$p_{\text{osmotic}} = cRT \tag{5}$$

where c is the concentration of solute. This means that the difference in the osmotic pressure across a membrane that transports solvent, but not solute, is approximately proportional to the difference in the concentrations across the membrane.

If the pressure applied to the high pressure side of the membrane is constant, the permeation rate will decrease with time in a batch cell or with distance down the channel in a continuous cell because the concentration increases on the high pressure side of the membrane. The increase in osmotic pressure with concentration places a practical limit on the degree of concentration that can be achieved by reverse osmosis, and this limit is usually a few molar. In continuous cells, there could also be significant pressure drop along the flow channel; then the pressure will be the least at the end of the channel where the opposing osmotic pressure is the greatest. Thus, the applied pressure in a reverse osmosis cell must (1) overcome the osmotic pressure, (2) drive the solvent through the membrane at an acceptable rate, and (3) move the fluid through the flow channels in the reverse osmosis cell. As the concentration of solute in the reject stream is increased, a greater osmotic pressure will have to be overcome, and with sufficiently high solution concentrations the applied pressure required to maintain a practical permeation rate may become too costly. Although the osmotic pressure is a thermodynamic phenomena that represents the minimum energy required to make a separation, the limitation on the practical solute concentration is likely to result from the need for sufficient additional pressure to maintain a practical permeation rate. The economics of reverse osmosis become less attractive at concentrations above a few molar, but there are often other limitations, such as the precipitation of solids, that could limit the retentate concentration to significantly lower values.

Concentration Polarization

The behavior of a membrane can be affected by mass transfer resistances; in membrane technologies, this is usually referred to as "concentration polarization." This mass transfer problem arises because of the different fluxes of solute and solvent through the membrane. The flux of solvent through the membrane, F_w, may be expressed as

$$F_w = -P \, \Delta p \tag{6}$$

where Δp is the pressure drop across the membrane, after correcting for osmotic pressure, and P is the permeability of the membrane, that is, the ratio of the flux of solvent through the membrane to the pressure drop driving force. In some cases, P will be approximately a constant with little dependence on pressure. In other cases, P could be strongly dependent on the pressure or the pressure difference. This dependence can result from compressibility of the membrane. Because the opposing osmotic pressure depends on the concentration, the permeating flux depends upon the concentration of solute at the membrane surface.

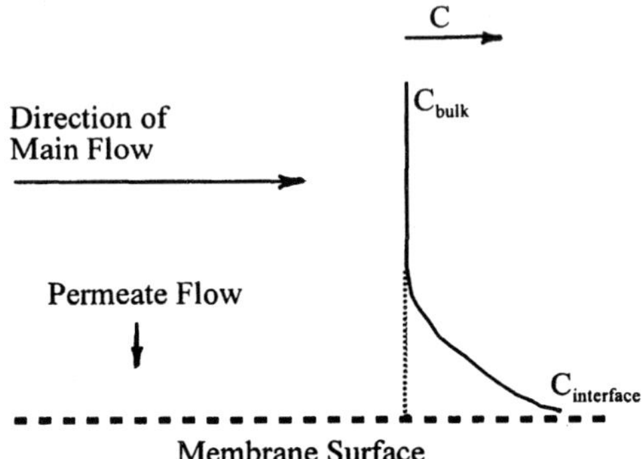

FIGURE 2 Schematic of concentration polarization on a membrane surface from convection of rejected solute carried to the membrane surface by the permeate flow that passes through the membrane. The accumulation of rejected solute is opposed by mass transfer (diffusion and eddy convection) of solute back into the main stream.

Concentration polarization arises because solvent flowing toward the membrane also carries solute toward the membrane. If the membrane rejects this solute, the excess solute must be transported away from the membrane by diffusion. That mass transfer process is usually called concentration polarization. If mass transfer from the membrane surface can be expressed in terms of a simple mass transfer coefficient, concentration polarization can be described as

$$u_w(c_i - c_p) = k(c_i - c) \tag{7}$$

The flux of solvent through the membrane (left side of the equation) is expressed in terms of the velocity of solvent toward the membrane at the membrane surface multiplied by the difference in the concentration at the interface and the concentration in the permeate. The concentration of solute is c in the bulk solution, c_i at the membrane surface, and c_p in the permeate. This equation simply equates the flux of solute toward the membrane by convection to the flux of solute from the membrane by diffusion of solute from the membrane surface and assumes that there is no accumulation of solute at the membrane surface. If the membrane totally rejects the solute, c_p will be zero. That is often a fair approximation for highly rejecting membranes.

The important result of concentration polarization is the increase solute concentration at the membrane interface (Figure 2). As Equation (7) has just shown, the concentration c_i at the surface will be greater than the concentration c in the bulk or main part of the solution. The higher concentration at the membrane surface is important because it affects the effective rejection and the osmotic pressure (and thus the flux through the membrane). Even if the fraction rejected were independent of concentration, the apparent rejection based upon the bulk phase concentration would change because the actual rejection is based upon the concentration at the membrane surface rather than in the bulk liquid. If the fraction rejected is independent of concentration, that would mean that the flux of solute is proportional to the concentration (the same fraction is rejected). Then the concentration of solute in the permeate (the stream going through the membrane) would be proportional to the concentration at the membrane interface not to the bulk concentration. Since the concentration at the membrane surface is always greater than the concentration in the bulk solution, the apparent rejection based upon concentrations in the bulk solution will always be less than the actual rejection, which should be based upon the concentrations at the membrane surface.

Similarly, the higher concentrations at the membrane surface mean higher osmotic pressures that must be overcome and lower fluxes for the same applied pressure, so the apparent permeability based upon the bulk concentrations could be less than the actual permeability based upon the osmotic pressure at the interface. The difference between the apparent and actual rejections will decrease as the mass transfer coefficient is increased further. Concentration polarization needs to be taken into account when designing reverse osmosis membrane systems and when interpreting data taken from them.

Another important effect of concentration polarization is the potential for precipitating solids on (near) the membrane surface. As noted earlier, reverse osmosis can only produce a concentrated solute, not a solute free solute. If a dry solute is required, it is necessary to use a second separation method for that purpose, perhaps evaporation. If the solutes in the feed stream are all highly soluble, the limited concentration that can be reached by reverse osmosis is likely to be set by the buildup of osmotic pressure as the concentration in the reject stream increases. As the osmotic pressure increases, the pressure required to drive the solvent through the membrane increases and eventually reaches impractical values. However, if some of the components in the feed stream have limited solubilities, they may precipitate before the osmotic pressure becomes unacceptably high. In those cases, the precipitation is likely to occur in the concentration polarization film on or near the membrane where it could foul the membrane. Because of the limited solubilities of common salts such as calcium carbonate, precipitation on the membranes is not an uncommon problem. It can be an important contribution to limiting membrane life because solids on the membrane surface can affect the membrane life significantly, and reducing the concentration polarization can reduce this problem.

When electrolytes are being rejected, it is useful to remember that diffusion of charged specie, such as ions, generate electric fields that act to maintain local electroneutality. Diffusion of ions is governed by the Nernst-Planck equation rather than the simpler Fick equation, and the effective diffusion coefficient of one ion is affected by the concentration gradients of the other ions. This effect is not likely to be large [3], and the effect may not be obvious if film resistance is measured experimentally. However, estimates of the effects of concentration polarization should be based upon the diffusion of salts (not individual ions) under similar conditions. The Nernst-Planck equation is described in more detail in Chapter 2 because it is more important in ion exchange than in other separation processes, including reverse osmosis.

The mass transfer coefficient can be increased by increasing the flow rate across the membrane. When special efforts are made to increase the mass transfer coefficient by using very high cross-flow over the membrane surface, the operation is frequently called cross-flow reverse osmosis. Of course, this is only a quantitative difference since there is always some flow over the membrane surface. The effects of cross-flow on concentration polarization are similar to the effects of cross-flow on most mass transfer operations (or even similar to the effects on heat transfer operations) in similar geometries, but with high membrane fluxes there can be some differences, because in reverse osmosis systems the new flow of fluid toward the membrane can alter the flow in the boundary layer near the membrane and thus affect the mass transfer coefficients. A significant flux toward the membrane will reduce the thickness of the hydrodynamic boundary and may stabilize it and delay transition to turbulent flow, but in most systems this effect will be small. These changes in the hydrodynamic boundary layer will thus have an effect on the mass transfer boundary layer, which is usually much thinner and affected largely by the velocity gradient at the membrane surface. When the net flux toward the membrane is sufficiently small (the most common case), these effects can be neglected, and the mass transfer coefficients and the resulting effects of concentration polarization can be estimated from standard correlations for flow in the channel shape of interest and, thus, neglecting the effects of the permeation flux on the mass transfer coefficient (but not neglecting the effect of permeation on transport of solute toward the membrane surface).

Structure of Membranes

Membranes can be made of numerous materials in many different ways, but the user is not likely to have to know details of their preparation. However, it is desirable to know at least the general structure of several membranes to aid in their selection and to anticipate their performance. Two structures have already been mentioned: porous membranes and solid (or essentially solid) membranes that transport solvent by dissolution and diffusion. These need not be distinct since membranes can have pores and still dissolve solvent. The possibility of adsorbed solvent molecules blocking the pores from transporting solute was also mentioned. If the adsorbed solvent molecules diffuse along the surface of the pores, the behavior can be much like dissolution of solvent in the solid membrane, except the permeability will depend upon the surface area of the pores.

Membranes can be constructed of polymers or inorganic materials, but we will consider three groups of membranes: porous materials with micro-pores, "crystalline" polymers, and "rubbery" polymers. Only the first two will be considered in detail. Rubbery polymers are discussed principally because their performance needs to be contrasted with the behavior of crystalline polymers.

Micro-porous membranes can be made of inorganic materials or polymers. Polymer membranes can be "solid" or porous, but almost all inorganic membranes of importance to waste and environmental processes will be micro-porous materials. (The metal membranes that are used for purifying hydrogen are notable exceptions—inorganic membranes that do not need pores. However, these metal membranes are less likely to have important applications in waste and environmental problems.) Micro-porous membranes are not likely to separate small molecular materials by simple filtering action because the pores are likely to be much larger than any of the molecules to be separated. Nevertheless, micro-porous materials can still separate some larger molecules. As noted earlier, macromolecules can often be separated according to their size, and this is usually called ultrafiltration or micro-filtration rather than reverse osmosis. Solvents or solutes can adsorb on the surface of membrane pores and enhance permeation of that component by surface diffusion.

A membrane can also be nonhomogeneous. This means that the membrane properties may be different at various depths below the membrane surface. In fact, it is usually desirable to have the membrane much different on one surface than on the other. There can be significant advantages in having the membrane properties vary with depth in the membrane, but, of course, there is no merit in having variation along the plane of the membrane. Variations of properties with depth into the membrane can be very important and is a characteristic of many commercial membranes, perhaps even of most new membranes. The difference between membrane properties on surfaces, or even throughout most of the membrane thickness, can result from the drying step or other steps in the method used to prepare the membrane. The difference also can be created intentionally by preparing the surface and the base support structure in separate steps with the explicit purpose of improving the properties of the membrane.

For small pores or even selective dissolution of solvent, it is usually desirable for polymer membranes to be relatively dense. However, dense membranes are usually not very permeable, and more porous or highly hydrated membranes are likely to have better permeabilities. The trade-off between greater permeability (higher throughput) and greater rejection (higher performance) is common in selecting membranes for reverse osmosis .

The way to achieve high selectivity and still maintain high permeability is to make the membrane very thin. However, there are practical limits to how thin a membrane can be made and still be effective. The first limit in making the membrane thinner usually is the need to maintain sufficient mechanical strength and avoid tears, holes, and other damage during even normal careful use and handling. As noted earlier, even a very small number of imperfections in a highly selective membrane can greatly reduce its effective selectivity. To make the membrane even thinner without losing mechanical strength, one can fabricate the ultrathin membrane on a much stronger substrate. The substrate can be strong and relatively thick but highly porous since no separation performance is required from the subtract. All of the selectivity can come from the thin film of "active" membrane. Some membranes, such as cellulose acetate membranes, which initiated the strong interest in reverse osmosis, are produced by "drying" the membrane on a flat solid surface. During the drying process, a thin film is formed on the surface facing the air or gas, and the rest of the membrane remains relatively porous. The procedure creates a thin film of active membrane on a stronger, but more porous, substrate. This very thin "skin" on the cellulose acetate membranes was one factor that made them such a significant advance in reverse osmosis, and similar thin films are usually preferred for other membrane applications. The same effect can be achieved by intentionally forming a very thin layer of active membrane on a preformed porous structure. The thin film could be made of the same polymer as the supporting porous structure, or it could be made from a second material. One is not limited to the use of only two layers of material; it may be desirable to have a thick and strong layer of highly porous material for the basic strength, a thinner layer of material with smaller pores, and a final ultrathin layer of the active film. It is also possible to form an active layer of organic polymer, and perhaps a layer of intermediate porous support material, on a highly porous and strong inorganic support.

It is probably obvious that thin supported membrane composites are directional, and the thin film should be on the high pressure side of the membrane. The more porous support side of the membrane should be on the low pressure side. The performance and the stability of the membrane can depend upon the orientation of the membrane. Placing too much pressure difference in the wrong direction could separate the thin film from the substrate. Note also that permeation in the opposite direction, through the highly porous layer first, could result in greater concentration polarization because the cross-flow would not be able to penetrate effectively into the highly porous, but not separating, region of the membrane.

One final membrane structure will be mentioned, which has found significant applications in environmental applications of reverse osmosis. These are often called "dynamic" membranes. This term describes membranes that are formed on a porous surface during, or just prior to, operation in reverse osmosis. This technique was pioneered by Johnson and co-workers at the Oak Ridge National Laboratory several years ago [4,5]. In this system a highly porous strong structure is used which has little or no power to reject the solute of interest. Then a suspension of unconsolidated polymer (usually a polyelectrolyte) is added to the solution, and pressure is applied across the porous material. As water is pumped through the porous material, the polymer is carried by convection to the porous surface, but the polymer molecules are too long to pass through the porous medium. Then a "film" of polymer is formed on the porous medium. This can be viewed as a profitable use of concentration polarization (polarization of the polymer not the solute) or the creation of a precoat much like that used in filtration. If the film of polymer can reject the solute, this "dynamic" membrane can offer performance much like other externally prepared membranes. It has the additional advantage that the membrane can be easily replaced or repaired. In most cases, repair will be almost automatic. The permeability and the solute rejection can depend upon the thickness of the dynamic membrane, and the thickness is set by the cross-flow rate and the permeation rate (determined by the applied pressure).

In principle, the membrane thickness of a dynamic membrane is governed by Equation (7), which gives the concentration polarization at a membrane surface in terms of the mass transfer coefficient, but some care is needed in estimating the mass transfer rate for the polymer, which

TABLE 1 Commercial Reverse Osmosis Membrane Manufacturer

Desalination Systems	USA
Dow Plate and Frame (Dow Chemical)	USA
Dupont & Co.	USA
FilmTec Corporation (DOE Chemical)	USA
Fluid Systems Corp (Allied Signal)	USA
Hydranautics, Inc. (Nitto Denko)	USA (Japanese owned)
Kitto Denko Corp.	Japan
Osmonics, Inc.	USA
Rochem	USA
Separem, SPA	Italy
Sumitomo Chemical	Japan
Toray Industries	Japan
Toyobo Co.	Japan
TriSep Inc.	USA

Source: List derived partially from Brandt, D. C., in *Reverse Osmosis: Membrane Technology, Water Chemistry, and Industrial Applications* (Z. Amjad, ed.), (1993), p. 1.

is likely to be elongated in the film with preferential alignment with the microporous membrane surface. Such diffusion coefficients could be significantly different from those measured for the same polymer molecules in more random arrangements. Once a film of polymer is collected on the porous surface, it is not rapidly displaced as long as continuous flow of permeate is maintained. The limitations of dynamic membranes are the selection of polymers that can be used, the thickness of the films needed, and the limited number of membrane shapes that can be used. Most testing of dynamic membranes has involved moderate sized tubes of porous materials. The high flow rates needed to control the thickness of the dynamic membranes probably eliminates the use of hollow fiber or some spiral wound systems which have extremely high membrane areas per unit volume of cell, but which may have too much flow resistance in one side of the membrane to allow sufficient control of the cross-flow rate. One important advantage of dynamic membranes is the ability to remove and replace the membrane without undesirable disassembly of the equipment. That could be an important advantage when the membrane is particularly prone to fouling. These dynamic membranes have found application in the textile industry to recover valuable dyes and to avoid discharge of the dyes to the environment.

A list of some example reverse osmosis membranes available commercially is given in Table 1.

Concentration Driven Processes

In many respects, there may be little difference between pressure driven and concentration driven membrane processes since both are based upon a difference in the concentration or activity of the permeating component across the membrane. Note that for gases the concentration and partial pressure are approximately proportional to each other. Many gas separation processes discussed here as concentration driven processes could have been discussed as pressure driven processes. Increasing the total pressure increases the partial pressure and the concentration of all components. In pressure driven permeation, the chemical activity of the permeating component on the upstream side of the membrane is increased by the applied pressure. In concentration driven membrane processes, the chemical activity on the upstream side of the membrane remains the chemical activity of the permeable component in the solution or gas mixture, but the activity on the downstream side can be reduced sufficiently for significant permeation. However, except for osmosis, the permeable component in these processes is more likely to be a solute and less likely to be the major component of the solution or mixture as in reverse osmosis. As noted, it is usually desirable to permeate the minor component(s) when that is possible and practical.

One of the most important applications of membranes, and one that seems likely to grow, is in separation of gases. Membranes can separate hydrophilic gas molecules, such as CO_2, from hydrophobic molecules, such as hydrocarbon molecules. Membranes have been found to be capable of separating common gases such as nitrogen and oxygen. Membrane processes can even compete with other methods (cryogenic separations) for small capacity units and where ultrahigh purity components are not required (but pressure swing adsorption also becomes competitive for very small units). A large fraction of the costs of membrane equipment is in the membranes themselves, and further improvements in the performance or in the cost of membranes could extend the range of applications or the range of facility capacities over which membranes can compete in these important separations.

The most common membrane applications for separating gases in waste and environmental problems may be for treatment of off-gas systems to remove selected pollutants. In these cases, the pollutant is likely to be present at low concentrations, and it will be desirable to have a membrane system that will pass the pollutant and retain the bulk air. One group of such operations could be the removal of hydrophobic gases such as hydrocarbons or solvent vapors from air. A variety of hydrophobic membranes will selectively favor permeation of organic vapors over air. Similar membranes may be useful in removal of organic components from water, but in those cases it may be necessary to raise the temperature to obtain a significant vapor pressure of the hydrophobic organic/solvent pollutant. If the temperature is raised to approach the boiling point of water, such operations are often referred to as "pervaporation," and will be discussed later in the chapter.

In gas systems, the downstream partial pressure (or chemical activity) of the permeating component must be maintained lower than the partial pressure in the upstream side of the membrane. This can be achieved by reducing the total pressure on the downstream side or by flushing another gas across the downstream side of the membrane. Of course, it is possible to pressurize the upstream side of the membrane as well as reduce the pressure on the downstream. Remember the similarities in gaseous pressure driven and concentration driven membrane processes since partial pressure and concentration of gases are approximately proportional.

In liquid systems, membranes are available that can selectively pass either hydrophilic components or hydrophobic components from solutions. It is even possible to selectively pass certain metal ions through membranes. In liquid systems, the mechanisms of any of the processes are more likely to involve some type of dissolution or attachment of the solute to the membrane and diffusion of the component or the complexed component through the membrane. The downstream side of the membrane is usually another liquid. Liquid in the downstream side of the membrane must be "flushed" from the cell. Because concentration driven processes depend upon the difference in the activity of the permeating solute on the two sides of the membrane, the concentration of solute on the downstream side can affect the permeation rate. Some of the most effective membrane systems for removing contaminants from liquids involve converting the permeating contaminant largely into a non-permeating component on the downstream side of the membrane. This can involve neutralizing acidic or basic contaminants or other chemical reaction with the permeating contaminant.

Permeation/Diffusion within Membranes

Gases can permeate through the pores of membranes or become dissolved in the membranes and diffuse through the solid membrane materials. Gas molecules can be adsorbed in the pores of the membrane, block the permeation of other components, and permeate through the membrane by surface diffusion. These are essentially the same mechanisms described for reverse osmosis membranes.

Membranes can be made of organic polymers or porous inorganic materials, and their performance for specific applications can be enhanced by incorporating highly selective reagents for separating specific components. These can be especially effective for transporting specific metal ions in liquid systems because highly specific ligands are available for several ions. These ligands can be grafted chemically to the membrane polymer, introduced into the polymer by a solvent that penetrates and swells the membrane (called a supported liquid membrane and discussed later), or trapped within the membrane polymer [6]. Moderately large molecules containing ligands, such as crown ether groups used in this example, can be added to the polymerization mixture. If the resulting polymer is sufficiently tight, the entrapped ligand bearing molecules will not be able to diffuse out of the membrane polymer.

In gaseous systems, the molecules will diffuse through a porous membrane by Knudsen diffusion, provided the diameters of the pores are significantly smaller than the mean free path of the molecules in the gas. The flux of all components will be proportional to the difference in the partial pressure of that component across the membrane. The diffusivity of each component of the gas will be inversely proportional to the square root of the molecular weight of that component. This relation comes from the ideal gas law, but it is a reasonably good assumption for most Knudsen flow situations because there are no significant molecule-to-molecule interactions. Since these interactions can be ignored, Knudsen flow has the simple feature that the flux of any component is independent of the concentration, or the presence, of any other molecules.

Thus, microporous membranes operating under Knudsen diffusion will separate molecules strictly according to molecular weight. The ratios of the permeabilities will be inversely proportional to the square root of the molecular weight of the components, and the lighter (smaller molecular weight) components will diffuse more rapidly. The separation factors will be the square root of the inverse ratio of the molecular weights of the two components of interest. Although this discussion has been limited to gaseous systems, one should remember that it could apply to some liquid systems such as pervaporation where the components act like gases within the membrane.

Knudsen diffusivity is also proportional to the diameter of the pores, so the permeability of a microporous membrane under Knudsen flow will increase linearly with pore diameter. However, once the diameter of the pores approach the mean free path of the gas, molecule-to-molecule interactions will begin to become important, and the permeation will begin to take on the properties of viscous flow—the case when interactions with the wall can be ignored and molecule-to-molecule interactions dominate the gas behavior. There is no separation of molecules on the basis of molecular weight in viscous flow, and the separation factors based upon Knudsen flow for a microporous membrane will decline and approach unity as the pore diameters are increased and exceeds the mean free path of the molecules.

Microporous membranes can also separate molecules because of preferential adsorption of one or more components on the surface of the pores. In some cases, there can even be condensation of a component (multiple layers of a component adsorbed) in the pores. The adsorbed material may then diffuse along the pores faster than Knudsen flow can transport nonadsorbed molecules through the pores. The "adsorbed" molecules can reduce the volume of the pores available to nonadsorbed molecules and block or reduce the flux of those molecules through the pores. Remember that Knudsen diffusion coefficients are proportional to the pore diameter, and the cross-section is proportional to the square of the pore diameter. This type of separation is not necessarily dependent upon the size of the molecules, but is more dependent upon chemical properties of the molecules and the surface. The effect of condensation in pores is also dependent upon the boiling points of the permeating components because the degree of "condensation" in the pores increases as the partial pressure and temperature approach the condensation/boiling conditions. (Of course, condensation in and out of the pores becomes complete when the condensation conditions are met.) The degree of adsorption/condensation increases as the pore size decreases and as the condensation conditions are approached.

Perhaps the most common case of selective adsorption on microporous membrane pores is in the separation of hydrophobic molecules from hydrophillic molecules. This can be very useful in separating hydrocarbons from water, but there can be a variety of pore surface and permeating molecules that affect selective permeation of components in microporous membranes. It is generally desirable for the adsorption to be selective, but the adsorption usually should not involve sufficient bonding strength that surface diffusion will be slow or even negligible. Thus the surface properties of the pores (such as the wettability to water or organic contaminants) can be particularly important.

Microporous materials for membranes can be produced using the same methods often used to make microporous adsorbents. That is, they can be made from inorganic materials by compressing and sintering powders of very fine particles. If uniform sized particles are used in the preparation, relatively uniform effective pore sizes can be obtained. Of course, to produce microporous membranes with pore sizes sufficiently small for many membrane applications, it is necessary to start with extremely small particles. Microporous carbon membranes can also be produced in the same way that carbon molecular sieves are produced for adsorption, by carbonizing a suitable polymer. Carbon molecular sieve membranes have been reported to have good selectivity for the important low molecular weight gas separations, but traces of hydrocarbon impurities in the gas streams can accumulate in the highly hydrophobic molecular sieve membranes and severely degrade membrane performance. Similarly, water can accumulate in the pores of membranes. Jones and Koros [7] reported that accumulated higher molecular weight hydrocarbons can be removed from carbon molecular sieve membranes by passing propylene through the membranes to remove the accumulated higher molecular weight hydrocarbons and restore the membrane approximately to its initial performance. Apparently propylene has an intermediate molecular weight, so its vapor pressure lets it permeate the membrane at a practical rate; yet it is able to "dissolve" or otherwise move the higher molecular weight hydrocarbons through the membrane. Although several common hydrocarbon contaminants were tested, it is possible that not all contaminants will be removed so easily by propylene.

Organic membranes can also be solid materials with no obvious pores for gases or liquid to pass. In these cases, the permeating molecules must, in effect, dissolve in the membrane surface. These polymers can be classified as "crystalline" or "rubbery." The crystalline polymers retain their structure as the permeating gases enter the membrane, but the rubbery polymers appear to swell, at least slightly, as components enter the polymer. Crystalline polymers generally give low permeabilties, but they can have high selectivities for permeation of different components. The permeation of a particular component through crystalline polymers can show little dependence on the concentration (or partial pressure) of the permeating component, or even on the presence of other components, until the loading of permeating components becomes relatively high. However, permeation rates through rubbery polymers can be highly dependent on the presence and concentration of all components. Often the permeation rate will increase with increasing concentrations, suggesting that the permeating components "expand" the polymer and allow higher permeabilities.

Because of their higher selectivities, crystalline polymers are more likely to be desired for many waste and environmental applications, as they are for so many other applications. However, because of their low permeability, it is desirable to make the membranes very thin. Since the mechanical strength of the membrane also declines with thickness, it is usually necessary to provide structural support for such thin membranes. This problem has been discussed earlier in terms of reverse osmosis membranes, and the same solutions to the problem are available for membranes used in concentration driven processes. There are also other parameters, other than strength, that would limit how thin the membrane could be made. For permeation to behave as just described, the membrane should be sufficiently thick that it is bulk diffusion within the membrane that determines the permeation rates and the separation factors (the permeation rates of all components) and not surface effects. This is not likely to be a limitation for most solid membranes, such as those likely to be used in most environmental and waste application, but surface effects are important in some extremely thin films such as soap bubbles.

The thin active layer on the surface can be produced naturally on some membrane materials simply because drying membrane materials, such as cellulose acetate in air, produces such a thin and dense film on the surface exposed to air. As noted, most reverse osmosis membranes and a large fraction of the newer membranes developed for other applications are "asymmetric" and consist of a thin active film supported by a more porous substrate. These include the more recent aromatic polyamide films made by Dupont and the composite membranes offered by UOP [8]. The thin active region of the membrane can be as thin as 0.1 μm. In other cases, the thin film may be created separately and attached to or formed on a more porous but thicker and stronger support membrane. The film and support material do not even have to be made of the same material as long as a suitable bond between the materials can be achieved. In some cases more than two layers of materials have been chosen. If an extremely porous support structure is needed, it may be desirable to insert a material with intermediate porosity between the then dense film and the highly porous support. The immediate layer may help the very thin, but weak, film span the larger pores of the base supporting substrate.

When gases are separated, the porous support material may have pores small enough for Knudsen diffusion to occur. In such cases, the porous support can contribute to the separation achieved, but this contribution is likely to be a relatively small fraction of the total separation for a highly selective membrane. In all cases, Knudsen diffusion favors permeation of the component(s) with the lower molecular weights. (This was discussed earlier in terms of the porous membranes.) If the thin surface film on a membrane is more permeable to the low molecular weight components, the Knudsen flow may contribute further to the separation; if the thin film is more permeable to the higher molecular weight components, the Knudsen flow may subtract, but perhaps only slightly, from the separation achieved by the "active" portion of the membrane.

Membranes can be made of inorganic materials as well as organic polymers. Although organic polymer-based membranes are probably the more common, there is a growing interest in inorganic membranes. These offer much greater structural strengths, and they can be used in higher temperature operations. High temperature applications were not important in reverse osmosis, but the new developments could result in new materials that will become important in reverse osmosis as well as in other environmental applications, especially in gas separations and off-gas treatment. Inorganic membranes can be made of metals, ceramics, or glasses. They do not have to be homogeneous; thin films can be added to inorganic materials just as with polymer membranes. The thin films can be other inorganic films or (more often) organic polymer films.

Pervaporation

Pervaporation can appear to be a cross between evaporation and membrane permeation [9]. In this method the fluid to be separated is a liquid mixture, and the fluid on the low pressure side of the membrane is a gas (or vapor). The liquid is heated to, or near, the boiling point, and the vapor is passed through the membrane rather than through the liquid-vapor interface as in normal evaporation. The membrane is selective and passes one component more readily than the other. As with many membrane processes, it is often preferable if the membrane passes only the more dilute component, but the important aspect of pervaporation is the ability to shift the effective vapor-liquid equilibrium from the normal equilibrium by including the selectivity of the membrane along with the selectivity of the vapor-liquid equilibrium. Thus, pervaporation can "break" azeotropes, reduce the water vaporized along with moderately volatile organic compounds, or assist in difficult separations in other ways by shifting the effective vapor-liquid equilibrium.

Pervaporation has been studied for several decades [10], but interest started to accelerate quickly during the 1970s. Before 1992, pervaporation was only a secondary subject in *Chemical Abstracts*, but since then it has been a primary subject. Pervaporation is currently one of the fastest growing parts of the membrane industry [11]. Although it is sometimes viewed principally as a potential substitution for distillation, it can be used in some cases to supplement distillation. The largest application of pervaporation at the present time is believed to be for removal of water from oil (organic liquids). Such operations use hydrophilic membranes that transport water preferentially over hydrophobic organic liquids.

There are numerous possibilities for using pervaporation in waste and environmental processing, but those applications have not been fully developed. The removal of trace volatile organic compounds from water is a prime example. That would be the opposite of the pervaporation process that is more common at this time—removal of water from organic compounds. Pervaporation should be especially attractive for cases where the organic component has insufficient volatility to remove by gas stripping at near ambient temperatures. When the temperature of the solution has to be raised significantly to increase the vapor pressure of the organic compound, perhaps a contaminant, the higher temperature will also increase the vapor pressure of water. Of course, when the temperature has to be raised to values near the boiling point of water, one is essentially vaporizing or distilling the water-organic mixture, and a multiple stage distillation system is required to recycle the water and produce a concentrated product of the organic compound. For instance, pervaporation using a membrane with a high permeability for the organic compound and a low permeability for water would reduce the evaporation/distillation of water and allow the organic compound to be concentrated in one or a few stages. A helpful review of pervaporation and pervaporation membranes was published recently [12].

Separations by pervaporation utilize the separation factor from evaporation, which appears in the vapor pressure driving force, and the permselectivity of the membrane. In some cases, pervaporation can increase the change in composition many-fold more in a single stage than could be achieved by distillation (or evaporation). That would mean, of course, that the membrane is the major contributor to the separation, not the evaporation. If the diffusion coefficients within the membrane are significantly different, they can also contribute to (or hinder) the separation. It is possible and usually desirable for these effects to be additive, but they can also act in opposite directions. Of course, for the cases where pervaporation is attractive, the two effects are certainly likely to be additive. Permeation of organic compounds through membranes usually involves a "dissolution" of the organic in the membrane material or a strong adsorption or wetting of the organic compounds on highly hydrophobic surfaces of the membrane, and the organic compounds are then transported across the membrane by diffusion of the dissolved or adsorbed materials. Solubility/adsorption and diffusion can behave differently than diffusion with changes in temperature or pressure; so the contribution of permeation to the separation can be different at different temperatures or pressures. That is, the contribution of the diffusion process to the separation may change with pressure of the transported specie(s).

Pervaporation offers the potential for greater separation in a single stage than ordinary evaporation or single stage distillation. As noted earlier, this can be especially important when azeotropes are formed that prevent a distillation operation from reaching the desired separation using ordinary fractional distillation methods. Pervaporation can be used to separate the azeotropic mixtures into two streams, one on each side of the azeotrope composition. This then allows further distillation to produce any compositions desired. This feature of pervaporation is interesting and important, but azeotropic mixtures are not common in environmental problems.

The ethanol dehydration facility built in France in 1988 may have been the first large-scale application of pervaporation [13]. More than 90 units were in operation in 1992. Even a slight selectivity in a pervaporation step can "cross" the azeotrope and allow the rest of the separation to be done by distillation. High selectivity by pervaporation could even complete the separation. Azeotropes are not necessarily common problems in waste and environmental processes, but the potential for removing moderately volatile contaminants at elevated temperatures with no more than limited vaporization of water makes the potential for pervaporation in environmental and waste treatment obvious.

One potential advantage of pervaporation over distillation is the possibility of evaporating only the minor component in the solution, for example, traces of volatile organic compounds in water. Without using pervaporation, this is difficult if the minor component does not have relatively high vapor pressure in the solution. If only the minor amount of volatile organic contaminant is evaporated (passes through the membrane), considerably less energy will be needed than that required to evaporate relatively large quantities of water along with the trace organic material. Of course, much can be gained even from pervaporation systems that are not completely selective, but do concentrate the minor concentration of organic contaminant significantly, if not completely. If evaporation carries too much water into the vapor along with the organic compound, several stages of distillation may be required and relatively large quantities of energy handled to maintain the required reflux required at each end of a fractional distillation system. However, with a highly selective pervaporation membrane, in principle it would be possible to evaporate little more than the organic compound. If that compound is only present in small or trace concentration, the amount of material that has to be evaporated can be very small. Of course, any membrane gives some resistance to permeation, and considerable energy may be required to heat the liquid to the proper temperature for acceptable pervaporation rates, but the advantage of reducing the vaporization of water can more than compensate for the energy required to force the permeating component(s) through the membrane.

Description of Pervaporation

Pervaporation involves dissolution or other uptake of selected components preferentially in the membrane, and diffusion of those components through the membrane to the low pressure side [14]. In principle, there is no difference between pervaporation and any other pressure or concentration driven process. The principal difference results when the membrane is only partially "wet" by the liquid. The region wet by (or filled with) the liquid behaves much like a liquid separation system, and diffusion of "dissolved components" can control the permeation in that region. However, part of the membrane on the low pressure side may be essentially free of liquid, and permeation in that region may be more like separation in a gas system. Thus, to describe the performance of a pervaporation membrane with pores, it may be necessary to describe two regions of the membrane, If the membrane is modeled as a single phase solid in which the permeating specie dissolve, there would be only one membrane region to model. But if the membrane is heterogeneous, diffusion of dissolved permeate in the dense film on the upstream side of the membrane could behave much differently for Knudsen diffusion of gases through the pores on the downstream side of the membrane.

Even if the membrane is a solid material without pores, the liquid saturated region (liquid side) of the membrane can have significantly different properties than the "dry" membrane without liquid present. Since the permeability of the wet and the dry portions of the membrane can be different, the concentration gradient in the membrane can be different in those two regions. Of course, the region where liquid penetrates the membrane may not end suddenly, leaving two homogeneous, but different, regions of the membrane, wet and dry. There can be a gradient of liquid concentration in the wet region of the membrane, and diffusion of solute in this region can depend upon the liquid concentration. In such cases, the concentration profile in this region may not be linear as expected for steady permeation through a homogeneous material [15].

Since the pervaporation rate for the different components of a mixture are driven by the partial pressures of the components (the vapor pressure at the pervaporation temperature) and the permeability of the two components through the membrane, the overall separation will be determined by both the evaporative separation factor and the membrane separation factor:

$$\alpha_{\text{overall}} = [\alpha_e][\alpha_P]$$

where α_e is the separation factor for evaporation and α_P is the separation factor for permeation of the two components through the membrane. If the membrane is modeled in terms of two regions of the membrane, each region could contribute to the separation factor, and the contribution from evaporation would occur between the two regions.

Obviously one prefers to have both (or all) α's with values greater than unity so that the contributions to the overall separation are additive. Otherwise, permeation and evaporation will not complement each other in the separation. In the special case where pervaporation is used to separate azeotropic mixtures, $\alpha_e = 1$, so the entire separation will result from the membrane performance.

Because of the latent heat involved in pervaporation, temperature effects as well as mass transfer can be important in analyzing and predicting pervaporation performance. Temperature effects become increasingly important as the mass of material evaporated per unit area of membrane increases. If only a trace component (a contaminant) were permeating the membrane, the heat effects could be small. Schofield and co-workers [16] analyzed the heat and mass transfer effects in pervaporation. Gryta and co-workers [17] recently analyzed pervaporation operations in laminar flow.

Membranes Used in Pervaporation

A wide variety of membranes can be used for pervaporation, in fact, most permselective membranes can be used for pervaporation providing they can be used at the temperature of interest and can withstand the pressure difference usually required. The temperature used in pervaporation is that necessary to achieve satisfactory vapor pressure of the permeating component(s), and thus sufficient permeation fluxes. Table 2 lists several membranes and their applications, most of which were mentioned by Zhang and Drioli [18]. Note that most of the membrane materials have been studied for separating water-ethanol mixtures. This is an important application since it avoids problems of azeotrope formation in ordinary fractional distillation. Furthermore, most of the membranes developed or tested for this application favor permeation of water over the ethanol.

There are relatively few established applications mentioned that are clearly important in waste and environmental processing, but important applications are expected. The current applications probably reflect the great interest in the azeotropic separation problems, not the lack of potential environmental and waste treatment applications. For the most likely environmental applications—removal of volatile organic compounds from water—one would prefer a hydrophobic membrane that is more permeable to organic compounds. A few such membranes are mentioned in Table 1. Silicone rubbers are relative inexpensive membrane materials that

TABLE 2 Pervaporation Membrane Materials and Applications

Material	Application
Cellulose materials	Separates water for ethanol and separates aromatic mixtures
Chitosan	Separates water from ethanol
Collagen	Separates water from ethanol
Nafion	Separates water from ethanol
LDPE	Separates aliphatic isomers
Nylon-4	Separates water from ethanol
Polyamide	Separates water from ethanol and/or acetic acid
Polyacrylonitrile	Separates water from ethanol
Polyetheramide (block polymer)	Separates alcohols and phenol from water
Fluoronate polyester	Separates ethanol from water
PVC	Separates water from ethanol
Silicone rubber	Separates organic compounds from water

seem to transport organic compounds more readily than they transport water, especially polydimethylsiloxane [19]. The polymers can be modified by adding cross-linking agents, plasma or radiation curing, or even entrapping adsorbents. Other hydrophobic membranes have been made of other polymers that do not contain silicon. Less hydrophobic membranes have been studied to separate alcohols and other components from organic liquids. Most membranes that can be used to permeate VOCs from water could be considered for removing less volatile organic (hydrophobic) compounds by pervaporation at higher temperatures provided the membranes are sufficiently stable and selective at those temperatures.

Pervaporation membranes can be constructed in any of the common membrane shapes—flat, spiral wound, and tubular (hollow fiber). The advantages of high surface area of spiral wound and (especially) hollow fiber membranes apply to pervaporation as to other membrane processes. Also, the high permeability of ultrathin "active layers" is as desirable for pervaporation as for other membrane processes, so composite membranes with thin dense layers on porous support layers are common.

Equipment for Pervaporation

The equipment used in pervaporation systems can be similar to that used in other membrane processes. The liquid is likely to be heated in a separate vessel and pumped through an appropriate membrane cell. This can be a spiral wound or hollow fiber membrane cell. It is important that the membrane be able to withstand the temperature and pressure of the liquid. This often limits the operation to temperatures less than 100°C for most conventional membranes, often as low as 60°C, but newer membrane materials can handle temperatures significantly higher.

It is always necessary to have the partial pressure of the permeating component on the downstream side of the membrane lower than the partial pressure on the upstream side, and this can be a problem if the partial pressure of the solute (or contaminant) to be removed is relatively low. This may mean that a reduced pressure or even a moderate vacuum will be used on the downstream side. If a high vacuum must be pulled on the downstream side of the membrane, it is necessary to provide as little flow resistance as practical for removing the permeate product from the cells. That will eliminate the use of narrow channels or the inside of hollow fiber membranes for the downstream side. A more suitable arrangement for such cases could be the use of liquid on the inside of hollow fibers and the low pressure side in the more spacious region outside the fibers.

Suitable vacuum systems and permeate recovery systems are also needed. If the permeate to be recovered can be condensed effectively, low temperature condensation systems are likely to be adequate. This is especially likely when pervaporation is used to "break" an azeotrope, and the partial pressure of the permeating component is relatively high at the high temperature/pressure side of the membrane but much lower at lower condensation temperature. It could be more difficult if the permeating component is a trace component with only a very low partial pressure. Adsorption or chemical destruction of the vacuum system exhaust may be applicable in such cases. One should note that adsorption can be useful in such systems, but high selectivity from the adsorbent may not be needed if the membrane is sufficiently selective that little besides the desired component permeates the membrane; the pervaporation membrane can be responsible for the selectivity. When removing semivolatile organic compounds, it is probably desirable for the membrane to remove all of the organic materials, and selectivity over permeation of water may be all that is required. Selectivity among the organic compounds may not be important.

Separation Methods Likely to Compete with Pervaporation

Pervaporation competes with distillation, gas stripping, adsorption, chemical destruction, and biological destruction. When to use pervaporation will depend upon the selectivity of the available membranes, the relative volatility of the contaminant (relative to the volatility of water), and the concentration of the contaminant. For removal of organic compounds with sufficient volatility at normal ambient temperatures, gas (air) stripping is likely to be more competitive, but other factors, such as the difficulty in recovering concentrated VOCs, could also be a factor. When the concentration of the contaminant is sufficiently low, adsorption is likely to be a serious competitor. It is usually less expensive to add surface area in adsorption beds (by increasing the volume of the bed) than to add membrane surface. (Note that adsorption may be considered for the "final" or "polishing" step in removal of a contaminant.) Pervaporation is more likely to be preferred when an organic contaminant has sufficient volatility only at higher temperatures where the vapor pressure of water is also significant. In those cases, distillation is likely to be the competing separation method.

Of course, pervaporation will be favored when a membrane is available that sufficiently enhances the separation that it can remove most of the contaminant in a single membrane stage. Chemical and biological destruction may be favored over pervaporation when there is no merit in recovering the contaminant and when suitable destruction methods are available (but, of course, these are not separation methods). There are often competing chemical destruction methods sufficiently aggressive (UV or catalyzed oxidation) that they are effective for a wide range of organic compounds, including many chlorinated compounds. Biodegradation can be more specific, and a few compounds such as highly chlorinated compounds (including PCBs), are difficult to degrade biologically. Some chlorinated compounds, such as the higher cogimers (PCB compounds with a higher number of chlorine atoms) of PCBs, are more easily degraded by first reducing them to lower cogimers using anaerobic organisms and oxidizing the resulting compounds with aerobic organisms. Although destruction processes will not be discussed in detail, their general performance should be understood because they often compete with separation methods. Thus, some understanding of all competing approaches is needed to make the best applications of separation methods.

Liquid Membranes

Separations using liquid membranes combine many features of separations using solid membranes with features of separations using liquid-liquid extraction, absorption, and stripping. Liquid membranes may be used to separate gases or components in liquids. In the following discussion, separations from liquid systems, usually aqueous solutions, will be stressed, but, with the exception of separations using micro-emulsion membranes, these concepts can, in principle, be used with gaseous as well as liquid systems.

A liquid membrane consists of a film of liquid that separates two regions of different phases, whether gases or liquids. Thus the active membrane material is a liquid rather than a solid. The membrane could be an organic solvent that separates two different aqueous solutions, an aqueous phase that separates two organic phases, or an aqueous or organic liquid that separates two different gases. The liquid membrane material must have a low solubility in the other liquid phase or a low volatility in the gas phases because the volume or mass of liquid in a liquid membrane is usually very small in comparison to the volume of fluid to be processed. Loss of small quantities of the liquid that could sometimes be easily accepted in liquid-liquid extractions or absorption operations could deplete the liquid in the membrane and terminate the separation.

Obviously a liquid membrane does not separate components by the size of pores in the liquid. A permeate (solute) to be separated is soluble in the liquid used in the liquid membrane and permeates through the membrane as a dissolved or extracted component. The permeate dissolves in the membrane and is transported by diffusion to the other side of the membrane where there is another aqueous solution or gas. Note that this mechanism is similar to one of the mechanisms discussed earlier for solid organic membranes, that is, solid membranes without pores. With liquid membranes, it is possible to use a number of highly specific solvents that dissolve and transport only selected components. This is an important property of liquid membranes because the variety of liquid membranes and the higher potential specificity of the liquid membranes for individual permeates offer numerous opportunities for applications in environmental and waste treatment as well as other chemical processes. Especially when separating metal ions, it is often easier for the liquid membrane to dissolve extractants that could be more difficult or costly to incorporate chemically into solid membranes. The corresponding disadvantage or limitation of liquid is the need to use solvents and extractants with very low solubilities in the liquid phases. Similarly, when using liquid membranes in gas separations, it is necessary for all components of the liquid membrane to have very low vapor pressures. Of course, chemically bonding an extractant to a solid membrane medium leaves it with essentially no solubility or vapor pressure.

There are several ways to construct liquid membranes, and some of the general classes are illustrated in Figure 3. The membrane can be freestanding (Figure 3a). This arrangement uses large amounts of solvent and long diffusion paths and is not likely to be used in actual separation

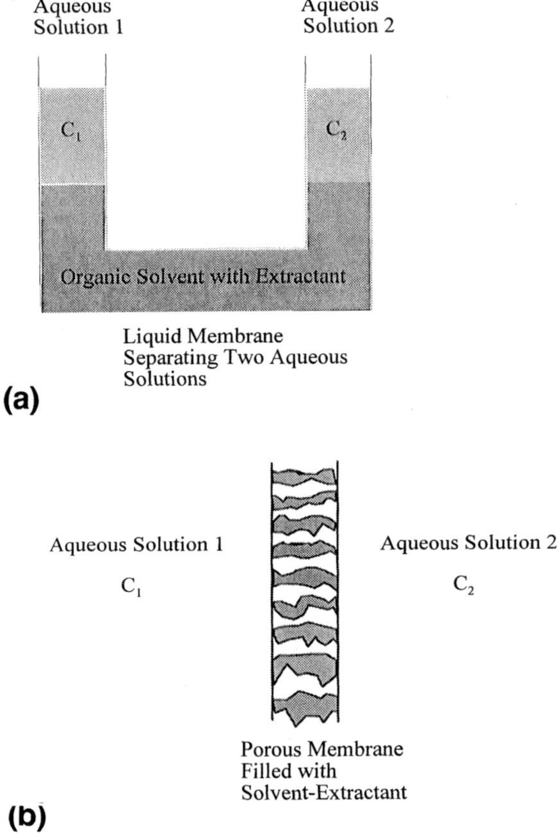

FIGURE 3 (a) Sketch of a simple "U" tube type arrangement that can be used to test solvents/extractants that could be useful in liquid membranes. This is not a practical arrangement for production systems, but it illustrates the chemical mechanism involved in liquid membrane operations. (b) Schematic of a supported liquid membrane. The solvent/extractant wets and fills the voids of a porous solid sheet, tube, or hollow fiber. The porous solid holds the membrane in place and can allow large membrane surface areas with relatively little solvent/extractant uses.

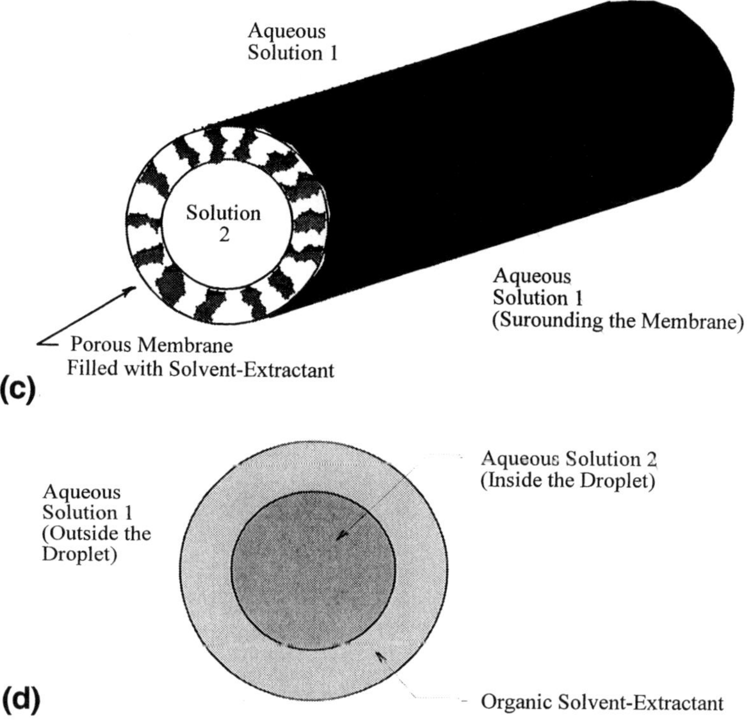

FIGURE 3 (c) A porous tube or hollow fiber used to support a liquid membrane. (d) An unsupported liquid membrane with a droplet of strip solution (downstream side of the membrane) surrounded by a layer of solvent/extractant membrane. The droplet is used in a much larger aqueous solution to extract selected solutes. These emulsion membranes are more likely to be practical when the volume of strip solution needed on the downstream side of the membrane is much smaller than the volume of solution to be treated.

systems, but it is a simple and convenient way to test or screen liquid membrane systems.

The liquid membrane can also be "supported" by a film of porous solid, as illustrated in Figures 3b and 3c. In these cases, the porous solid does not need to play an active role in the separation; its purpose is to provide structural support for the solvent, which is the active component in the separation. The solid support must be wet by the membrane solvent, and the solvent is usually held in the solid by capillary forces. Figure 3b shows the membrane supported in a porous solid sheet, and

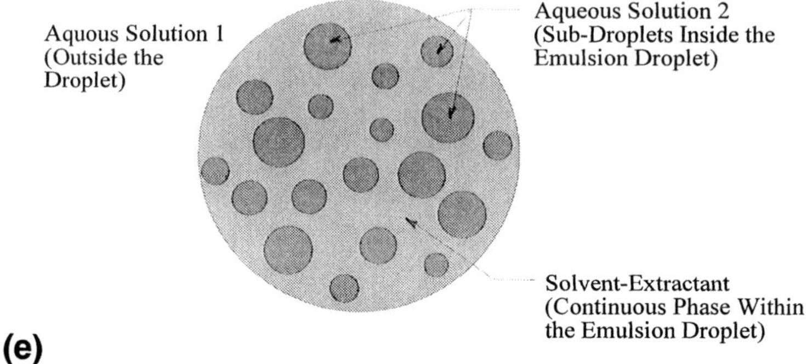

FIGURE 3 (e) A micro-emulsion unsupported liquid membrane. Rather than having a single drop of strip solution, the strip solution is in a series of smaller droplets dispersed within the solvent/extractant droplet. This is likely to be the more common type of emulsion membrane.

Figure 3c shows the solvent supported in a tubular-shaped porous solid. Note that these are the principal membrane shapes discussed for solid membranes. Later other possible liquid membrane shapes illustrated in Figures 3d and 3e will be discussed (micro-emulsions). The liquid membrane can also be supported between two solid membranes. One such approach places the liquid between two ion exchange membranes [20]. The solid membranes protect the solvent from entrainment (as in emulsion membranes) and can contribute to (or detract from) the selectivity of the overall membrane system.

Since most organic solvents have a finite solubility in aqueous solutions (and, for gas separations, all solvents have finite vapor pressures), there can be a slow loss of solvent from the membrane unless it is replenished. When the solvent consists of an extractant and a diluent, and that is often the case, the loss of performance could result from loss of extractant and/or diluent. To minimize such losses, it is usually desirable to saturate the feed stream with the membrane solvent, but there is only limited experience that indicates the extent to which presaturation can prevent loss of solvent and membrane performance. Use of highly insoluble solvents, saturation of the feed solution with the solvent, and avoidance of exceptionally thin liquid membranes with too little solvent capacity to control or minimize the loss of solvent often results in a suitable membrane life. Small concentrations of certain components in the fluid phase can affect solvent loss. For instance, small concentrations of surfactants in an aqueous phase can result in solvent loss by micelle formation, losses that can far exceed losses to solubility. Supported liquid membranes can sometines be regenerated relatively easily by cleaning the support and replacing the solvent. However, regeneration of the membrane by adding fresh solvent can be limited, if the aqueous phase has wet the membrane surface; wettability is not always easily reversed.

Since the solvent is held in the membrane by capillary forces, there is a limit to the pressure difference that can be applied across a supported liquid membrane without forcing the solvent from the pores of the support. One must consider the applied pressure difference and any osmotic pressure difference that may occur in the system. The force is inversely proportional to the diameter of the pores and directly proportional to the interfacial tension of the interface between the feed solution and the solvent.

Because the chemistry of liquid membranes usually resembles the chemistry of solvent extraction (or absorption for gaseous systems), many of the same parameters need to be considered with liquid membranes and solvent extraction. The solvent must dissolve (extract) the desired solute (permeate) selectively, and a suitable strip solution must be available for the other side of the membrane.

It is usually desirable to concentrate the permeate. For liquid systems, this means use of an effective downstream (strip) solution is desirable, and that usually means that the feed stream and downstream solutions will be significantly different. When both solutions have essentially the same properties, the driving force for moving the permeate across the membrane is only the difference in the concentration on each side of the membrane, and this always results in dilution of the permeate. However, if the solubility of permeate in the membrane solvent is strongly affected by such solution parameters as pH or redox conditions, it may be possible to achieve permeate concentrations in the downstream solution much higher than those in the feed. This is much like the possible concentration of solutes in extraction (back-extracting) operations. Of course, it is important that the component(s) in the strip solution, which can make a great difference between permeate solubility on different sides of the membrane, not be soluble enough in the membrane solvent that components of the strip solution are transported across the membrane to the feed solution at a significant rate.

For environmental applications, the solubility of the solvent and the toxicity of the solvent can be important considerations, just as they are in solvent extraction. Of course, many solvents can be removed from treated aqueous streams by adsorption on carbon, but this is another operation and the resulting carbon could be another solid to handle. Entrainment of solvent is not as likely to be as serious a problem with liquid membranes as it is with more conventional solvent extraction systems, and the expense from losses of solvent by entrainment can be one reason for considering or selecting liquid membrane systems over more conventional liquid-liquid extraction.

Supported (Immobilized) Membranes

As noted earlier, the support structure needs to be preferably wet by the solvent; if the separation is to be from an aqueous solution, the porous support needs to be hydrophobic. For separations of gases, the membrane may be hydrophobic or hydrophilic depending upon the solvent used. For liquid membranes using water or an aqueous solution as the solvent, ion exchange membranes are hydrophilic and can be used to support water films. For instance, a cation membrane was used in tests of carbon dioxide removal from methane [21]. The ionizable groups on ion exchange materials can make them very hydrophilic.

Sheets or tubes of porous support materials can be used. Membranes are usually available in sheet form or tubes. These support materials are often the same materials that are used to support thin film organic or inorganic membranes, but in this case the membrane materials go into the pores and do not just rest on the surface of the porous materials. The high surface area per unit volume of cell make hollow fiber membranes particularly attractive for liquid membrane separations as well as other membrane separations. A variety of hollow fibers are available and have been tested for use with liquid membranes.

If greater strength is desired, larger structures such as porous tubes can be used. Of course, porous organic polymer support materials for liquid membranes do not need thin, dense, or crystalline material on the upstream side of the membrane. However, note that the allowable pressure difference across a supported membrane will be limited by the available capillary forces holding the solvent-extractant in the pores as well as by the strength of the support itself, and mechanical strength of the support is less likely to be the principal limitation.

There will always be pressure differences between the two sides of membranes. Not only must the porous structure withstand that pressure, but the pressure also must not force the solvent from the pores of the porous support. If the porous support is rigid, the pressure required to force solvent from the pores can be estimated from the size of the pores and the interfacial tension. For circular pores

$$\Delta P = \frac{2\sigma}{r_p} \cos \phi$$

where r_p is the radius of the pores, σ is the interfacial tension, and ϕ is the wetting angle. Since the porous structure needs to be strongly wet by the solvent, cos ϕ is usually close to unity. This is, of course, an idealized expression because pores are not all circular or uniform. The situation is even more complicated for gel-like membrane support systems such as the ion exchange support membranes mentioned earlier. In those cases, it is difficult to define a realistic pore size. Nevertheless, this equation does provide insight into how the solvent properties (interfacial tension) and approximate pore structure affect the ability of the porous support to retain solvent.

The stability of supported liquid membranes can also be affected by osmotic effects and by transport of water through the membrane with the solute, especially when the solute is an ion. Water of hydration appears to enter the membrane with metal ions, at least to a limited extent, and water may be able to accumulate in the membrane if it is not transported through the membrane. Eventually there can be enough water in the membrane to form channels which make the membrane less selective. This tendency can be minimized by using highly hydrophobic porous support materials.

Extractant Transport

The solvent used in liquid membranes (like those used in solvent extraction) need not be a single compound. There are merits for using a diluent solvent and an extractant. This is especially desirable when extracting metals or other inorganic specie from aqueous solutions. It is unlikely that ions will have high solubilities in solvents that are highly insoluble in water. However, one can add a second component to the solvent that is strongly distributed to the solvent phase but which coordinates or otherwise interacts strongly with specific ions in the feed solution. The extractant can have large hydrophobic groups which make it highly insoluble in the aqueous phases, but also contain selective groups which interact with specific ions.

The extractant increases the concentration of the component of interest in the upstream side of the liquid membrane. Since the driving force for transport of permeate through a liquid membrane is the difference in concentration of the permeate in the solvent on different sides of the membrane, the driving force will be low if the concentration on the upstream side is low. If addition of an extractant increases the solubility of the solute in the solvent or the concentration on the upstream side of the membrane, there is a potential for significantly increasing the permeation rate. To take full advantage of the increased concentration on the upstream side, it is still necessary to maintain low concentrations of permeate in the solvent on the downstream side. Often with effective strip solutions in the aqueous downstream phase, it is possible to keep the concentration on this side of the membrane near zero, even with the addition of an extractant. Extractant transport can be used in essentially any form of liquid membrane, support membranes, or emulsion membranes.

Permeation Rates

For gaseous solutes, the driving force for permeation through liquid membranes is the difference in the effective partial pressure of the solute on both sides of the membrane. This is similar to the driving force for other types of membranes. For solutes in liquids such as water, the driving force is the difference in the activity of the solute of both sides of the membrane.

The other factors affecting permeation rates are the diffusion coefficient of the permeate within the membrane, the thickness of the membrane, and the structure of the pores (diameter and tortuosity). These effects are usually lumped into the permeability of the membrane. The diffusion coefficient within the membrane will include the effects of the extractant. If the permeate (for instance a metal ion) is firmly attached to the extractant, the diffusion will involve the entire complex containing extractant and solid. If the complex of extractant and metal ion is relatively large, the diffusion coefficient may be smaller than one would expect for the permeating solute alone. Note also that the extractant molecules that diffuse across the liquid membrane with the permeate must also diffuse back across the membrane after they lose the permeate molecule (or ion) to be ready to receive another permeate molecule to transport across the membrane. This "back diffusion" of extractant can contribute to the total resistance to permeation.

Of course, permeation rates, or effective diffusion coefficients, are affected by the porosity of any support structure used. Low void fractions in the porous solid mean that there is less volume of solvent or cross-sectional area for diffusion. Furthermore, the pores are not all parallel to the direction of average diffusion, so a tortuosity factor will have to be applied. There are also potentially important effects of pore size. These effects of pore size become especially important as the size of the diffusing complex approaches within small multiples of the effective pore diameter. The diffusion rate can be decreased considerably as wall effects become important. Since most porous materials do not have uniform size pores, the diffusion rates are likely to be determined by the size of constrictions or "necks" in the pores. These effects are generally the same as those discussed elsewhere for diffusion in solid membranes or in solid adsorbents. An important point here is that the size of the diffusing specie can be increased considerably if large extractant molecules are used.

Equipment for Liquid Membranes

Membrane cells for supported liquid membranes may differ little in appearance from those used for other membrane separations. The principal options are the choice of sheet or tubular forms. Spiral wound and micro-fiber shapes appear likely to gain in importance over other systems because they can accommodate more surface per unit volume of cell.

As mentioned several times in this section, liquid membranes can be considered a modified form of liquid-liquid extraction as well as a special type of membrane, especially emulsion membranes which will be discussed next. However, there are at least two characteristics of supported liquid membranes that are important and affect their selection for particular separation tasks.

First, one should note that a single supported liquid membrane cell incorporates the extraction and stripping operation into a single unit. Next note that the liquid is stationary, but in conventional liquid-liquid extraction systems the two phases must flow in countercurrent directions. Thus, conventional solvent extraction equipment is limited to relatively modest flow ratios to maintain sufficient interfacial area or to maintain the desired continuous phase. These issues never appear with supported liquid membranes because the interfacial area is set by the membrane geometry, and neither phase is dispersed in the other.

Thus supported membranes can use high capacity solvents as well as solvents with high selectivity, that is, solvents with distribution coefficients that could not be utilized effectively in conventional liquid-liquid extraction equipment that can operate effectively only within a relatively narrow range of volumetric flow ratios. This feature of supported liquid membranes allows one to use extractants that are highly specific and have strong affinities for the solute (high distribution coefficients). This can be particularly important for waste and environmental problems where one is more likely to deal with dilute streams, but sometimes with large streams.

Another difference in supported liquid membranes and conventional solvent extraction is the almost complete absence of entrainment in supported liquid membrane systems. Although solvent can still be lost by solubility, the elimination of entrainment can reduce solvent losses considerably. There is more incentive to select solvents with extremely low solubilities for liquid membranes because this can be the only significant factor in solvent loss; there is less incentive in conventional solvent extraction systems to reduce solubilities to extremely low levels if the solvent and extractant losses to entrainment are difficult to avoid. Solvent losses are particularly important when the solvent is toxic and cannot be released with the processed stream.

Finally, supported liquid membranes require very small quantities of solvent, usually much less than is required for conventional solvent extraction. This need for only small quantities of solute when coupled with the significant lower solvent loss, because of the elimination of entrainment, means that solvent costs are much lower or that much more expensive solvent can be practical for supported liquid membranes. This can be important for situations where highly specific and effective solvents, such as crown ether compounds, are relatively expensive to manufacture. Although one may not consider the use of such costly solvent in conventional solvent extraction systems, they may be practical or, perhaps, even optimal for supported liquid membrane applications.

Emulsion Membranes

There is one type of liquid membrane that has no common equivalent solid membrane form. Emulsion membrane cells are likely to look much different from traditional membrane cells using solid membrane materials. Emulsion membrane operations are much like solvent extraction and could be discussed under that heading. However, they are discussed here under membranes because such systems are usually described as "membrane" systems rather than as "extraction" systems. In this author's opinion, either heading would be appropriate. Because of the small volume of downstream (strip) solution, emulsion membrane systems are likely to be used principally when the removal can be essentially irreversible. In those cases, there is no need for countercurrent flow of the feed and the extracting fluids.

The emulsion is stabilized by adding a surfactant to the system. This stabilizes both the aqueous droplets in the organic membrane material (solvent) and the larger drops of the solvent in the bulk liquid. The use of higher viscosity solvents can also add to the effective stability of the emulsion drops [22]. This probably does not affect the ultimate or thermodynamic stability of the emulsion, but it can slow the breakup of the emulsion considerably, and that can be adequate for most applications. Increased viscosity of the solvent can also slow diffusion rates in the solvent (membrane), but mass transfer rates in emulsion membranes are usually relatively good. A reduction in the mass transfer rate may be readily accepted if it adds sufficiently to the emulsion stability.

The simplest emulsion contactors can be stirred tanks with the settled emulsion skimmed from the top of the contactor to recover the permeate and recycle the solvent. However, it may be possible to achieve higher removal rates per unit volume of contactor in a cell with approximate plug flow. In such cases, the feed solution would travel approximately in plug flow along a tubular contactor. Most likely the feed would enter the top of a tower-shaped cell and flow down the tower while the emulsion would flow up the tower. This would be much like a liquid-liquid (solvent) extraction unit, but the emulsion removed from the top of the tower would have the permeate already stripped into an aqueous phase. No additional strip tower would be needed, only an emulsion breaker.

Emulsion membranes are likely to be droplets of solvent in aqueous solutions. The solvent droplets surrounds a core of the aqueous "strip" solution, the downstream aqueous fluid into which the permeate is being transferred. These droplets can be placed into an aqueous solution from which the permeate is to be removed (Figures 3d, 3e). Figure 3d shows the solvent film surrounding a single droplet of the downstream (strip) solution, and Figure 3e shows a number of fine droplets of downstream solution enclosed within an organic drop (the liquid membrane). This second form is often called micro-emulsion membranes. The emulsion can be prepared in a separate vessel and introduced into the membrane contactor. In these cases, the membrane contactor (membrane cell) may be a column (tower) or a mixed tank. These may have appearances considerably different from those of other membrane cells.

It is necessary to separate the emulsion from the feed solution, so the emulsions must have different densities from the feed solution if the emulsion droplets are to be separated from the bulk fluid by gravity. Usually the emulsion droplets are less dense than the aqueous feed solution, so the emulsion droplets rise and are collected at the top of the cell.

Emulsion preparation can take place in a separate unit. Droplets of strip solution can be dispersed within the solvent and the solvent then formed into small droplets in another aqueous phase. With proper use of surfactants, the emulsions can be relatively stable. In some cases, they have been stable enough to use in medical applications where the emulsion can even be taken orally to release a drug at a controlled rate or to remove a toxic material from the digestive system. After contact with the feed solution, the emulsion must be broken, the strip solution from within the emulsion separated from the solvent, and the solvent recycled to prepare new emulsion. After the emulsion has accumulated the extracted solute, it settles to the top of the membrane unit (assuming that the liquid membrane system is less dense than the water feed) and is removed. Usually a weir or a "skimming" system is sufficient to remove the emulsion with only small quantities of the feed solution (with solute concentration greatly reduced). To recover the solute and recycle the solvent to the emulsion maker, it is necessary to break up the emulsion. Although any of several methods can be used to break up the emulsion, electrostatic and acoustic methods appear to be the most common approaches. Since most emulsions that are likely to be used in environmental systems are droplets of solvent with micro-droplets of aqueous strip solution within them, electrostatic methods have been reported to be the desired method [23]. The most effective way to impose the electric field is with a high frequency AC field onto a steady (DC) potential gradient. The explanation of the mechanism in which this application of the electric field is so effective is not certain, but one can speculate that the DC component works to bring the aqueous droplets suspended in the organic solvent closer, and the high frequency AC component of the field disturbs the interface of the water (strip) droplets, possibly creating new surface area which is temporarily devoid of stabilizing surfactant and thus more likely to coalesce with other droplets. The DC component of the field can bring the droplets together by dielectric effects, and the force should depend upon the square of the field strength.

Perhaps the most attractive feature of emulsion membranes is the high surface areas that can be obtained. For some applications, there can be other important advantages for emulsion membranes. For instance, with highly irreversible systems, the permeate can be concentrated greatly in a small volume of strip solution. In a supported liquid membrane or in most conventional solid membranes, the volume of fluid on the downstream side of the membrane often must be small enough to be incorporated within the solvent droplets, so large increases in concentration are essentially a requirement for this type of operation. Note that with such small volumes of solution on the downstream side of the liquid membrane, that is, inside the emulsion, very little time is required for such systems to reach steady state and the concentration factor is excessively high. Remember that it can be difficult to operate most solvent equipment with such extremely high (or low) flow ratio for the two phases. On the other hand, emulsion membranes may be less attractive when large volumes of strip solution are necessary.

Electrodialysis

Electrodialysis is an electrical potential driven membrane separation method. It is used to remove electrolytes from aqueous solutions. Although some selectivity can be achieved among cations (or among anions), electrodialysis is usually used to remove all electrolytes, not for separating electrolytes from each other. Electrodialysis can also be used to "split" water and thus produce an acid and a base from a salt. Although that may not be considered strictly a separation process, it is convenient to discuss water splitting to make the treatment of electrodialysis somewhat more complete. The key roles of electrodialysis in waste and environmental processing are as an alternative to evaporation to remove electrolytes and to recycle acids and bases.

The key element in electrodialysis is the ion selective membrane. These are membranes that selectively pass only cations or anions. Such membranes can be fabricated in the same manner as ion exchange particles; the only major difference may be in the shape, being sheets rather than spherical particles. The selectivity for electropermeation of only cations or anions results from Donnan exclusion of anions (or cations). As described in Chapter 2, ion exchange materials, often polymers, have ionizable sites on the molecules or surface. The ions (cations for a cation exchanger and anions in an anion exchanger) are free to migrate (or exchange). These ions are neutralized by fixed ions of opposite charge that are chemically bound to the solid ion exchange material and are not free to migrate.

A typical cation exchange material may be a polymer with sulfonic acid groups incorporated throughout the polymer. The sulfonic acid group remains with the polymer. However, the hydrogen ions are free to diffuse within the polymer, and other cations can diffuse into the polymer and be "exchanged" with the hydrogen ions. These materials are called cation exchangers because cations can diffuse into the polymers and be exchanged with cations that were there previously. Note, however, that there can be no significant net removal of electrical charge from the polymer, so sufficient cations must remain within the polymer to neutralize the electrical charge of the sulfonic acid groups. The concentration of sulfonic acid groups is then the "capacity" of the cation exchange materials. Other ionizable groups can be incorporated with polymers to make different cation exchange materials. The ion exchange performance will depend upon the ionizable group incorporated within the polymer, the concentration of the ionizable groups in the polymer, and the nature of the polymer itself. The selectivity of ion exchange membranes for cations or anions will be discussed briefly here, but more details on ion exchange materials and their behavior are given in Chapter 2.

If the ionizable group is an amine and has free anions, the polymer will be an anion exchange material. Quaternary amine groups make strong anion exchange materials, and ternary and secondary amine groups make successively weaker anion exchange materials. If the ionizable group is a weak acid group like carbonic acid, the exchanger will be a weak acid exchanger and generally favor having hydrogen ions on the resin rather than most metal ions. The concentration of ionizable groups affects the exclusion of ions with charges opposite to those of the exchanging ions and the change in the selectivity of the resin for divalent ions and monovalent ions as a function of concentration.

Donnan Exclusion

This topic was discussed in Chapter 2, but it is mentioned again briefly here because it is particularly important to the function of ion exchange membranes in electrodialysis. If the concentration of ions in the exchanger is significantly greater than in the electrolyte solution, ions with charges opposite to those of the exchanging ions will be excluded from the ion exchange material. Donnan exclusion from a membrane of ion exchange material results because at equilibrium the chemical activity of any electrolyte (salt, acid, or base) must be the same inside the membrane as outside in the solution. The chemical activity of the electrolyte is approximately proportional to the product of the concentration of cations and the concentration of anions raised to powers equivalent to the number of ions in the (defined) formula for the salt. Deviations from this proportionality represent changes in the activity coefficients of the ions in one or both of the two phases. For instance, with a 1–1 salt such as sodium chloride, the activity would be proportional to the concentration of sodium ions multiplied by the concentration of chloride ions. (One could define the activity as the square root of this product, but since we will only equate the activities at two different conditions, it is only necessary to use the same basis for both conditions.) For a 1–2 electrolyte such as calcium chloride ($CaCl_2$), the activity would be proportional to the concentration of calcium ions multiplied by the square of the concentration of chloride ions.

For simplicity consider a dilute solution of a 1–1 electrolyte such as sodium chloride in equilibrium with a cation exchange membrane. Ignoring activity coefficients (assuming that they will be approximately constant and of order unity), the concentration of cations and anions outside the membrane and within the membrane will be related by

$$[C^+][A^-] = [\underline{C^+}][\underline{A^-}]$$

where $[C^+]$ is the concentration of the cation in the solution, and $[\underline{C^+}]$ is the concentration of the cation in the ion exchange membrane. Similarly, $[A^-]$ is the concentration of anions in the solution, and $[\underline{A^-}]$ is the concentration of anions in the membrane. Of course, if activity coefficients are taken into account, these products will not be exactly equal, and a better approximation can be made by inserting a constant (on the right side, for instance) that incorporates the products of the activity coefficients. However, to illustrate exclusion of anions from cation membranes and cations from anion membranes, one need look only at the magnitudes of these terms.

Note that for a cation membrane the concentration of cations in the membrane must be equal to the capacity of the membrane plus the concentration of anions in the membrane. For dilute solutions, the capacity of the membranes will be much higher than the concentration of the dilute solution. For instance, cation exchange membranes may have capacities as high as 5 molar. Then $[\underline{C}^+]/[C^+]$ is very large, and $[\underline{A}^-]/[A^-]$ must be very small. Then the concentration of anions is very low in cation membranes. Similarly, the concentration of cations in anion membranes is very low. Note that ion selectivity increases as the concentration in the solution decreases, and the membrane can become less selective as the concentration in the solution approaches the concentration of active sites in the membrane material.

With anions "excluded" from cation membranes, an applied electric field across a cation membrane will result largely in transport of cations through the membrane and relatively little transport of anions in the opposite direction. Similarly, an applied electric field across an anion membrane will result principally in transport of anions through the membrane. This makes the membranes "ion selective." This is the key to understanding the performance of ion selective membranes. For many purposes, it is sufficient to remember that cation membranes primarily allow only cations to pass, and anion membranes allow only anions to pass. When the concentration in the solution becomes sufficiently high, of the order of the concentration within the ion exchange material, the membranes will lose some of their selectivity.

Description of an Electrodialysis System

The way these properties can be used to remove electrolytes from dilute aqueous solutions is illustrated in Figure 4. Note that the electrodialysis cell consists of a series of membranes with flow channels between them. The membranes are alternately cation and anion membranes. When an electrical field is applied across the cell, the resulting electric current will be carried largely by cations passing through the cation membranes and by anions passing through the anion membranes. This results in depletion of electrolyte in alternate flow channels and concentration of electrolyte in the other channels. In the end channels with membranes on only one

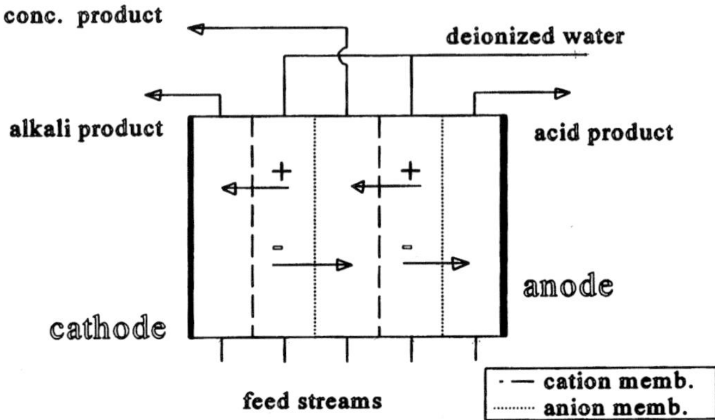

FIGURE 4 A few compartments of an electrodialysis unit. The units are constructed of alternating cation and anion membranes with cathodes and anodes at the end of the series of cells. The cell arrangement consists of electrolyte depleted cells and electrolyte concentrated cells.

side, acid and oxygen will be produced at electrodes on one side, and a base and hydrogen gas at at the electrode in the channel on the other end of the cell. Thus water is "split" at the end electrodes unless the cation can be reduced at the cathode or the anion can be oxidized at the anode.

Any current carried by anions through cation membranes or by cations through anion membranes represents a loss of current. Thus the current efficiency is likely to be higher with relatively dilute solution, that is with solution concentrations well below the concentration of ionizable groups in the membrane (the membrane capacity). On the other hand, the conductivity of the solutions in (half) the flow channels decreases as salt (ions) is (are) removed from the solution. This means that the voltage required and, thus, the energy consumed per mole of electrolyte removed increase as lower concentrations are sought in the deionized product.

Although the simplified description of electrodialysis did not mention transport of water through the membranes, significant transport can occur. This can be particularly important when large amounts of electrolyte are to be removed and when extremely low concentrations are needed in the product. There is always considerable water in most ion exchange membranes. The charged groups incorporated into the polymer structure make the polymer hydrophilic, even if the other portions of the polymer would otherwise be hydrophobic. In fact, water penetrates the membranes and can "swell" them significantly. This water is free to diffuse through the membranes, and there will be an osmotic pressure driving force once the concentrations in the adjacent channels becomes different. The osmotic pressure will drive water from the dilute channels from which the electrolyte is being removed into the more concentrated channels into which the electrolyte is being transported. However, additional water transport can result from the electrodialysis itself; this is usually called electro-osmosis. Electro-osmosis results because water molecules are transported through the membrane with the ions. Hydration of ions within the polymer membrane may not be the same as that in the channel solution, and the association of water with ions within the membrane may be complex and not even understood. Electro-osmosis usually should be viewed simply as the water that is "pulled" across the membranes by the flux of ions across the membrane without suggesting details of the mechanism.

Electrodialysis to Concentrate Salt Solutions

Electrodialysis cannot be used to remove electrolytes to extremely low concentrations. If it is necessary to reduce a contaminant concentration even lower, electrodialysis may not be adequate because of low conductance of the waster. Other methods such as adsorption or ion exchange can be used as polishing steps to reduce the concentrations further. The electrolyte is recovered at a higher concentration. If the electrolyte is used elsewhere in the process, the concentration may be sufficient to recycle the material if few impurities were added. Although electrodialysis usually does not remove specific electrolytes, it can be important in environment and waste management, even if it is only indirectly useful. For instance it can be used to treat water to be used in other processes to minimize introduction of electrolytes.

Salt Splitting

With only a slight modification, electrodialysis systems can be used to convert salts to the respective acids and bases of the anions and cations incorporated in the salt. Although this is not likely to be a common approach for treating many waste and environmental streams, it offers an important opportunity in a few cases to reduce salt effluents when they are formed in the process from the acid-base reactions. If the concentration of other contaminants is not excessive, the acid and/or base produced may be recycled to the system. This is often called "waste minimization," and is viewed by some as even more beneficial than waste treatment.

The modification required for generating acid and base by electrodialysis rather than a concentrated salt stream is illustrated in Figure 5. The important difference between this arrangement and that in

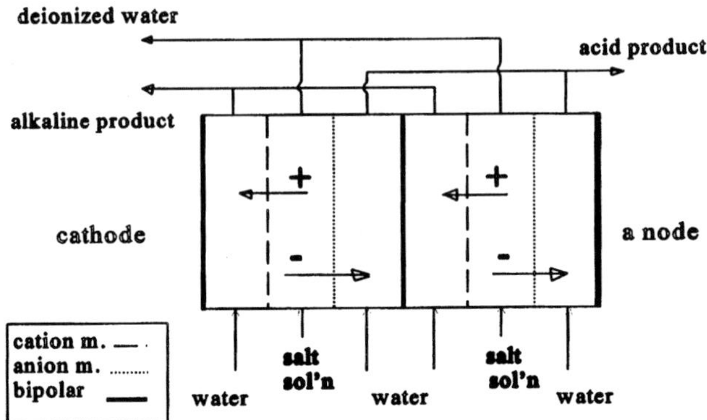

FIGURE 5 A salt splitting unit. This is similar to an electrodialysis unit, but each portion of the unit consists of a central cell with a cation membrane on one side and an anion membrane on the other side. The central cell is surrounded on both sides by cells with bipolar membranes that act as cathodes on one side and anodes on the other side. Electrolyte salt is depleted in the central cell, acid is formed in the cell on one side, and base is formed in the cell on the other side.

Figure 4 for conventional electrodialysis is the inclusion of bipolar electrodes. These electrodes act as cathodes on one side and as anodes on the other side. Thus hydrogen gas is usually generated at one side of the bipolar electrodes and oxygen is generated at the other side. These are electrodes at which water is split to form hydrogen and hydroxide ions of the respective sides of the membrane. This makes the electrodialysis cell into a series of small units that deionize the central stream and generate acid and base on the adjoining streams. No water passes through the bipolar electrodes that separate these channels. Note that the region between each bipolar electrode is divided into three channels separated by a cation membrane or an anion membrane. The water solution to be treated (deionized) flows through the center channel, and small water streams flow through the other two channels that are bounded by one of the bipolar electrodes. The cations are transferred to the channel between the cation membrane and the cathode side of a bipolar membrane, and anions transfer to the third channel, the channel between the anion membrane and the side of a bipolar membrane that acts as an anode. Then the channel between the cation membrane and the cathode gives off hydrogen gas and accumulates the base (hydroxide) solution, and the channel between the anode and the anion membrane gives off oxygen and accumulates acid. The central channel becomes depleted of salt, both cations and anions; that is, water in the central channel becomes deionized.

There is, of course, a price to pay for generating acids and bases, because the voltage drop at the bipolar electrode faces must be sufficient to split water. This increases the energy consumption to levels that can be significantly greater than that possible in conventional electrodialysis. The usefulness of generating acids and bases depends upon their potential uses, usually within the plant. If the acid and/or base are not used within the plant, it is not likely to be practical to market the material elsewhere. Also, all electrolytes in the waste stream will be converted to acids and bases. These will contaminate the main acid and base and may make it unsuitable for reuse. Even if the contaminants are relatively low in concentration, continuous recycling of acid and base may result in buildup of contaminants to unacceptable levels. A "bleed" stream may be required to limit the concentration of contaminant buildup.

Outline of Design Considerations for Electrodialysis

Design of electrodialysis units usually involves determining the amount of electrolyte to be removed, the practical current density for the membrane to be used, and the membrane area needed. If flow rates are low and current densities and electrical resistance are high, heat generation in the electrodialysis cells can be important and require cooling. Interstage cooling can be used if the electrodialysis is carried out in several units with cooling between the units. One additional advantage of performing electrodialysis in more than one unit is the potential for better control of the voltage gradient and the current density in each section of the system. If simple straight paths are used in all channels (Figure 4), the current densities will be highest at the end of the cell with the higher concentration and lowest at the end with the lower concentrations (and lower electrical conductivity). One can flow the solutions in opposite directions in some channels and/or run the depleting streams through the cell several times rather than straight through single channels for their entire residence time in the electrodialysis unit. These actions will all allow some smoothing of the current densities over different regions of the cell.

There are other important considerations with the geometry of the flow channel, but they are more likely to be the concern of the supplier of the electrodialysis unit. Of course, if one chose to design an electrodialysis unit from scratch, all of these factors would have to be considered. As long as the electrical resistance of the depleted stream is significant compared to the resistance in the membrane, the increasing resistance of the electrolyte cells that results from decreasing concentration should be taken into account. Each unit of an electrodialysis system could provide the entire separation, or only part of the electrolyte could be removed in the first unit, and the stream from the first unit could be sent to a second unit for removal of more electrolyte. Alternatively, some cells in the unit could receive fresh electrolyte solution, and others could receive the depleted electrolyte from these cells. The same electric current passes through all cells in an electrodialysis unit since each chamber (or cell) is in series, the electric resistance will be the highest in those cells with the lowest electrolyte concentration, and the voltage drop across the cells in the unit will not be the same. As noted, using separate units, one can have different current densities in each unit, and that could be useful when large changes in concentration are needed.

To minimize dispersion (mixing) of the solution (particularly the stream being depleted of electrolyte), each cell usually contains a "spacer" which channels the flow through specific paths (often in zigzag patterns) to create high local velocities and reduce nonuniform flow and mixing. These spacers also support the membranes and minimize flexing from the high pressure cells to the low pressure cells. (Pressure can be different in adjoining cells because of the flow patterns and flow resistance.) Of course, spacers can cover part of the membrane surface, and this must be taken into account when sizing electrodialysis cells and units. These internal details of the electrodialysis unit may be determined by the manufacturer if the equipment is purchased in units.

CONSTRUCTION OF MEMBRANE EQUIPMENT

Most membrane equipment using solid membranes can be divided into three general classes. The first involves flat membrane surfaces with a high pressure (or high concentration) on one side and a low pressure (or low concentration) on the other. Since there are generally better arrangements for most environmental application, this is not used in many environmental systems. However, there may be some applications where the ease of cleaning and membrane replacement makes this type of arrangement desirable. It is usually most economical to have as much membrane area as possible in a given volume of membrane equipment, and simple flat membrane cells usually do not give the highest area per unit volume of equipment. It is important, however, to consider such systems. For instance, simple systems that illustrate the operation of membrane systems may be used to test new membrane materials (Figure 6). In the figure, the solution to be treated can be introduced in the upper chamber of the cell and pressure applied to force the solute through the membrane. When a pressure difference is applied, the membrane can be mounted on

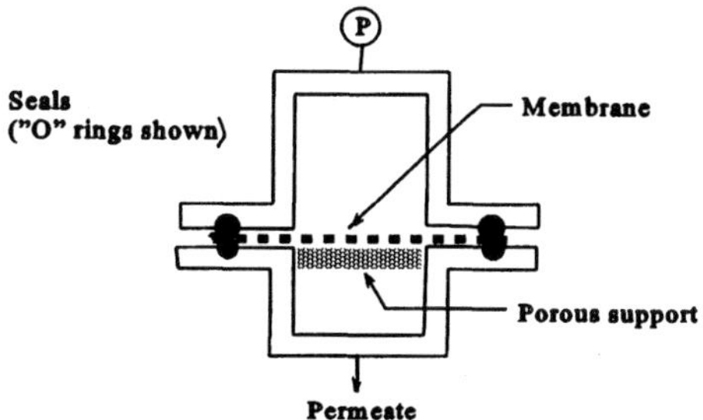

FIGURE 6 Simple pressure cell showing a membrane sealed by a pair of membranes and supported by a porous frit or screen. Convenient for testing membranes, it is not usually practical for large-scale operations.

a support screen or frit to provide strength against the applied pressure. The upper chamber can be stirred to reduce concentration polarization, or the solution can be pumped over the surface. With the upper chamber stirred and no pumping of fluid through the chamber, the system will not operate at steady state, and samples must be taken from both chambers after several time periods to determine the rejection. If solution is pumped through the cell, the conditions can at least approximate steady state. It usually is desirable to keep the volume of the low pressure/concentration chamber small to approximate the ideal non-steady-state operations more closely or to reach steady-state operation more quickly.

In production units, flat membranes can be more effective when mounted in filter press type arrangements with flow across the membrane (Figure 7). In such units, the area of membrane and thus the productivity per unit of volume of cell is much greater than could be achieved in a single membrane test unit and is determined by how closely the membranes can be installed. The principal disadvantage of this type of unit is the difficulty in packing as much membrane surface into a given volume of cell as can be achieved using some of the alternative membrane cell designs that will be discussed next. The principal advantage of the plate-and-frame membrane cells is the ease of disassembling and cleaning. Unlike the other cell designs that will be mentioned next, the plate-and-frame cells can be taken apart so that each membrane surface can be cleaned or replaced. That makes this type of cell especially useful in those industries and applications where frequent cleaning is necessary. Frequent cleaning is required in the food and pharmaceutical industries, and plate-and-frame cells are used there. Such frequent cleaning is less likely to be necessary in the waste and environmental

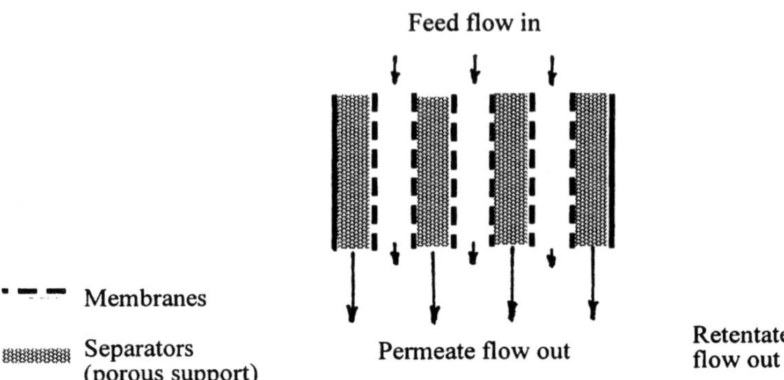

FIGURE 7 Simple schematic of a filter press type unit using several membranes. Pressure is applied to alternate cells in the unit. Permeate passes through to the adjoining cells. A wire or other porous structure is needed to support the pressure applied on the "upstream" side of the membranes.

applications, so plate-and-frame cells are not as likely to be as common in these applications.

The difficulty of getting high membrane surface area into a given volume of cell can be addressed by the second type of membrane cell to be discussed, the spiral wound membrane cell (Figure 8). A "stack" of several membranes is bound together and wound into a spiral arrangement. The membranes in the stack are separated by a loosely woven mesh spacer, usually made of plastic. Sections between the membranes alternate from high to low pressure portions of the cell. If the membrane is directional,

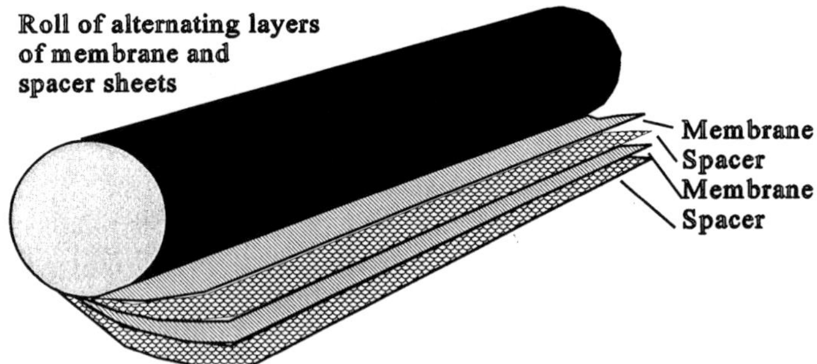

FIGURE 8 Spiral wound membrane cell. This arrangement allows the use of much larger membrane areas than the filter press arrangement in a unit volume of cell.

the high pressure or high concentration side must always be facing the high pressure or high concentration sections. The spacers are important especially in the permeate (low pressure) sections because they prevent the membranes on opposite sites of the low pressure section from coming together and blocking flow of the permeate fluid. In the retentate (high pressure or concentration) side of the membrane, the spacers keep the flow channel open. There can also be additional benefits from the spacers if they promote turbulent flow and thus reduce concentration polarization. On the other hand, poorly designed spacers can create "dead spaces" that promote precipitation or growth of microorganisms that degrade membrane performance. Fluid can flow through the mesh along the membrane surfaces to the permeate exit. It is usually easier to obtain high membrane surface areas per unit volume in this type of equipment because the spacing between the membranes can be very small, usually smaller than can be maintained in a stack of membranes in a filter press (plate-and-frame) type arrangement.

By winding these membranes in a spiral manner, the membranes also can be placed in a convenient cylindrical vessel. The membranes are joined by casting the joined edges in a "potting" plastic. This is the most popular type cell to use with membranes that are manufactured in sheets. Several hundred square meters of membrane surface can usually be incorporated into a cubic meter of spiral wound membrane cell. Although the area per unit volume of cell depends upon the thickness of the membrane and the channel width required for flow through the cell, the area in spiral wound membrane cells is almost always greater than that in the same volume of flat membrane cells Moderately high pressures can be used as long as the membrane spacers are not deformed or the membrane is not damaged from being pressed against the spacers with too much force. Spiral wound cells usually require relatively strong membranes because it is not as easy to construct spacers which can protect fragile membranes as effectively as fine or graded porosity frit membrane supports. Nevertheless, much support for thin films can be built into the membrane itself (as noted earlier).

The third popular arrangement is illustrated in Figure 9. In principal, this is a tube-in-shell arrangement much like those used in many heat exchangers. Like heat exchangers, these cells can be operated with single or double pass "tubes." In a single pass cell, the tubes of membranes simply go from one end of the shell to the other end. In double pass cells, the tubes go the length of the cell and then are reversed and return to the same end at which they entered. The membranes are shown quite small, to illustrate the common "hollow fiber" membranes. A variety of membranes can be fabricated in this shape. The hollow fibers can be small,

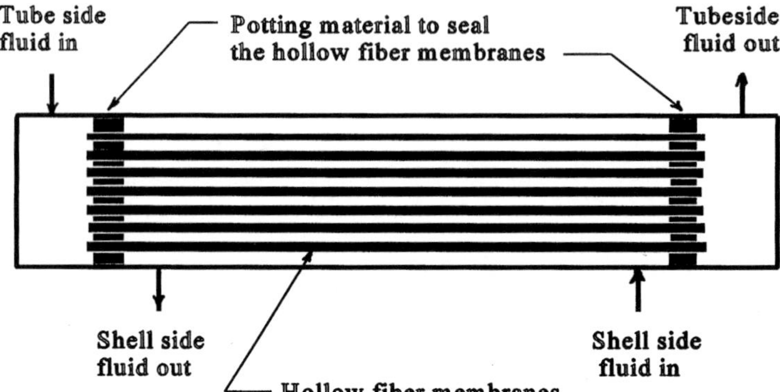

FIGURE 9 Micro-capillary type membrane cell. With the use of extremely small hollow membrane fibers very large membrane areas can be packed into a small volume of cell.

and sizes of 0.1 mm diameter are common. It is probably obvious that with such small membrane fiber diameters, the membrane surface area per unit volume of cell can be very high. It is often possible to have thousands, or even a few tens of thousands, of square meters of membrane surface area per cubic meter of hollow fiber membrane cell. Although the maximum pressure difference that can be applied to such systems is limited by the strength of the hollow fibers, the small diameters of the fibers allow them to withstand considerable pressure. The high pressure can be on the inside or outside of the fiber, but it is more common to be on the outside. Liquid pumped into or around such small fibers would need to be free of particles that could plug the fibers or the spacing between fibers. This is certainly a limitation, but the presence of particles is a serious problem for most membranes because particles can be transported to membrane surfaces and damage the surface; this is one form of "fouling" of the surface and will be discussed in a separate section. Nevertheless, hollow fiber membranes are usually more susceptible to plugging by small concentrations of particles than the other membrane cell designs discussed because the channels are so small that channel blockage as well as membrane fouling can become a problem.

In a few cases, tube-in-shell membrane systems are operated with much larger tubes. Some of the dynamic membrane cells were operated with tubes having diameters of 1/8 to 1/4 in, closer to the diameters of common heat exchanger tubes. Although these systems are not as susceptible to plugging, they do not allow the extremely high packing

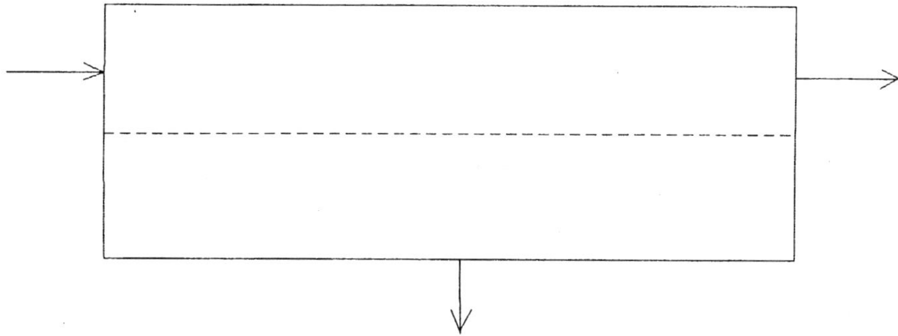

FIGURE 10 (a) Flow diagram for systems where the downstream composition has little effect.

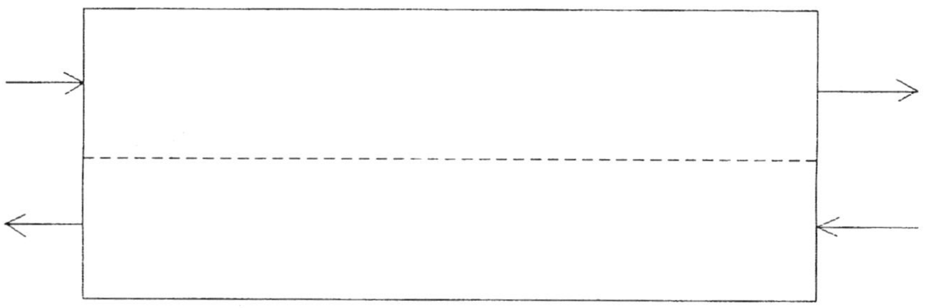

FIGURE 10 (b) Countercurrent flow. The most efficient arrangement when fluids flow on both sides of the membrane.

of membrane area into a unit volume of cell that can be achieved with hollow fiber membranes.

Although tube-in-shell membrane cells are similar in general shape to heat exchangers, one should note a few important differences. For many systems such as reverse osmosis, there is no inlet stream on the low pressure side and thus no exact analogy to countercurrent flow (Figure 10). There can, however, be a similar situation because the permeate could be withdrawn from the inlet or outlet side of a single pass cell. We have assumed in the discussions thus far that the permeate is removed immediately as it comes from the low pressure side of the membrane. However, from Figure 10 the permeate has to flow for a finite distance to exit the cell. If the concentration of solute in the high pressure side of the membrane is varying and the permeate concentration is a function of the concentration on the high pressure side of the membrane, the concentra-

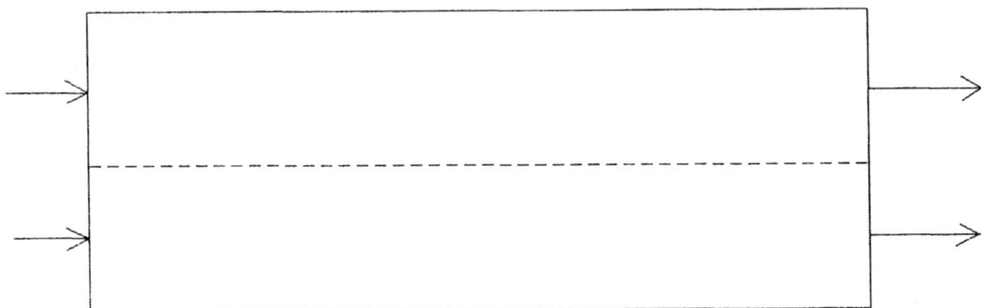

FIGURE 10 (c) Cocurrent flow. Often minimizes the pressure differences across membranes in concentration driven systems.

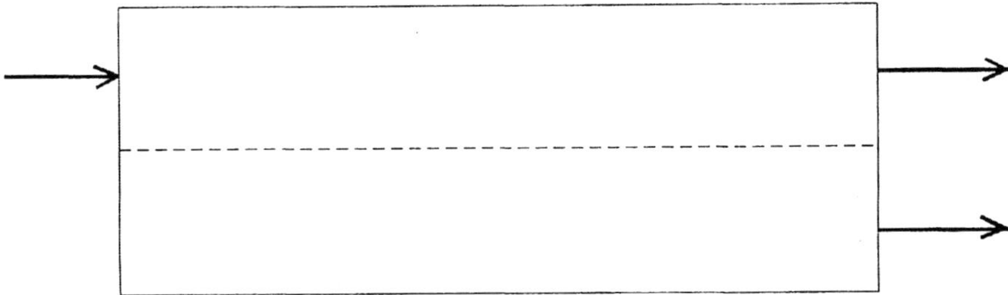

FIGURE 10 (d) Withdrawal of permeate from a pressure driven system from one end of the cell, in this case the downstream end.

tion of solute in the permeate can be different at different positions in the cell. This means that the concentration of the permeate is a composite of all of the material being permeated through the membrane. If the membrane is extremely effective in rejecting the solute, the concentration may be so low that the consequence of averaging the concentrations may be of little importance. However, if the rejection of solute is far from complete, this averaging may be important and could even limit the degree of concentration that can be allowed in the high pressure (reject) fluid.

The effects of the flow pattern on the membrane's low pressure side can depend upon the relationship between the fluid composition on the membrane's low pressure side on the concentration just inside the low pressure side of the membrane. Particularly, it depends upon whether the flow pressure face of the active portion of the membrane is in chemical equilibrium with the fluid on the low pressure side of the membrane. The options for arranging the flow patterns in the low pressure/concentration side of the membrane were discussed by Bansal et al. [24]. In some studies, the fluid on the low pressure side is assumed to be in equilibrium with the low pressure side of the membrane. However, there can be mass transfer resistance on the low pressure side of the membrane just as there is on the high pressure side. Since the fluid on the low pressure side comes directly from the membrane, it will be in equilibrium with the membrane surface unless there is cross-flow as the fluid is removed and significant differences in the composition coming from different portions of the membrane.

The flow pattern on the downstream side of the membrane is important only when the permeating flux of one or more components is affected by the concentrations on the downstream side of the membrane. In reverse osmosis and in some other pressure driven processes, there may be little effect from the downstream concentration, but for concentration driven systems, there will usually be a significant effect. There are several ways to flow the downstream fluid through the cell (Figure 10), and there can be different behavior depending upon the flow pattern. If the membrane is a once-through system, the downstream flow can be in the same direction as the upstream flow (cocurrent) or in a direction opposite to the upstream flow (countercurrent). Of course, as noted earlier, the upstream or downstream flow could involve multiple passes through the cell. In single pass systems, countercurrent flow gives the greater average "driving force" and a more nearly uniform flux through all portions of the membrane.

If the pressure drop from flow through each chamber of the cell is significant, the pressure difference between the two sides of the membrane will change less with cocurrent flow. That is, the pressure in both chambers of the membrane cell will be higher at the fluid inlet end of the cell. Note that the pressure drops for the upstream and downstream flows are likely to be much different because of the differences in flow rates and for tube-in-shell systems, with different channel sizes and shapes on the membrane sides. The concentration driving force for permeation will usually change less over the length (or surface) of the membrane with countercurrent flow, but the total pressure difference across the membrane (a limit set by the structural strength of the membrane) is likely to be more nearly constant for cocurrent flow. The smaller variation in pressure difference across the membrane may be more important when high pressure gradients across the membrane must be avoided because of structural problems.

In discussing concentration polarization on the high pressure side of the membrane, the role of the net flux or fluid velocity toward the membrane was noted to affect the hydrodynamic boundary layer near the membrane, and thus the mass transfer boundary layer. When there is a significant flux "from" the membrane, the effect is just the opposite, and the boundary layer becomes less stable and more likely to become turbulent. When only the trace component is being transported through the membrane, as in many concentration driven and gas separation systems, the flux or velocity of fluid to or from the membrane is likely to be much less, and probably less important. The presence of a relatively thick and highly porous supporting substrate on the downstream side of the membrane is likely to be of more importance. As noted earlier, the separation occurs at the thin active layer of the membrane, and the more highly porous supporting part of the membrane can hamper the role of cross-flow in reducing the effects of concentration polarization. That is, the fluid leaving the thin surface film on the membrane may be in equilibrium with the "surface of the thin film," but the fluid outside the porous support portion of the membrane may not be in equilibrium with the thin active portion of the membrane. The support structure can have significant mass transfer resistance and hinder equilibration of the bulk permeate stream with the surface of the active portion of the membrane.

Membrane Fouling

One common problem with all membrane processes is fouling of the membrane surface. Fouling can refer to a number of phenomena that degrade the membrane or block the membrane surface. The membrane could be degraded by some materials in the feed stream, gas, or solution. This component could be the solvent itself, the solute, or a third component that otherwise would be of no importance in the separation process. For instance, dissolved oxygen in water can eventually affect some organic polymer membranes, but dissolved oxygen is not usually listed as part of the feed solution composition.

Suspended solids will usually damage the performance of any membrane and may be particularly damaging to pressure driven membrane processes such as reverse osmosis where the flux of fluid toward the membrane carries suspended solids to the membrane surface. Considerable effort is usually devoted to removing essentially all particles from the feed to reverse osmosis or other pressure driven membrane processes. Since no particulate removal system is absolutely effective, some particles will always reach the membrane surface. The permeate flow creates a flux toward the membrane that carries the particles to the membrane surface. Rapid flow of fluid across the surfaces (cross-flow) is helpful in reducing the effects of particles accumulating on the surface just as it is effective in reducing the buildup of filter cake in cross-flow filtration. The accumulation of particles at the membrane surface and the convective/diffusive mass transfer of the particles back to the bulk stream are also similar to concentration polarization of soluble specie at the membrane surface. In this case, however, it is eddy motion and Brownian motion rather than molecular diffusion that removes the particles from the surface. Since the buildup of particle concentration at the membrane surface (and concentration polarization) is (are) driven by the flux toward the membrane, the buildup is likely to be more serious for cases with high membrane fluxes, usually those cases such as reverse osmosis where the bulk component (the solvent) rather than the minor component (the solute) permeates the membrane. The most significant damage to the membrane probably results when the particles actually interact with the membrane and become imbedded in it or erode its surface.

In some cases, concentration polarization can result in damage to the membrane by concentrating solutes at the membrane surface components that damage the membrane. In all cases, polarization results in higher concentrations of rejected components at the membrane surface. If the membrane is affected by the nonpermeating solute (in cases such as reverse osmosis), the higher concentration of solute at the surface will increase the damage. The membrane life should then be estimated using the time of exposure to the concentration actually on the surface, not to the concentration in the bulk solution.

Sometimes, concentration polarization can even result in precipitation of a solute on the surface; this would then have obvious undesirable effects, much like those effects from particulates in the feed stream that reach the membrane surfaces in a reverse osmosis cell. Since high cross-flow rates reduce concentration polarization as well as filter cake buildup, high cross-flow rates will reduce the rate of membrane degradation from concentration polarization both by reducing the concentration of the sparsely soluble components and by sweeping away more of the precipitates that may form on the surface. To prevent solids from fouling a membrane it is necessary to maintain the concentrations throughout the cell, including the liquid film next to the membrane (concentration polarization) below the saturation level. Remember that the concentrations of solutes that do not permeate the membrane increase with position down the flow channel as more of the permeating solvent or other permeating components are removed. Thus for reverse osmosis, the concentration of the nonpermeating solute increases down the flow channel, and precipitation from concentration polarization may be more likely down the retentate flow channel. To prevent precipitation from concentration polarization, one must consider the concentrations and polarization conditions throughout the system, not just at the inlet.

In addition to the major components in the feed liquid, one must also consider some trace components. In many reverse osmosis systems treating groundwater or wastewater, the concentration of cations, such as calcium or magnesium, and anions, such as carbonate or sulfate, rather than the principal component or even the contaminant, may limit the degree of concentration in the reject stream. This is because of the very limited solubilities of calcium or magnesium carbonate or sulfate. It is common practice to add small concentrations of acid to the feed for reverse osmosis operation to lower the pH and reduce the carbonate concentration. Of course, this will have little effect on the formation of insoluble sulfates on the membrane surface. Reducing the carbonate concentration by adding acid to the feed also produces carbon dioxide gas, which usually permeates most reverse osmosis membranes with the water. At the relatively high pressure of the feed stream, much of the carbon dioxide may remain dissolved in the water, but it may be necessary to degas the product permeate stream to remove the excess carbon dioxide. Silica can be an important fouling material if its total concentration is much greater than approximately 100 ppm in the feed stream. The solubility of silica will decrease as the pH is lowered, and changes in the pH after the feed stream is filtered can affect the importance of silica fouling.

In aqueous systems such as reverse osmosis, membranes can also be fouled by biomass filtered by the membrane or that grows on the membrane surface [25]. The importance of biofouling is a function of the composition of the feed stream and the membrane material used. If the feed stream contains significant micro-organisms and/or nutrients, biofouling is more likely to be a problem. Groundwater is likely to contain organisms and nutrients, especially if it comes from near the soil surface; wastewater could be sterile or could contain essentially as many organisms as groundwater. Carbon beds and other upstream equipment [26] can be the source of organism growth as well as the original source of the feed. It is also desirable to reduce the volume of "dead spaces" upstream of the membrane where organisms can grow without being rapidly flushed from the system. Dead spaces are reservoirs with only small flow, thus regions with long residence times for organisms to grow. Some membranes such as cellulose acetate are more prone to biofouling than some of the newer reverse osmosis membranes. Other membrane may be more susceptible for organism attachment or even contain nutrients for the organisms to use for growth.

The effects of biofouling usually appear first as a decline in the permeability of the membrane, and this can be followed by a decline in rejection. The decline in permeability is to be expected because biofouling adds a film on the membrane surface and increases flow resistance. The decline in solute rejection may be less expected. A decline in rejection can result from additional concentration polarization. The biofilm on the membrane can allow the solute to penetrate the film to the membrane surface, but with the biofilm on the membrane surface, there will be little cross-flow to flush the solute from the membrane surface. Of course, in some cases, the biomass could even degrade the very thin membrane surface film that is responsible for the rejection.

Biofouling can be reduced or minimized by careful pretreatment of the feed solution and by regular treatment of the membrane surface. There is some evidence that there is an initial accumulation of a porous coating of the membrane surface well before the effects of biofouling is evident. This could be an accumulation of humus or other micro-solids that do not alone hinder membrane performance, but the humus film may provide a suitable surface for further biomass growth that does degrade membrane performance [27]. Although one would expect biofouling to be affected strongly by the cross-flow rate which would carry the biomass from the membrane surface, there is not much evidence of this. This may result because the biomass is bonded more firmly to the membrane surface than particulates or solute concentrations that can be swept from the membrane surfaces by higher cross-flow rates. There is also evidence that, in some cases, exposure of the feed stream to light (through glass or plastic piping or open tanks) can result in increased organism (bacteria) growth and biofouling.

Pretreatment of the feed to remove traces of solids such as humus may be helpful even before the effects of biofouling are seen. Frequent and regular treatment of the membrane should begin well before biofouling is detected. These treatments usually involve biocides, such as chlorine, but most of the more common biocides are not totally effective; they do not totally disinfect the system. Sulfite may be more effective. If all of the organisms are not killed, the biomass may grow back more rapidly than it was formed originally. Failure to kill all of the organisms can result from failure to expose all of the fluid and biomass to the disinfectant or from the greater tolerance to the biocide of some of the organisms. Surviving organisms may be more tolerant to the biocide, and future applications may be less effective in reducing biofouling. Existing biomass may still remain on the membrane surface even if the organisms that form the biomass are killed. This remaining biomass can be a convenient place for relatively rapid growth of surviving organisms. Thus repeated applications of biocides are often less and less effective, and the membrane must eventually be replaced—that is, the biomass must be completely removed from the system.

Although most membranes used in waste and environmental applications are not expected to be in plate-and-frame cells where mechanical as well as chemical cleaning can be used, it is still possible to use in situ chemical cleaning (as well as biomass treatment) without disassembling the cells. Carbonate fouling can be cleaned by flushing the cells with an acid stream, usually HCl or citric acid, and such treatment is likely to be quite successful. This usually involved lowering the pH of the feed to approximately 2 for 30 to 60 min [28]. The acid concentration that can be tolerated depends upon the susceptibility of the membrane to damage by low pH solutions, and the limit of pH 2 is approximately the limit of cellulose acetate membranes. Sulfate fouling is far more difficult to remove, but solutions of citric acid or other complexants such as EDTA (ethylene diamine tetracetate) can help. Metal oxides can sometimes be removed by citrate ions or other chelating agents. EDTA is usually a better chelating (complexing) agent than citrate ions for ferric iron and most metals other than citrate, but it is more costly. Ferrous citrate has a very limited solubility, and this can hamper the use of citric acid for removing iron deposits from membranes. However, better results can often be obtained by using ammonium citrate because ferrous ammonium citrate has a significantly higher solubility. Oxalate ions are also effective chelating agents, but oxalic acid has a relatively low solubility, and it is difficult to apply high concentrations of oxalate ions.

Silica and silica polymers can also be serious fouling agents, and such fouling is very difficult to clean. Alkaline solution (phosphates, carbonates, or hydroxides) cleaning can be beneficial. If there are other components such as calcium or magnesium carbonates in the fouling scale, it may be necessary to alternate the alkaline treatment with acid treatment to remove calcium phosphate and other metal components of the scale.

Biofouling can usually be removed by treatment with solutions of oxidants or biocides such as sodium bisulfite or chlorine. In such treatments, the membrane must resist the treatment. Because biofouling is such a common problem in treating aqueous streams, it is usually highly desirable for the membrane to be resistant to oxidation, and this is one of the criteria considered in developing and selecting membrane materials. Both of the biocides mentioned above are oxidizing agents. Because the active portion of a membrane can be very thin, often as thin as practical, very little damage to the membrane can be allowed without degrading the membrane performance considerably.

The frequency of membrane "cleaning" will depend upon the fouling problem, the cost of the membrane, and the application, but cleaning approximately every six months is common [29]. More frequent treatment may be needed if biofouling is a problem. One can select the cleaning agents mentioned above, but these agents are also available in commercial preparations and marketed specifically for membrane cleaning.

Membrane degradation can occur from chemical changes as well as from physical fouling. Chemical degradation can occur during cleaning and production operations. Although many major and trace components in feed streams could degrade membrane materials, the most common type of degradation probably results from oxidation. In aqueous systems, chlorine is a common oxidant in the feed streams because chlorine is used so often to treat process water. Polyamide membranes, some of the more attractive reverse osmosis membranes, and probably most other polymer membranes can be degraded through incorporation of chlorine into the polymer structure. Of course, the "process" that produces a wastewater can introduce other oxidants that could degrade some membranes. In treating wastewaters, it is possible to reduce the problem of membrane degradation from dissolved chlorine by first passing the feed stream through an activated carbon bed before the stream goes to a reverse osmosis cell. Of course, the carbon bed will remove other components as well as chlorine. The lifetime of the carbon bed can be relatively long if chlorine removal is its only purpose. Since a carbon bed is likely to be considerably less expensive than a membrane cell, one would want to monitor the carbon bed output to ensure that the bed is replaced before chlorine breakthrough reaches a level that would significantly damage the membrane cell.

In pressure driven membrane separation systems, the membrane can be degraded physically without the presence of chemical changes, such as oxidation or chlorination, or fouling by solids. The sustained pressure applied across some membranes can result in compression that is not totally reversible. Compression usually results in a decline in flux through the membrane. The amount of compression is usually a function of the maximum pressure drop that has been applied, and this can occur relatively early after a membrane is placed into service. There is usually no way to repair such damage.

Membrane lives can range greatly from a few hours to a few years. The cost of membrane replacement is always an important consideration in reverse osmosis systems, but recent membrane and cell designs generally minimize fouling and membrane degradation. These longer membrane lifetimes of a few years can only be achieved by careful design of the membrane cells and operating them to minimize membrane fouling. Commercial membrane cells have addressed fouling problems, and there usually is sufficient cross-flow to provide help in reducing concentration polarization and sweeping particles from the surface. Adequate filters or other means must be provided to prevent a significant amount of solids from the membrane cell. It may also be necessary to reduce the concentration of selected components that would be particularly damaging to the membrane. The lifetime of a membrane can affect the economics of a membrane process considerably, and poor estimates of the membrane lifetime can result in very poor estimates of the treatment cost.

Considerations in the Design of Membrane Systems

The first consideration in design of membrane cells is identification of the potential membranes and the crude estimates of the membrane areas needed. The performance parameters of specific membranes must come from the manufacturer and/or from experimental data on the particular problem and fluid streams. The first goal is to determine if the membrane can provide sufficient selectivity to meet the desired concentrations in the permeate and retentate streams on a differential basis. That is, can acceptable concentrations be achieved with the feed composition? In the case of reverse osmosis, the question is likely to be whether the membrane can reduce the concentration of solutes in the permeate to sufficiently low levels. In the case of concentration driven systems, the question could be whether the concentration of the component to be removed can be recovered at sufficiently high concentrations in the permeate.

The design methods for membranes are not as standardized as those for distillation, absorption, and some other separation systems. This results only to a limited extent from the relatively recent growth in membrane processes. The principal reason for less standardized methods is the wide variety of different types of membrane processes of interest and the different types of equipment used. Only an example of approaches to a few applications is given here, but it is hoped that the readers will find information on a system similar to the ones they are considering and be able to make the minor adjustments that may be needed.

One must remember, however, that because in practical systems it is necessary to allow a significant quantity of material to permeate the membrane, the conditions at the feed composition represent only one point in the membrane cell. In reverse osmosis, one will want to allow a significant portion (usually most) of the water to permeate the membrane, and the concentration of solutes in the retentate stream will then increase significantly. In concentration driven membrane processes, the concentration of the component permeating the membrane will decrease significantly across the membrane cell. One generally wants to remove large fractions of a contaminant, so the concentration could be severalfold lower than that in the inlet feed. Thus in both types of processes, the concentration of the contaminant and the flux of the permeating components may decrease greatly over the membrane cell. This means that membrane performance needs to be understood over a range of solute concentrations.

After the membrane is selected, a major effort in the process design of membrane systems is to determine the outlet composition of both streams leaving the cell and to determine the membrane area required. Both depend upon the flow arrangement selected for the membrane cell. If the flux through the membrane and the concentration in the permeate depends only upon the concentration in the high pressure/concentration side of the membrane, the calculations can be relatively straightforward, and this will be approximately the case for many systems such as many reverse osmosis systems.

The change in the flux of solute on the high pressure side of the membrane is

$$dQ = VdC + C\,dV \tag{8}$$

where C is the concentration at any point on the membrane surface (along the flow channel), Q is the rate at which solute is flowing down the flow channel, and V is the volumetric rate at which solution is flowing down the flow channel. In reverse osmosis if there is no permeation of solute, Q is constant and the equation reduces to the simpler form

$$\frac{dC}{C} = -\frac{dV}{V} \tag{9}$$

When solute does permeate the membrane, the rate of solute permeation can be expressed in terms of the rejection coefficient, $R = 1 - C_p/C$. Then

$$dQ = C_p\,dV = (1-R)C\,dV \tag{10}$$

and

$$dC = (1-R)C\left(\frac{dV}{V}\right) - C\left(\frac{dV}{V}\right) \tag{11}$$

The change in concentration as the solution passes through the membrane cell can then be estimated by integrating this equation to obtain

$$\int \frac{dC}{C} = -R\int \frac{dV}{V} \tag{12}$$

If the reaction is constant, R can be moved outside the integral, so

$$\log_e\left(\frac{C_{\text{out}}}{C_{\text{in}}}\right) = -R\log_e\left(\frac{V_{\text{out}}}{V_{\text{in}}}\right) \tag{13}$$

or

$$\frac{C_{\text{out}}}{C_{\text{in}}} = \left(\frac{V_{\text{out}}}{V_{\text{in}}}\right)^{-R} \tag{14}$$

Thus one can estimate the concentration in the high pressure side of the membrane as a function of the volume of solution remaining in the retentate. If the solute is rejected totally ($R = 1$), the fractional change in solute concentration in the retentate is inversely proportional to the fraction of the water (solute) remaining in the retentate.

The concentration in the fluid permeating the membrane at any point in the membrane cell is simply $(1 - R)C$. However, one is more likely to be interested in the average concentration of all of the permeate because the permeate is likely to be mixed. This average concentration can be estimated by integrating (averaging) the concentration in the permeate, but it is probably easier to simply estimate it from the differences in the solute and water (solvent) that enter and leave the retentate side of the membrane:

$$C_{p,\text{avg}} = \frac{C_{\text{in}}V_{\text{in}} - C_{\text{out}}V_{\text{out}}}{V_{\text{in}} - V_{\text{out}}} \tag{15}$$

The required membrane area is determined from the amount of permeate that needs to pass through the membrane. The rate of permeation is proportional to the applied pressure or concentration difference if the permeability is constant, but for reverse osmosis one may need to account for the resisting osmotic pressure. It is also often necessary to account for the pressure loss through the retentate (high pressure) side of the membrane. For instance, for reverse osmosis, one can write the expression for the area required as

$$dV = P(p - p_{\text{osm}})\, dA \tag{16}$$

or

$$A = \int \frac{dV}{P(p - p_{\text{osm}})} \tag{17}$$

If the permeability, P, and the applied pressure, p, are constant and if the osmotic pressure, p_{osm}, can be neglected, this expression can be easily integrated. In the more general case, p will decrease with distance down the channel. Because the flow rate on the retentate side of the membrane will change as more water has permeated the membrane, so the pressure down a reverse channel (retentate side) will not decrease linearly with distance down the channel. To perform the integration accurately, one would need to take into account the change in p, with distance down the channel, that is with the change in area, A. (Distance down the channel can be related to the area of membrane "used" or "involved" in the cell up to that position. In most cell designs, distance and involved membrane area will be proportional.) Also remember that the concentration increases as more water is removed by permeation, and this raises the osmotic pressure. If the osmotic pressure become significant relative to the applied pressure, it will be necessary to account for that in Equation (17).

The effects of concentration polarization appear in both the concentration of solute in the permeate and in the resisting osmotic pressure. The rejection should be specified in terms of the concentration at the membrane interface. However, the rejection given in Equations (12)–(14) is based upon material balance and refers to bulk concentration. One can relate the concentration at the membrane p, C_i, to the mass transfer coefficient, k, the rejection, R, and the resisting osmotic pressure, $p_{i,\text{osm}}$.

$$RC_iP(p - p_{i,\text{osm}}) = k(C_i - C) \tag{18}$$

Solving for the concentration at the membrane gives

$$C_i = \frac{C}{1 - R(p - p_{i,\text{osm}})/k} \tag{19}$$

Of course, the concentration at the membrane surface approaches the concentration in the bulk retentate when $k \rightarrow \infty$ when $P \rightarrow 0$, or when $R \rightarrow 1$ (no rejection). The concentration at the membrane interface should be used in estimating the resisting osmotic pressure. If the osmotic pressure is proportional to the concentration of solute (the case for dilute solutions), the value for $p_{i,\text{osm}}$ can be replaced with the constant of proportionality. Then Equation (13) can be solved directly for the concentration at the membrane surface and the resisting pressure needed to calculate the permeation rate.

The selectivity of the membrane, which has been described in terms of the fraction of solute that passes through the membrane, R, can also be affected by concentration polarization. In general, the fraction of solute going into the permeate can be a function of the concentration in the retentate side of the membrane. However, if R does not change much between C and C_i, its apparent value can be written as

$$\frac{1 - R'}{1 - R} = \frac{C_i}{C} \tag{20}$$

Calculations of the membrane area required for concentration driven processes can be complicated by the relatively complex flow pattern that can be used and by the mass transfer resistance that can occur. If the membrane is a solid film, the mass transfer resistance on the films on opposite sides of the membrane can be estimated and treated as in other mass transfer operations. The only difference is that the mass transfer resistance in the membrane must also be taken into account.

With concentration driven membrane processes, one is less likely to be concerned with removing the high concentration of solvent from a relatively small concentration of solute. In waste and environmental problems, concentration driven processes are more likely to be used to remove relatively low concentrations of a solute (often a contaminant) from the bulk fluid. In these cases, the mass flux through the membrane is likely to be small, and there is not likely to be much net flow to or from the membrane. This makes relatively standard mass transfer coefficients based upon no flow at the surface (interface) likely to be good approximations.

When the permeation rate or performance depends upon the concentration on the permeate side of the membrane, one must take into account the flow pattern on that side as well as on the retentate side. Although there can be a wide variety of flow patterns in concentration driven membranes (cross-flow, single-pass flow in the shell, and double pass in the tube, etc., were described earlier; see Figures 10a–d), we will consider only cocurrent flow, countercurrent flow, and complete mixing on the permeate side of the membrane. This last case may be a crude approximation for some shell-side systems without good baffles, and it represents one intermediate case between the cocurrent and countercurrent cases. Most other arrangements will also be intermediate between these two cases. When the flow patterns are neither cocurrent nor countercurrent, it may be possible to break up the membrane cell area into parts where each part is approximated by one of these three cases, or one can pick the case that is believed to best fit the flow pattern of interest.

For all cases, the mass transfer through the membrane can be estimated from the resistances on both sides of the membrane and the resistance to mass transfer across the membrane. Since there will be no significant accumulation of solute in the membrane or in the thin fluid films on each side of the membrane, the fluxes through each film and through the membrane will be equal:

$$\text{Flux} = k_R(C - C_{Ri}) = P(C_{Ri} - C_{Pi}) = k_P(C_{Pi} - C_P) \qquad (21)$$

The subscript Ri refers to the condition at the membrane surface on the retentate side, the subscript Pi refers to conditions at the membrane surface on the permeate side, and the subscript P refers to the bulk fluid on the permeate side of the membrane. The concentration in the bulk retentate side is C, and the concentrations at other positions are noted by appropriate subscripts. The mass transfer coefficients are k_R on the retentate side of the membrane and k_P on the permeate side of the membrane. The permeability of the membrane for the solute is P. Note that when the driving force for permeation is written in terms of concentrations, the permeability is much like a mass transfer coefficient for the membrane.

As discussed in earlier sections, the permeability can incorporate more than one factor. These could include the dissolution or adsorption of the solute on one side of the membrane, diffusion of the solute through the membrane, and desorption of the solute on the other side of the membrane. There could also be a significant contribution to the diffusion resistance from the porous support part of the membrane. (This resistance will be assumed to be incorporated into the overall resistance of the membrane.) The permeability will not necessarily be a constant, but it is a useful term because there are many cases where the permeability does not vary greatly with concentration. Note, however, that if the adsorption or dissolution of the solute into (or onto) the membrane is not linear (a constant distribution coefficient), the effective permeability may not be independent of concentration.

For cocurrent flow on both sides of the membrane,

$$V_R\, dC = -V_p\, dC_P \tag{22}$$

That is, the solute (or any component) that transfers from the retentate stream appears in the permeate stream. If dispersion in the retentate and permeate streams is neglected, the change in concentration is related to the rate of mass transfer through the membrane. If the solute is being transferred from the retentate, the concentration in the retentate will decrease with distance into the retentate flow channel. That is,

$$V_R\, dC = -k_R(C - C_{Ri})\, dA \tag{23}$$

$$\frac{dC}{C - C_{Ri}} = -\frac{k_R}{V_R}\, dA \tag{24}$$

The term dA is a measure of the differential area of membrane. For a given flow channel, the area of the membrane will increase as the length of the flow channel is made longer. Thus, for a given channel cross-section, the area will be proportional to the distance down the channel, so the total area is another measure of channel length or membrane cell size. A similar equation could be written for the concentrate in the permeate side.

Integration of the left side of Equation (24) should extend from the initial concentration in the retentate to the desired final concentration in the retentate. Integration of the right side should extend from an area of zero to the final area (the area of membrane needed), and the result of the integration is simply $-(k_R/V_R)A$. The integration of the left side must include values for the concentrations at the membrane surfaces.

The resulting concentration of solute in the permeate stream is the integral of the solute flux over the membrane surface divided by the flow across the permeate side of the membrane.

$$C_p = \int \frac{\text{Flux}_p \, dA}{F_p} \tag{25}$$

The term Flux_p refers to the flux of the permeating component of interest through the membrane, and F_p is the flow through the chamber on the permeate side of the membrane. When the permeate side of the membrane is well mixed, Flux_p and F_p can be taken outside the integral, and the concentration in the retentate is an accumulation of the permeate that occurs from permeation over the entire surface area on the retentate side. Equation (25) ignores the volume of fluid permeating through the membrane, but for concentration driven processes, that is usually not large.

When to Consider Membranes

As noted throughout this discussion, membrane processes seem to be gaining attention in many areas, and environmental and waste operations are no exception. Of course, the first consideration is whether a membrane system is available for a particular use. But there are also a few guidelines that one can consider before spending excessive time looking for and considering a membrane process for a given waste or environmental problem. There are four major groups of environmental applications that should be considered: (1) separation of dissolved (or even dispersed) organic liquids in water, (2) separation of gases, (3) removal of selected metal ions from water, and (4) reverse osmosis to separate water from waste solutions.

Reverse osmosis can compete with ion exchange, evaporation, and other approaches that remove essentially all of the dissolved components in an aqueous waste or environmental stream. Because reverse osmosis is usually not highly specific in the dissolved components that are removed, it is not likely to be preferred when a trace component is to be removed from relatively high concentrations of other dissolved components. Other membrane processes (especially concentration driven membrane processes)—adsorption, ion exchange, or liquid-liquid extraction—may prove to be more desirable for such application. Reverse osmosis is also more likely to be the choice operation for removing dissolved material at immediate concentrations. When the concentration is sufficiently high, evaporation may be preferred. Evaporation can concentrate the dissolved materials to dryness (for the nonvolatile components), and it may be desirable to couple reverse osmosis with evaporation. Reverse osmosis can be more economical to concentrate the dissolved materials up to a point where the osmotic pressure makes evaporation more attractive.

Reverse osmosis is generally a more energy efficient approach than evaporation to removing water from solutions of low concentration. Considerable savings in energy costs can be achieved by multiple effect evaporation systems, but at the expense of higher capital costs. However, if suitable reverse osmosis membranes are available for the solution of interest, they usually should be considered. Remember, however, the reverse osmosis will be attractive only for concentrating the solution to the point where the osmotic pressure becomes excessive; for electrolytes, this limit is likely to come at concentrations of 1 to 4 molar. Evaporation may still be needed to concentrate the solute further or to take it to dryness.

At the other end of the concentration scale, reverse osmosis will have to compete with adsorption and ion exchange, which are competitive at low concentrations. Reverse osmosis can eventually become less competitive as the concentration decreases, and at extremely low concentrations ion exchange or adsorption processes are likely to be more cost-effective. These processes remove the trace solutes from the solvent rather than the solvent from the solutes. In addressing the role of concentration in the competitiveness of reverse osmosis and other competing processes, it is convenient to consider the concentration of the feed and the desired concentration of the product permeate. Since reverse osmosis passes the solvent and rejects the solute, at low concentrations large volumes (and masses) of material are being permeated to remove a small quantity of solute. The cost of reverse osmosis at low concentration will be almost independent of the solute concentration. The cost of adsorption and ion exchange operations will be highly dependent upon solute concentration since it determines the size or operating time (between regenerations) of an adsorption or ion exchange bed. Thus the cost of adsorption or ion exchange operations will decrease as solute concentration decreases, and eventually those methods are likely to become less expensive than reverse osmosis.

Several important variables determine when or if adsorption or ion exchange will become more attractive, the most important of which are the performance and price of the membranes, adsorbents, and/or ion exchange materials. Thus the range of feed concentrations where reverse osmosis is most likely to be competitive has a lower bound at those concentrations suitable for adsorption and ion exchange and an upper bound at the concentrations where evaporation is just becoming competitive.

However, the exact positions of those range boundaries depend upon the application and the available membranes.

The required concentration of the permeate product can be an important consideration even when the feed concentration is relatively high or low. When extremely high decontamination (removal of large fractions of the contaminant) is required, adsorption or ion exchange may prove to be more attractive than reverse osmosis because it is usually so much easier to incorporate many adsorption or ion exchange stages into a system, usually into a single bed. Many reverse osmosis membranes will not reject more than 90% to 99% of the dissolved solids. Although this is adequate for many applications, perhaps even for most applications, it may require two or more membrane stages to reach the extremely low contaminant concentrations required for some problems. In some such problems, adsorption or ion exchange systems may be able to combine sufficient stages in a moderate size bed to meet those needs more economically than can a series of reverse osmosis cells. However, one should not overlook the opportunity to use both reverse osmosis and adsorption or ion exchange over the concentration range for which they are most economical. The best approach for some applications may be to use reverse osmosis to reduce the concentration of the contaminant (or all dissolved solids) as much as possible in a single stage (or possibly in a couple of stages), and then use adsorption or ion exchange to reduce the concentration below the final desired level. In such operations, the use of reverse osmosis would allow the adsorption or ion exchange beds to be operated with much longer periods between regeneration because the bulk of the dissolved solids would already be removed. The economic optimization can then be made between the cost of the reverse osmosis (and possibly an additional evaporation step) and more frequent regeneration of the adsorption or ion exchange beds. The nature of the regeneration and the regeneration product will also play a role in this optimization. Remember that it may be optimum to combine reverse osmosis with evaporation at the other end of the concentration range, with reverse osmosis doing much of the concentration of the retentate and with evaporation taking the retentate concentration even higher, perhaps even to dryness.

When an organic contaminant is volatile, concentration driven membrane systems for aqueous streams are likely to compete with gas stripping processes. Obviously as the volatility of the contaminant declines, membrane systems are more likely to become competitive. With only moderately volatile contaminants, distillation may become the competing process. This is one area where membrane processes have become increasingly competitive. Although distillation is a well established and trusted separation method, membrane systems can often operate with significantly less energy consumption. The presence of trace components that could damage a membrane or trace surfactants that would create foaming problems in gas stripping could have major effects on the choice of the better technology for removing volatile organic compounds.

The scale of the operation can also affect the relative competitiveness of membrane processes. The change in competitiveness with scale results principally from the difference in the way the capital cost for membrane systems and competing systems change with the scale of the operations. Because a significant fraction of the cost of membrane systems is often the cost of the membrane, the cost of membrane systems depends strongly upon the scale of the operation (i.e., the production rate). Because the other parts of the membrane systems will not depend exactly upon the production rate, the total capital costs will not be strictly proportional to the production rate (i.e., membrane area needed), but will be much more so than the cost for some of the competing separation methods. The capital cost of distillation systems can be approximately proportional to the 0.6–0.7 power of the production rate, but the capital costs for membrane systems will be proportional to a high power of the production rate. This means that membrane systems may not compete well for applications that require a very high production rate. However, when the capital costs are scaled down to smaller facilities, the stronger dependence of capital costs on the production rate becomes an advantage. This different dependence of capital costs is at least partially responsible for membrane systems to gain over cryogenic distillation for air separation (to oxygen and nitrogen) for small facilities, but not for very large facilities. Although air separation is not a waste or environmental application, similar relative behavior of membrane and competing separation methods can be expected to be important in waste and environmental facilities.

Hydrophobic membranes are often selective for dissolved hydrocarbons and many other organic compounds. Of course, to be applied to a practical system, it is necessary to have an even lower activity of the organic material on the downstream side of the membrane. This is likely to mean that the most practical downstream condition is a vapor phase and the membrane is operated as a pervaporation membrane. Pervaporation is more effective for volatile organic compounds. Even for volatile hydrocarbons, the membrane processes are more likely to be the choice for concentrations that are moderate to high in organic materials. At sufficiently low concentrations, adsorption processes using materials like activated carbon may become more competitive.

Selective removal of specific metal ions using liquid membranes appears to have a good chance to become more common. The chemistry of liquid membranes systems is similar to that of solvent extraction processes. A potentially important advantage of liquid membrane systems is the ability to utilize very small volumes of solvents. This can be important when the selective extractant is relatively expensive. High selectivity is more likely to be important when a component is being removed from a concentrated mixture. However, extremely high specificity may not be as important for groundwater and dilute wastewaters that contain relatively small concentrations of other ions. There are often opportunities to incorporate the active ligands from many extractants (including crown ether groups) onto solid substrates that can be used as adsorbents. Such systems can also operate with relative small quantities of expensive extractants (or the extracting ligand) and may compete effectively with liquid membranes unless the diluent plays an important role in the liquid membrane (extraction) operation.

Specific Environmental Applications of Membrane Processes

Membrane processes have received considerable interest during recent years, and many new research programs have begun during the last 10 to 20 years. Numerous papers of membrane separation processes have been written, and new journals have been started that are devoted largely or even exclusively to membrane development and membrane separations. This growth in interest in membranes has resulted largely from improvements in membrane technology. Interest in membrane technologies seemed to jump significantly with the discovery (or development) of effective reverse osmosis membranes in the 1960s for desalination [30]. The cellulose acetate membranes developed by Loeb and Sourirajan [31] were unusual for the time because the active portion of the membrane was limited to a thin section of the membrane surface. That is, the membrane was heterogeneous with a thin surface layer responsible for most of the membrane performance. The rest of the membrane was very porous and acted principally as a support for the thin active surface rather than contributing directly to the separation process.

Reverse Osmosis

Jacobs et al. [32] formed very thin poly-2-vinylimidazoline membranes on micro-porous polysulfone substrate membranes in tubular shapes and obtained high reverse osmosis fluxes and NaCl rejection often better than 99%. Yang and Chu [33] prepared a range of thin membrane films from mixtures of sulfonated polysulfone and polyvinyl alcohol composites and obtained relatively good fluxes and salt rejection. As with so many reverse osmosis membranes, high rejection is obtained only with high fluxes, that is, with high pressure differences. They investigated their membrane performance over a range of temperatures and found fluxes increasing significantly with temperatures up to 50°C (the range tested), but there was a corresponding drop in rejection efficiencies.

Because reverse osmosis is particularly effective at low concentrations where the opposing osmotic pressure is low, it is proposed and used for cleaning up residual contamination in wastewaters. It could be used to treat contaminated groundwater, which is also usually relatively dilute. One requirement of essentially all reverse osmosis systems is a good clarification (removal of solids) of the feed water. The treated water is often of higher quality than most fresh intake water to water treatment plants; it may even be of higher quality than the process water used in the plant. Thus wastewater processed by reverse osmosis is often a suitable candidate for reuse. The advantage of reducing feed water treatment and the merits of reducing pollutant discharge can be additive.

Although reverse osmosis generally shows little selectivity among the solutes, there are many cases where the concentrated product—the reject stream—can be largely recycled to the plant process. These are cases where the reagents are diluted but not significantly degraded or contaminated. Concentration of waste from the pulp and paper industry and cooling tower "blowdown" are two relatively well established applications for reverse osmosis. Two significant examples are the metal plating industry [34] and the textile industry [35].

Reverse osmosis is also a major player in "recovering" materials in the food industries [36]. Although these may not be considered waste treatment applications, one should remember that not too many years ago materials such as cheese whey were at least partially discarded, and recovery of these materials for use in human or animal food can be considered a waste treatment operation. It would be better if more of our waste treatment operations could be considered recovery operations to claim a valuable and useful product.

Membranes can be used in some applications to remove contaminants such as organic vapors from gases, including off-gases. The principal limitation may be the difficulty in removing some contaminants to the level needed. In one case, a membrane has been recommended for removing unreacted monomers and other volatile hydrocarbons from polymerization reactors [37]. In that case, the monomer and other hydrocarbons can be recovered and reused while previous installations have usually sent such streams for incineration in "flares."

Ceramic membranes (titania) have been studied for removing proteins and other valuable dissolved solids from cheese wheys by reverse osmosis [38]. This is an example of the possible use of a separation method for removing a contaminant and turning it into a potentially valuable product, especially if the components of the whey are later separated into selected components.

Gas Separations

Gas separations are among the most extensively used membrane processes, mostly with air (oxygen-nitrogen) separations, not a major environmental separation. However, separations of acid or basic gases from air can be achieved by suitable membrane systems, and in gaseous systems molecular size can be a more useful basis for separation than in some liquid systems. For instance, phosphazene polymer membranes have been tested for separation of methylene chloride or carbon tetrachloride from nitrogen (air) [39]. Such separations could be used on effluent streams which contain such contaminants or from gas pulled from gas stripping or soil venting operations. Note that if the gas is largely recycled, either to the stripping operation or to the soil for removing additional contaminant, the removal efficiency needed may not be as high as that needed for discharge of the gas.

Pervaporation

The main advantage of pervaporation over distillation is the potential increase in the separation factor of the membrane over the relative volatility of the components of the mixture. This advantage, of course depends upon the relative voaltility of the components and the availability of a suitable membrane. Although there would appear to be several potential applications in waste and enviromental processing, few, if any, applications for dilute waste streams or groundwater have been reported. The principal reason for so few applications is believed to be the cost. The construction of pervaporation membrane cells does increase the cost, and there are usually competing separation processes available. The successful applications of pervaporation currently appear to be associated with relatively high value products, not with low value wastewater or groundwater. We will discuss briefly some conditions when pervaporation is more likely to become practical for environmental and waste applications and give a few examples of proposed uses of pervaporation that have been tested in laboratories.

If the organic contaminant is a highly hydrophobic solvent, it is likely that a suitable hydrophobic membrane can be found that will have a significant separation factor. The higher separation factor can result in less energy consumption since less material has to be evaporated. For environmental problems with dilute contaminants, evaporation of only the contaminant could result in very significant energy savings.

The advantages of pervaporation are partially offset by the increased mass transfer resistance imposed by the membrane, by the additional cost and complexity of the membrane over a single stage of distillation (but not necessarily greater complexity than a multi-stage distillation column), and the additional pumping and/or condensation required to collect the permeate. It is possible to have multiple stages of pervaporation, but the additional stages are not constructed as easily as additional distillation column lengths. If numerous pervaporation stages are required, distillation may still be the more attractive option. That is, the pervaporation membrane may not have sufficient selectivity to displace distillation.

Pervaporation is most likely to be used to remove organic contaminants from water when the vapor pressure of the contaminant is not many times greater than the vapor pressure of water. Thus it is more likely to replace distillation than gas stripping, which is more likely to be selected when the vapor pressure of an organic contaminant is relatively high and it is not necessary to raise the temperature to obtain sufficient contaminant pressures and thus useful removal rates.

It is desirable to remove the organic contaminant sufficiently from the water to allow reuse, release, or retreatment with another method, such as adsorption, that can remove traces of contaminant more economically but is usually not cost effective at higher concentrations. Thus pervaporation is more likely to be effective at moderate concentrations, but may not necessarily need to remove the last traces of contaminant. The permeate stream also need not be totally free of water. For instance, if the permeate is enriched in the organic contaminant sufficiently that condensation of the permeate produces two phases, an organic phase saturated with water and a water phase saturated with organic, it is possible to recycle the small water stream back to the feed stream to the permeation equipment without affecting the capacity of the unit significantly. Such operations would have only two products: an aqueous phase that meets the requires for low contaminant concentration, and a contaminant stream in the form of an organic phase.

Chlorinated hydrocarbons have been removed from wastewater by pervaporation through a polyphosphazene membrane [40]. Separation factors of approximately 10^4 were observed for carbon tetrachloride from water, and moderate separation factors (10 to 100) were obtained for carbon tetrachloride and methylene chloride from nitrogen and carbon dioxide. Note, however, that both of these compounds are relatively volatile and can often be removed effectively by gas stripping. Pervaporation of trichloroethane, trichloroethylene, and tetrachloroethylene were studied using composite membranes with poly(acrylate-*co*-acrylic) acid [41]. There seems to be a growing interest in the use of pervaporation to remove hydrocarbons and chlorinated hydrocarbons from waste water [42,43].

Pervaporation can be used to separate acetone from water or water from acetone depending upon whether a hydrophobic or hydrophilic membrane is used. An acrylamide-acrylic acid copolymer formed in the pores of a ceramic membrane gave a very high separation factor, approximately 2000, for water over acetone [44]. On the other hand, polypropylene and ethylene-propylene copolymer membranes were slightly selective for acetone over water [45]. More recently, Hollein et al. [46] found a composite silicone membrane and polydimethylsiloxane (PDMS) membranes had more substantial selectivities for acetone over water, values of 30 to 50. Phenol-water separation by pervaporation has also been studied [47].

Pervaporation can also be useful in cases where foaming would make distillation or gas stripping difficult. This can occur in treatment of waste streams used to remove nonaqueous phase liquids (NAPLs) from soils by surfactant flushing. These highly insoluble liquids can be especially toxic, and their low solubility means that they can contaminate large volumes of soil and remain for long periods as small pools or droplets of insoluble liquids. By flushing surfactants through the soils, NAPLs can be removed with micelles formed by the surfactants. Volatile NAPLs can normally be removed from the flush liquid by gas stripping, but the presence of surfactants makes foaming highly likely. By using pervaporation rather than gas stripping, there is no foaming problem. With a hydrophobic membrane that transports the NAPLs in preference to water or surfactant (the larger surfactant molecule and the probable low vapor pressure of the surfactant make it likely that surfactant will be retained very well), the NAPL can be removed and recovered. It may be possible to reuse the NAPL-free water-surfactant stream for further flushing.

The most interesting current applications of pervaporation are "breaking" azeotropes, which cannot be separated with distillation alone, and drying solvents. Although azeotropes are not usually found in effluents from processing plants, which one typically associates with environmental processing, there are important solvent recovery operations within chemical processing plant systems where pervaporation can enhance the performance sufficiently to affect the release of contaminants to effluent streams. For instance, recovery of tetrahydrofuran from water is hindered by an azeotrope at 94%(weight), which can be "broken by a polyvinylalcohol membrane that is hydrophilic and passes water selectively" [48].

Drying solvents is essentially the opposite of the operation most likely to be of use in wastewater treatment. When pervaporation is used to remove traces of water from a solvent before it is used or recycled, the product is a dry solvent that may have value, and a significant or modest expense can be used to enhance the value of that solvent. When less costly alternatives are not available, pervaporation certainly has a chance to gain applications. The next section will review some of the alternatives to pervaporation that are likely to be available.

Liquid Membranes

Although there is much current research on liquid membranes, relatively few applications are known at this time. The principal reason why supported liquid membranes have not found more applications is believed to the the lack of stability of the membranes, or the lack of confidence in the stability of the membranes. The major sources of membrane stability have been discussed, but more details of the causes of membrane stability and possible ways to reduce the instabilities is given by Kemperman et al. [49]. However, this should be interpreted to mean that this is a relatively new field with applications likely to grow. Since liquid membranes involve liquid solvents, they also have many features of solvent extraction. If the solvent has significant solubility in water, there will obviously be solvent in the processed water. If the solvent is toxic, it will have to be removed, probably by adsorption on carbon, before the water is discharged. If the water is reused, the solubility may be sufficient to not be a serious problem, depending upon the use of the recycled water. However, supported membranes have two important advantages over conventional solvent extraction (liquid-liquid extraction) systems: essentially no entrainment, and the need for only a very small inventory of extractant. In many pieces of solvent extraction equipment, the major loss of a highly insoluble solvent may be by entrainment rather than by solubility. The used of supported liquid membranes essentially eliminates entrainment. Furthermore, essentially all of the extractant is used at all times, and the quantities of extractant required can be minimized. This can make it practical to use highly selective extractants, even when those extractants are too expensive to be practical for use in conventional solvent extractant systems.

The French CEA is investigating the use of crown ethers to remove cesium from their high level radioactive wastes stored as concentrated acid solutions in stainless steel tanks [50]. Cesium is the principal source of penetrating radiation in these wastes, and removal of the cesium makes further treatment of the wastes less difficult. They have found derivatives of 21-Crown-7 to be highly selective for cesium from these wastes. The same chemical system could be operated as a solvent extraction process, but they are investigating the use of supported liquid membranes, probably in part because of the high cost of crown ether compounds.

Although entrainment should be expected for unsupported liquid membranes (micro-emulsions), it may be less than that found in some solvent extraction equipment. Remember that the intense mixing to form micro-emulsions comes in the dispersion of strip solution in the solvent, not in forming the droplets of emulsion that contacts the water feed.

Another attractive feature of liquid membranes is the ability to use very small quantities of solvent and very high water to solvent volumes. This can be an important advantage in dilute systems that are believed to be more common in environmental and waste processing. The small volume of solvent required makes it possible to use relatively expensive solvents that would not be practical in conventional solvent extraction operations. Crown ether extractants are one good example of highly selective extractants that are relatively costly and not likely to be practical for many large volume solvent extraction operations. Note also that the lack of significant entrainment is also particularly important when an exceptionally costly solvent is used.

Conventional solvent extraction equipment usually operates best when the ratio of the flow rates of aqueous to solvent flow rates is not far from unity, or generally within the range of 0.1 to 10. When a contaminant is to be removed, a very dilute solution, maintaining a flow ratio of even 10, would mean that the concentration of contaminant in the solvent would not become more than 10 times the concentration in the original feed. Furthermore, a flow ratio as low as 0.1 can be used, and the concentration of contaminant in the solvent cycle, extracting and stripping, is only a factor of 100. Some adsorption and ion exchange processes can concentrate a contaminant more, and that is one reason why those methods become more attractive than liquid-liquid extraction for very dilute systems. Liquid membrane methods that utilize chemistry much like liquid-liquid extraction can sometimes overcome this limitation when high extraction capacity and highly selective extractants are available along with good re-extraction methods.

If the solvent in a liquid membrane is not flowing, there are no flow ratio restrictions, and high contaminant loadings in the solvent can be allowed, and they are even desirable. Again note that this makes it possible to consider even highly expensive solvents provided their solubility losses are acceptable. Highly selective solvents that have high contaminant concentrations in equilibrium with the water feed stream are also more likely to have good mass transfer through the membrane because the driving force for diffusion through the membrane can be high.

These considerations may have played a role in the study of strontium removal from nuclear waste with supported liquid membranes containing the relatively costly crown ether extractants reported by Dozol, Casas, and Sastre [51] and the cesium removal mentioned earlier. The use of a highly selective extractant such as a crown ether can remove specific contaminants, but their study does not show permeabilities of likely competing ions like calcium for comparison with the strontium permeability.

Supported liquid membranes have also been studied for removal of organic contaminants from wastewater. Zaha, Fane, and Fell [52] investigated the removal of phenol with *n*-dodecane in kerosene solvent in several porous substrates. The phenol was stripped from the membrane with caustic solutions. The fraction of phenol removed from 0.5 to 1 g/L solutions was approximately 60% to 90% in batch experiments. The data were not scaled to commercial flowing equipment, and there is no information on whether this recently reported study is likely to be applied.

It is only fair, however, to point out that there are limitations to the solvents that can be used in liquid membranes. Solubilities must be very low in water because there is so little inventory of solvent that the membrane would quickly become devoid of solvent if its solubility in the feed water were too high. It is possible to saturate (or approximately saturate) the feed with the solvent before it enters the liquid membrane cell to minimize, if not eliminate, solvent loss. It is also necessary to have a highly suitable strip solution to use on the downstream side of the membrane, usually one that removes the contaminant almost totally from the solvent. Fortunately for metal ion removal membranes, pH or redox changes can often meet these conditions. There may be fewer opportunities for stripping organic contaminants from liquid membranes so effectively, and this is one reason that liquid membranes have been studied more extensively for removing specific ions from water.

Emulsion membranes are even closer to liquid-liquid extraction and have somewhat different advantages and disadvantages. Since the "membrane" consists of dispersed droplets, it is possible to lose solvent by entrainment. Although, in principle, this is essentially the same problem as in solvent extraction, it usually is not a problem of the same magnitude. The emulsion droplets are usually significantly larger and probably more uniform in size than the droplets desired for liquid-liquid extraction, but this is not necessarily true. In solvent extraction, it is often desirable to continuously create fresh surface area and enhance mass transfer rates, and many liquid-liquid extraction devices will involve continuous coalescence and formation of new surface area. However, that is not required if the initial surface area is adequate for suitable mass transfer rates. With emulsion liquid membranes, one is not likely to want to break up the emulsion droplets into smaller droplets. The surfactants used in liquid emulsion membranes are likely to stabilize the emulsion droplets as well as the strip droplets within the emulsion.

Emulsion membranes are not likely to be highly tolerant to particulates in the feed stream, and highly effective filtration systems will be necessary to prevent particles from accumulating at the interfaces, probably on the outer interface of the emulsion droplets. However, this is not a unique problem for emulsion membranes since particles are usually harmful to any membrane system. It is not obvious that particulates are more harmful to emulsion membranes than they are to solid membranes or to supported liquid membranes. Exact comparisons are difficult because of the different tolerance of different membranes to particulates. Likewise, it is not obvious that emulsion membranes are less tolerant to the presence of particulates than are the corresponding liquid-liquid extraction systems.

Perhaps the principal attractions of emulsion membranes are their ability to combine extraction and stripping into a single unit and the resultant gain in efficiency from continuously stripping the solute within the extraction vessel. The only additional equipment required is for a separate emulsion breaking step. The additional gain in efficiency from the continuous stripping of the solute results because there is no accumulation of solute in the solvent; it transfers to the strip solution as it is extracted. This last advantage is reduced if the strip solution is not able to reduce the concentration of solute in the solvent essentially to zero.

Several emulsion membrane systems have been reported for removing specific ions from dilute solutions, but there is not strong evidence that emulsion membranes are used extensively to solve environmental problems. One system for removing zinc from a wastewater produced in a viscose rayon plant has been in operation since the 1980s [53]. Several studies have investigated the removal of metals from dissolved ores or contaminated water [54–56]. Tutkun and Kumbasar [57] reported using LIX 860 (5-nonylacetophenone oxime) in kerosene with Span 80 as the surfactant. The Cu was selectively separated from Zn, Mg, Fe, Co, and Ni. The initial Cu concentration was approximately 175 ppm. The internal (strip) solution within the emulsion was concentrated sulfuric acid. There is strong interest in emulsion membranes, and it is a popular topic in the R&D community. One example study reported using D2EHPA extractant in *n*-hexane diluent to remove lead from wastewater. Span 80 surfactant was used to stabilize the emulsion, and 0.1 *M* sulfuric acid was used as the strip solution in the core of the emulsion [58]. Much of the work on emulsion membranes up to 1990 is reviewed by Ho and Sirkar [59]. It is likely that the relatively strong R&D currently under way will result in more applications of the technology, and several could be in environmental and waste processing.

Electrodialysis

Electrodialysis has potential applications in separating electrolytes from water and in salt splitting. In separating salts from water, electrodialysis must compete with other methods such as reverse osmosis and even evaporation. Although there have been significant advances in electrodialysis in recent years, this author feels that the advances in reverse osmosis have generally been greater, so the interest in applying electrodialysis may be a little less now than it was a few years ago. However, there are still potential applications for electrodialysis, and these should not be ignored. It is less likely to be preferred for very dilute solutions because reverse osmosis is so effective. It is also not likely to be preferred for very high concentrations because Donnan exclusion is not effective as the concentration approaches that of fixed charges in the membrane. For concentrated solutions, evaporation may be the preferred method.

The applications for electrodialysis are most likely to occur in the immediate concentration range. Although so many environmental applications dealing with aqueous solutions are relatively dilute, this does not mean that there will be little or no potential applications for electrodialysis. For instance, reverse osmosis may be the preferred method for removing a salt from a very dilute wastewater, and one may want to concentrate the salts removed by reverse osmosis using electrodialysis. However, as the concentration increases, the osmotic pressure increases and eventually makes reverse osmosis less desirable. Of course, if one wanted to concentrate the salt sufficiently, perhaps even to dryness, evaporation would also have to be considered. Thus, two steps using different methods may be preferred to take a salt from a dilute solution to dryness; there may be cases where it would be even better to use three methods.

The one application where electrodialysis offers a completely different service is in salt splitting. Salt splitting is the basis for the chloralkali industry, a very substaintial segment of the chemical industry where sodium chloride (salt) is converted to chlorine and sodium hydroxide (caustic soda), but salt splitting is not commonly used in waste and environmenal processing, although important potential applications are likely to exist. If the acid and base can be recycled, salt splitting offers the potential for eliminating a waste, not just concentrating it. One interested application involves the recovery of waste acid from metal finishing and waste battery acid [60]. These can be acid streams with significant concentrations of metals, some of which can be toxic. The acid is likely to be more dilute than the original acid used in the process, and electrodialysis can be used to remove the metal contaminants and to concentrate the acid to the desired level. Both functions can be carried out in the single unit. Furthermore, the metal contaminants can be concentrated in the other, alkaline or less acid, stream. In any process system where the process results in neutralizing an acid and base, salt splitting could be considered to reverse that aspect of the process to reuse the acid and base. The most likely limit that would discourage salt splitting could be the buildup of impurities in the salt stream that would become contaminants in the acid and base.

The most common limitation to deployment of membrane systems is probably the cost of the membranes themselves. The cost of membrane systems scale with a relatively high power of the system capacity (a power near unity), and this results because the cost of the membranes is a significant/important part of the overall system cost. This means that significant reductions in membrane costs would have important effects on the overall system costs and could allow opportunities for additional applications of membrane systems, including environmental and waste treatment.

REFERENCES

1. Lahiere, R. J. "Environmental and Process Applications for Membrane Technology in the CPI." Preprints of the First Separations Division Topical Conference on Separation Technologies: New Developments and Opportunities. 1992, pp. 788–792.
2. Loeb, S. and S. Sourirajan. *Adv. Chem. Sci. 38*, 117 (1962).
3. Melton, J. H., et al. *Recent Developments in Separation Science*. Vol. V, CRC Press, Cleveland (1979), pp. 1–10.
4. Marcinkowsky, A. E., et al. *J. Am. Chem Soc. 88*, 5744 (1966).
5. Johnson, J. S., et al. *J. Electroanal. Chem. 37*, 267 (1972).
6. Lamb, J. D., R. T. Peterson, and A. Schow. "Novel Polymer Inclusion Membranes Incorporating Macrocyclic Cation Carriers." Presented at the Ninth Symposium on Separation Science and Technology for Energy Applications, Gatlinburg, TN, Oct. 22–26, 1995; submitted to *Sep. Sci. Technol.*

7. Jones, C. W. and W. J. Koros. "Regeneration of Carbon Molecular Sieve Gas Separation Membranes Following Organic Exposure." Paper presented at the 1995 Spring National Meeting of the American Institute of Chemical Engineers, Houston, TX, March 19–23, 1995.
8. Cadotte, J. E., et al. *Desalination 32*, 25 (1980).
9. R. Y. M. Huang (Ed.), *Pervaporation Membrane Separation Processes*. Elsevier, Amsterdam (1991).
10. Zhang, S. and E. Drioli. *Sep. Sci. Technol. 30*, 1 (1995).
11. R. J. Lahiere. "Environmental and Process Applications for Membrane Technology in the CPI." Preprints of the First Topical Conference on Separations Technology: New Developments and Opportunities. AIChE, 1992, pp. 788–792.
12. Zhang, S. and E. Drioli. *Sep. Sci. Technol. 30*, 1 (1995).
13. Rapin, J. L. "The Betheniville Pervaporation Unit—The First Large-Scale Productive Plant for Dehydration of Ethanol." *Proceedings of the Third International Conference on Pervaporation Processes in the Chemical Industry* (R. Bakish, ed.), p. 364. Bakish Materials Corp. Englewood, NJ (1988).
14. Kujawski, W. *Sep. Sci. Technol. 31*, 1555 (1996).
15. Zhang, S. and E. Drioli. *Sep. Sci. Technol. 30*, 1 (1995).
16. Schofield, R. W., A. G. Fane, and C. J. D. Fell. *J. Memb. Sci. 33*, 299 (1987).
17. Gryta, M., M. Tomaszewska, and A. W. Morawski. *Sep. Purification 11*, 93 (1997).
18. Zhang, S. and E. Drioli. *Sep. Sci. Technol. 30*, 1 (1995).
19. Zhang, S. and E. Drioli. *Sep. Sci. Technol. 30*, 14 (1995).
20. Wodzki, R. and G. Sionkowski. *Sep. Sci. Technol. 31*, 1541 (1996).
21. Noble, R. D., J. J. Pellegrino, E. Grosgogeat, D. Spersy, and J. D. Wat. *Sep. Sci. Technol. 23*, 1595–1609 (1989).
22. Stevens, G. W., et al. *Sep. Sci. Technol. 31*, 1025 (1996).
23. Draxler, J. and R. J. Marr. "Preparation and Breaking of Emulsions: Applications Including Emulsion Liquid Membranes." Paper presented at the 1995 Spring National Meeting of the American Institute of Chemical Engineers, Houston, TX, March 19–23, 1995.
24. Bansal, R., V. Jain, and S. K. Gupta. *Sep. Sci. Technol. 30*, 2891 (1995).
25. Flemming, H. C. in *Reverse Osmosis: Membrane Technology, Water Chemistry, and Industrial Applications* (Z. Amjard, ed.). Van Nostrand Reinhold, New York (1993), p. 163.
26. Geldreich, E. E., et al. *J. Am. Waterworks Ass.* 77, 72 (1985).
27. Winfield, B. A. *Water Res. 13*, 561 (1979).
28. Amjard, Z., et al. in *Reverse Osmosis: Membrane Technology, Water Chemistry, and Industrial Applications* (Z. Amjard, ed.). Van Nostrand Reinhold, New York (1993), p. 210.
29. Pittner, G. A. in *Reverse Osmosis: Membrane Technology, Water Chemistry, and Industrial Applications* (Z. Amjad, ed.). Van Nostrand Reinhold, New York (1993), p. 90.

30. Merten, U. (Ed.), *Desalination by Reverse Osmosis*. MIT Press, Cambridge, MA (1996).
31. Loeb, S. and S. Sourirajan. *Adv. Chem. Ser. 38*, 117– (1962).
32. Jacobs, E. P., M. J. Hurndall, and R. D. Sanderson. *Sep. Sci. Technol. 29*, 2277 (1994).
33. Yang, M.-H. and T.-J. Chu. *Sep. Sci. Technol. 28*, 1315 (1993).
34. Nirmal, J. D., V. P. Pandya, N. V. Desai, and R. Rangarajan. *Sep. Sci. Technol. 15*, 2083–2098 (1992).
35. Brandon, C. A. *Ind. Water Eng. 12* (6), 14 (1976).
36. Koseogula, S. S. and G. J. Guzman. in *Reverse Osmosis: Membrane Technology, Water Chemistry, and Industrial Applications* (Z. Amjad, ed.). Van Nostrand Reinhold, New York (1993), p. 300.
37. Wijmans, H., et al. "Hydrocarbon Recovery from Polyolefin Vent Streams Using Membrane Technology." Paper presented at the 1996 Annual Meeting of the AIChE, Chicago, IL. Nov. 12, 1996.
38. Peterson, R. A., M. Anderson, and C. Hill. *Sep. Sci. Technol. 28*, 327 (1993).
39. Peterson, E. S., et al. *Sep. Sci. Technol. 28*, 271 (1993).
40. Peterson, E. S., et al. "The Removal of Organic Chemicals from Waste Streams Using Polyphosphazene Membranes." *Membrane Processes: Water Treatment-Pervaporation*. Proceedings of Euromembrane-92, Paris, France, Oct. 1992 1992, p. 381.
41. Nakagawa, T., M. Hoshi, and A. Higuchi. *Proceedings of the Fifth International Conference on Pervaporation Processes in the Chemical Industry* (R. Bakish, ed.), Bakish Materials Corp. Englewood, NJ (1991), p. 88.
42. Lipski, C. and P. Cote. *Environ. Prog. 9*, 254 (1990).
43. Clement, R., Z. Bendiama, Q. T. Nguyen, and J. Neel. *J. Membrane Sci. 66*, 193 (1992).
44. Sakaoara, S., F. Muramoto, T. Sakata, and M. Asaeda. *J. Chem. Eng. Jpn. 23*, 40 (1985).
45. Featherstone, W. and T. Cox. *Br. Chem. Eng. Process Technol. 16*, 817 (1971).
46. Hollein, E., M. Hammond, and C. S. Slater. *Sep. Sci. Technol. 28*, 1043 (1993).
47. Matsumoto, Y., et al. *Proceedings of the Sixth International Conference on Pervaporation Processes in the Chemical Industry* (R. Bakish, ed.), Bakish Materials Corp. Englewood, NJ (1992), p. 55.
48. Mencarini, J., R. Coppola, and C. S. Slater. *Sep. Sci. Technol. 29*, 465 (1994).
49. Kemperman, A. J. B., et al. *Sep. Sci. Technol. 31*, 2733 (1996).
50. Dozol, J. F. and J. Casas. *Sep. Sci. Technol. 30*, 435 (1995).
51. Dozol, J. F., J. Casas, and A. M. Sastre. *Sep. Sci. Technol. 28*, 2007 (1993).
52. Zaha, F. F., A. G. Fane, and C. J. D. Fell. *Sep. Sci. Technol. 29*, 2317 (1994).
53. Ruppert, M., J. Draxler, and R. Marr. *Sep. Sci. Technol. 23*, 1659 (1988).
54. Hayworth, H. C., et al. *Sep. Sci. Technol. 18*, 493 (1983).
55. Li, N. N., et al. *Hydrometallurgy, 9*, 227 (1983).
56. Weiss, S., et al. *J. Memb. Sci. 12*, 119 (1992).
57. Tutkun, O. and R. S. Kumbasar. *Sep. Sci. Technol. 29*, 2197 (1994).

58. S. Kaur and D. K. Vohra. in *Emerging Technologies for Hazardous Waste Management: 1992 Book of Extended Abstracts*. American Chemical Society (1992), pp. 201–204.
59. Ho, W. and K. Sirkar. *Membrane Handbook*. Van Nostrand Reinhold, New York (1992).
60. Asano, B. and S. Kok. "Treatment of a Dilute Acid Waste Stream Using Electrodialysis: A Process Evaluation." *Proceedings of the 49th Industrial Waste Conference*, Purdue University, May 9–11, 1994, p. 311.

5
Leaching/Extraction

Leaching and extraction will be defined here as removal of a contaminant from a solid by a fluid, usually a liquid. There is a possibility for confusion because the word "extraction" is also used in Chapter 6. In fact, there are important similarities because some of the leaching operations discussed here will involve use of organic solvents that could also be used in liquid-liquid extraction separations. There could be even closer chemical similarities when leaching operations are used to remove contaminants that are dissolved in (or linked though hydration to) water that is attached to solids such as soils. Nevertheless, the distinction used for including a problem in this chapter is the presence of the contaminant in or on a solid, regardless of the chemical form of the contaminant in/on that solid. Note that there can also be a chemical similarity with desorption of contaminants from adsorbents when the contaminant is an adsorbed specie.

The choice to discuss extraction of components from solids separately from the discussion of extraction of components from liquids or from adsorption/desorption is not a universal practice in separation texts, but it is a logical choice because the nature of solid-liquid extraction (or leaching) equipment is significantly different from liquid-liquid separation equipment or from most adsorption/desorption separations, and many leaching operations discussed in this chapter would not fall conveniently into the other chapters. Leaching operations are discussed together almost regardless of the mechanism by which the contaminant is attached to the solid; in many cases, the mechanism of attachment is not known.

To minimize confusion, only the term "leaching" is used, it is unlikely that anyone using this term will confuse his or her audience. However, the reader should remember that the term "extraction" is also used for such operations in other publications and be prepared to recognize this use of the term when necessary. Also remember that when the contaminant is adsorbed on the solid or dissolved in a liquid within or on a solid, the chemical (thermodynamic) equilibria can be similar to those discussed in the other chapters (e.g., Chapters 2 and 6).

Note also that many solid-liquid leaching operations involve chemical reactions and may be considered reactions rather than separation processes. Although this discussion will focus on leaching operations that are not principally chemical reactions, no great effort will be made to exclude leaching operations that involve, or could involve, chemical reactions.

Leaching operations are traditionally involved in recovering valuable products from ores or other solid sources of the desired products, and those are probably the largest and most common applications of leaching. Although ore processing is the largest application of leaching operations, there appears to be more growth in interest in leaching for environmental and other applications. Examples of nonenvironmental application range from leaching of dyes or drugs from plants or other natural products to large-scale leaching of minerals from ores. Even very small scale leaching operations can have great economic value to society. In environmental applications, leaching/extraction is more likely to involve removal of a toxic or undesirable product from a solid such as soil, sludge, or contaminated equipment.

Leaching operations can be described by several names. For instance, removal of toxic materials from equipment or structural materials could be called "decontamination," and removal of toxic materials from soils could be called "soil washing." Nevertheless, the reader should have little difficulty in recognizing any process step that involves removing a component from a solid by contacting the solid with a liquid or gas and realize that the process could be described as leaching (or extraction).

The state of the solute or contaminant (or any components that are leached from the solid) is extremely important to the leaching operation, and only a few examples were given earlier of the form of the contaminant in/on the solids. The contaminant could be coated as a slightly soluble solid on the surface of particles, adsorbed on the surface of solids, physically entrapped within the solid, or even chemically bonded to the surfaces. Physically, the component to be leached could be located on the solid surface, dispersed uniformly in the solid, or encapsulated within the solid. The appropriate leaching method can be different for each of these situations. The removal mechanism could involve dissolving only the component to be leached or dissolving a portion of the solid itself into the fluid. Of course, it is likely to be undesirable to dissolve too much solid material other than the contaminant. Grinding the solid is often desirable to increase the surface area and thus increase the leaching rate, but grinding could be essential for removal of solids that are physically entrapped within the solid. That is, it may be necessary to fracture the particles to expose the contaminant to the leach fluid.

The rate of leaching can involve rates of dissolution, diffusion of leaching reagent to the solid surface, diffusion of the leaching reagent into the solid, diffusion of the leached component from within the solid, or diffusion of the leached component from the solid surface to the bulk fluid. In some cases, it may be necessary to grind or crush the solids first to expose the solute to the leach fluid so that these mechanisms can begin to have an effect, and the nature of the particle fracture can be a crucial parameter in determining the leach rates. Note that the rate limiting steps in leaching can be similar to the steps considered for desorption operations, even when the method of attaching the contaminant to the solid is much different from adsorption. However, there can be additional mechanisms involved, especially those involved in transport and reaction of the leaching agent and fracturing of the solids. Like desorption operations, leaching operations can involve multi-stage contacts with a fluid to desorb sufficient fractions of the contaminant and/or to achieve better utilization of the leaching reagents. In some cases, the contaminant could be physically removed by hydrodynamic and/or surfactant interactions with the surface [1]; it is not always necessary to use exotic, corrosive, or expensive leaching reagents.

GOALS OF LEACHING OPERATIONS

The goal of more traditional leaching operations is to recover the leached component, preferably at a high concentration, which is likely to be a metal in an ore or a valuable component in a treated solid biological material. On the other hand, the goal of environmental leaching operations is more likely to be removal of the toxic component to a concentration below some desired level. The difference is important. There is a tendency to assume that leaching technologies developed for more traditional industries, such as ore leaching, can be applied with little difficulty to environmental applications, but the different goals can make the technologies of one group of applications difficult or even impractical to use for the other group. Leaching operations that remove 60% to 90% of the component of interest may be considered highly successful for recovering valuable components from ores; there may be little incentive to expend further efforts to remove the relatively small fraction of the remaining component.

However, for environmental applications, a much higher percent removal may be (and often is) needed, and leaching operations that do not meet the requirements can be unacceptable.

It may be impractical, or even impossible, to reach the degrees of leaching needed for environmental applications although apparently similar leaching operations are considered highly successful for product recovery operations. Note, however, that United States laws link the toxicity of "nonlisted" toxic solids to the ease of leaching the toxic components from the solids under standardized conditions that are intended to simulate leaching under environmental conditions, not necessarily to the residual contaminant concentration. These are wastes that are toxic by characteristics and not listed as toxic. It may be likely that such toxic solids after leaching the more soluble components with an aggressive leach solution may no longer lose additional toxic components to the milder solutions used in a standard leach test. That is, even when high removal efficiencies are not achieved, the remaining contaminants on/in the solid may be sufficiently difficult to remove that the remaining solid will pass leach tests. However, leaching with an aggressive leach solution provides no assurance that sufficient fractions of the toxic components can no longer be leached.

Although there can be many reasons why high removal efficiencies can not be achieved (slow leach rates, marginal solubilities of the contaminant in the leach fluid, etc.), a relatively common reason is the presence of two or more forms of the contaminant in the solid. Some contaminant could be on the outside of the solid particles, and the other portion could be in small pores that can hardly be reached by the leach liquor. Some contaminant could even be trapped in the solid and completely out of reach of the liquor. The contaminant can also be in different chemical states, such as soluble salts and insoluble salts of metal contaminants. There can also be different types of solid materials in a solid mixture; contaminants could be easily leached from some solid particles and not leached from other particles.

ALTERNATIVES TO LEACHING OPERATIONS

When a solid contains an unacceptable concentration of a contaminant, there are usually only three alternatives: leaching, destruction of the contaminant (usually destruction of organic contaminants), and incorporation of the entire solid into a suitable waste form from which the contaminant cannot be leached by the standard tests. If the contaminant cannot be removed sufficiently by a practical leach operation, one must consider one or more of these alternatives. Incorporating small volumes of solid waste into a final disposal form may be inexpensive, and the added cost of setting up and operating a leaching operation may not be justified. However, it may be extremely costly to incorporate large volume waste streams directly into the disposal form. Current United States laws do not give much credit for removing even significant fractions of the contaminant unless the concentration is reduced sufficiently to move the waste from the hazardous category (or move the waste to a lower category, for nuclear wastes), but it may be worth testing leaching processes and the resulting leached solid to see if it can then pass the standard environmental leach test (the TCLP in the United States). If the contaminant exists in more than one form, even removal of a small fraction may be sufficient if it removes essentially all of the component that can be removed in the environmental leach test.

When essentially complete removal of the contaminant from the solid is not practical but leaves only contaminant that is not removed by the environmental acceptance tests, it is important that the waste not be listed. That is, leaching of the easily removed contaminant would not change the classification of a listed waste; a formal delisting procedure would be required, which is slow and expensive. However, the classification of nonlisted wastes is is relatively easily altered by showing that there is no significant release of contaminant to the environment.

MATERIAL BALANCES

Although the basic material balances and rate expressions for all leaching operations are essentially the same, there are enough differences in the chemical and physical operations involved that we will discuss some important examples separately. Experimental material balances are highly desirable to check data and test results, and, as seen earlier for other separation methods, material balances are also one of the key tools for analyzing and designing separation equipment.

Leaching/extraction operations can often achieve their goals in single stage operations, but some solids may require several stages of contact with the leach solution. Both cases will be considered. As noted, in some ways leaching can resemble the regeneration of adsorbents with a fluid, usually a liquid or a supercritical fluid, and, in some cases, the same techniques could be used. However, when irreversible changes occur in the solid phase from desorption or chemical reaction affecting the bulk solid, the equilibrium relation used to describe leaching may not be simple, or even relevant.

Single Stage Operations

The material balance for single stage operations is simple. In steady-state leaching, the amount of contaminant removed from the solid must be equal to the contaminant accumulated in the leach liquor. That is,

$$Q(q_{in} - q_{out}) = L(C_{out} - C_{in}) \tag{1}$$

where Q is the mass of solid leached, q is the concentration of contaminant in the solid expressed in mass of contaminant per unit mass of solid, L is the volume of leach liquid used, and C is the concentration of contaminant in the solution expressed in mass per unit volume of liquid. The subscripts indicate the initial and outlet concentrations of contaminant in the solid and liquid. For highly efficient leaching operations, the final concentration of contaminant in the solid may be insignificant relative to the initial concentration, and the final concentration in the solid would have very little bearing on the material balance. However, the concentration of contaminant remaining in the solid will still be of utmost practical importance as long as the toxicity of the treated solid is significant. This material balance can apply to a single stage or to an overall leach system with multiple stages.

Equilibrium Stage Operations

Single stages of continuous leaching may be sufficient if equilibrium does not restrict the contaminant removal significantly. In many or most cases the degree of removal is limited by rate processes, not by equilibrium. This is often the case because one is likely to use aggressive leaching agents that shift the thermodynamic equilibrium strongly toward transfer of the contaminant to the fluid. In such cases, leach time and hydrodynamic/mechanical conditions may be the principal factors determining the degree of leaching. When the degree of leaching is controlled by equilibrium, the outlet solid and liquid could be in equilibrium, and, of course, no single equilibrium stage leaching operation can remove more solute than one would achieve from equilibrium conditions.

For those operations with highly favorable equilibrium (and thus controlled by rate processes), high contaminant removal from the solid can be achieved even with a significant concentration of contaminant in the leach liquor. It may even be possible to recycle the liquor with no regeneration or after relatively simple regeneration steps that remove no more than a fraction of the contaminant from the liquor, and it may be possible to recover the contaminant at a relatively high concentration from the leach solution, a concentration more suitable for reuse or concentrated into a small volume more suitable for disposal or destruction.

Recycling of leach liquor can be highly desired to reduce the volume of waste for disposal or that must be destroyed by further processing, such as incineration. Maximum recycling of the leach liquor is especially attractive when the leach liquor is corrosive or otherwise presents disposal problems or high disposal costs.

However, equilibrium can be an important consideration when the concentration remaining on the solid is a strong function of the concentration in solution; those are the cases when multistage leaching may be necessary to achieve high removal efficiencies and when removal of the contaminant from the leach liquor is most desirable if the leach liquor is recycled. Note that for highly porous solids, the effective equilibrium has to account for the volume of leach liquor in the pores of the solids. In such cases, some multiple leach stages may be necessary just to wash the contaminant laden leach liquor from the pores of the solid. These "wash" stages do not always have to use full strength leach solutions and may be able to use water alone.

To estimate the maximum removal that can be achieved from an equilibrium leaching stage (with excellent kinetics and mass transfer rates), one can combine the material balance equation with the equilibrium that relates the final (outlet) q and C. This curve is much like an "isotherm" used in adsorption, and desorption could even be a mechanism for retention of the solute in/on the solid. If the equilibrium is expressed in graphical form with q plotted as a function of C, one can calculate the final concentration in the solid and liquid in the following way. First note that the material balance relates the final concentration in the liquid, q_{out}, to the final concentration in the liquid, C_{out}, and the relationship is linear with a slope of $-L/Q$. Thus this line represents all of the possible combinations of q_{out} and C_{out} and goes through the point (q_{in}, C_{in}). This plot of the material balance can be called a "single stage operating line." The simultaneous solution of the material balance and the equilibrium relation occurs when the operating line intersects the equilibrium curve (Figure 1). If the leaching does not reach equilibrium, the final concentrations will still lie along the operating line between the original composition and the intersection of the operating line with the equilibrium curve.

If there is a simple expression for the equilibrium curve, one may be able to calculate easily the maximum degree of removal that is possible in a single equilibrium stage. For instance, the concentration of contaminant in the solid could be proportional to the concentration in the liquid; that is, the equilibrium curve could be linear. This is not an unusual case, especially for washing of contaminants in the pores of solids, even when physical adsorption is not a significant factor. Leaching systems with

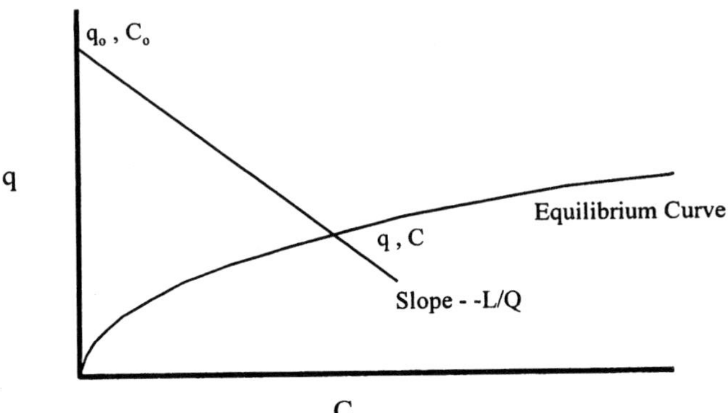

FIGURE 1 Equilibrium leaching. Equilibrium curve and operating line for a single stage equilibrium leashing operation.

highly favorable equilibrium (where the remaining contaminant adsorbed or precipitated in the solid is essentially zero after reaching equilibrium with the leach liquor) can appear to behave somewhat like a system with a linear equilibrium curve if the contaminant dissolved in the liquid inside the particles (usually in the pores) is included with the contaminant in the solids. The concentration of contaminant in the pores will be approximately the same as the concentration in the leach liquor after equilibrium; so the contaminant that appears to remain in the solid will be proportional to the concentration of contaminant in the leach liquor. In other cases, the contaminant could be adsorbed on the solid, and in dilute systems the adsorption isotherm could be approximately linear. In both cases, the equilibrium curve would appear to be approximately linear, at least at low concentrations when the pore volume is the principal source of contaminant in/on the solid. For a linear equilibrium curve,

$$q_{\text{sat}} = Kc_{\text{sat}} \tag{2}$$

where K is the slope of the equilibrium curve, sometimes called the distribution coefficient. This relation can then be used to eliminate either q_{sat} or C_{sat} in the material balance, and the equation can be solved for the other variable. If one eliminates C_{sat}, the solution for q_{sat} becomes

$$q_{\text{out}} = \frac{Qq_{\text{in}} + LC_{\text{in}}}{Q + L/K} \tag{3}$$

When equilibrium strongly favors removal of the contaminant, K will approach zero, and q_{out} will approach zero.

Comments on Mass Transfer Resistance in Leaching

Although there can be significant diffusion resistance in the liquid film, the most serious mass transfer resistance is likely to be in the solid itself, probably diffusion through pores or from internal surfaces of the solid particles. For soils or other solids that have been in contact with flowing water or other leach fluid for long periods, solutes that can be leached easily from the solids are likely to be gone unless the contamination occurred very recently. For instance, clays have layered structures, and some contaminants can be located within the layers; soluble materials that were located between the particles may have been washed from the solids long ago. Other solids may have three-dimensional pore structures from which the contaminant must be removed. The longer the solid has been exposed to the contaminant, the more time has been available for the contaminant to diffuse into the smaller pores.

In some cases such as soil washing, attrition washers are needed to break up larger clay particles into smaller particles to give shorter diffusion paths during the removal (wash/leach) process. It is even possible that all or a portion of the contaminant can be "trapped" within the solid particle and not be reached by the leaching reagent nor have a suitable path to exit the particle. Usually the solid particles must be formed or altered in the presence of the contaminant for true entrapment to occur. Of course, contaminant would not diffuse into such regions, but they could have been trapped there when the solids were formed. The opportunity for contaminant to be trapped within the solid depends upon the history of how the solids were formed. In some cases, contaminants can precipitate within preformed solids. Sometimes, aggregate particles trap contaminants and can be broken in attrition "wash units" or in separate grinding operations, but very small and/or very strong aggregates sometimes can be difficult to break. Considerable energy may be required if the solids have to be ground to extremely small sizes.

When leaching involves chemical reactions or actual dissolution of some of the particles, one may question if that should be considered truly a "separation" process rather than a chemical "reaction," but we will leave that decision to the reader and proceed to mention such systems. Including such reacting systems with separation operations can be justified by the goal of the operation if not by the mechanisms involved. Chemical dissolution can attack principally the contaminant that may coat the surface of the solid, or it may attack the interior of the solid itself to free the contaminant. Leaching can open pores that allow the contaminant to escape to the leaching fluid. Leaching can sometimes attack only a portion of the solid that is responsible for adsorbing the contaminant or whose presence interferes with leaching, perhaps by blocking the path for diffusion of the contaminant from the particle and/or diffusion of the leaching agent to the contaminant.

The local rate of leaching can be controlled by any number of mechanisms, and it may not even be practical to determine and confirm the exact mechanism(s) involved, but it is worthwhile to keep in mind some of the mechanisms that could contribute. Ideas about the leaching mechanism(s) can provide guidance in selecting changes to promote improved leaching, but without proof of the mechanism it will be necessary to run tests to show that the expected improvements can be achieved.

There are several reasons why leaching performance, even in well stirred systems, may not be as simple as most analyses suggest. Many solids are not contaminated uniformly, and this can also affect the performance of leaching systems. For instance, when removing contaminated soils for treatment, it is usually necessary to remove some additional uncontaminated soil to ensure that all of the contamination is removed. Limited knowledge of the location of the contamination often makes if necessary to remove extra soil, and the poorer the information on contamination location, the more clean soil is likely to be removed. It is also possible that the contamination will not all be in a single form. One fraction of the contaminant can be in a relatively soluble and easily leached form, and the other fraction(s) could be in a nonleachable, or inaccessible, form.

It is likely that contamination will be spread throughout the treated soil during the initial stages of treatment, and for completely mixed leach systems it may sometimes, but not always, be acceptable to assume that the contamination is uniform. The distribution of contaminant in the soil is more likely to have an effect on the leach rate in nonstirred systems such as trickle beds, but it is not practical to present much detail on this question since the effects are strongly dependent upon where the contamination is distributed within the bed and flow channels and upon the method of contacting the solids (soil) with the fluid.

Nonuniform particle size can also affect the leach rate, especially if the contaminant resides deep in the particles and the leach rate is controlled by diffusion within the particles. The contaminants are likely to be leached more rapidly from the smallest particles and slowest from the larger particles, if the properties and diffusion rates in the particles are identical. This tends to "spread" the leaching rate, with high initial rates corresponding to the smaller particles and slow rates later resulting from leaching from the larger particles. However, contaminants often are concentrated in the "clay" fraction or the organic (humus) fraction of soils, and the leach rate may be affected by only a portion of the particle size distribution.

Nonequilibrium Single Stage Operation

Since single stage operations are often (perhaps even more likely to be) used when the degree of removal is not controlled by equilibrium condition, the rate expressions are likely to be a dominant concern. These are single stage operations, but it would not be appropriate to call them equilibrium stage operations. These include cases where the equilibrium conditions would correspond to essentially complete removal, perhaps greater removal than is likely to be obtained in a single pass through standard or practical size equipment. The equilibrium curve in Figure 1 will then be horizontal and approach the x-axis. The degree of removal in these cases is then controlled by rate processes, not by equilibrium conditions. For systems that do not involve chemical reactions (such as desorption of physically adsorbed contaminants or liquids trapped in pores of the contaminated solids), the leaching or decontamination rate could be diffusion controlled, but equilibrium limitations can still affect the leaching rate by affecting the driving force for diffusion. That is, the rate at which the leached component(s) can diffuse from the remaining solid substrate would depend upon the difference between the concentration in the pores (or adsorbed on the solid) and the concentration in the bulk. When chemical reactions play a major role in the leaching process, the rate at which the leaching agent diffuses into the solid substrate can contribute to, or control, the leaching rate. In some cases, chemical kinetics could control it. In diffusion (or kinetic) controlled systems with equilibrium conditions corresponding to essentially complete removal, there may be no significant incentive to operate multi-staged systems other than the need to keep individual pieces of equipment to a practical scale (size). When multiple pieces of equipment are used, there may be no incentive to operate in countercurrent flow; cocurrent flow may be just as effective. However, when the leach rate depends upon the concentration of contaminant in the bulk fluid, multi-stage systems and countercurrent flow should be considered. Cocurrent operations can provide as much time as necessary for the solids and fluid to approach as close to equilibrium as needed. Cocurrent operations can be carried out continuously, but it is difficult to construct and operate continuous countercurrent fluid solid contacting systems. (Such systems are possible and are discussed in more detail in Chapter 2.)

There are two extreme classes of nonequilibrium continuous single stage leaching operations that need to be considered. In one class, the solid and liquid flow are not well mixed and proceed through the leaching vessel in an approximately cocurrent plug flow manner that was described above and can be represented by the limiting case of a plug flow leaching system. The second class of single stage operations includes highly mixed systems that can be represented by the extreme case of the continuous stirred tank leaching system, a form of a continuous stirred tank reactor (CSTR). Multi-stage cocurrent systems, such as those with a series of leaching vessels, can give performances between these two limiting cases. When both phases reach equilibrium, there is no difference between the performance of these two systems, but there can be important differences when mass transfer or chemical kinetics of leaching reactions are important. Single stage systems are more likely to be highly stirred systems, but there are cocurrent systems with "screw drives" that move the solids in the same direction as the liquid flow. Perhaps more importantly, the cocurrent systems behave much like batch leach systems when time, rather than position in the leaching equipment, is considered the independent variable, especially when the liquid and solids are transported at the same velocity and in plug flow. As the flow of either phase differs more and more from plug flow because of mixing or dispersion, the performance of cocurrent system will deviate toward that of a CSTR. Thus the plug flow and CSTR leaching systems correspond to limits for the range of operating performances of nonequilibrium single stage leaching systems.

Cocurrent and Batch Leaching Systems

The following discussion will stress batch or cocurrent operations, but the reader should not consider these to be general cases. The behavior of idealized cocurrent leaching systems and batch leaching systems will be discussed together by considering the behavior of both systems as a function of time. The idealized cocurrent system will have the solid and liquid both in plug flow and traveling at the same rate. Thus at any point in the leaching system, the solid will be considered to be in contact with the same liquid throughout its trip through the leaching equipment. Of course, that is exactly what happens in a batch system in which the solids remain in contact with the same liquid throughout its residence in the leach equipment.

There are various opportunities for a cocurrent system to differ from this ideal system. It is possible to have significant mixing in the liquid and/or in the solid stream, and that would make the performance of such a system deviate from the ideal plug flow behavior and become more like the performance of a mixed system. It is also possible for the residence time of liquid in the equipment to be different from the residence time of the solid. If the reader has difficulty in imagining this situation, it may help to look at the volume fraction of the solid and the volume fraction of the liquid at any point in the cocurrent leach equipment. If the ratio of the flow rates of the two phases are not the same as the ratio of the fraction of the equipment occupied by each phase, the residence times of the two phases will not be equal. The effects of different residence times can resemble the effects of mixing one or both phases and thus make the system perform more like a mixed system. When the leaching process is completely irreversible and not dependent upon the fluid concentration(s), the residence time of the solids will be the important term.

Although different "resistances," or leaching mechanisms, can control the rate of leaching, consider first the simple cases where the leach rate can be expressed as a mass transfer resistance in the solid or liquid phase. In such cases, the controlling resistance could result from diffusion of the leaching reagent from the bulk liquid to or into the solid or diffusion of the leached component (usually the contaminant in environmental and waste application) through the solid or from the surface of the solid to the bulk of the fluid. In such systems, when the leaching rate is controlled by the rate at which leaching reagent enters the solid, the mass transfer rate (or the leaching rate) can be expressed as

$$\frac{dC_r}{dt} = -\left(\frac{a}{V}\right) k_s(q_{ri} - q_i) \tag{4}$$

or

$$\frac{dC_r}{dt} = -\left(\frac{a}{V}\right) k_f(C_i - C_{ri}) \tag{5}$$

When the leaching rate is controlled by diffusion of the leached component (usually the contaminant),

$$\frac{dC}{dt} = \left(\frac{a}{V_f}\right) k_s(q - q_i) \tag{6}$$

$$\frac{dC}{dt} = \left(\frac{a}{V_f}\right) k_f(C_i - C) \tag{7}$$

These equations are similar to those used in describing absorption, stripping, and liquid-liquid extraction processes. The concentration of the contaminant is given as C (in solution) or q (in the solid). The concentration of these components at the solid-liquid interface is denoted by the subscript i. The mass coefficient is k_f or k_s, and the surface of solid per unit volume of the leaching equipment and the total volume of liquid in a

TABLE 1 Some Mechanisms That Affect Leaching Rates

Liquid film mass transfer
Diffusion of leachate to the solid surface
Diffusion of contaminant from the solid surface
Solid pore diffusion
Diffusion of leachate in the solid pores
Diffusion of contaminant in the solid pores
Dissolution of the solid
Dissolution of the contaminant
Dissolution of solid to allow access to the contaminant
Fracturing of the solid
Increased surface area and shorter diffusion paths
Exposure of trapped contaminant

unit volume of the equipment are a and V_f, respectively. Note that the mass transfer coefficients for the leaching reagent and the contaminant will usually be different. The two equations can be coupled if a given mass (or chemical equivalent) of leaching reagent is required to leach a given mass (or chemical equivalent) of the contaminant.

In many systems, there may be no simple equivalence; there may be no leaching reagent at all (other than water) or the concentration of leaching reagent may be much higher than that required for the leaching. Leaching agents can be consumed in leaching or dissolving far more materials than the contaminant alone. Such cases are probably more common. One should remain aware that there could be other mechanisms that control the leaching rate. For instance, the role of the leaching agent can be to dissolve a third component of the solid, which allows the contaminant to be easily dissolved by water alone. There is only an indirect coupling of the reagent consumption and the release of the contaminant. Of course, the leaching reagent can participate in dissolution of the contaminant as well as contribute to dissolution of other components of the solid.

Some examples of possible controlling "resistances" are given in Table 1. The possibilities are too numerous to cover them all in this book. In many cases, one may not know what mechanism controls the leaching rate, even if there are some experimental data on the leaching. Unless the leaching experiments are specifically designed with sufficient variation in the proper parameters, it may not be possible to verify that any particular mechanism controls the leaching rate. As noted, it is often highly desirable to test leaching rates carefully in laboratory experiments, but practical considerations may limit the extent of testing. If the leaching conditions in the production equipment are simulated accurately, one can usually design effective leaching equipment without fully understanding the leaching mechanism, but the designs may not be optimal. Temperature, concentrations, etc., are usually easily simulated in laboratory experiments, but complex flows, unusual residence time distributions, or attrition of the solids may be more difficult to simulate in laboratory equipment. Although understanding the leaching mechanisms (as with most separation processes) is always desirable because it can tell the investigator or designer what changes are most likely to improve the performance, it may not always be practical to devote that much effort to do so and to evaluate the parameters involved.

Since the contaminant leached from the solid goes to the fluid, the concentrations of contaminant in the solid and fluid phases are always related:

$$\frac{dC}{dt} = -\left(\frac{V_s}{V_f}\right)\frac{dq}{dt} \tag{8}$$

The volumes of solid and liquid are denoted by V_s and V_f, respectively. For a batch leaching system, these can be viewed as the total volume of each phase in the equipment if the solid and fluid phases move through the equipment with the same velocity. The same assumption can be made for cocurrent leach equipment, but there are greater possibilities for the ratio of the volumes of the two phases to change with position in the equipment and/or for the velocities of the two phases to be different. This can cause the performance to deviate from this idealized case. It may be convenient to express this relation as

$$\frac{dq}{dC} = -\frac{V_f}{V_s} \tag{9}$$

Note that V_s will not be constant if a significant portion of the solid is dissolved, and these changes will have to be taken into account if the concentration of target contaminant in the solid is to be expressed per unit volume or unit mass of the solid.

For mass transfer resistance in the liquid phase, Equation (5) can be written as

$$\frac{dC}{C_i - C} = \frac{k_f a}{V_f}\,dt \tag{10}$$

This expression is much like that for countercurrent systems with a number of transfer units, the height of a (cocurrent) transfer unit, and length. However, in this case, the problem is dealing with time rather than length. A similar expression can be written based upon concentration changes in the solid, when the controlling resistance is in the solid phase. It is then usually more convenient to use the concentrations in the solid phase rather than the concentration in the liquid for the driving force when most of the mass transfer resistance is in the solid phase. Since the accumulated concentration of the contaminant in the liquid phase is related to the concentration in the solid phase, the results can be described either way, and the results of experiments can be observed by following the concentration in either phase with the appropriate correction for the volumes of the two phases.

Remember that when the concentration of contaminant in the solid can be viewed as adsorbed on the solid or otherwise related by an equilibrium expression to the concentration in the liquid, the problem can be treated much like a batch desorption process and countercurrent leaching may be preferred. (Such operations are described in more detail here than in Chapter 2.) An expression for the equilibrium concentration of the contaminant in the solid as a function of the concentration in the liquid is much like an adsorption isotherm (and would be an adsorption isotherm if that were the mechanism for retention of the contaminant in the solid). If the equilibrium condition corresponds to essentially zero concentration of contaminant in the solid, the equilibrium concentration would be essentially zero. Then for each concentration of contaminant in the bulk liquid, there is a corresponding concentration in the bulk solid determined by Equation (11), which notes that the contaminant leaving the solid appears in the liquid (Equation 9).

If all of the contaminant in the solid is free to diffuse toward the interface, the rate expression can be approximated in terms of the average concentration in the solid phase as the driving force. This expression is only appropriate when all of the contaminant can diffuse and when leach times are moderately long to long:

$$\frac{dq}{q - q_i} = \frac{k_s a}{V_s}\, dt \tag{11}$$

Then the concentration of the contaminant at the solid-liquid interface is obtained from the equilibrium curve, that is, from the isotherm. In such a procedure, the change in concentration in the liquid can be calculated by integrating Equation (11). The corresponding concentration of contaminant in the liquid is obtained from Equation (8), which relates the change in concentration in the liquid to the change in concentration in the solid.

Batch or plug flow leaching rates usually should be evaluated experimentally, and if (as usual) the resistance to leaching is principally in the solid phase, it is likely to be most practical to use the results of leaching tests directly. With far more complex leaching mechanisms possible, it may not be practical to determine the exact mechanism or even prove that a simplified driving force mechanism is a reasonable approximation. In such cases, it may be sufficient to measure the degree of leaching as a function of time in small test equipment. This is an empirical approach that can be scaled to larger equipment, but cannot be scaled to longer leach times. Fortunately, batch leaching tests with control of leach rates in the solid phase are likely to behave in small test equipment essentially the same as in larger test equipment. The principal differences are likely to occur when the mixing or agitation changes the particle size; the agitation may not be the same in the test equipment and the process equipment. When control of the leaching rate is in the liquid phase, the hydrodynamics and mixing can affect leaching rates greatly, and it is desirable for the hydrodynamics of the test equipment to be as similar as possible to that of the production equipment.

Continuous Stirred Tank Leaching

Stirred tank leaching can also be carried out in continuous operations, but where the effluent liquid and solids are not necessarily in equilibrium. Thus this is a type of continuous (single stage) operation that occurs when equilibrium cannot be achieved. This type of operation still can be effective because leaching operations often employ reagents that are thermodynamically so strong that the contaminant concentration in the leach liquid has no significant effect on the leaching rates, which are sufficiently high that enough contamination is removed from all but an acceptably small fraction of the solids. If the system is completely mixed, there will always be a small fraction of solids that have a very short residence time in the leacher, so if one must remove very large fractions of the contaminant from the solids, it may be necessary to make the stirred leacher differ at least partially from a CSTR.

CSTR (batch or continuous) leaching usually offers high leach rates per unit volume, but continuous operations have potentially significant disadvantages in terms of the utilization of leaching reagents and in reaching extremely high contaminant removal. In a highly mixed tank, the solids and the liquid can be assumed to be completely mixed. This means that the liquids will all have the same concentration, but it does not mean that all of the solids will have the same concentration. Although there will be a constant average concentration in the solids, because the leaching operations will be at steady state, there can be considerable variation in the contaminant concentration in the different solid particles. If the leaching operation affects the solid in other ways other than dissolving the contaminant, there will also be similar variations in the other leaching effects in different solid particles. These differences result because the solids have different residence times in a stirred vessel, and that results in different concentrations of contaminant and other changes that take place during leaching.

When both phases are completely mixed, the probability of a particular particle of solid or volume of liquid leaving in the solid or liquid outflow is always the same. This means that some of the solids and some of the liquid that flows into the tank will leave the tank almost instantly. This problem can sometimes be reduced sufficiently by adding baffles between the solids inlet and the solids outlet to eliminate or reduce the fractions of solids that have very short residence time in the leach vessel. Of course, such baffling will prevent the system from behaving exactly like a CSTR, but the changes are likely to be favorable. The baffling can affect only a small portion of the leacher vessel volume, so vigorous mixing and high mass transfer performance can still be maintained in the bulk of the leaching volume. Single vessel CSTR leaching is not usually preferred if leaching times are moderately long and very high contaminant removal is required.

In some cases, it is necessary to stir a leach tank vigorously (often to attrite the solids, and sometimes to reduce liquid phase mass transfer resistance), and such stirring results in mixing and an approach to CSTR conditions.

Solids are fed to a continuous mixed tank leaching operation and are continuously removed. In a completely mixed system, some of the particles will statistically be removed essentially immediately after they enter the tank, and some particles will remain in the tank long after the mean particle residence, which is the volumetric or mass holdup of particles in the tank divided by the volumetric or mass rate in which particles are fed to the tank:

$$t_{\text{mean}} = \frac{V_s}{F_s}$$

The volume of solids in the tank is V_s, and the volume of solids fed to (or withdrawn from) the tank per unit time is F_s. A similar expression could be written in terms of the mass of particles in the tank and the mass feed rate to the vessel. Remember that for steady-state operations, the feed rate and the withdrawal rate of particles and liquid are the same, after any necessary corrections are made for the change in volume or mass of the particles due to the leaching operation.

For a completely mixed vessel (a CSTR), there is an equal probability that any particle will be removed during any differential time increment, the residence time of particles in the tank has a first order or logarithmic distribution. Thus the particles in the vessel have a range of residence times. For a CSTR, the range is very broad; for a plug-flow vessel, the range is very narrow.

The range of residence times in completely mixed leaching vessels makes continuous single stage leaching ineffective when high removal efficiencies are required, especially when long residence times are required for effective leaching. This problem results if the fraction of particles with short residence times in the leaching vessel leaves with unacceptable quantities of contaminant remaining. In some cases, it may be better to consider using batch leaching or multistage CSTR leaching. By breaking the volume of the CSTR leach tank into two or more CSTR tanks in series, the performance will improve and change toward that of a plug flow or batch reactor, and in the limit (an infinite number of CSTRs in series) the performance will be the same as that of a plug flow or batch leach system.

Multi-stage Leaching

When the extent of leaching/extraction is limited by equilibrium conditions, it will usually be necessary to resort to multiple stage leaching to achieve both high leaching efficiency and high concentrations of the solute (contaminant) in the leach liquid. The most effective use of the leach solvent occurs with countercurrent leaching. In this case, the solids move countercurrent to the leach solvent much like the flow pattern in countercurrent absorption, liquid-liquid extraction, or distillation. The problem here is that it is more difficult to move solids countercurrent to the liquid, the same problem that occurs in adsorption operations in which it would be desirable to operate in a countercurrent mode. Countercurrent operations can involve keeping the solids in the same container, flowing the liquid from one container (possibly a mixing tank), and simulating countercurrent operations by changing the location of the leach liquor feed and withdrawal point in a stepwise manner.

For countercurrent operations, the material balance is essentially the same as that used in absorption or liquid-liquid systems, but one of the phases is a solid. The material balance is

$$Q(q_{\text{in}} - q) = L(C - C_{\text{out}})$$

This balance applies from one end of the countercurrent system (the end where the solid enters) to any point in the system, and it can be extended to the opposite end of the system and thus incorporate the liquid feed and the solids leaving the system. The analysis of countercurrent equilibrium stage leaching is essentially the same as analysis of countercurrent liquid-liquid extraction or absorption. Let one phase (usually the solid phase) be the heavy phase, and the other (usually the liquid phase) the light one, and the analysis will resemble gas stripping operations.

Packed Bed Leaching

One way to utilize some aspects of countercurrent leaching is to carry out transient leaching operations with a packed bed of solids. This is only effective when the solid particles are sufficiently large that adequate liquid flow rates can be achieved with acceptable pressure losses. The liquid can be pumped through the bed of solids as a continuous liquid phase just as in adsorption or ion exchange operations.

Alternatively, the liquid can be pumped to the top of the bed and allowed to "trickle" down through the bed, leaving much of the void space in the bed filled with air. Trickle bed leaching is sometimes called "heap leaching." The leach solution can be sprayed on the top of a tank filled with solids or even a "heap" or pile of solids on a plastic drainage system (or a drainage system made of another material). Open heap leaching has been used in the ore industry and considered for environmental leaching, but this author is not aware of any current environmental open heap leaching operation. Heap leaching in an open pile is likely to be limited to relatively arid regions where sudden heavy rainfall can be prevented from flooding the system and returning contaminants to the environment. However, in-tank or covered trickle bed leaching can be used essentially anywhere. Analysis of the performance of trickle bed leaching is essentially the same as the analysis for the more common "flooded" bed leaching, but the mass transfer parameters needed can be significantly different.

Cross-Flow Leaching

With leaching operations, it is sometimes practical to operate a cross-flow system (Figure 2). In cross-flow operations, the solid is always contacted with fresh liquor. This offers the potential for achieving essentially any level of contaminant removal, and it makes better use of the solvent than single stage operations; that is, it gives higher contaminant loading in the solvent. However, note that the concentration in the solvent is highest after the first leach step and then decreases with each succeeding leach step. Although there are usually significant merits for using countercurrent operations, the cross-flow arrangement may be easier to construct and can bring down the contaminant level in the solids more quickly because the

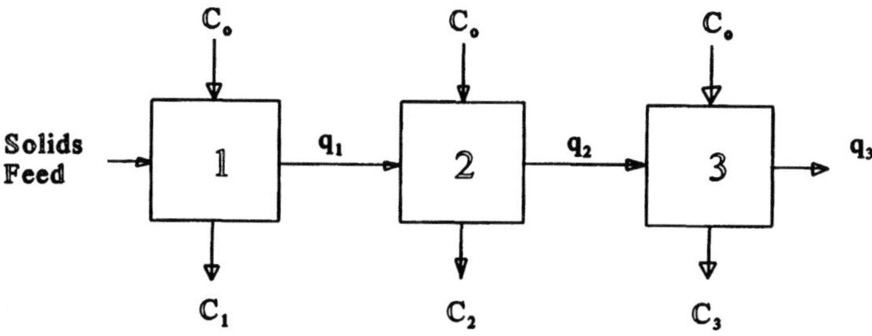

FIGURE 2 Schematic of cross-flow leaching operation. Fresh leach solution is used in each stage, but the solid moves from one stage to the next.

driving force for mass transfer is greater. The importance of the disadvantage of greater solvent use and lower solvent loading depends upon the difficulty in removing the contaminant from the solvent and subsequent reuse of the solvent. If that separation is easy and not costly, there may be less incentive to adopt countercurrent systems over the cross-flow alternative.

Each stage or step in the cross-flow leacher is a single stage leach operation. The composition of the feed solution is the same for each stage, but the composition of the solids is different in each stage. The leached solid product from one stage is the feed solid to the next stage. Calculation of the compositions from each leach stage is illustrated in Figure 3.

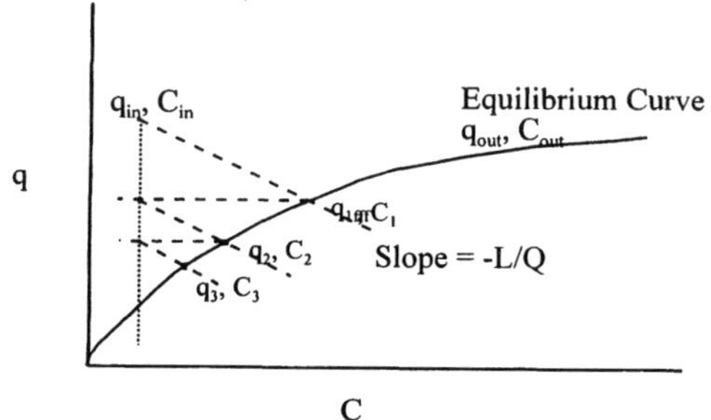

FIGURE 3 Equilibrium cross-flow leaching. Calculation of the concentrations from each stage based upon the equilibrium curve and operating lines for each stage.

An important variation of cross-flow is the use of continuous leach liquor feed to a batch leaching operation. That is a semicontinuous operations. The solids are loaded into the leaching vessel as a batch at the start of a leaching run. The leaching liquor (extractant) is fed continuously to the leaching vessel during the run. If the leaching operation is efficient enough that the leaching cell approximates an "equilibrium stage," the liquid leaving the leaching vessel will be in equilibrium with the solid. The material balance for any instant would be

$$Q\frac{dq}{dt} + V\frac{dC}{dt} = W(C_{\text{in}} - C)$$

Here, Q is the mass of solid in the leach vessel, V is the volume of liquid in the vessel, W is the volumetric flow rate of liquid through the vessel, q is the contaminant concentration in the solid in mass fraction, and C is the contaminant concentration in the liquid in units of mass of contaminant per unit volume of liquid. The subscript "in" denotes the concentration of contaminant in the inlet leach liquid, usually zero or a very low value; and the concentration with no subscript represents the concentration of contaminant in the liquid in the completely mixed leaching vessel, and thus concentration in the liquid leaving the vessel.

In some cases, it will be possible to neglect the mass of contaminant in the liquid within the vessel if it (the second term) is sufficiently small in comparison with the mass of contaminant in the solid (the first term). When the equilibrium relation between the solid and liquid is linear, the first two terms can be combined to get

$$Q\left(\frac{dq}{dt}\right) + V\left(\frac{dC}{dt}\right) = \left(Q + \frac{1}{K}\right)\frac{dq}{dt} = W(C_{\text{in}} - C)$$

where K is the ratio of the equilibrium concentration of contaminant in the solid to the concentration in the liquid, that is, the slope of the linear equilibrium curve.

Continuous fed batch leach systems are likely to be used when large volumes of liquor are required to supply sufficient leaching agent such as when it is necessary to dissolve significant quantities of the solid. With vigorous stirring and control of the leaching agent concentration (feed rate), it is possible to control the reaction rate and thus the heat or off-gas generation rates. Without mixing and with the solids compacted as a packed bed, continuous fed batch leachers become like the fixed bed leaching operations described earlier.

APPLICATIONS OF LEACHING TO ENVIRONMENTAL PROBLEMS

Comments on the Remaining Leached Solids

Often one is tempted to think of leaching operations as a simple case of removing a contaminant (or a valuable component) from a solid and to think of the decontaminated solid as a material much like the original solid without the contaminant. However, because leaching agents can be so aggressive, it is possible that the solids will be much different, and perhaps degraded in value. Remember that aggressive leaching agents can dissolve some of the bulk of the solid particles. Soil washing operations can result in a soil free of contaminants, or with greatly reduced contaminant concentrations, but sufficient other materials may also be removed that the soil loses some or all of its important properties that are useful for supporting plant life. A washed soil may no longer support vegetation or may do so significantly less effectively. When aggressive leaching agents are used, it is often better to think of the process only as an operation to remove contaminant, not as a way to reclaim the soil or other solid for its former use, such as to support the growth of plants. Washed soil will have important components removed so that it can no longer support plant life, but, in some cases, nutrients and organic matter can be added to restore some of the ability to support plant life. However, it still remains likely that the quality of the soil for such purposes will be significantly degraded. Even the restoration of some of the components may be inadequate for the soil to be useful. Similar changes can occur with any solid during aggressive leaching, and reuse of the solid can be affected by the leaching.

Since complex solids such as soils contain many components besides the contaminants that could be removed by leaching processes, this often must be taken into account when determining the reagent consumption and when estimating the properties of the solids after leaching. In many cases, these normally harmless, and even useful, components of soils can consume the major portion of the leach agent. For instance, acid leaching will dissolve any limestone in a soil, and with many soils limestone is likely to consume far more acid than the actual contaminant. (Note that the use of sulfuric acid leach would also result in the formation of fine particles of insoluble calcium sulfate that could be a problem in handling the soil later.) Organic solvents or highly oxidizing leach solutions are likely to remove all or much of the organic content of the soils. In many cases even the removal of the organic components from the soil will leave a residue soil that is not nearly as useful for growing plants. Similar degradation can also result from thermal treatment of soils. When the temperature required to remove a volatile contaminant from soil is moderately high, the other organic components in the soil may be burned off. Of course, on an industrial site that is to be used as a new construction site, reduction in the agricultural properties of a soil may not be a problem, but it could be a problem for other areas such as adjoining fields if off-site soils must be treated.

In summary, the final state of the soil or other solid waste should be a consideration in selecting soil or waste decontamination methods, and the stakeholders who participate in the selection/approval of methods to be used in removing contaminants from soils should not necessarily expect the soil to have the same value after treatment as it had originally. This can be an especially important consideration when removal of a highly insoluble contaminant is being considered. If there is very little risk from highly insoluble contamination, the cost of expected or possible soil degradation should be taken into account in deciding the merits of soil/solid treatment.

Leaching Fluids for Organic Contaminants

Although it is possible to remove some organic contaminants with water or water-surfactant mixtures, many organic materials may require organic solvents to remove them effectively. Since many organic solvents have some toxicity, this presents some difficulty in their use because one does not want to contaminant a soil or solid waste with one toxic material while removing another contaminant. There are two ways to approach this problem. One is to use a nontoxic solvent, perhaps a vegetable oil that is a natural, nontoxic, and biodegradable material, possibly even an edible product; the other approach is to use a solvent that is more easily removed from the solid and from the contaminant, even if the solvent has some toxicity. One class of solvents easily removed from the solid and easily separated from the contaminant are highly volatile solvents. Light hydrocarbon solvents may remove many hydrophobic contaminants from soils or other solid, and the solvent often can be removed from the solid relatively effectively by simply heating it, perhaps under vacuum. The solvent can also be captured by condensers and reused, leaving little residual waste other than the contaminant and the other nonvolatile materials removed from the solid. A highly volatile solvent is thus likely to be easily separated from significantly less volatile contaminants by distillation. Although it is possible to leach volatile contaminants from solids, one should remember that often it is also possible to remove such contaminants by evaporation or air stripping, and those options should be considered when the contaminant is highly volatile. Another option is to use a solvent that is particularly easily degraded biologically in the soil or waste. Easily degraded solvents, even if they have some immediate hazard, may pose less long-term hazard, and there may be fewer limitations on the degree of solvent recovery than one would require for a solvent that would have a long lifetime in the "cleaned" soil or waste.

Perhaps the most notable solvents that can be easily removed from the solid and easily separated from a contaminant are supercritical fluid solvents. (Although supercritical fluids are not strictly liquids, their use will be included in this discussion.) As gases approach their critical point, their densities increase, and their ability to dissolve selected materials increases greatly. The extracted contaminant can be recovered by simply reducing the pressure, and the solubility of the contaminant in the low pressure gas is simply its vapor pressure. This provides an exceptionally powerful method for selectively removing those components that interact with the supercritical fluid. Supercritical fluids are particularly effective in penetrating small openings and are thus effective for leaching some porous solids and solids with small particles such as soils. Furthermore, some of the most attractive supercritical fluids are not toxic.

The supercritical fluid most used for extraction is carbon dioxide, but its uses are not (yet) common for environmental applications. Carbon dioxide has a critical point that is near ambient temperature, and thus high temperatures that would destroy temperature sensitive solutes are not required. The critical pressure is also not excessively high. This makes supercritical carbon dioxide important in leaching type operations in the food and pharmaceutical industries. Although carbon dioxide itself may not be effective for leaching many important organic contaminants, the addition of other components (usually called "modifiers") to the supercritical carbon dioxide can improve the leaching of many organic contaminants. Additions of 0.2% to 5% methanol can increase the solubilities of polyaromatic hydrocarbons in supercritical carbon dioxide [2]. Other modifiers such as benzene, toluene, and even water have been studied. Water is a particularly important modifier because it can be present in soils and does not have to be added to the solvent. Small concentrations of water can increase the solubility of some contaminant such as naphthalene, but as the water content in the carbon dioxide increases to 20% or above, the mass transfer appears to decrease significantly [3].

Environmental applications have been studied for supercritical leaching, but the number of cases where supercritical extraction has been used economically for soil washing appears to be very limited. There are significant costs associated with the high pressures required even for supercritical carbon dioxide, and even higher pressures are likely to be needed for most other potential solvents. The loss of carbon dioxide can be a significant cost, especially when there is no local low-cost source of carbon dioxide. Also the effectiveness of supercritical leaching with real soils may not be as good as leaching of test simulant soils. In one study [4] soils artificially contaminated with diesel fuel were easily decontaminated with supercritical carbon dioxide, but only 21% of the diesel fuel from soil taken from an old spill could be removed. The difference in the results with freshly contaminated and aged soils could have resulted from diffusion of the aged fuel into micro-pores in the soils or from degradation of the fuel into a different spectrum of constituents.

Zhou et al. reported on the use of supercritical carbon dioxide to remove polychlorinated biphenyls (PCBs) from soils and sediments [5]. These are especially important environmental contaminants that are difficult to remove. They have low volatilities, and they are not easily destroyed by oxidation, except at very high temperatures and moderately long residence times. PCBs are also not easily degraded by microorganisms, but evidence of two-step biodegradation (an anaerobic dechlorination followed by aerobic degradation of the lower chlorinated materials) has been discovered. Oxidation of PCBs is made especially difficult because partial oxidation can result in the formation of dioxins that are also toxic. Those investigators reported both moderately high PCB removal and a higher removal efficiency for the more highly chlorinated coggimers (those PCBs with more chlorine atoms). The PCBs with higher chlorine content are both more toxic and less susceptible to biodegradation.

Supercritical carbon dioxide has also been studied for removal of mineral oils from glass cutting wastes [6]. The mineral oil removed from this waste is not particularly hazardous, but it is a waste with significant disposal cost, and it can be recovered in a form suitable for reuse.

Although supercritical extraction with other leaching agents has received considerable attention for removing contaminants from solids (and regeneration of adsorbents), the applications have usually not developed, principally because of the costs of capital equipment (high pressure systems) and/or operations at high pressures. One would like to work with an extractant that has a critical temperature that is not too high and a critical pressure that is low enough to be contained in relatively conventional equipment with moderate cost. Those conditions limit the practical supercritical fluids to only a modest number of possibilities and are a major reason why interest in supercritical extraction have centered on the use of carbon dioxide. Of course, to be practical, the supercritical fluid also must extract the contaminant of interest. Supercritical water is a powerful extractant and with modest concentrations of oxygen can be a powerful oxidizing agent that can destroy essentially all organic contaminants relatively quickly. However, the critical pressure for water is very high, and supercritical water is such that equipment costs can be raised even higher. There is no further discussion of supercritical water oxidation in this book because oxidation is not usually included as a "separation process," but the applications of supercritical water oxidation could become significant if the cost of high pressures, moderately high temperatures, and very corrosive fluids can be accepted. Without sufficiently oxidizing conditions, the "extraction" (or separations) aspects of supercritical water could be utilized.

In some cases, it is not necessary to resort to the high pressures usually required for supercritical extraction, but one can operate with liquefied gases such as propane. Such operations maintain the benefits of a highly volatile extractant that can be removed from the solid residue and the contaminant by simply reducing the pressure. However, it forgos the benefits of enhanced extraction that occurs near the critical point of the fluid. Thus condensed gases such as liquid propane will behave more like highly volatile organic solvents and not like supercritical fluids. Reducing the pressure (or increasing the temperature) will volatilize the solvent, but it is not likely to alter the solubility of the contaminant in the solvent significantly.

Leaching of Inorganic Contaminants

Inorganic contaminants are likely to be removed by aqueous leaches, usually water, acid, or solutions of specific ligands for removing the contaminant. Caustic solutions may be more effective for some contaminants. By including surfactants, it may also be possible to use aqueous solutions to leach organic contaminants from solids, including some soils. Obviously, water is likely to be the choice for leaching highly soluble materials from solids. When the contaminant is insoluble in water, it is sometime possible to add leaching/washing reagents that allow the contaminant to be or behave as if it was soluble in water. For instance, inorganic contaminants, especially small particles, can "stick" to oils and other organic contaminants on solid surfaces. Surfactants can disperse the organic contaminants and any associated solids in water. Soaps are one group of surfactant. Just as soaps are commonly used to remove grease and oils from many surfaces (including hands), similar materials can be used to remove organic compounds from solid wastes and soils. This can make the operations appear even more like "washing" operations.

As one begins to use more aggressive reagents, the term "washing" may become too much of a simplification. Special ligands are particularly interesting because they can sometimes remove specific metallic ion contaminants more effectively than more aggressive acid systems that could dissolve significant quantities of other materials as well. For instance, carbonate ions in alkaline solutions can remove uranium contamination from contaminated metal surfaces, and the metal surfaces often will not be attacked by alkaline solutions [7]. Acid solutions could remove some uranium, but they are likely to dissolve much of the metal as well. Selected ligands such as citrate ions or natural ligands such as sederaphores [8] remove a number of metal contaminants from metal surfaces, including highly toxic plutonium, without significant attack of some metal substrates.

Soil Washing to Remove Organic Contaminants

Soil washing is the removal of contaminants from soil by washing, leaching, or extracting with a liquid. Although soil washing has a significant following in this country, it appears to be practiced more widely in Europe. It may be helpful to consider the washing of organic contaminants and inorganic contaminants separately. First, note that if the contaminant has sufficient volatility, it is unlikely that soil washing will be able to compete with vapor extraction either in situ or ex situ. That is, soil washing is less likely to be the choice for removing volatile organic compound compounds (VOCs) than for removing higher molecular weight contaminants that are less volatile. Even when the vapor pressure of the contaminant is only moderately high, it may still be preferable to heat the soil and use thermal extraction or to apply both heat and vacuum. The advantages of vapor extraction or even moderate thermal treatment are the relative ease of recovering the VOCs. The VOCs can be condensed and collected in a low-volume concentrated form. Even when moderately low temperatures are required for condensing highly volatile VOCs, one can use packed beds of activated carbon to adsorb any trace of contaminant that may pass through the condenser. There is even the option of incinerating the organic contaminants in the soil. Incineration is not necessarily a poor choice, but it is not usually considered a "separation" process. Note that incineration usually has two possible important problems. First, incineration is not viewed favorably by many sectors of the public, a situation that could change with time. Second, incineration (or incomplete incineration) could result in other toxic contaminants. These two problems are, of course, closely coupled since it is the fear of airborne releases of toxic components that drives the public dislike for incineration.

Incinerators usually generate serious concerns from residents because of fears that some contaminants will be volatilized but not destroyed. The degree of combustion can usually be improved to essentially any level desired by increasing the temperature and residence time in the incinerator. When chlorinated compounds such as PCBs of trichloroethylene are burned, dioxins can be formed if the temperature becomes much greater than approximately 400°C. Even when the chlorinated compounds are destroyed, dioxins can form as the combustion products cool toward ambient. As important as the chemistry of combustion is the perception of important segments of the public. The difficulties with using incineration may result as much from the fear that operations will not or cannot be monitored effectively as from the performance of specific incinerators themselves.

Thus, soil washing is more likely to be applied to removal of relatively nonvolatile contaminants, especially for compounds that form more toxic products when partially burned or heated to moderately high temperatures. PCBs are obvious contaminants of this type, but heavy oils may also be in this category since they are often difficult to burn and their exact composition may not be known. (Even small "droplets" of PCBs trapped in soils can contaminate such large volumes of water, and dense nonaqueous liquids such as PCBs are easily spread by disturbing soils.) Four types of washes could be considered: aqueous washes with surfactants, solvent washes with "light" organic solvents, liquefied gas extraction, and supercritical extraction.

The more common approaches to aqueous soil washing (and probably to aqueous cleaning of solids in general) rely principally upon surfactants. This is not greatly different from most people's view of washing. Generally, surfactants are used to remove organic materials from solids (soils) [9], but removal of the coating of organic bearing materials can release other contaminants as well [10]. Just as in any use of water and soap to remove grease or any other organic compound, the surfactant (soap is one type of surfactant) can mobilize the organic contaminant by dispersing it into the water. This approach has the advantage that it does not necessarily involve the use of a toxic or hazardous material. The disadvantages are that it may be a slow process, require relatively large volumes of wash water, and/or produce a fluid from which it is difficult to separate the contaminant from the water effectively. By recycling the wash water, the discharge of contaminated water to a waste treatment system can be minimized, but some water cleanup problems are always likely. The primary water-oil (contaminant) separation can involve little more than breaking the emulsion and decanting the organic contaminant, or the separation could involve more efficient systems such as micro-porous membranes that will not pass the emulsion. (Note, however, that most membrane systems cannot tolerate significant concentrations of solid such as soil particles.) Final cleanup of discharge wash water may have to involve secondary treatment systems such as adsorption beds of activated carbon.

Most soil washing with surfactants is done ex situ in washing vessels, usually stirred tanks or "attrition" mixers with intense mixing. However, it is also possible to do surfactant washing in situ, but this operation may be called by other names, such as "surfactant flushing," Perhaps the largest scale in situ operations of this type are carried out by the petroleum industry to remove desirable crude oil, rather than contaminants, from reservoirs. The aim may be different, but many aspects of the operations are similar.

This is one way to remove dense nonaqueous liquids (such as PCBs) from soils. This treatment does not require disturbing the soil and thus risking spreading the contaminant by releasing trapped droplets and letting them move toward the bottom of the disturbed or excavated region. However, surfactant flushing does mobilize the contaminant with the aim of flushing it to a collection well. However, such operations would have to be done most carefully to ensure that the contaminant is collected and not just spread further. The extremely low solubility and the high toxicity of contaminants such as PCBs makes spreading of the contaminant especially undesirable. Even minute quantities of liquid PCBs can contaminate vast volumes of groundwater to concentrations exceeding acceptable limits.

Light solvents often remove many heavy organic contaminants more quickly, but the solvents themselves present some problems since they are also not likely to be allowed in the remaining soil. The reason for suggesting a "light" solvent is the relative ease of removing the residual solvent from the soil and recovering the solvent from the less volatile contaminant. It is usually desirable to dry the soil (remove the water) before applying the solvent, especially when the solvent is highly hydrophobic. Solvent washing with volatile solvents requires relatively "tight" systems to prevent fugitive gaseous emissions. Obviously the solvent needs to be selected to have a high solubility for the contaminant.

The ultimate light solvents are probably those that are normally gases at the usual ambient temperatures and atmospheric pressure, such as propane, carbon dioxide, butane, or even methane or ammonia. The use of such highly volatile solvents obviously makes removal of the solvent from the soil and separation from many contaminants quick, but it also complicates the equipment requirements. Higher pressure equipment is required, and solvent containment is likely to be more costly. Economic as well as environmental considerations are likely to require efficient recovery of the wash/leach/extraction solvents. The extremely high volatility of the liquefied gases at atmospheric pressure means that residual solvent concentration in the soil is usually very low. The gases just mentioned are not particularly toxic, so the degree of containment may not need to be as high as for more toxic solvents. Compressors will also be needed with most systems.

As with (normally) liquid solvents, nonpolar liquefied gases such as propane are likely to be most effective in removing highly hydrophobic contaminants such as PCBs from soils. One company reports that most of its work with solvent soil washing uses propane [11]. The contaminant can be adsorbed on one or more of the components of the soil, such as the humus materials, so removal efficiency can be much lower than one would estimate from only the solubility of the contaminant in the solvent.

Supercritical fluids are used extensively to remove relatively valuable components, such as caffeine from coffee, perhaps the most profitable application, but there have been difficulties finding practical systems for most environmental applications where the volumes of solids to be leached can be large and the funds available for the treatment are limited. This author know of only one large-scale application of supercritical extraction for soil cleaning, but there have been several studies of the technology. The performance does not seem likely to be as important a factor as the cost. Capital costs and operating costs are increased by the high pressures required. Even the loss of the supercritical fluid can be important, and many applications assume significant losses per cycle. High pressure equipment large enough for high throughput systems can also be expensive. There seems to be a growing interest in liquefied gas washing, at the expense of interest in supercritical extraction. One advantage of supercritical extraction over most liquefied gas is the nonflammability and nontoxicity of carbon dioxide, the most popular supercritical fluid. However, significant research into supercritical extraction is continuing on contaminants that are the most difficult to remove from soils such as PCBs [12]. With the addition of chelating agents, toxic metals could be removed by supercritical fluids [13].

Perhaps the most critical aspects of in situ washing with surfactants are the determination, or estimation, of the flow of surfactants as well as the effects of the surfactant on the organic contaminants. In the petroleum industry's "surfactant flushing" to remove petroleum crude from reservoirs, the oil originally in the large pores may have been removed initially by the initial pumping, by later water washing, or even by thermal treatment. This can mean that the remaining petroleum is located in the smaller pores ore cavities in the soil or in other regions from which it is more difficult to remove the oil. However, fresh spills or releases of contaminants may follow the larger channels in the soil and thus be more easily washed from the soil by the surfactant-water flow. However, older spills or dense liquids may be retained in the smaller cavities of the soil and be more difficult to reach with the surfactant. The long time since the spill may allow contaminant in the larger channels to have been swept away by continuous or periodic groundwater flow. In general, the more heterogeneous the soil, the more difficult it will be to predict the flow paths and the ability of the surfactant to reach the contaminant. Adsorption of the contaminant by the clay or humus portions of the soils can also hinder surfactant (or any form of) washing.

Removal of inorganic contaminants by soil washing is complicated by the different forms the contaminant can take. Although much can be inferred from knowledge of the chemistry of the contaminant, there are enough uncertainties in real problems that it is almost always necessary to perform treatability tests before considering actual washing operations [14]. Since the chemistry of the different size fractions of the soil can be quite different, the contaminant could be concentrated in one or more fractions of the soil. The smallest size group, the clay fraction, is often the source of adsorbed contaminants because of its greater surface. The organic fraction, usually humus, can be the source of organic and even inorganic contaminants that are adsorbed on humus or are easily reduced to less soluble forms by the humus. Remember that the humus fraction of the soils will be greatest in soils from the surface or shallow positions and can be small or absent in deep soils. Before planning a soil washing operation it is usually desirable to locate the source of the contaminant and limit the relatively costly soil washing to the smallest volume of soil necessary. Not only can this reduce the cost of the washing operation, but it can also reduce the volume of secondary wastes produced by the operation.

Treatment goals for leaching soils are likely to be set by general or specific regulations. Some contaminants, especially those from very old spills, may be largely in low solubility forms, and such soils may pass the Toxic Characteristic Leaching Procedure (TCLP). However, if the regulations require removal of the contaminant, it will be necessary to use a leaching agent that will solubilize the contaminant. This is where the chemical form and chemical properties of the contaminant must be known. For instance, some contaminants such as chromium and technetium are relatively insoluble in their lower valance states, so oxidation is likely to be necessary to wash them from soils effectively. Although the contaminant may pose little risk to the environment as long as it remains insoluble, the danger of subsequent oxidation and solubilization must be considered by regulators. Some contaminants such as arsenic can be relatively soluble in acid or alkaline leach solutions depending upon the valance of the contaminant [15]. As(V) can be more soluble in alkaline solutions, but As(III) is more soluble in acid solutions. Since humic acid is extracted from soils by alkaline washing, arsenic associated with the humus is also likely to be removed by alkaline washing.

Decontamination of Surfaces (Equipment, Facilities, etc.)

Leaching of contaminated solids may be called "decontamination." However, soils are only one common set of solids that routinely become contaminated with toxic materials. Decontamination can involve a physical dissolution of a contaminant or chemical dissolution of either a contaminant containing coating on the surfaces of the solid or part of the surface of the solid itself. Again the solids can be contaminated with organic or inorganic materials. For instance, machine cuttings can be contaminated with cutting oils, or solid materials from a nuclear facility could be contaminated with inorganic radioactive materials. The organic contaminants can be removed by washing with water and surfactants, with light organic solvents, or with liquefied gases or supercritical fluids [16]. Inorganic contaminants are likely to need water or even acids to "dissolve" the contaminant. Surfactants are often helpful for removing even inorganic materials because there can always be some traces of oil on parts with inorganic contaminants. Acids or oxidants may be needed to dissolve some contaminants, and specific ligands may help dissolve specific contaminants.

Most aqueous leaching agents are not selective for the contaminant of interest, but there are a few exceptions. For instance, selective ligands can be added to the wash/leach liquors that are specific to the contaminants of interest. EDTA has looked promising for assistance in removal of lead from contaminated soils [17] and probably can be useful for metals and other solids as well. Generally one would like to remove the contaminant while removing as little of the base solid as possible. For instance, acids would dissolve contaminated concrete relatively rapidly, perhaps more rapidly than they would attack the contaminant; so one may prefer to try water or even alkaline washes on concrete before resorting to acids. On the other hand, some metals can withstand significant acid treatment before excessive corrosion removes too much metal.

Contaminants that reach the solid as a fluid can enter pores or cracks in the solid, and diffusion of those materials from the pores or cracks may be rate controlling factors. Some solids that appear to be relatively solid can have high porosity. Concrete is one such material, and contaminants from aqueous solutions can enter concrete via the pores. The regions inside the concrete are highly alkaline, and many metallic contaminants will precipitate in the alkaline media and thus not deeply penetrate the concrete. Because it is not practical to use acids to dissolve material inside concrete, one may prefer to not use alkaline leaching, or any separation method, to remove such contaminants. A few inorganic contaminants that are insoluble in alkaline media such as concrete can still be leached with specific ligands.

For some contaminated concrete, physical spallation methods may be preferred over direct leaching. This involves removing the outer layer of the concrete, the region that is likely to contain all or most of the contamination. The surface layer of concrete can be removed by abrasion using water, ice, or solid carbon dioxide (dry ice) or by thermal stress generated by microwaves. However, even when the surface of concrete is removed, it may still be desirable to leach the solid residue of surface concrete to concentrate the contaminant and reduce waste volume.

In extreme cases, where substantial quantities of solid are dissolved or when additional driving forces such as electrolytic corrosion are applied to remove portions of the contaminated surfaces, the decontamination activities may not resemble ordinary leaching operations.

Electrokinetic Methods

Electrokinetic methods have been studied for removal of contaminants from soils and porous solids such as concrete. These are methods for moving contaminants within the soil or porous medium toward one of the sets of electrodes. This motion can result from the application of an electric field across the porous material. The electric potential gradient can induce motion to the contaminant itself if it is electrically charged (that is, an ion), or the motion can result from movement of water within the pores of the solid (electro-osmosis) that sweeps the contaminant with it, either as ions or nonionic dissolved contaminants.

Electrokinetic leaching may appear to be a simple operation, but it can be relatively complex, especially when the structure of the porous solid and the chemistry within the pores are complex. These processes are not yet fully developed and evaluated. One recent report described the problems with application of electrokinetics to a particular contaminated concrete and did not find the approach promising. Work on electrokinetic leaching of soils and concrete continues, and cost effective applications may be found and used. However, the most detailed current studies seem to indicate that electrokinetic leaching is more likely to be used for specific applications and is less likely to become a general solution to decontamination of most soils or porous solids.

REFERENCES

1. Grant, C., et al. *AIChE J. 42*, 1495 (1996).
2. Andrews, A. T., R. C. Ahlert, and D. S. Kosson. *Environ. Prog. 9*, 204 (1990).
3. Pratte, T. S., et al. "Influence of Water on Supercritical Fluid Extraction." *Proceedings of the 51st Industrial Waste Conference*. Purdue University, May 6–8, 1996, p. 95.
4. Firus, A., W. Weber, and G. Brunner. *Sep. Sci. Technol. 32*, 1403 (1997).
5. Zhou, W., P. Chen, and L. L. Tavlarides. "Supercritical Fluid Extraction of Polychlorinated Biphenyls from Sediments." Paper presented at the Ninth Symposium on Separation Science and Technology for Energy Applications, Gatlinburg, TN, Oct. 22–26, 1995; submitted to *Sep. Sci. Technol.*
6. J. Schön, et al. "Supercritical Fluid Extraction of Glass Grinding Waste Contaminated with Mineral Oil." *Sep. Sci. Technol., 32*, 883 (1997)
7. Francis, C. W., A. J. Mattus, and L. L. Farr. "Selective Leaching of Uranium-Contaminated Soils: Progress Report 1," ORNL/TM-12177 (1993).
8. Wilson, J. H., et al., "Carbonate and Citric Acid Leaching of Uranium-Contaminated Soils: Pilot-Scale Studies (Phase II)," ORNL/TM-12960 (1995).
9. Bourbonais, K., et al. "Comparison of Surfactant Properties with Washing Effectiveness for Soil from a National Priorities List Site." Paper presented at the AIChE 1995 Summer National Meeting Boston, MA, July 30–Aug. 2, 1995.
10. Moussavi, M. "Removal of Biphenyls and Some Heavy Metals from Soils by Imediation of Surfactants." Paper presented at the AIChE 1995 Summer National Meeting, Boston, MA, July 30–Aug. 2, 1995.
11. Markiewicz, J. and D. Driscoll. "Treatment of Organic Contaminated Soils with a Liquefied Gas Solvent Extraction Process." Paper presented at the Summer Meeting of the AIChE in Boston, MA, August 1995.
12. Chen, P., W. Zhou, J. Zhang, and L. L. Tarlarides. "Supercritical Extraction of Polychlorinated Biphenols from Lab-Spiked Soils/Actually Contaminated Sediments." Paper presented at the AIChE 1995 Summer National Meeting, Boston, MA, July 30–Aug. 2, 1995.
13. Yazdi, A., M. Ataai, and Eric Beckman. "Development of Highly CO_2-Soluble Chelating Agents for Removal of Lead, Arsenic, and Mercury from Contaminated Soil." Paper presented at the AIChE 1995 Summer National Meeting, Boston, MA, July 30–Aug. 2, 1995.
14. Legiec, I. A., et al. *Environ. Prog. 16*, 29 (1997).
15. Legiec. I. A., et al. *Environ. Prog. 16*, 29 (1997).
16. Adkins, C., et al. "Oil, Grease, and Solvent Removal from Solid Wastes Using Supercritical Carbon Dioxide." Paper presented at the 1995 Summer National Meeting of the AIChE, Boston, MA, July 30–Aug. 2, 1995.
17. Peters, R. W., et al. "Remediation of Metal Contaminated Soil by Soil Flushing Using Chelating Agents and Enhanced Pretreatment Techniques." Paper presented at the AIChE 1995 Summer National Meeting. Boston, MA, July 30–Aug. 2, 1995.

6
Liquid-Liquid Extraction

Liquid-liquid extraction, sometimes called solvent extraction, involves removal of a solute from one liquid to another liquid. The two liquids must be different liquid phases. It is usually desirable that the two liquids be almost completely insoluble in each other, but solvent extraction operations can be carried out as long as there are two separate liquid phases. Very often one phase is an aqueous solution and the other is an organic solvent, but, in principle, any two immiscible liquid phases could be used.

The role of liquid-liquid solvent extraction in waste management and environmental separations can be for removal of organic contaminants (solvents, insecticides, etc.) or toxic ions (usually heavy metals or radionuclides). Solvent extraction is a powerful separation method that can be very selective and effective for removing toxic materials from wastewater and, possibly, even from groundwater. One major problem in many environmental liquid-liquid extraction operations is the finite solubility of solvents in aqueous solutions. Many solvents are or may be considered toxic and not suitable for release to the environment. One usually wants to avoid the situation where one toxic component is removed from a waste water while another toxic material is added, even if the toxicity of the added solvent is lower than the toxicity of the original contaminant. However, there can still be merit in the use of solvent extraction with its potentially high selectivity for the solute(s) of interest, especially if the solvent is not toxic, has a very limited solubility in the water, and/or if the solvent can be easily removed from the wastewater.

For instance, if a low volatility pollutant can be removed effectively by a light solvent with high volatility, the residual trace solvent could be removed from the wastewater by air stripping or distillation more easily than the low volatility pollutant could be removed. It is usually not desirable to have more operations than necessary, but here is an example where

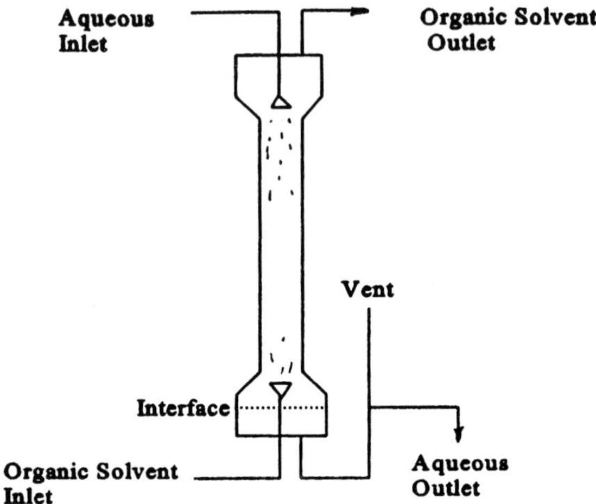

FIGURE 1 Liquid-liquid extractions tower/column showing expanded sections for de-entrainment and hydraulic control of the interface position.

the solvent extraction approach at least improves the situation. Perhaps an extreme case of solvent volatility would be a supercritical fluid, which could be grouped either with gas stripping or liquid extractions since a supercritical fluid is technically neither a liquid nor a gas. Supercritical CO_2 was suggested as an extractant for removing organic pollutants from water [1]; residual CO_2 would certainly be easily removed and would not be toxic. More recently, there has been interest in using liquefied gases, like liquid CO_2, instead of the more common liquid solvents or supercritical fluids.

In appearance and analysis, liquid-liquid solvent extraction looks much like absorption and stripping. The equipment can be a packed tower or a tray tower, not obviously different from those used in absorption/stripping operations (Figure 1). As in absorbers, the light phase enters at the bottom of the column, and the heavy phase enters from the top. This sketch shows a distributor for the heavier aqueous phase and one for the organic solvent phase. There is also an interface control system to maintain the column continuous in the light phase; alternatively, the interface between the two phases could have been fixed near the top of the column, making the heavier aqueous phase continuous throughout the column.

However, some additional equipment types, usually called "mixer-settlers," are sometimes used in solvent extraction operations. These are not likely to be used in gas absorption or gas stripping operations. There are several variations in the mixer-settler concept, and a couple of these variations will be discussed.

DESIGN OF LIQUID-LIQUID EXTRACTION SYSTEMS

In the simplest case where the two liquids are completely insoluble in each other, the height of column (or the number of stages) required for a given liquid-liquid extraction separation is determined in the same way as the height required for an absorption or gas stripping column. A simple sketch of a liquid-liquid extraction column looks much like those for a gas stripping column/tower (Figure 2). In this case the solute is removed (stripped) from the heavy phase (the aqueous phase) and extracted into the solvent phase. All of the analyses of the number of stages or the number of transfer units required are the same as those used in gas stripping. One only needs to substitute the heavy (aqueous) phase in solvent extraction for the liquid phase in gas stripping, and the light (solvent) phase in solvent extraction for the gas phase in gas stripping, and the equations given in Chapter 3 can be used directly. One can simply

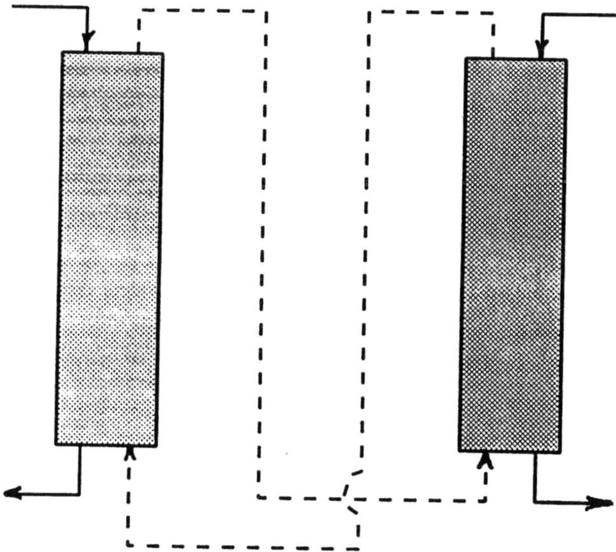

FIGURE 2 Two column extraction-stripping system with solvent recirculating between the extraction column on the left and the strip column on the left.

consider the solvent phase as the gas phase and the aqueous phase as the liquid phase in gas stripping.

To operate solvent extraction operations, it is only necessary for there to be two liquid phases with different compositions. It is not necessary for the two phases to be totally insoluble in each other, i.e., immiscible. When the two liquids have significant solubilities in each others, the miscibility of the solvents can be a function of the concentration of the third component, the solute being extracted. There are reasonably simple procedures for calculating the compositions in the two streams from the different stages. The graphical methods are based upon triangular diagrams relating the compositions of all three components (aqueous, solvent, and solute), the immiscible concentration range, and the tie lines connecting equilibrium compositions in the two phases. However, a decision was made not to include such analyses in this book because it is unlikely that systems with significant mutual insolubilities will play important roles in environmental and waste processing. Generally, high solubility of a solvent in an aqueous environmental stream would be highly undesirable because it would require a major effort to remove the solvent to the levels needed for discharge. However, there certainly plenty of ways to use solvent extraction within a process system to purify streams and eventually reduce waste volumes. If the reader needs to use such systems, they should study more detailed standard textbooks on solvent extraction [2].

If more than a single solute is to be removed by a liquid-liquid extraction step, the system can be called "multi-component, and the use of computer assisted calculations is highly recommended. However, if all of the solutes are sufficiently dilute, and the concentrations of no solute in either phase affect the concentration of the other solutes in the other phases, the system can be treated as several different single component liquid-liquid extraction operations carried out simultaneously.

When several solutes are extracted, one cannot size the tower (column) for the specified removal of *all* of the components or select the optimum solvent rate (ratio of the solvent rate to the aqueous rate) for all components. One can only select one solvent rate and one number of stages. These two conditions can be selected to meet the removal requirements and the optimum solvent rate and number of stages for only one component. The removal of the other components then can be calculated with the solvent rate and the number of stages fixed. It is usually best to select the solvent rate and the number of stages for the solute that is most difficult to remove and meet the specified removal requirements. In many cases, the solute that is most difficult to remove is obvious; it is usually either the solute that distributes less strongly to the solvent phase or the solute that must be removed with the greatest efficiency. The most difficult solute to remove may not be obvious when one component has a smaller distribution coefficient (ratio of the concentration in the solvent to the concentration in the aqueous solution) and another solute has the highest removal requirements. When the solutes appear to behave independent of the presence of the other solutes, one can calculate the solute rate and the number of stages required to remove each solute to the degree specified and then use both the largest solvent rate and the largest number of stages. That would be a conservative approach, but would not be far from optimum if the difficulty of removing one solute is far greater than the difficulty of removing the other solutes. Note, however, that in any case, it is necessary to have a solvent rate that exceeds the minimum rate for each solute.

Once a solvent rate and number of stages are selected usually based upon the removal of the most difficult solute, one can calculate the removal of all other solutes that will occur for those conditions. Even when the solutes behave independent of each other, such calculations are not straightforward and involve trial-and-error procedures. With the solute rate specified, the slope of the operating line is specified, but the location of the line is not fixed. In single component examples, one could locate the operating line from the concentration in the aqueous stream leaving the tower/column. That was possible because the concentration in the effluent stream was specified, and the goal was to calculate the number of stages required. When the number of stages is specified, one must try several positions of the operating line and calculate the number of stages involved, by the procedure used before. If the number of stages calculated is greater than the number of stages specified (for the other component), the operating line is too close to the equilbrium, and the effluent concentration of this solute is lower than can be achieved with the specified solvent rate and number of stages. Conversely, if the number of calculated stages is less than the number specified, the operating line is too far from the equilibrium curve, and the effluent concentration of this solute is higher than can be achieved with the specified conditions. Remember, that since the solvent to aqueous flow rates have been sepcified, the slope of the operating line is fixed. The trial-and-error calculations are made by trying a series of parallel possible operating lines. Although such a calculation can proceed quickly, one will probably prefer to use computer codes if they are available.

As with several other separations methods, the complications of multi-component systems become important only when the concentrations of some, or all, solutes affect the equilibrium distribution of the other components in the two phases. As additional solutes are to be considered in multi-component systems, the need for extensive equilibrium information increases, and suitable methods for organizing the data and interpolating the data are needed. Liquid-liquid extraction equilibrium for a single phase can be described in a single graph with the equilibrium concentration of the solute in the solvent phase plotted as a function of the concentration of the solute in the aqueous phase because the concentration of the solute in the solvent phase is only a function of its concentration in the aqueous phase. When there is more than one solute and the extraction equilibria for each solute is independent of the concentration of the other solutes, the system can be described with a series of equilibrium curves for each solute, and the number of stages, solvent rate, and the removal of the different solutes can be estimated as just described. However, for multi-component systems, the concentration of a solute in the solvent phase can be a function of the concentration of all solutes in the aqueous phase. For two solutes, one can make plots of the concentration of one component in the solute as a function of its concentration in the aqueous phase for a fixed concentration of the other solute, but one would need to have several equilibrium curves, each corresponding to a different fixed concentration of the second solute. Where there are several solutes, it becomes impractical to show the equilibrium in any simple graphical form.

Most solvent extraction operations have an associated "stripping" or "back-extraction" operation. For instance, if a solute is removed from an aqueous solution, it is often desirable to recover the solute as a purified compound or in another purified aqueous solution. In some cases where volatile solvents are used, especially when organic compounds are being removed from water, the solvent can be distilled from the solute economically. When solvents with limited volatility are used, this is usually not an option. For instance, most solvents that extract metal ions from aqueous solution contain at least one component that has little volatility. In those cases, it is often desirable to re-extract the solute into another aqueous phase (Figure 2). The first column is an extraction column, and the second column is a back extraction of a strip column which returns the solute to an aqueous phase. The recovered solute could be more concentrated in the back-extract product, or it could be at a comparable or even more dilute concentration. However, the recovered solute is usually separated from all or most other components in the original feed solution. The recovered solute can be at a much higher concentration if suitable modifications can alter the equilibrium relation between the solvent and the aqueous (back extraction or strip) phase. Such alterations could include changes in the pH or the redox condition in the strip solution. Even when the concentration of the recovered solute is relatively low, it may be practical to concentrate the solute by evaporation or another method. The important point can be that none of the other solutes are present at significant concentrations. The solvent is recirculated between the two columns. It removes the solute from the aqueous phase in the first column and gives up the solute to the aqueous strip solution in the second column. The solvent leaves from the top of one column and is pumped into the bottom of the other column. In some cases, solvent reconditioning or even cleaning may be needed before the solvent is returned to the first (extraction) column.

It is traditional to plot single component equilibrium curves with the concentration of the solute in the light phase on the vertical axis and the concentration of the solute in the heavy phase on the horizontal axis. Note that this convention offers the potential for confusion. Using the usual convention, solvent extraction is similar to gas stripping in gas-liquid systems. That is, the solute is removed from the heavy phase and transfers to the light phase. Thus, one can view extraction as stripping of solute from the aqueous solution, but a light liquid is used rather than a gas. The possible confusion arises because removal of solute from the solvent into another aqueous stream by back extraction is also called stripping. Although the author apologies for the possible confusion, the reader is encouraged to understand that different and conflicting terms are used in absorption and in liquid-liquid extraction so that the literature for both methods can be studied. Using the conventional definitions of heavy and light phases, this would be like absorption, not stripping, in gas-liquid systems. Although the reader may want to reduce the confusion by referring to removal of solute from a light solvent to another aqueous phase by another term such as back extraction, it is wise to be aware of the different uses of "stripping" for back extraction in the solvent extraction literature. To minimize the confusion in this book, transfer of the solute from the light phase into another heavy phase will be called "back extraction," and "stripping" will be added only occasionally to remind the readers that they may find it in the literature.

SOLVENT EXTRACTION EQUIPMENT

Packed Towers

Almost any type of packing can be used in solvent extraction. Rings, saddles, and other random packing shapes have been used. Structured packing is of interest in solvent extraction, but the interest in these new packing materials does not yet appear to be as evident as it is in absorption and distillation. Sieve trays have been used in several solvent extraction systems, especially in the nuclear facilities. The sieve trays usually involved "pulsed flow," and the towers were usually called "pulsed columns" (towers). The pulse was imposed by a diaphragm located near the bottom of the tower and increased the shear of droplets that were forced up and down through the sieve trays. This gave enhanced surface areas and mass transfer coefficients. Other contactors use mixer paddles or rotating disks operating between baffles to generate interfacial area and thus enhance mass transfer. The baffles reduce axial mixing in such devices. These mechanically agitated towers are not likely to be used in gas-liquid systems.

Most of the behavior of packed towers of all types can be discussed together. One should note, however, that the quantitative performance of these different types of packing can be quite different. Furthermore, the behavior of pulsed columns can be quite different depending upon pulse rate, pulse amplitude, and spacing geometry of the sieve trays, and the behavior of stirred towers depends upon the stirring rate. The choice of packing materials can depend upon several parameters. Structured packing usually allows very high liquid flow rates and seems to be attracting growing interest. However, structured packing may not give the best mass transfer coefficients, that is, the shortest heights of transfer units or stages. Better mass transfer may be possible with one of the agitated columns. Note that when the extraction process is highly favored and/or extremely high extraction efficiencies are not needed, the columns needed may be relatively short. With extremely short columns, the cost of the end sections may be so significant that making the columns somewhat longer may not add greatly to the overall cost. When that is the case, and very high throughput is needed, it may be better to consider a packing suitable for handling high flow rates, even if that means that a somewhat taller column may be required. On the other hand, when large numbers of transfer units are needed, one may need to consider the height of a transfer unit more carefully. Other considerations may be required for some applications. The nuclear industry adopted columns with pulsed sieve plates because of the ease of removal and decontamination of column intervals and because of the relatively short height unit heights of those columns. Space in shielded nuclear processing cells was very expensive, and there were strong motives to keep the equipment volume as small as possible.

There are two factors to consider in selecting a packing material: the mass transfer performance (stage height or height of a transfer unit) and the throughput capacity. Note that these are the same factors to consider in selecting packing materials for most countercurrent separation operations such as absorption, gas stripping, or distillation. The best mass transfer rates are achieved by increasing the interfacial surface area, and usually the interfacial area will be higher when smaller size packing is used. However, decreasing the packing size will also usually increase the pressured drop and lower the maximum throughput through the tower. At high flow rates, smaller packing may not always give higher interfacial area when the flow rates start to approach flooding rates. In a properly operating solvent extraction tower, the dispersed phase will be dispersed into small droplets. It is desirable to have small droplets and to have all of the dispersed phase in droplet form. However, since it is also necessary to separate the two phases before they leave the tower, one does not want the droplets to be too small. A dispersion with approximately uniform size droplets is usually desirable. In particular, a wide distribution of droplet sizes with some sizes becoming very small is undesirable. Fine droplets may be entrained and carried out with the continuous phase. The ability of a packing material to generate large surface areas and minimize the formation of fine drops that are difficult to coalesce are major measures of the performance of a packing material; the ability to approximate plug flow in both phases is the other important measure.

Either liquid can be the dispersed phase. There are ways to affect which phase is continuous. For instance, in Figure 1 the interface is maintained near the bottom of the tower. The interface is the position in the tower above or below which the phases are separated. Note that the interface is maintained near the bottom of the tower by bringing the liquid withdrawn from the bottom of the tower to approximately the position where the interface is to be located before removing it to the next process step. When the interface is near the bottom of the tower, the light phase (usually) is continuous; that is usually the solvent phase. When the interface is near the top of the tower, the heavy phase is usually continuous; that is usually the aqueous phase. The piping arrangement shown that establishes the position is often called a "jack-leg." To be effective, the jack-leg must have a vent to avoid siphoning off fluid from the tower. The position of the interface in the tower will not be exactly at the top of the jack-leg because of pressure drop and buoyancy of the light phase droplets in the tower. By having a much higher jack-leg, the interface could have been maintained at the top of the tower. The same results can be achieved by controlling the pressure at the bottom of the tower by using a pressure transducer and an appropriate pressure control system.

It is usually desirable to have the phase with the higher volumetric flow rate be the dispersed phase. If the residence time of each phase is approximately the same, having the dispersed phase be the phase with the highest volumetric flow rate results in a higher holdup of dispersed phase than if the phase with the lower volumetric flow rate were dipsersed. Then, if the dispersed phase droplets are approximately the same size, there will be more interfacial area when the holdup of dispersed phase is greater. This is usually the major factor in determining which phase should be dispersed. However, when the holdup of the dispersed phase becomes too large, the droplets can coalesce and become the continuous phase. This can occur over the entire height of the column or only over a portion of the column. Reversals of the continuous phase are usually undesirable and degrade mass transfer performance.

With the interface near the top of the tower, the continuous phase in the tower is more likely to be the heavy phase. If a light organic solvent is extracting a solute from a wastewater, the water phase would be more likely to be the continuous phase. Conversely, if the interface were maintained lower, the solvent (lighter) phase would be more likely to be the continuous phase. This is usually an adequate way to set the continuous and dispersed phases for low flow rates, but other factors become important at higher flow rates. Unfortunately, maintaining a continuous proper phase is not always easy. In some cases, especially near flooding flow rates, local reversals of which phase is continuous can occur as mentioned earlier. When the ratio of the flow rates of the two phases is far from unity, there may be difficulties in maintaining the liquid with the lower flow rate continuous. The packing can also affect which phase is continuous and which is dispersed. There is usually a greater ease in making a phase continuous if that phase wets the packing or internal structure of the tower. It is usually undesirable to operate a liquid-liquid extraction system near the flow rates where reversal of the continuous phase is likely to occur. This often means that one should not operate with flow rates close enough to flooding rates for reversal of the continuous phase to be likely.

How far one wants to operate below the flooding rates probably should depend upon how accurately one knows the flooding rate. For instance, if one designs a liquid-liquid extraction system and estimates the flooding rate from correlations, one probably should be very cautious and not try to push the operations to flow rates much greater than 50% to 70% of the flooding rate, and it may be necessary to operate at even lower rates. This is simply a recognition that flooding rates are not easily measured, that the available correlations of flooding rates show scatter from the available experimental mesurements of 30% or more, and that reliable flooding measurements are not available in quantity for all of the packing material and liquids that could be of interest. On the other hand, when good measurements are available for the same packing and fluid to be used in the facility, one could consider trying rates somewhat closer to the known flooding rates. In such cases, the uncertainties in fluid properties and measured flow rates may be more important than the uncertainty in the flooding rate. Some caution is still recommended because when operating near the flooding rate, small upsets in the operating conditions can cause flooding to start. Once flooding begins, it is difficult to stop it without reducing flow rates drastically, that is, to values far below the flooding rates. There is a large hysteresis in the transition between flooding and nonflooding conditions.

Reversal of the continuous phase may or may not be evident in the performance of a liquid-liquid extraction column. If it is the initial stage of flooding, one is likely to see a steady increase in the holdup of one phase in the column, and that increase will eventually lead to flooding. The change in holdup may be evident if one measures a material balance on the two phases leaving the column and/or from changes in the pressure drop across the column. In large towers, the change can occur relatively slowly, and the onset of flooding may only be obvious when looking back at data taken prior to detection of two phases leaving one end of the column, the definitive sign of flooding.

Flooding and Tower Capacity

As just mentioned, the capacity of packed columns/towers is determined by the flooding rate. That is the rate at which the tower is no longer able to separate the phase, and at least one phase begins to exit from both ends of the tower. It may take some time for the full effects of flooding to be apparent, but the onset of flooding can usually be estimated from transient measurements. As the flow rates of either or both phases are increased, the holdup of dispersed phase in the tower will increase. Up to a point, this is desirable because it increases the interfacial area and thus the mass transfer performance. However, the improved performance ends when the tower floods, and then the performance falls abruptly because of mixing that occurs when the phases are not completely separated. This situation is similar to the behavior of packed absorption or gas stripping towers, but with liquid-liquid extraction, the difference in the densities of the two fluids is usually only modest, and it is notably more difficult to separate the two phases. Thus there are quantitative differences in the onset of flooding between gas-liquid and liquid-liquid towers, but qualitatively the behavior is similar.

The flooding rate for a given tower, packing material, and liquids can be explored experimentally and correlated. A typical flooding rate can be expressed in graphical form. The positions below the flooding curve represent allowable operating conditions without flooding, and the conditions above the curve correspond to flooding conditions. Such curves are specific for operating the column with a certain phase continuous; another curve would be required if the interface were at the other end of the column, so the other phase would be continuous. Flooding rates are usually presented as superficial velocities, that is, as the volumetric flow rate of a phase divided by the cross-sectional area of the tower/column. The superficial velocity is calculated without considering the portion of the tower volume occupied by either the packing material or the other phase. Note that the allowable rate for one phase depends upon the flow rate for the other phase.

There are several correlations available for flooding rates in packed column. Empirical correlations can be represented by a somewhat older correlation by the author [3] that included data from high density fluids as well as the more conventional organic and aqueous liquids. Although this is a relatively old correlation, most empirical correlations are even older. The inclusion of flooding data with high density fluids could be important in extrapolation of the correlation. Although most correlations include the density of the fluids and the difference in the density of the two fluids as variables, essentially all of the data are for aqueous solutions and organic solvents with densities that are within 10–20% of the density of the aqueous fluid. This can result in uncertainties in the effects of those variables in the final correlation.

Many of the more recent empirical and semi-theoretical correlations (as well as mass transfer rate correlations) are summarized by Seibert and Fair [4]. New experimental data are reported for ceramic Raschig rings, metal Pall rings, ceramic Intalox saddles, and structured packing made from corregulate metal and from metal gauze. The experimental results and the correlation (proceedure) are compared with data from earlier well known work, such as that of Dell and Pratt [5], and Nemunatis et al. [6]. The prediction of flooding rates is based upon estimations of the droplet size, the slip velocity of the droplets, and the holdup of dispersed phase. Semi-theoritical or empirical expressions for these parameters are presented and explained. Flooding is assumed to occur when the slope of the holdup of dispersed phase with increasing velocity of either phase approaches infinity.

The Siebert and Fair approach to flooding is included in a recent paper by Siebert and Humphrey that explains all steps in design of liquid-liquid extraction systems [7]. The Sauter mean drop diameter is first estimated from the properties of the fluid:

$$d_{sm} = 1.15\eta \left(\frac{\sigma}{g\Delta\rho}\right)^{1/2}$$

The coefficient η is 1.0 when the solute is transferred from the continuous to the dispersed phase and 1.4 when the solute is transfered from the dispersed to the continuous phase. This difference results from circulation that can occur near the interface when the interfacial tension is a function of the concentration; this is usually called the Marangoni effect and can be especially important to mass transfer coefficients. The interfacial tension is σ, the gravitational acceleration is g, and the difference in the density of the two phases is $\Delta\rho$.

These drops have to move countercurrent to the continuous phase and the driving force for the gravitational force of the particle. First note that the actual drop velocity will have to be greater than the apparent superficial velocity of the dispersed phase because much of the column volume can be occupied by the packing material and the continuous phase. The actual drop velocity will be

$$V_{\text{drop}} = \frac{V_d}{\varepsilon_v \varepsilon_d}$$

The velocity of the drops is V_{drop}, the superficial velocity (volumetric flow rated divided by the column cross-section) is V_d, the void fraction in the column (fraction of the column volume not occupied by packing material) is ε_v, and the fraction of the volid volume (not the fraction of the entire column volume) occuped by dispersed phase is ε_d.

Because the continuous phase is also moving, the rate at which the two phases pass each other (the slip velocity) is

$$V_{\text{slip}} = V_{\text{drop}} + \frac{V_c}{\varepsilon_v(1 - \varepsilon_d)}$$

where V_c is the superfacial velocity of the continuous phase. To obtain the actual velocity of the continuous phase, the superficial velocity was divided by the fraction of the column ocupied by the continuous phase.

The net force moving the two phases countercurrent to each other is the difference between the gravitational force and the buoyancy force on the droplets, and this force can be equated to the drag force on the droplets:

$$F_{\text{drag}} = \frac{\pi}{6} d_{sm}^3 g(\rho_d - \rho_c) = \frac{1}{2} C_D \rho_c \left(\frac{\pi}{4}\right) d_{sm}^2 V_s$$

The densities of the dispersed and continuous phases are ρ_d and ρ_c, respectively, C_D is the drag coefficient for the droplets, and V_s is a characteristic slip velocity. This equation can then be solved for the characteristic slip velocity.

Note that this equation does not account for interaction of the droplets with the packing material or for flow of the dispersed or continuous phases in directions other than vertical, and the failure to describe the characteristic slip velocity, V_s, and the actual slip velocity, V_{slip}. The characteristic and actual slip velocities were related by Siebert and Fair by the equation

$$V_{slip} = V_s \exp(-1.92\varepsilon_d) \cos\left(\frac{\pi\zeta}{4}\right) + \left[1 - \cos\left(\frac{\pi\zeta}{4}\right)\right] \frac{V_c}{1 - \varepsilon_d}$$

The factor ζ accounts for many of the packing effects:

$$\zeta = \frac{a_p d_{sm}}{2}$$

Since the drag coefficient is a function of the slip velocity, the determination of the characteristic slip velocity is not straightforward but can require trial-and-error calculations. Grace et al. [8] offered an alternative direct calculation approach.

Once the characteristic slip velocity is established, the flooding rate for any flow ratio of continuous and dispersed phases can be determined as suggested by Siebert and Fair:

$$1.08V_{cf} + \left[\cos\left(\frac{\pi\zeta}{4}\right)\right]^{-2} V_{df} = 0.192V_{so}$$

The subscripts "cf" and "df" refer to flooding conditions.

Scrubbing and/or Wash Stages

When removing inorganic contaminants, it is helpful to include a few stages above the aqueous feed point to wash or scrub some of the components that are not to be removed from the solvent. This is a problem largely in multi-component systems where more than one solute can be extracted, but a few wash stages can also reduce the effects of entrainment of aqueous droplets in the solvent phase that may carry other components that would not normally be extracted. In some cases, two or more solutes are extracted, but small changes in the aqueous phase will cause the other components to be essentially stripped from the solvent. If this is done with a small aqueous stream that then flows down the column and joins the feed stream, this is usally called a "scrub" solution. If the other solutes are stripped in a separate column and the strip solution is removed before it reaches the feed stream, this could be called a two-step stripping operation, and the other solutes would appear in a second aqueous solution, not returned to the feed.

Mass Transfer Rates in Packed Towers

The number of transfer units can be calculated for liquid-liquid extraction systems in the manner described for absorption or stripping. One only has to define one phase as the liquid phase and the other phase as the gas phase and the same mass balance, and even the notation used to design absorption and/or stripping towers, can be applied to liquid-liquid extraction columns. This is evident when one considers the streams entering and leaving a liquid-liquid extraction column. The heavy liquid phase in a liquid-liquid extraction column is equivalent to the liquid phase in a gas stripping tower, and the light liquid phase in the liquid-liquid extraction column equivalent to the gas phase. The equations and methods used to calculate the number of stages or the number of transfer units required are essentially the same as those used for gas stripping when the solute is being transfered from the heavy liquid phase to the light liquid phase. If the transfer is in the opposite direction, the liquid-liquid extraction operation is equilivent to a gas absorption operation.

Siebert and Humphrey [9] suggested that the mass transfer coefficients for the continuous phase be estimated from the equation by Siebert and Fair [10]:

$$k_c = 0.698 \left(\frac{D_c}{d_{vs}}\right) N_{\text{Re},c}^{0.5} N_{\text{Sc},c}^{0.4} (1 - \varepsilon_v)$$

where

$$N_{\text{Re},c} = \frac{\rho_c V_s d_{sm}}{\mu_c}$$

and

$$N_{\text{Sc},c} = \frac{\mu_c}{\rho_c D_c}$$

The expression recommended for the mass transfer coefficient in the dispersed phase depends upon a modified value for the Schmidt number of the dispersed phase:

$$N_{\text{Sc},d} = \frac{\mu_d}{\rho_d D_d}$$

and

$$N_{\text{Sc,mod.}} = \frac{N_{\text{Sc},d}^{0.5}}{1 + \mu_d/\mu_c}$$

When $N_{\text{Sc,mod.}}$ is less than 6,

$$k_d = 0.023 V_s N_{\text{Sc},d}^{-0.5}$$

When $N_{sc,mod.}$ is greater than 6,

$$k_d = \frac{0.00375V_s}{1 - \mu_d/\mu_c}$$

Using these parameters, one can calculate the number of transfer units required for a liquid-liquid extraction operation in the same way that was described for absorption or gas stripping operations. In many liquid-liquid extraction cases, the equilibrium curves are linear, and the resistances in the two phases can be added:

$$\frac{1}{K_{oc}} = \frac{1}{k_c} + \frac{1}{K_{d,d}k_d}$$

or

$$\frac{1}{K_{od}} = \frac{1}{k_d} + \frac{K_{d,d}}{k_c}$$

where $K_{d,d}$ is the distribution coefficient (slope of the equilibrium curve/line) when the concentration in the dispersed phase is plotted versus the concentration in the continuous phase.

The material balance used to describe the operating line assumed plug flow of both phases. Two functions of packing material are to reduce axial mixing of the phases and to increase the interfacial area and mass transfer performance. In most solvent extraction or gas absorption/stripping towers, the axial dispersion (also called axial mixing) in the dispersed phase is moderate to low, but one is more likely to find dispersion effects in the continuous phase. The dispersion effects are usually ignored because they are incorporated into the effective mass transfer coefficients (or the effective stage heights). All mass transfer measusrements involve some degree of axial diffusion, so the use of standard correlation such as those discussed above is likely to include an accounting for axial diffusion effects as long as the operations are sufficiently similar to those used in the measurements that the importance of axial diffusion will be essentially the same. However, if one is selecting packing material and estimating what changes in packing materials would improve performance, it can be important to know what factors are contributing to or limiting the performance of a packing material. This is more likely to be important in cases where the operating conditions are expected to be significantly different from those used in estimating the mass transfer coefficients. This could include cases where the concentration gradients in the column (concentrations changes per stage or transfer unit) are much greater than in the cases where the coefficients were measured. The higher axial concentration gradients could make dispersion more important in the "new" system.

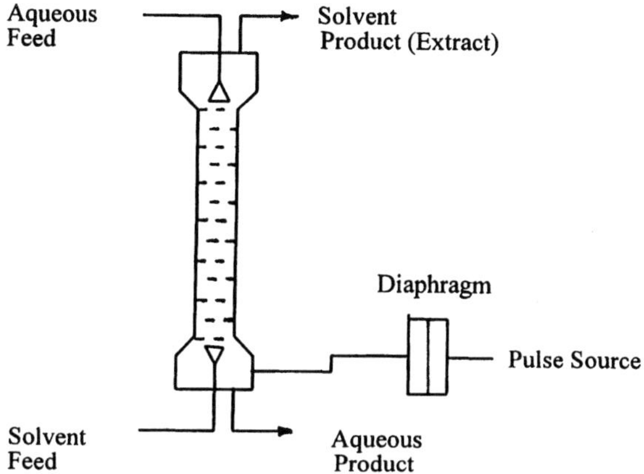

FIGURE 3 Pulse column showing a diaphragm pulse source, possibly in a remote location.

Plate Filled Towers

In principle, any plate type packing used in distillation or other countercurrent fluid-fluid separation system can be used with liquid-liquid extraction. However, two types of plate based towers will be mentioned. Simple sieve plate towers were used with pulsing of the fluids in the nuclear industry for processing nuclear fuels and have been tested for a number of nuclear waste treatment applications (Figure 3). The fluid was pulsed in a sine wave manner, so the flow through the sieve plates included the average flow rates of the two phases added to the pulse velocity. The purpose of the pulsing action was to force the continuous and dispersed phases through the openings in the sieve plates several times during the residence of each phase in the tower and create additional and new interfacial area for mass transfer. The pulse action could be imposed by a piston or diaphragm located at a removed position (outside the heavy shielding required for nuclear fuel processing) and transmitted pneumatically to the tower. The increased mass transfer rates resulting from the pulsing action are particularly important in applications such as nuclear fuel or waste processing, where the space inside shielding can be extremely expensive. Almost any effort that will reduce the size of equipment in such systems will be economical. Details of design and performance of pulse towers are given by Beyer and Edwards [11].

Another type of plate contactor used relatively frequently in liquid-liquid extraction systems involves periodic stirrers placed along the tower. The stirrers are driven from a central shaft as shown in Figure 4. The mix-

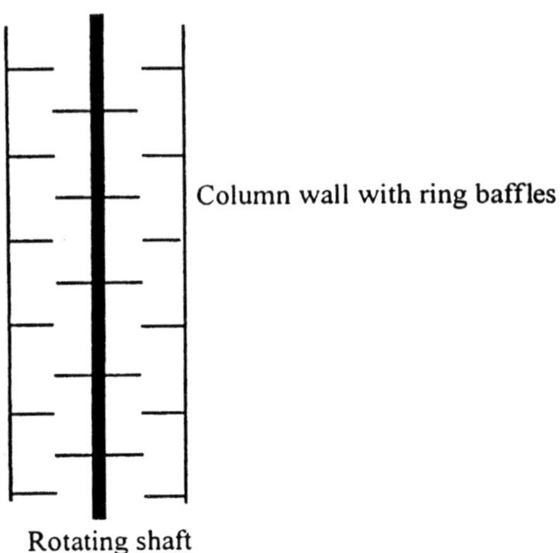

FIGURE 4 Stirred and baffled column. The example shows disk stirrers, but paddles can also be used on the stirrers.

ers are usually separated by annular baffles as shown. The mixers can be small paddles or simple disks. These also can create good interfacial area and mass transfer rates. Interfacial area can be created at the mixers, and some phase reseparation can be created by centrifugal forces. The hydrodynamics within such systems can be relatively complex and depend upon the fluid properties. An important limitation on the performance of such towers can be the axial mixing that occurs from all of the turbulent motion. The effective dispersion in rotating disk type contactors was studied recently by Moris and co-workers [12]. Quantitative use of dispersion coefficients in analyzing is described by Miyanchi and Vermeulen [13], but the analyses are relatively complex, especially when dispersion is significant in both phases.

Mixer-Settler Extraction Equipment

Non-column (tower) equipment is more likely to be used in liquid-liquid extraction than in gas absorption or gas stripping. Some solvent extraction equipment consists of discrete units in which the two liquids are mixed aggressively to create a large surface area and subsequently separated. The arrangement of these units into a countercurrent system can be just like the "stages" discussed for gas absorption towers, but in that case, the stages may have been only a certain length of the tower. There are several reasons for selecting such a system and several important variations in

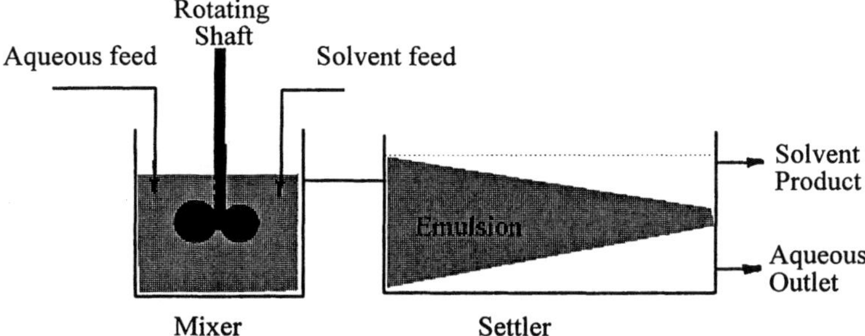

FIGURE 5 Schematic of a mixer-settler system. The two liquid phases are mixed vigorously into an emulsion in the mixer, and the emulsion is separated in the settling tank.

design, but these will all be called "mixer-settler" devices to describe the two operations that must take place in each stage. That is, the two liquids must be mixed to create a large interfacial area, provide sufficient residence time for the two liquid phases to approach close to equilibrium, and separate the two liquids so that the two liquid streams from the stage can be sent to the appropriate next stages.

A mixer-settler unit is illustrated in Figure 5. The two liquids are fed into a tank with a mechanical mixer. The agitation creates a large interfacial area and may even create an emulsion. The emulsion (or mixture of the two liquids) flows over a weir into a settling tank. This tank could be a long device through which the mixture flows as the two phase separate. The light phase can then be withdrawn from the top of the settler, and the heavy phase can be withdrawn from the bottom of the settler. Small baffles can protect the withdrawal lines from fluctuations in the emulsion level. This type of arrangement is be more likely to be used in large throughput systems such as those used to treat ore leach liquor in the hydrometallurgical industries.

There are several potential reasons for choosing mechanical mixer-settlers such as these instead of towers/columns for liquid-extraction applications. The stirred tank and gravity settlers illustrated in Figure 5 can handle large throughput, and their operations are relatively flexible. The tanks can be any size required, so they can provide sufficient residence time for even relatively slow extraction systems. It is also important to note that they can handle solids which are often inadvertently introduced into or formed in the system. Solids tend to concentrate at the liquid-liquid interface and always are harmful to all liquid-liquid extraction systems, especially in separating the phases. Thus, they are not desirable in either mixer-settlers or towers, but solids usually have more obvious effects on the operation of towers, probably largely because larger "settler" systems can be incorporated into mixer-settler systems if necessary. The presence of solids is an important consideration in using liquid-liquid extraction to remove metal values from leach liquors in hydrometallurgical applications. It could also be important if liquid-liquid extraction is selected for processing wastewaters or groundwater. These devices are probably more attractive when the number of stages required is not too large, but they have been used with significant numbers of stages.

Another arrangement for a mixer-settler is illustrated in Figure 6. This class of device does the mixing and settling along a rotary shaft. Some authors do refer to these devices not as mixer-settlers but as "centrifugal contactors." Mechanical energy is supplied to the mixer region to create the desired interfacial area, and then in another portion of the device the rotating shaft is used to introduce centrifugal forces to separate the two phases. These devices are relatively small, but can have exceptionally high throughput for their size (volume).

The centrifugal mixer-settlers are more likely to be selected when the volume of space available for the equipment must be limited. For instance, in the nuclear industry, it is desirable to limit the time that the organic solvent is exposed to aqueous solutions with high levels of radiation that can speed degradation of the solvent. It is also desirable to keep the volume of the equipment as small as possible to minimize the high cost of shielded "cells" that must contain the equipment. One center of recent development in centrifugal contactors is the Argonne National Laboratory [14,15], and the nuclear applications were of obvious interest. However, one could imagine that such devices may also be useful in portable units or anywhere compact units are desirable. There could be such needs for treating environmental problems, but they are less likely to be common.

Another class of "centrifugal" contactors behaves much like packed columns but uses the centrifugal forces to increase the flow rates through the packing [16,17]. In these devices the packing material (sieve trays, random packing, etc.) is placed in an annular space of a rotating device. The heavy phase is introduced near the center of the annular "column," and the light phase is introduced near the outside of the device. The centrifugal field acts much like the gravitational field for ordinary liquid-liquid extraction columns, but, of course, the centrifugal forces can be made much greater than the normal gravitational forces. That means that significantly greater flow rates can be used in such devices before flooding occurs. The efficiencies of this type of centrifugal contactor depends upon the packing used and how effectively the flow of both phases is introduced into the required regions of the device and distributed over

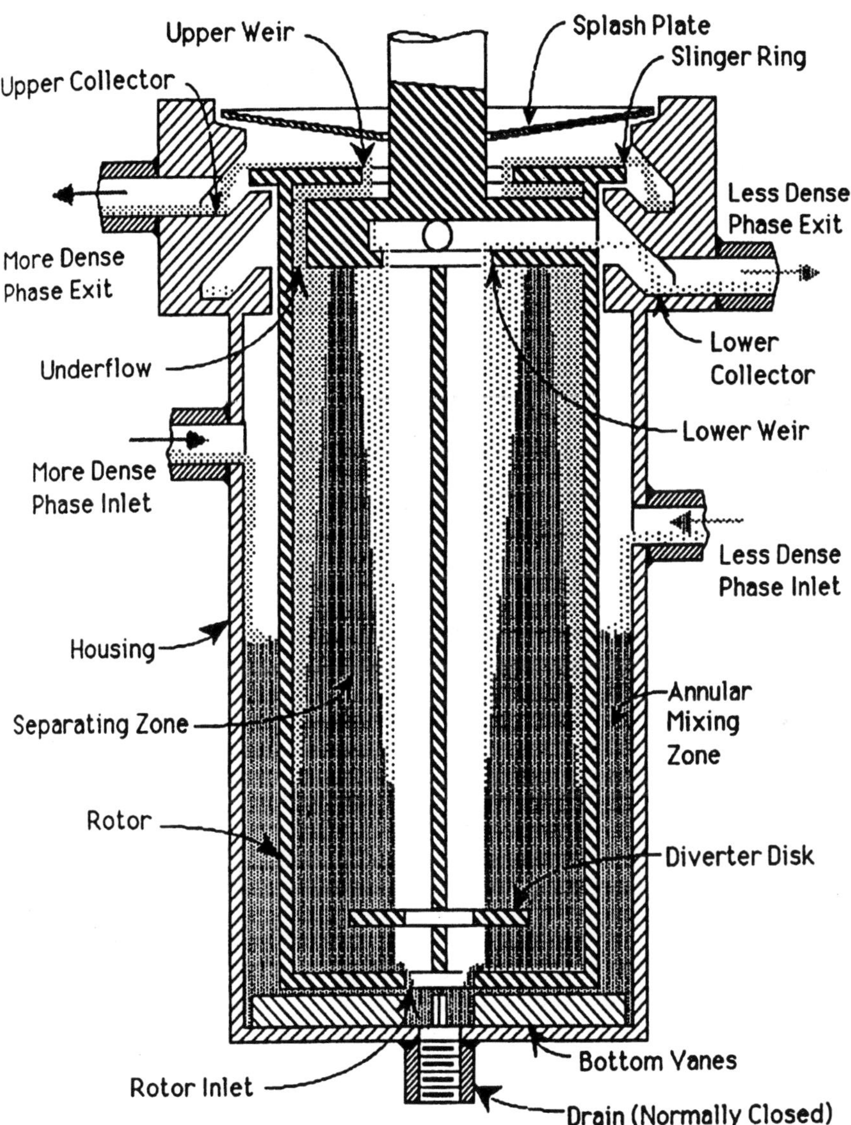

FIGURE 6 Annular mixer settler. (From Ref. 23.)

the cross-section of flow. Although the flow path across the annulus may be relatively short, the high centrifugal forces involved allow one to use somewhat smaller equipment (packing) volumes than would be practical if only gravity were available to force the two phases through the packing in opposite directions. Thus several stages or transfer units can be obtained in a single contactor of this type.

Flooding can occur in mixer-settler equipment just as in solvent extraction towers. Again, the ultimate result of flooding is the presence of one phase in both effluent streams, the inability of the system to effectively separate the two phases. Flooding is largely a failure of the settling part of the mixer-settler units, but that failure can occur because of breakup of the dispersed phase into excessively small droplets in the mixer part of the equipment. Local flooding can occur in either tower or mixer-settler equipment and degrade the performance, although pure phases are still leaving the overall system. The role of local flooding is easier to understand when considering staged type systems like mixer-settlers. When the phases are not completely separated, some portion of one or both phases travels in the wrong direction with the bulk of the other phase. This is called "back mixing" and can severely degrade the effective stage efficiencies, that is, decrease the separation achieved in each stage. The material balances that described the operating line assumed that the phases were separated completely between stages. When the concentration is changing greatly with each stage, even backflow of an apparently small fraction of either phase can reduce the separation considerably.

Liquid-Liquid Extraction with Membranes

Other systems use the solvent in the form of a "liquid membrane," and it would be appropriate to discuss such equipment as a form of liquid-liquid extraction or as a form of membranes. The choice has been made to discuss such systems in Chapter 4, principally because they are more commonly recognized as membrane systems. As pointed out there, such systems have advantages of low solvent inventory and little or no loss of solvent by dispersion.

However, one membrane related type of solvent extraction equipment will be discussed in this chapter because it more closely resembles conventional solvent extraction. This is sometimes called a "nondispersive" contactor (Figure 7). In these systems, the two phases flow countercurrent, but one is on one side of a membrane and the other phase is on the other side. In these cases, the operation closely resembles a common liquid-liquid extraction, and the advantage of the low solvent volume is not present. The membranes are likely to be "hollow fiber"

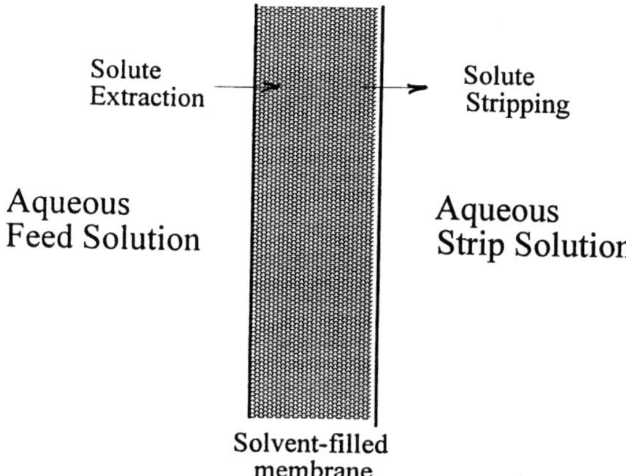

FIGURE 7 Schematic of a supported liquid membrane that acts much like a liquid-liquid extraction and stripping system.

type membranes (see Chapter 4) because one can obtain high membrane surface area (interfacial areas) per unit volume of equipment. The membranc must be preferably wet by one of the two phases (one does not want the membrane to be filled with air), and the pressure of that phase is maintained a little lower than the pressure of the other phase to prevent it from escaping through the membrane. This author knows of no current application for this device, but it has been studied and appears likely to find applications [18]. The advantages of this type of equipment are the essential elimination of solvent loss by dispersion, the high interfacial area, when using hollow fibers, and the ability to maintain that area over a wide range of flow rates and flow ratios. The disadvantages are an additional mass transfer resistance from the membrane and the (probable) additional capital costs from adding the membrane.

APPLICATIONS OF LIQUID-LIQUID EXTRACTION TO ENVIRONMENTAL PROBLEMS

Recall that liquid-liquid extraction can be used to remove organic compounds or metal ions from aqueous solutions. Despite the problems with solvent solubility and entrainment in water, there are important roles for liquid-liquid extraction in waste and environmental processing. Consider possible applications of liquid-liquid extraction first for removal of organic contaminants and then for removal of toxic ions.

Liquid-Liquid Extraction to Remove Toxic Organic Compounds

A hydrophobic solvent can remove many hydrophobic contaminants dissolved or (sometimes) even suspended in aqueous streams. The hydrophobic contaminants are likely to favor the solvent phase and be removed with considerable efficiency. Since recycling of the solvent is essential, it is potentially as important to have a suitable method for removing the solvent from the contaminant, and, if the solvent has a significant solubility in water, it may be necessary to have a suitable method for removing the trace residue of solvent from the "cleaned" water.

Of course liquid-liquid extraction always competes with other separation methods, but no generalization can be made about the preference for solvent extraction until information is available on specific solvent for specific applications, the recovery of the solvent from the contaminant, and the toxicity of the residual solvent in the water. However, a few general guidelines can be considered that will indicate when solvent extraction is most likely to be the separation method of choice. These guidelines are based as much upon the suitability of alternative separation methods as on the merits of solvent extraction. For instance, a highly volatile organic contaminant is likely to be removed more easily by gas stripping, and solvent extraction is not likely to be the best method to use. "Highly volatile" means "much more volatile than water."

When the volatility of the contaminant is relatively low, but still greater than the volatility of water, distillation will be a possible competing method. When the volatility of the contaminant is essentially the same as that of water, such as when an azeotrope is formed, distillation generally will not work. These are the conditions for which solvent extraction is likely to become a separation method of choice. An exception would occur when the azeotrope contains relatively little water, and there is no objection to carrying a little water with the contaminant. Furthermore, a low boiling azeotrope (high vapor pressure azeotrope) could make the distillation easier since the separation factor between the water and the azeotrope could be significantly greater than what one would expect for the pure components alone, based upon their vapor pressures. A sufficiently low boiling azeotrope could even make air stripping practical for a contaminant whose vapor pressure would suggest that it could not be removed effectively by gas stripping.

When the volatility of the contaminant is significantly less than the volatility of water, it generally will be less practical to use distillation, but solvent extraction could be a better choice for many systems. Since most contaminants are present at relatively low concentrations, distillation would have to distill the bulk water, but it is usually preferable to distill off the minor component(s). Thus solvent extraction could be preferred over distillation for relatively low volatility contaminants.

Organic contaminants that form hydrogen bonds with water are likely to be difficult to separate from water by air stripping or by distillation, and liquid-liquid extraction may be better able to overcome such bonding. Liquid-liquid extraction is commonly used to recover and purify acids such as acetic acid and formic acid. A number of contaminants that form hydrogen bonds that hinder contaminant volatility may also be candidates for removal by liquid extraction.

The concentration of the contaminant is also important in determining if solvent extraction is likely to be the preferred method for removing a contaminant. The preference for distillation for higher concentration of the less volatile component has already been mentioned. However, more important may be the preference for adsorption at very low contaminant concentrations. (For further description of why adsorption is so attractive for extremely low contaminant concentrations, see Chapter 2.) Thus solvent extraction is most likely to be the separation method of choice when the contaminant volatility is near to or below that of water and when its concentration is moderately high, too high for adsorption and not high enough for distillation to be preferred.

These conditions are not greatly different from those likely to favor some membrane separations. In fact, solvent extraction is likely to compete with membrane systems for applications, and some liquid membrane systems can be viewed as special forms of solvent extraction. It is difficult to say which separation system should be used, without investigating the solvents and membranes available. There is a great variety of solvents available for extraction operations, but the number of commercial membranes available may be more limited. However, there are also restrictions on solubility, viscosity, density, etc., that may reduce the number of solvent options several-fold.

Although solubility of the solvent in water and entrainment of the solvent are always problems with liquid-liquid extraction, the problem may not be sufficiently serious to eliminate the use of liquid-liquid extraction for environmental applications if the solvent is not toxic, not costly, and/or can be removed relatively effectively from the aqueous product. For instance, if a highly volatile solvent can be used, it may be possible to remove the trace of dissolved or entrained solvent from the aqueous product by gas stripping or even distillation. Although any additional operation increases the costs of the overall system somewhat, a sufficiently effective liquid-liquid extraction system may justify the additional step.

Cusack [19] summarized the use of liquid-liquid extraction for removing organic contaminants from wastewater and focused upon the removal of phenol as an example. Phenol is an excellent example of a contaminant that is difficult to remove by gas stripping or even by distillation. Several possible solvents (toluene, benzene, isopropyl ether, *n*-butyl acetate, and methyl isobutyl ketone) were considered. Methyl isobutyl ketone (MIBK) was selected as the most promising solvent. The basis of this selection included consideration of the distribution coefficient of phenol between MIBK and water, the low solubility of MIBK in water (2.7%), the low toxicity of MIBK, the chemical stability of MIBK, and the ease of recovery of MIBK for reuse.

Liquid-Liquid Extraction to Remove Toxic Metal Ions

One class of contaminants that cannot be removed by gas stripping or distillation are the totally nonvolatile contaminant such as heavy metal ions. (Of course, one could separate these materials by evaporation of all the water, a possible but expensive option that would not select among the ions present.) Liquid-liquid extraction and membranes are likely to be the more attractive method for separating toxic ions at moderate concentrations, and ion exchange processes are likely to become more competitive at lower concentrations. Removal of ions by membranes may utilize liquid membranes, usually supported membranes, but possibly emulsion membranes, and all of these membrane processes can be viewed as different types of liquid-liquid extraction devices.

Solvents for removing ions from solutions usually consist of at least two components. One component is the actual extractant that bonds with the metal ions of interest. With care, one may find extractants for selectively removing the ions of interest, often heavy metals. The other component is the diluent, often a relatively inexpensive and nontoxic hydrocarbon. Actually the diluent can play an important role in determining how a solvent will perform even though it is not the principal extractant. Even when the diluent does not interact strongly with the metal ion, it can perform important services by controlling conditions such as the hydrophobicity or water content of the region around the extractant molecules, and these conditions can affect the amount of the contaminant metal ion extracted and the selectivity for that ion by controlling the extraction of other ions.

Highly selective solvent extractants were developed for the nuclear industry, and these processes are responsible for a large fraction of the uranium recovered from ores and for essentially all of the plutonium recovered from irradiated nuclear fuels for defense programs. Although these processes have been modified and adapted for recovery of the last traces of transuranium elements in nuclear wastes [20,21], the modified processes still operate at high salt concentrations, like those in concentrated nuclear wastes and at high acid concentrations, and like the concentrations of some nuclear wastes (those that have not been neutralized) or the concentrations of dissolved precipitated nuclear wastes. The TRUEX process [21] uses a highly selective solvent [octyl(phenyl)-*N*-diisobutylcarbamoylmethylphosphine oxide, better known as CMPO] with the more common tributylphosphate (TBP) used as a phase modifier. Even more complex processes have been proposed that add other extractants to the solvent to remove combinations of contaminants. The TRUEX-SREX process [22] adds a crown ether to the TRUEX solvent to remove strontium as well as the actinides because strontium also needs to be removed from most or many high level nuclear wastes.

Because of the importance of toxic (and radioactive) metal ions in, e.g., mining wastes and radioactive wastes, there has been significant research devoted to finding better and more specific extractants, or extractant-diluent combinations. Some of the newer processes for nuclear wastes were mentioned above. There have also been extractants developed for lead, copper, and numerous other toxic metals. Some extractants are effective only in moderate to high ionic strengths, so some of them could not be used for groundwater or many wastewaters, but they could often be used earlier in a process system before the contaminants reach the wastewater or groundwater.

Generally, the "neutral" extractants (those that do not ionize) require high ionic strengths in the aqueous phase to extract the metal ions. This results because the metal ion must also be accompanied by an anion to retain electroneutrality. That is, the neutral extracts remove the entire metal salt. The use of high ionic strengths to increase the metal extraction is sometimes called "salting out" the metal salt. The neutral extractants include some of the crown ethers and similar compounds. Such extractants are likely to be less attractive for treating dilute waste waters.

Extractants that can be ionized, such as amines or sulfonic acids, are more likely to be attractive for removing metals from very dilute solutions. These solvents are similar to the structures inside some ion exchange materials (Chapter 2). The use of these extractant has sometimes been called "liquid ion exchange." The reason why these extractions are more likely to be attractive for dilute systems is evident when one notes that, like ion exchange resins, these extractants must always have their ionizable groups neutralized by a cation (or an anion for the amines). Thus extraction of a metal ion is likely to involve exchange of one cation for another cation, often the exchange of hydrogen ions from the extractant for the metal ions from the solution. The reason why this type of solvent is likely to be attractive for dilute solutions is similar to the reason why ion exchange is often attractive for dilute solutions. The total loading of cations in the organic solvent (extractant and diluent) remains essentially the same regardless of the concentration in the solution. That means that the distribution coefficients (the ratio of the concentration of metal ions in solvent to the concentration of metal ions in solution) can be very high when the solution concentration is very low. One potential advantage of liquid-liquid extraction over ion exchange is the ability of the diluent to play a role and add to the selectivity for the metal ions of interest. There are also similar opportunities to alter the interior structure of ion exchange resins, but the degrees of freedom for changing resins are often somewhat more constrained than the opportunities for changing the diluent in a liquid-liquid extraction system.

Back extractions of ionic contaminants can sometimes be achieved by changing the ionic strength (often at lower ionic strengths), the pH, or the redox condition. Often water, acid molecules, or other components may extract or be released by the solvent in the liquid-liquid extraction column, and the effective equilibrium curve can be difficult to determine. In effect, the column can behave like a multi-component system with a number of components extracting, and all of those components may not be metal ions.

REFERENCES

1. Moses, J. M., Z. Altiparmakov, and D. Weber. "Startup and Operation of a Commercial Critical Fluid Wastewater Extraction Plant." *Emerging Technologies for Hazardous Waste Management: 1992 Book of Extended Abstracts*. American Chemical Society (1992), pp. 36–39.
2. Treybal, R. E. *Liquid-Liquid Extraction*. McGraw-Hill, New York (1963).
3. Watson, J. S., L. E. McNeese, John Day, and P. A. Carroad. *AIChE J. 21*, 1080 (1975).
4. Siebert, F. A. and J. R. Fair. *I&EC Res. 27*, 470 (1988).
5. Dell, F. R. and H. R. C. Pratt. *Trans. Inst. Chem Engr. 29*, 89 (1951).
6. Nemunatis, R., et al. *Chem. Engr. Prog. 67* (11), 60 (1971).
7. Seibert, A. F. and J. L. Humphrey. *Sep. Sci. Technol. 30*, 1139 (1995).
8. Grace, J. R., T. Wairegi, and T. Nguyen. *Trans. Inst. Chem. Engr. 54*, 167 (1976).
9. Siebert, A. F. and J. L. Humphrey. *Sep. Sci. Technol. 30*, 1139 (1995).
10. Siebert, A. F. and J. Fair. *I&EC Res. 27*, 470 (1988).
11. Beyer, G. H., and R. B. Edwards. U.S. AEC Report ISC-553 (1959).
12. Moris, M. A., F. V. Diez, and J. Coca. *Sep. Purification 11*, 79 (1997).
13. Miyauchi, T. and T. Vermeulen. *I&EC Fund. 2*, 113 (1963).

14. Leonard, R. A., et al. *Sep. Sci. Technol. 28*, 177 (1993).
15. Leonard, R. A., "Centrifugal Contactors for Laboratory-Scale Solvent Extraction Tests." Paper presented at the Ninth Symposium on Separation Science and Technology for Energy Applications. Gatlinburg, TN, Oct. 22–26, 1995; submitted to *Sep. Sci. Technol.*
16. Todd, D. B. *Chem. Eng. 79*(16), 152 (1971).
17. Davies, G. "Centrifugal Solvent Extraction." Paper presented at the National Meeting of the AIChE, Chicago, IL, Nov. 15, 1996.
18. Schimmel, K. A., J. Williams, and S. Ilias. "Copper Removal and Recovery from Dilute Wastewater Using Nondispersive Solvent Extraction." Presented at the Ninth Symposium on Separation Science and Technology for Energy Application, Gatlinburg, TN, Oct. 22–26, 1995; submitted to *Sep. Sci. Technol.*
19. Cusack, R. W. *Chem. Eng. Prog.* 56 (April 1996).
20. Horwitz, E. P. and W. W. Schulz. in *New Chemistry Techniques for Radioactive Waste and Other Specific Applications* (L. Cecille, M. Casarci, and L. Pietrelli, eds.). Elsevier Applied Science, London (1991).
21. Schulz, W. W., and E. P. Horwitz. *Sep. Sci. Technol. 23*, 1191 (1988).
22. Horwitz, E. P., M. L. Dietz, and R. A. Leonard. in *Chemical Pretreatment of Nuclear Waste for Disposal* (W. W. Schulz and E. P. Horwitz, eds.). Plenum Press, New York (1994), p. 81.
23. Leonard, R. A. *Sep. Sci. Technol. 23*, 1473 (1988).

7
Distillation, Evaporation, and Steam Stripping

Fractional distillation is the most extensively covered topic in most textbooks on separations. That reflects the importance of distillation in the process industries. One recent report [7] states that most of the energy consumed by separations is in distillation towers. Since separations processes consume 30% of the energy used by the process industries, or 10% of the energy consumed by the United States, distillation is certainly an important subject that well deserves prominent coverage. Distillation is often selected for high throughput applications because the capital cost for distillation facilities usually scales with a relatively low power of the throughput capacity, often as low as the 0.5 power. Alternative systems are likely to scale with higher powers of the throughput. However, the coverage in this book is less prominent because distillation is not viewed by the author as having such a prominent role in the environmental and waste industries. This is, of course, a matter of opinion and definition. Since waste management should always consider waste minimization as well as waste treatment, distillation (or improvements in distillation) can be of major importance in waste minimization. This would certainly place distillation high in importance.

Steam stripping, a particularly important method for removing volatile organic compounds from water, is a special case of distillation where the reboiler which supplies energy to the system is replaced with steam and the emphasis is placed on stripping of the volatile organic rather than upon enriching the concentration of the more volatile organic "product." Steam stripping should not be confused with the very similar "air stripping," which introduces air rather than steam to remove the volatile organic component. The differences between single stage steam stripping and air stripping may appear to be relatively minor, and single stage steam stripping could be treated as either a variation of air stripping or a special case of distillation, as in this book. However, when reflux and additional stages are added to steam stripping, the system becomes much like fractional distillation, and the only unusual feature will be the use of steam at the bottom of the tower instead of a reboiler. When the distillation operation produces high purity water at the bottom of the tower, the difference between adding steam to the system and the use of a reboiler can be minor.

In both distillation and gas stripping, the volatile component is transferred to a gas stream (air or steam), usually flowing countercurrent to the aqueous stream. The major difference in the two operations is likely to be the quantity of material vaporized. Gas stripping is more likely to be used with relatively volatile components (such as contaminants) whose volatility is much greater than the volatility of the less volatile component (such as water in many environmental applications), and the concentration of the component to be removed is likely to be relatively small. This means that in analyzing gas stripping operations, it may not be necessary to account for vaporization of the heavier component; that will not be the case for distillation operations.

Distillation is covered here because of its prominent place in the process industries and because it will be needed for several waste treatment operations, especially for aqueous wastes with relatively high concentrations of volatile organic contaminants or for organic components whose volatilities are not greatly different from the volatility of water. However, it is not treated as extensively in this book as it is in most separations textbooks because the environmental and waste applications are not believed to be as prominent as they are in the current process industries. Distillation would be most likely to be chosen for environmental application when the relative volatility of the contaminant to be removed from a wastewater is greater than approximately 1.3 times the volatility of water. Distillation is less likely to be economical when the major component, usually water in environment applications, is the more volatile one. This means that distillation is not likely to be used to remove trace contaminants that are less volatile than water. On the other end of the scale, distillation can receive considerable competition from other separation methods, such as air or steam stripping, when the relative volatility of the contaminant becomes too high. It is not possible to place an upper limit on the upper value for the relative volatility since it depends upon how low the concentration of the contaminant must be reduced. As the desired concentration for the contaminant in the water effluent stream becomes lower, the relative volatility at which stripping becomes more economical increases. Steam stripping is usually used when the bottom water product is essentially pure water, and this is likely to be a common situation for many waste treatment operations. Remember that distillation with steam injection in place of a reboiler can be viewed as an extension of steam stripping that is capable of reaching a lower concentration of contaminants without sending too much water to the volatile product, but it is principally a slight variation of fractional distillation.

The term "distillation" is often used to mean the same as "fractional distillation," that is, multistage distillation involving stripping and enriching operations. Single stage distillation is more likely to be called evaporation or, for transient operations, "batch distillation."

EVAPORATION

Evaporation is a single stage separation of liquid and vapor phases after a portion of the liquid is "evaporated" and converted to vapor. Evaporation concentrates the more volatile material in the vapor and the less volatile components in the liquid. When there is a great difference in the volatility of the components to be separated, the separation from one stage of evaporation can be essentially complete. The two most obvious examples are when one component has essentially no volatility and when one component has a very high volatility. For example, nonvolatile inorganic salts can be separated from solutions by evaporation, also called "drying."

On the other extreme, highly volatile organic compounds can be removed, almost completely, by distilling only a small portion of the less volatile liquid. An important process question to be answered in designing evaporation equipment may be, "How much liquid needs to be evaporated to achieve the desired removal of the volatile component?" Of course, if the goal were to recover the volatile component, the question could have been, "How much liquid must be evaporated to recover a specified fraction of the more volatile component?" or "How much liquid can be evaporated to recover the volatile component at a concentration above some specification?" Those are more common goals for the conventional chemical process industries; they are less likely to be the goals of waste and environmental applications, but they can be important in efforts to recovery and recycle solvents.

The concentration of a more volatile component in the vapor will, in general, depend upon its concentration in the liquid. It should be obvious that the concentration of the more volatile component in the vapor will be zero when its concentration in the liquid is zero, and its concentration in the vapor will be 100% when the liquid is 100% volatile component (none of the lower volatile component present). However, the behavior of the vapor composition with intermediate liquid compositions can take many forms. Considerable vapor-liquid equilibria data have been published (e.g., Walas [1] or Palmer [2]). In general, it is desirable to have experimental data on vaporization to know the vapor composition at any liquid composition. There are several ideal cases such as Raoult's law liquids where the partial pressure of each component in the vapor is equal to the vapor pressure of that component (at the temperature of interest) multiplied by the mole fraction of that component in the liquid. Other relations can be used for some liquids over limited ranges of composition. For instance, the composition of the components in the vapor could be proportional to their mole fraction in the liquid, but the constant of proportionality may not necessarily be the vapor pressure of the components. There has also been considerable progress in simulating the behavior of liquid and vapor systems, and it probably will be possible in the near future to predict the behavior of more vapor-liquid systems from knowledge of fundamental interaction parameters for the molecules involved. Several computer codes for process analysis contain an extensive data base on vapor-liquid equilibria and use advanced thermodynamic methods and correlations to utilize the data. Progress in computer simulation and correlations is rapid, and is reducing the quantity of experimental data needed for many systems but has not completely eliminated the need for them. Measurements and empirical (usually graphical) information on vapor-liquid equilibria on new systems are still needed, at least to confirm the computer model results.

Vapor-liquid equilibria can be measured at constant pressure or constant temperature. More often, measurements at constant pressure are desired because most evaporation (or distillation) equipment operates at an approximately constant pressure. It is usually convenient to tabulate or plot the data as the mole fraction of the more volatile component in the vapor phase versus the mole fraction of the more volatile component in the liquid phase. The choice to base the plot on the more volatile component and the orientation of the graph to show the vapor composition on the vertical axis are traditional. This method will be used in presentation of vapor-liquid data throughout these discussions.

Continuous Evaporation

In continuous evaporation, the feed solution is fed to the evaporation continuously, and liquid and vapor products are withdrawn continuously (Figure 1). Two material balances can be drawn around the system, one for the more volatile component and another for the less volatile component

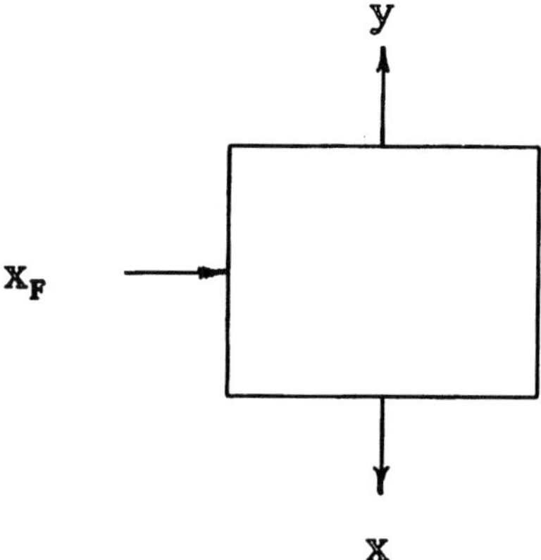

FIGURE 1 Distillation as the splitting of a liquid stream into a vapor stream and another liquid stream.

or for the total mass (both components). The balance for the more volatile component is

$$Fx_F = Vy + Lx$$

where F is the feed rate in moles per unit time, x_F is the mole fraction of the more volatile component in the feed, V is the vapor rate leaving the evaporator, y is the mole fraction of the more volatile component in the vapor, L is the liquid rate leaving the evaporator, and x is the mole fraction of the more volatile component in the liquid leaving the evaporator. The overall (both components) material balance is

$$F = V + L$$

These equations can be combined to eliminate one flow rate (usually L) and express the results in terms of the two product rates:

$$x_F = \frac{V}{F}y + \left(1 - \frac{V}{F}\right)x$$

Here, V/F is simply the fraction of the liquid evaporated. There are still two unknowns, x and y (x_F is assumed to be known since it is the composition of the liquid to be treated).

The additional information required to eliminate another variable comes from the vapor-liquid equilibrium data that was just described. One can usually assume that the vapor leaving the evaporator is essentially in equilibrium with the liquid leaving the evaporator. That is, one

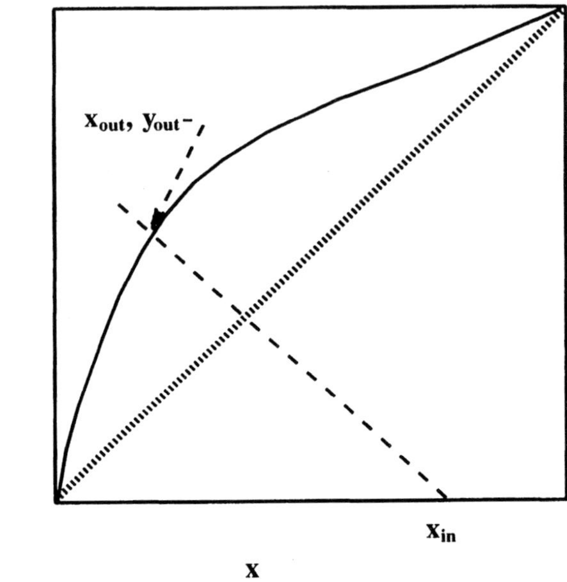

FIGURE 2 Operating line for a single stage evaporator relating the possible vapor and liquid phases from a material balance.

can usually assume that the evaporator behaves approximately as an equilibrium stage. Thus, x and y are related by the equilibrium curve. That is, the values of x and y will lie somewhere along the equilibrium curve. The liquid and vapor compositions are also related by the material balance equation given above. It may be easiest to solve the equation for y. Then

$$y = \left(1 - \frac{F}{V}\right)x + \frac{F}{V}x_F$$

The values for x and y satisfy both the equilibrium curve and the material balance equations, so the solution is the interaction of the material balance line with the equilibrium curve. Note that the material balance is a linear equation (gives a line when plotted on the y versus x diagram) with slope $1 - F/V$ and an intercept on the x axis at $[F/(F-V)]x_F$. The material balance is often called the "operating line" (Figure 2). It is a little more conventional to express the slope as $-L/V$, which is equivalent; see the overall material balance. The intercept can also be written in terms of the ratio of the liquid to the vapor rates, $(1 + V/L)x_F$.

The operating line in Figure 2 gives the range of liquid and vapor compositions that can satisfy the material balance, and the equilibrium curve gives the range of liquid and vapor compositions that can satisfy the requirement that the liquid and vapor be in equilibrium. Thus, the intersection between the operating line and the equilibrium curve represents the composition of the liquid and vapor from the single stage evaporator that satisfies both relations. If the equilibrium curve has a simple shape and can be expressed as a simple algebraic equation, it may be possible to solve the equilibrium equation and the operating line equation for the compositions. The solution is simple for a linear equilibrium curve, but the graphical approach is valid for any shape of equilibrium curve, including experimental curves for which there is not necessarily an acceptable algebraic equation. Note that for a given feed composition, a wide range of compositions can be obtained by using different ratios of the liquid and vapor rates, that is by having different operating lines. All possible operating lines converge at the feed composition.

Evaporation Equipment

Several forms of evaporators are in common use. The immediate image of an evaporator may be a simple pot. There are two approaches to evaporation. In one case the heat is applied to the liquid in the evaporator, and in the other the heat is applied to the fluid before the evaporator. The latter cases are called flash evaporators and are operated at reduced pressures.

A batch evaporator heated with steam and the evaporated product collected with a condenser is sketched in Figure 4. If the heat came from a wood fire, this could be the design for a "classical" moonshine whisky still. Although this author cannot confirm that such equipment is still used for making illegal whiskey, equipment somewhat like that in Figure ? is used or for small-scale batch distillation/evaporation. In this example, the heat exchanger is located inside the evaporator, and flow through the heat exchanger is by natural convection.

For larger heated evaporation systems it is usually desirable to have a high rate of heat transfer per unit volume (or unit cost) of the equipment. Two major groups of heated evaporator are most commonly used: film evaporators and forced circulation evaporators. A film evaporator is largely a heat exchanger with vertical tubes and a relatively small inventory of liquid in a lower chamber (Figure 4). Liquid is pumped to the top of the heat exchanger and flows down the exchange surface as a falling film. The relatively thin film provides good heat transfer to the vapor-liquid interface and gives a minimum inventory of liquid. The small inventory is

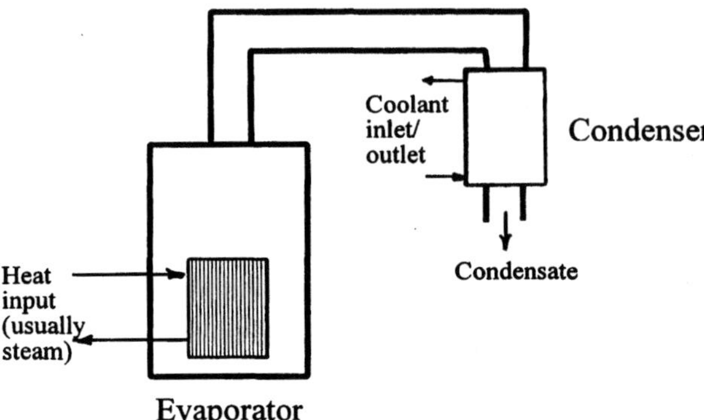

FIGURE 3 An evaporator with an internal heat exchanger.

important in some industries, but it will not necessarily be important in many waste and environmental processes.

One of the limitations or possible problems in falling film evaporators is the formation of solids on the surfaces. Although that can be a problem with any evaporator system, falling film equipment is often more susceptible to scale or other solid forming on the heat transfer surfaces because the inventory of water in the film is so low that large changes in the concentration of material in the water occur during a single pass of liquid down the heat exchanger surface. The hydrodynamics of the film is also particularly important, so falling film evaporators can be affected

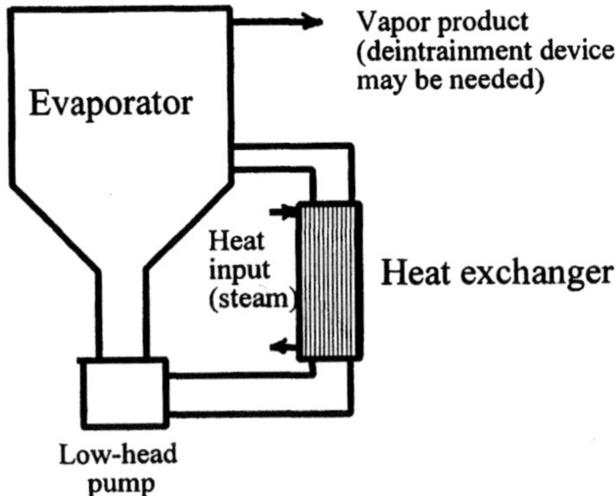

FIGURE 4 An evaporator with circulation through an external heat exchanger.

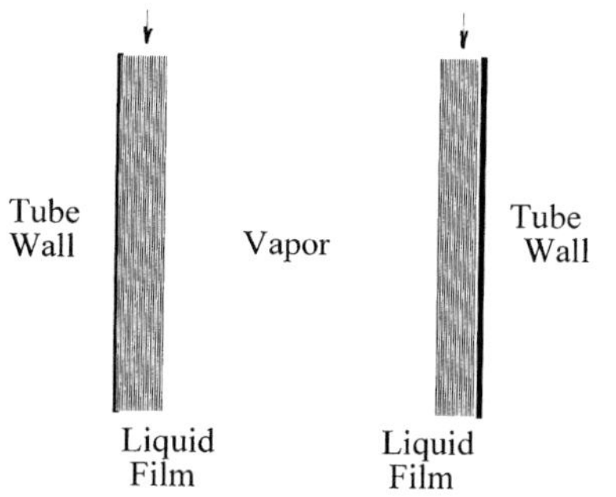

FIGURE 5 A falling film evaporator.

greatly by changes in the viscosity of the liquid with evaporation and may even be unsuitable for use with high viscosity liquids.

The evaporation rate is controlled by the rate of heat transfer to the liquid film, and heat transfer coefficients can be determined by testing or estimated from information supplied by some manufacturers. Falling film evaporators with waterlike liquids, believed to be the most likely liquids to be treated in waste and environmental systems, usually have overall heat transfer coefficients of a few hundred Btu/h/ft^2/°F.

Forced circulation evaporators transfer heat in liquid filled heat exchangers and the liquid is "forced" through the heat exchanger by an appropriate circulation pump. To keep the heat exchanger filled (largely) with liquid, it is necessary to limit the temperature rise in the heat exchanger by circulating the liquid relatively rapidly. High liquid velocities also reduce fouling on the surfaces due to higher heat transfer coefficients and lower temperature rise in the liquid. Gas-liquid separation then occurs in a separate part of the vessel, as shown in Figure 5. Many types of heat exchanger can be used in such evaporators. Standard tube-in-shell type exchangers seem to be common because of their standard design and modest costs. There are many options for the shape of the evaporator and even options for the location of the heat exchanger. The heat exchanger can be inside or outside the evaporator vessel (Figure 5). The enlarged upper region of the vessel creates lower vapor velocities and thus better deentrainment of liquid droplets from the vapor. Additional devices such as cyclones can also be used to reduce liquid entrainment. The deentraining device can be located within the evaporator vessel or outside the vessel with a liquid return line.

Flash evaporation involves heating the fluid using conventional heat exchangers and flashing the heated fluid in a chamber (the evaporator) at a lower pressure. Unless one wants to flash the liquid at extremely low pressures, it is usually necessary to heat the liquid and hold the liquid at a higher pressure so it will not boil before it reaches the lower pressure in the evaporator. A major advantage of flash evaporators for some applications is the reduced chances for forming scale in the evaporator itself or even the heat exchanger. However, salts, such as calcium carbonate and calcium sulfate, whose solubilities decrease as the temperature is increased, can still scale any of these heat exchanger systems.

In both systems, the ratio of liquid to vapor flow rates is determined by the amount of heat added to the liquid, that is, by the latent heat of evaporation. Since the circulating evaporators operate at approximately constant pressure, the relation between the heat added and the quantity of vapor created can be determined relatively simply. Information on the enthalpy of the fluid mixtures is needed at the pressure, temperature, and composition of both the original mixture after heating at an elevated pressure and at the vapor and liquid phases at the lower pressure (and lower temperature after the flash evaporation). If the environmental applications are limited to sufficiently dilute systems with relatively volatile contaminants, the enthalpy properties can be simplified somewhat by equating the sensible heat of superheated water to the latent heat of the vaporized contaminant and vaporized water. Although it may often be sufficient to assume that the sensible heat of very dilute mixtures is similar to that of pure water, the latent heat of the vaporized mixture often needs to be determined for the real mixture because a significant portion of the vapor could be the volatile contaminant because the latent heat from the solution may be significantly different from the latent heat from pure components.

Although one usually thinks of evaporation being driven by adding heat to the liquid, it is also possible to substitute mechanical energy for heat by using vapor compression. Vapor compression can have lower energy costs, but usually involves considerably more capital cost for the additional mechanical equipment. Thus, vapor compression is more likely to be chosen for evaporators that will operate with large streams continuously or for an extended time.

A single stage of vapor compression can be illustrated by a simple modification to the forced circulation evaporator. This is essentially a way to produce a liquid product from the overhead and apply work rather than heat to drive the evaporation. Mechanical vapor compression utilizes the increase in temperature of a vapor during (near) adiabatic compression. The hot compressed vapor is then used to evaporate more of the feed material (in the heat exchanger/evaporator), and the compressed vapor can be condensed in the process. There is usually some thermodynamic

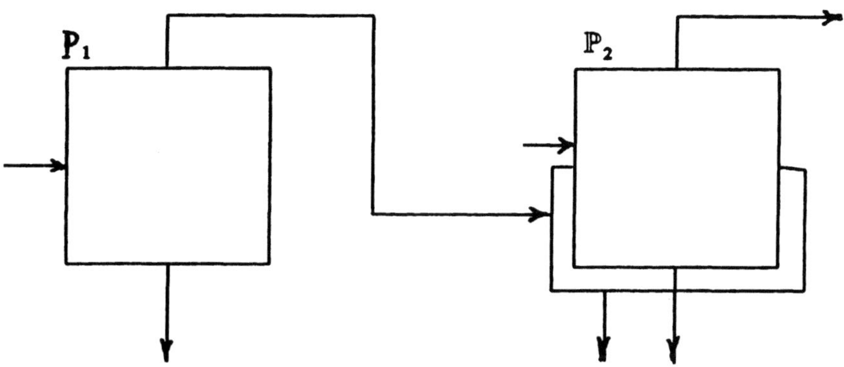

FIGURE 6 Multi-effect evaporation.

energy savings from such operations, principally because there is at least some utilization of the thermal energy in the vapor. Of course, the system will not be able to recover all of the energy because of inefficiencies in the compressor and heat exchanger; so vapor compression should be viewed as a way to reduce the need for fresh steam to run the evaporator. However, this is not the only way to recover a significant portion of the vapor. One could recover some of the thermal energy in the products by using them to help heat the feed stream with a recuperative heat exchanger. Another way is to use the thermal energy in "multiple effect" evaporators.

Multiple Effect Evaporation

The energy consumed by evaporation can be high and, thus, costly, so it is worthwhile to reduce the net energy used in the processes. One way to do this is to use multiple effect evaporators. Since it is more common to want both the liquid and vapor products as liquids, it is necessary to condense the vapor. Multiple effect evaporators use the heat removed from condensing vapor to evaporate more liquid (Figure 6). Because there is some resistance to heat transfer through the walls of the evaporator and film resistance in the liquid and condensing vapor, it is necessary to have a positive temperature driving force for each effect of the evaporation. This can be achieved by pressurizing the first effect and/or reducing the pressure on the last effect. In Figure 6 the liquid is fed to the evaporator as saturated liquid, possibly at elevated pressure, so the heat recovered from condensing vapor in one effect is used to vaporize liquid in another effect operating at a lower pressure and thus a lower boiling temperature.

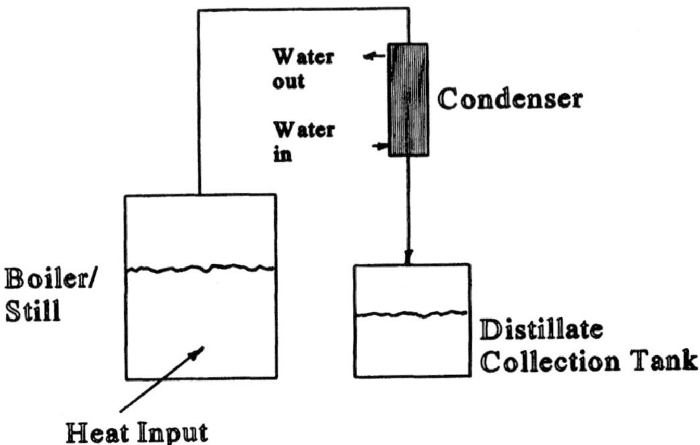

FIGURE 7 Batch distillation/evaporation system.

Thus much of the heat applied to the system can be used to evaporate liquid several times. The sensible heat required to heat the liquid to the boiling point may not be as great as the latent heat of vaporization, but much of the sensible heat can also be recovered by using recuperative heat exchangers where the exiting liquid products are used to heat the incoming liquid. However, even in well designed systems, evaporation will be an energy intensive process.

Figure suggests that the heat is added to each stage through a jacket surrounding the evaporation vessel, but this is only a sketch. The heat can be added through exchange tubes within the vessels as in other single effect evaporators.

Batch Evaporation

Batch evaporation is a transient operation used in the alcohol beverage industry (legal and, possibly, illegal), where it is usually called distillation. (In this book, "distillation" will be reserved for "fractional distillation," which involves use of multiple stages in the stripping and enriching towers. Other cases will be called evaporation.) In batch evaporation, a liquid feed solution is introduced to a batch vessel, and heat is applied to evaporate some fraction of the liquid. The composition of the liquid and the vapor change during the evaporation, so this is a transient operation. The vapor is usually condensed into another liquid product, one that is enriched in the more volatile component (Figure 7).

Again it is usually acceptable to assume that the vapor leaving the batch is approximately in equilibrium with the liquid remaining in the evaporator. Thus, at all times, the liquid and vapor compositions will lie along the vapor-liquid equilibrium curve. The final composition of the liquid will be the composition that remains in the liquid when the evaporation is stopped. If the "overhead" (the vapor) is condensed into a single container, the composition of that container will be the average of the composition of the vapor received during the evaporation. Since the evaporator will become depleted in the more volatile component, the concentration of that component in the vapor will be the highest at the start of the batch distillation and will decrease with time. The change in composition in the vapor and liquid can be followed by applying a material balance around the evaporation at any instant in time.

The rate of change in the liquid composition can be determined from the vapor composition. For a constant evaporation rate, V,

$$\frac{d(Q_L x)}{dt} = \frac{Q_L dx}{dt} + \frac{x dQ_L}{dt} = Vy$$

and

$$\frac{dQ_L}{dt} = V$$

Here Q_L is the moles of liquid in the evaporator at any time, and t is time. Note that Q_L will decrease with time as more liquid is evaporated. These equations can be combined to give

$$Q_L\left(\frac{dx}{dt}\right) = V(y - x)$$

Since y and x represent point on the vapor-liquid equilibrium curve, this equation can be integrated numerically or graphically for any shape of vapor-liquid equilibrium curve. For a constant distillation rate, V,

$$Q_L = Q_{L0} - Vt$$

The material balance can be written as

$$\frac{dt}{Q_{L0} - Vt} = -\log\left(\frac{Q_{L0} - Vt}{Q_{L0}}\right) = -\log(1 - F_v) = \int_{x_1}^{x_{\text{final}}} \frac{dx}{y - x}$$

where F_v is the fraction of the original liquid feed that is evaporated, and $1 - F_v$ is the fraction that is not evaporated. Although, for convenience, this discussion was based upon a constant evaporation rate, V, the last equation holds for any evaporation rate. If the vapor and liquid are in equilibrium, y is a function of x and can be evaluated from the equilibrium relation.

The integral on the right can be evaluated from any initial liquid composition to any desired final liquid composition (both expressed in terms of x), and the value of the integral between those limits will be the logarithm of the fraction of liquid remaining. The average composition of the condensed vapor can be evaluated by averaging the composition of the vapor over time or over the total moles of vapor produced. However, it is usually more convenient to obtain the average composition of the condensed vapor from a material balance on the liquid:

$$V_T = Q_{L0} - Q_L = Q_{L0}F_v$$

and

$$V_T y_T = Q_{L0}x_0 - Q_L x$$

In environmental applications, batch distillations are likely to be used when there is a great deal of difference in the volatilities of the two components, for instance where relatively volatile organic compounds are to be evaporated from water. In those cases, the concentration of the more volatile component can be reduced rapidly with evaporation of only a small fraction of the liquid. When the volatility of the two components are more similar, fractional distillation is the more likely choice. However, for those cases where only a modest fraction of the volatile component must be removed, evaporation could also still be considered. Batch distillation is also more likely to be used on smaller scale operations; high throughput systems are more likely to use continuous evaporation.

Multistage batch distillation can be used to treat batch quantities of liquid where it is desirable to recover the most volatile component at a higher concentration than could be achieved with simpler single stage evaporation/distillation. The arrangement for a multistage distillation is shown in Figure 8. Analysis and control of such systems has been described by Skogestad et al. [3] for the relatively complex cases of multiple product removal. To have multiple stages, it is necessary to condense some of the vapor and return it to the evaporator by cascading through the different stages, shown in Figure 8 as a hatched rectangle. The condensate can be produced by a condenser located above the stages or simply by heat losses from the walls of the distillation stages or piping above the stages. In the simplest case where the inventory of all components in the different stages is very small when compared with the inventory in the distillation vessel, such systems can be treated as quasi-steady-state operations and analyzed much like the enriching section of a fractional distillation system described next.

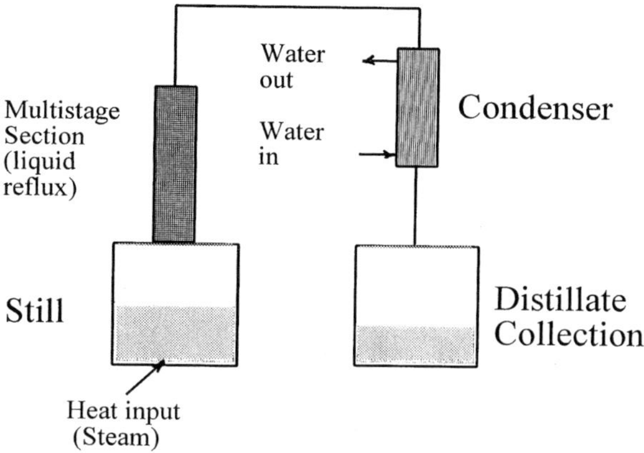

FIGURE 8 Batch multistage distillation system.

FRACTIONAL DISTILLATION

Fractional distillation involves two towers that operate much like absorption and stripping towers. It is possible to combine these two operations physically into a single tower, and this is the usual practice. However, the two functions take place in different sections of the tower, and the behavior of each section can be analyzed in the same manner described or for absorption and stripping.

The important difference between distillation and absorption is that the gas phase is a vapor containing the two components to be separated. In absorption, the gas phase was largely air or another gas which had little solubility in the liquid phase, and the liquid phase had little volatility at the temperature involved and did not enter the vapor phase to a significant degree. As in gas stripping, the objective is to remove a volatile component from the less volatile component that remains largely in the liquid. In stripping, however, the temperature is usually sufficiently low (sufficiently below the boiling point) that the volatility of the liquid itself could be ignored, but that is not the case in distillation.

Before describing the two individual portions of a fractional distillation tower, it is best to consider the overall arrangement of a fractional distillation system for separating two components (Figure 9). Note that there are essentially two sections of the system: one tower located above the point where the feed is introduced, and another tower located below the point where the feed is introduced. In both sections liquid flows down the tower, and vapor flows up the tower. The tower could consist of

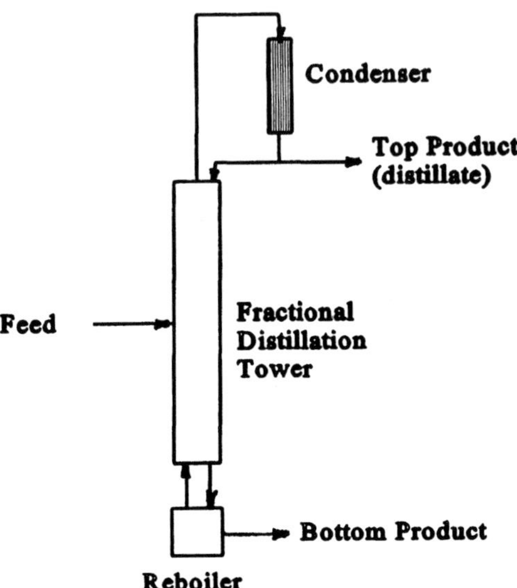

FIGURE 9 Fractional distillation.

trays (which resemble or approximate stages) or packing materials which allow continuous contact between the two phases. The process design of fractional distillation systems can be made on the basis of stage or transfer units, just as with absorption/gas stripping towers. The countercurrent flow of gas and liquid is achieved by refluxing some of the vapor which leaves the top of the tower by condensing it and feeding it back as liquid to the top stage of the tower. This provides a liquid phase to flow down the upper section of the system. Similarly, a portion of the liquid leaving the bottom of the system is refluxed by boiling it and sending the vapor to the bottom stage of the system (Figure 9). As will be shown, the fraction of the material refluxed determines the behavior of the separation. In general, greater reflux results in improved separation, but increased energy costs and lower throughput for a given size tower. An overall material balance will show that the reflux of condensed vapor at the top of the system will fix the boil-up rate at the bottom of the system; the reflux and boil-up rates are not independent.

The discussion in this book will focus upon the simple case of binary distillation for two reasons. First, binary distillation is considered more likely to be important in environmental and waste applications when the contaminant to be removed is present at relatively low concentrations. Then even when there is more than one contaminant, it is likely that each component can be treated separately as part of a binary distillation with water unless there are sufficient interactions between the contaminants that the presence of one contaminant affects the vapor pressure of another. This is likely to be the case in the stripping section, but it still may not be so in the upper stages of the enriching section. The second reason is the belief that most multi-component distillation problems should be designed with computer codes rather than by hand, as described here. Although the use of good computer codes is recommended for all process design when practical, the recommendation is especially strong when the design becomes more complex. Not only does it become less practical to design such systems by hand, but there can also be more chances for errors. However, an understanding of the principles involved in distillation is believed to be important to those using computer codes, but most of the understanding can be achieved by studying the simpler binary cases. There are opportunities to err when using computer codes as well as when doing calculations by hand, and the errors can be large if the codes are misused. Being able to even approximate the calculations by hand can be one way to spot such errors, and a basic understanding of the operations within the computer codes can provide the user with an understanding of how errors can occur and even of some possible limitations in the calculations.

Computer simulations are being used more frequently and more extensively to design chemical process equipment, and fractional distillation systems are among the most extensively covered by most simulation codes. When such systems are available and the reader is familiar with their use, computer simulations are highly recommended. In many cases, the simulation codes even incorporate data and correlations on vapor-liquid equilibria, so their use may save as much time and effort in the search for information as in the actual simulation calculations. However, for binary distillation, the calculations may be sufficiently simple that it is practical to consider doing them by hand. On the other hand, if one wishes to explore the effects of many design parameters, even calculations that can be done moderately easily by hand become laborious if they must be repeated many times. Multi-component calculations require trial-and-error calculation, and it is almost always desirable to use computer simulations for such calculations. The difficulty of trial-and-error calculations is likely to increase with the number of components to be considered, especially when many of them have similar volatilities; thus, the desirability of using computer simulations rather than manual calculations is likely to increase with the number of components involved. The following section contains a relatively simple description of manual calculations that are found in most introductory texts. Although they may

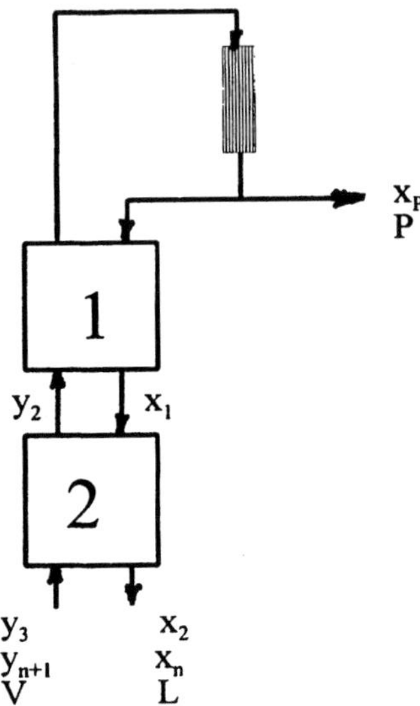

FIGURE 10 Material balance around part of the enriching section.

be practical only for binary distillations, the descriptive material should also be understood by those using computer simulations because they are similar to the calculations performed by simulators, even for multi-component systems.

The Enriching Section

The enriching section is the upper section of a fractional distillation system. Consider a material balance around the feed exit and an arbitrary point within the column (Figure 10). To follow the most common convention, the stages will be numbered starting from the top. Then there is only one stream entering this section of the column, the vapor coming from stage $n+1$. There are two streams leaving the section, the top product stream (sometimes called the distillate) and the liquid from stage 2. Because there are two components, two independent material balances can be made around the dotted envelope. It is conventional to use the total (overall) material balance on all materials and the balance on the more volatile component:

$$V_3 = P + L_n$$

$$V_3 y_3 = P y_p + L_2 x_2$$

or, in general,

$$V_{n+1} = P + L_n$$

$$V_{n+1}y_{n+1} = Py_P + L_n x_n$$

As in the analyses of absorption towers, the subscript describes the stage from which the fluid is leaving.

Here, P is the overhead product rate or the distillate rate. The distillate composition is given as y_P, which implies that it is a vapor. That is simply to indicate that it comes from the vapor rising up the tower, actually, the distillate product can be condensed to a liquid or left as a vapor. The only necessary condition is that a portion of the overhead vapor that is to be refluxed be condensed and fed to the top stage as a liquid.

It is best to express the flow rates in moles and the compositions in mole fractions rather than in mass (weight) and mass fractions. This is commonly done in gas absorption as well, but there mass fractions and other concentrations usually can be used almost equally well if desired. However, with distillation it is certainly preferable to use mole units. This is because in many mixtures, the molar heat of vaporization is often approximately constant, independent of the composition of the binary mixture. Since this is not a universally good approximation, the merits of using molar units may be somewhat reduced for a few systems.

As in the analysis of absorption towers, the material balances give the operating lines. The last equation which is the material balance on the more volatile component is the operating line. However, by using the overall material balance (the previous equation), the operating line can be simplified and contain fewer unknowns. Look first at the simplest system where the molar heat of vaporization can be assumed to be approximately constant, independent of composition. That would mean that the liquid rate will be the same throughout the enriching section of the system, and the vapor rate also will be the same throughout the enriching section if there is no heat loss. Some of the vapor will be condensed, and some of the liquid will be vaporized. If the heat of vaporization per mole of liquid is constant, the same number of moles will be vaporized as condensed. For these conditions, the subscripts can be dropped from the flow rates:

$$Vy_{n+1} = Py_P + Lx_n$$

$$V = P + L$$

Solving for y_{n+1},

$$y_{n+1} = \left(\frac{L}{V}\right)x_n + \left(\frac{P}{V}\right)y_P$$

This is the operating line that relates the concentration of the more volatile component in the vapor leaving the stage below the dotted envelope to the concentration of the more volatile component in the liquid leaving the stage just above the boundary of the dotted envelope. Where L and V are constant (equal molar overflow), the operating line is straight. However, if the heat of vaporization were not independent of composition, the liquid and vapor rates would not be constant, and the operating "line" would be curved.

This expression is often written in terms of the "reflux ratio," that is, in terms of the ratio of the liquid flow rate (the reflux) to the distillate product rate (P). Since the vapor leaving the first stage is divided between the refluxed liquid and the overhead product (distillate),

$$R = \frac{L}{P} = \frac{V - P}{P} = \frac{V}{P} - 1$$

Then the operating line can be expressed as

$$y_{n+1} = \left(\frac{L}{V}\right)x_n + \left[\frac{1}{R+1}\right]y_P = \left[\frac{R}{R+1}\right]x_n + \left[\frac{1}{R+1}\right]y_P$$

Calculation of the number of stages required in the enriching section of the fractional distillation system proceeds in the same manner used in the calculations for absorption or gas stripping towers (Figure 11). The operating line is easily constructed on the vapor-liquid equilibrium curve by noting first that the slope of the operating line is L/V. This is also $R/(R+1)$ if one wants to work in terms of the reflux ratio. This expression for the slope of the operating line is helpful because it shows that the slope must always be less than unity. The intercept is $[1/(R+1)]y_P$, but it is even easier to remember that the composition of the vapor leaving the first stage is the same as the liquid entering the first stage (the liquid from stage zero) because the liquid is simply a portion of the vapor from that stage that has been condensed and refluxed. This means that the operating line crosses the diagonal (line $y = x$) at the composition x (or y) $= y_P$. This is usually a convenient way to locate the operating line because the composition of the distillate product is usually specified in the problem definition.

Starting with the compositions at the first stage (the top stage in the tower), the compositions at all other stages can be determined as

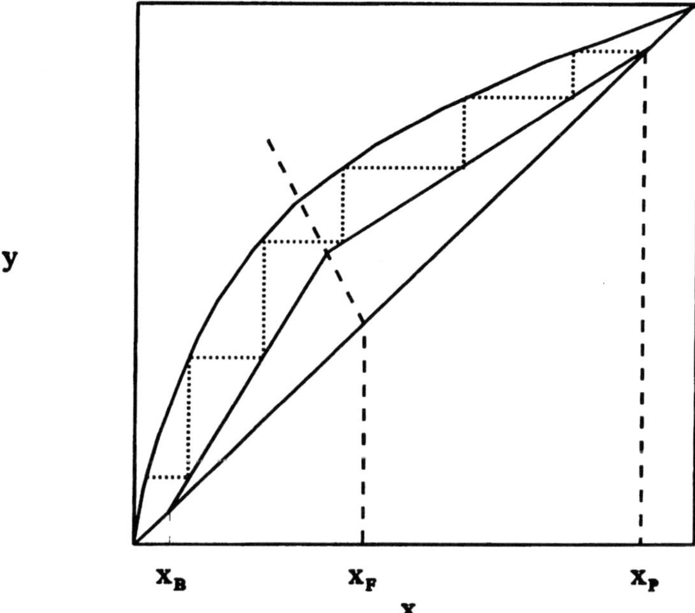

FIGURE 11 McCabe-Thiele calculations of stages in fractional distillation.

in gas absorption/gas stripping calculations. Since the vapor composition from the first stage is known (the distillate product composition), the liquid leaving that stage can be obtained from the equilibrium diagram. On the vapor-liquid equilibrium curve, the liquid composition is obtained by looking horizontally (constant vapor composition) to the equilibrium curve (Figure 11). This point represents the compositions of the two streams leaving the first stage. Then once the composition of the liquid leaving the first stage is known, the composition of the vapor leaving the next stage can be obtained from the operating line which relates the composition of the vapor from any stage to the composition of the liquid from the stage above. Thus one can look just below the position on the equilibrium line for the first stage and find the vapor composition from the next stage on the operating line. In a similar manner, once the vapor composition from the second stage is known, the liquid composition from the second stage can be obtained from the equilibrium curve.

This process can be repeated to determine the compositions of the vapor and liquid streams from each stage in the enriching section.

This method, known as the McCabe-Thiele method, quickly determines the composition of all streams in this tower section. The graphical approach resembles successive steps between the operating and equilibrium curves, but it is best to remember the reason for these steps—that is, those positions on the equilibrium curve simply represent possible compositions of vapor and liquid streams leaving any stage, and positions on the operating line (or curve) represent possible liquid compositions leaving one stage as a function of the vapor composition leaving the stage below. Such calculations should continue until the end of the enriching section is reached, that is, the point where the feed is introduced, usually when the composition of the liquid (or vapor) reaches the composition of the feed. Although one can introduce the feed at any position in the tower, the optimum location is where the composition of the liquid in the tower is close to the composition of the feed (liquid feed).

The feed can be introduced as a liquid, a vapor, or a mixture of the two. Consider how the operating line changes at the feed point. If the feed is saturated liquid (liquid at its boiling point), the vapor flow rate will be unchanged and will be the same in the lower portion of the fractional distillation system (usually called the stripping section) as in the enriching section. However, the liquid rate will be higher. In fact, the liquid rate in the stripping section will be equal to the sum of the liquid rate (the reflux rate) in the enriching section plus the feed rate. The slope of the operating line in the stripping section (below the feed point) will be L'/V', where the primes indicate the conditions in the stripping section, the lower part of a fractional distillation system. For a feed that is saturated liquid L'/V' will be $(L+F)/V$. If the feed is supercooled liquid (liquid below its boiling point), the liquid rate in the stripping section will be increased further by the condensation of vapor to heat the feed to its boiling temperature. Conversely, it the feed were vapor or superheated vapor, the vapor rate in the stripping section would be lower than the vapor rate in the enriching section be cause the feed vapor rate would add to the vapor rate in the enriching section.

In all cases, the slope of the operating line will be greater in the stripping section than in the enriching section. This is evident when looking at the material balance around the reboiler that supplies vapor to the stripping section. The balance will be made around the dashed envelope in Figure 12:

$$L'x'_n = V'y'_{n+1} + Bx_B$$

$$L' = V' + B$$

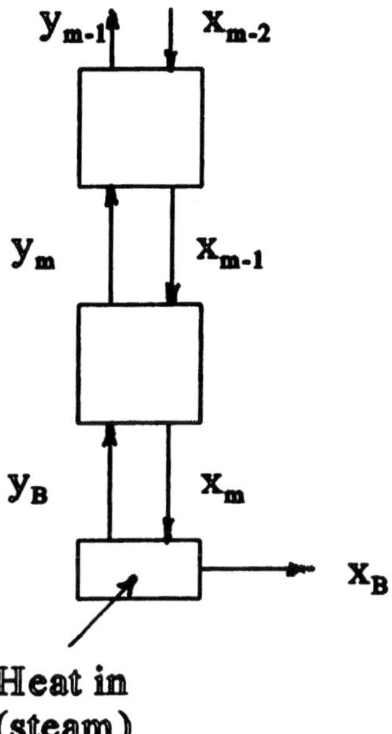

FIGURE 12 Material balance around part of the stripping section.

Solving for y'_{n+1} gives

$$y'_{n+1} = \left(\frac{L'}{V'}\right)x'_n - \left(\frac{B}{V'}\right)x_B$$

Since L' is always greater than V' (the difference is B), the slope of the operating line in the stripping section will always be greater than unity and, thus, greater than the operating line for the enriching section. This is illustrated for one case in Figure 12.

The slope of the operating line in the stripping section is coupled to the slope of the operating line in the enriching section. To see this coupling, one needs to remember that the slope of the operating line is the ratio of the liquid to the vapor rate and note the change that takes place at the feed point. The feed increases the liquid rate in the stripping section if it contains liquid and "reduces" the vapor rate in the stripping section if it contains vapor. Of course, the feed could contain both liquid and vapor. It could also be supercooled liquid or superheated vapor (vapor at a temperature above the boiling point). Thus, the thermal state of the feed and the reflux ratio at the top of the enriching section fix the slope of the operating line in the stripping section.

Not only is the slope of the operating line in the stripping section fixed, but the location of the line is also fixed. This results because one point on the operating line is already fixed, the location at the reboiler. If one assumes that the reboiler is not an efficient stage, or counts it as a stage, the composition of vapor coming from the reboiler is the same as the liquid from the reboiler; that is the same composition as the bottom product. This means that $y_{N+1} = x_N$. Here the subscript N represents the last stage in the system. The vapor composition y_{N+1} represents the vapor from the reboiler. Thus, the operating line crosses the diagonal ($y = x$) at the bottom product composition. With this point fixed, the entire operating line is determined once the slope is determined.

When working with McCabe-Thiele graphs, it is often easiest to draw the operating line for the stripping section from the point where it intersects the diagonal ($y = x_B$) and the point where it intersects the upper operating line. The point where the two operating lines intersect can be determined from material and heat balances around the feed stage. It is traditional to specify the thermal condition of the feed in terms of the parameter q which is the fraction of the feed that is saturated liquid. Then $1 - q$ is the fraction of the feed that is saturated vapor (some times called its "quality"). For feed that is saturated liquid, the value of q is unity, and for feed that is saturated vapor the value of q is 0. Obviously, for mixtures of liquid and vapor, the value of q lies between 0 and 1. For supercooled liquid feed, the value of q is greater than 1, and for feed that is superheated vapor, the value of q is less than 0.

For supercooled liquid feed,

$$q = 1 + \frac{C_L(T_b - T_F)}{H_B}$$

where C_L is the heat capacity of the liquid (at constant pressure), T_b is the boiling temperature of the feed solution, T_F is the temperature of the feed, and H_B is the heat of vaporization of the liquid feed. Both C_L and H_B should be expressed in molar units.

For feed that is superheated vapor,

$$q = -\frac{C_V(T_F - T_b)}{H_B}$$

where C_V is the molar heat capacity of the vapor, and T_b is, in this case, the condensing point of the vapor (the boiling point of the liquid that would produce this vapor mixture). To find the intersection of the two equilibrium lines, first consider the equations for the two lines. For the enriching section, the operating line can be written as

$$Vy = Lx + Py_P$$

For the stripping section, the operating line can be written as

$$V'y = L'x - Bx_B$$

Subtracting the second equation from the first gives

$$y(V - V') = x(L - L') + Py_P + Bx_B$$

This equation describes the locus of points at which the two operating lines can intersect. Note again that the relations between V and V' and between L and L' are determined by the thermal condition of the feed:

$$L' = L + qF \qquad \text{and} \qquad V' = V + (1 - q)F$$

These relations can be used to relate $V - V'$ and $L - L'$ to the feed condition:

$$-y(1 - q)F = -xqF + [Py_P + Bx_B]$$

From an overall balance on the more volatile component over the entire system,

$$Fx_F = Py_P + Bx_B$$

Note that the right side is the same as the term in brackets. Then

$$y = -\left[\frac{q}{1 - q}\right]x + \frac{x_F}{(1 - q)}$$

This equation expresses the locus of points where the two operating lines can intersect in terms of the feed composition and the thermal condition of the feed, q. That is, for a given feed and feed "quality," q, all operating lines for any reflux ratio will intersect somewhere on this line. For feed that is saturated vapor, $q = 0$, the slope of the locus of intersections is zero, and $y = x_F$. (Of course, if the feed were all vapor, it would have been more consistent to call the feed composition y_F instead of x_F but retaining x_F for the feed composition of any state is expected to create less confusion.) Thus one can locate the locus of intersections by drawing a horizontal line at $y = x_F$. Then by locating the intersection of the enriching operating line with this horizontal locus (often called the q line), the stripping section operating line can be located by simply drawing a line from this intersection to the position on the diagonal where $y = x_B$.

When the feed is saturated vapor, the q line has an infinite slope since the denominator of the coefficient for x is zero. The location of the horizontal q line is evident when the equation for the q line is multiplied by $1-q$. Then when $q = 1$ (feed is saturated liquid), the left side becomes zero and x must equal x_F. Then the q line or the locus of points where the two operating lines can intersect is a vertical line at $x = x_F$. Note that in both cases, the q line intersects the diagonal at $y = x = x_F$. This is true for all values of q, so the q line can be drawn from the diagonal position where $y = x = x_F$ with slope $-q/(1-q)$. When the feed is a mixture of liquid and vapor, the slope is negative and lies between 0 (saturated vapor feed) and $-\infty$ (saturated liquid feed). When the feed is supercooled liquid, the slope of the q line is positive, and the line lies to the right of the vertical line at $x = x_F$. Similarly, when the feed is superheated vapor, the slope of the q line is positive and the intersection of the two operating lines falls below the horizontal line at $y = x_F$.

The procedure for calculating the compositions of liquid and vapor streams throughout the distillation system (in both the enriching and stripping sections) proceeds as described or, for the enriching section, until the feed stage is reached. Although one could locate the feed stage at many positions, Figure 12 illustrates the location of the feed introduction at the stage where the compositions cross the feed composition, that is, when the calculations cross the point where the two operating lines intersect. Thermodynamically, this is the best location to place the feed. The merits of locating the feed at this point are illustrated by considering the consequences of locating the feed at some other point in the tower. The locations of the two operating lines are independent of where the feed is introduced. If the feed point were located several stages below the point where the two operating lines cross, one would keep calculating stage compositions by using the operating line for the enriching section, the upper operating line. Note that after the operating lines cross, the operating line for the enriching section lies closer to the equilibrium curve, and that makes the changes in composition per stage smaller than one would have by using the operating line for the stripping section, which lies below the enriching section operating line and thus farther from the equilibrium curve. On the other hand, if the feed point were located above the stage nearest the point where the two operating lines intersect, one would need to begin using the operating line for the stripping section earlier, and then the operating line for the stripping section lies above the operating line for the enriching section and thus closer to the equilibrium curve. This also would result in smaller concentration change per stages and more stages required for a given separation.

Thus, after the stage by stage calculations cross the intersection of the two operating lines, it is preferable to introduce the feed immediately so the calculations will begin to use the lower operating line. The reasons for not introducing the feed at this point would be rare; for instance, if the separation was to be made in an existing distillation system and changing the location of the feed point would be difficult.

Selecting the Reflux Rate

At very low reflux rates (reflux ratios) the slope of the operating line or for the enriching section would decrease and approach a horizontal line. At sufficiently low reflux ratios, the intersection between the two operating lines would occur "above" the equilibrium curve. Construct such a pair of operating lines and try making stage by stage calculations and it will quickly become evident that an infinite number of stages would be required to approach the point where the operating line first crossed the equilibrium line. That certainly would not be a practical system. This means that there is a minimum reflux required for a given separation, namely the lowest reflux ratio which will give two operating lines (the enriching section and stripping section operating lines) that will not "touch" or intersect the equilibrium at any point between the two ends of the system (the distillate product composition and the bottoms composition). For the relatively simple equilibrium curve in Figure 12, the minimum reflux ratio would occur when the intersection of the two operating lines touches the equilibrium curve, but for less ideal equilibrium curves which are far less symmetrical in shape the first intersection could occur along the enriching or stripping section operating lines.

Since it would take an infinite number of stages for the calculations just described to reach a point where the operating line(s) and the equilibrium curve intersect or touch, a fractional distillation system operating at the minimum reflux ratio would require an infinite number of stages. That would result in a tower of infinite cost. The number of stages required becomes finite as the reflux ratio becomes slightly greater than the minimum, and the number of stages required decreases as the reflux ratio is increased above the minimum.

Note that the operating lines both have slopes of unity when the reflux ratio is infinity. This means that with infinite reflux ratio, both operating lines lie along the diagonal ($y = x$), and the minimum number of stages required for the separation can be calculated by making the stage by stage calculations using the diagonal as the two operating lines.

With an infinite reflux ratio, all of the vapor at the top of the system is refluxed (total reflux), and all of the liquid from the last stage is boiled to make vapor. There is no product withdrawal and no feed; the location of the feed point is immaterial. This is also not a practical way to make a separation. The number of stages in the tower is a minimum, but the reflux rate for any finite feed rate would be infinity. That would mean that the tower would have an infinite diameter and consume an infinite amount of energy in the reboiler. Neither of those are economically attractive conditions.

The most practical reflux ratio obviously lies between these extremes. Increasing the reflux ratio decreases the number of stages required, but it also increases the diameter of the tower needed for the separation and the energy needed for the reboiler, as well as the cooling required for the condenser. The diameters of distillation towers will be described later, but it is important to note that the throughput of a tower of essentially any design type is proportional to the cross-sectional area. The liquid and vapor rates per unit cross section of the tower are set by the flooding conditions, that is, by the maximum flow rate that can be used before one phase becomes entrained in the other phase. Flooding is described in more detail in Chapter 3.

Exact calculations of the optimum reflux ratio require information on the capital cost of the tower as a function of diameter and height (number of stages), the cost of energy input to the reboiler, and the cost of reboiler and condenser surface areas. The optimum reflux ratio will depend upon the materials of construction and the tower packing type, but it often lies between 1.5 and 2.0 times the minimum ratio. This number can be used in preliminary cost estimates and may even be adequate for selecting the size of relatively small and inexpensive systems, but one may want to make more exact calculations for especially costly installations.

Alternative Ways to Specify Distillation Problems

MaCabe-Thiele calculations have been described in the most usual form, that is, when the feed composition and the desired product compositions are known. However, there are a number of other ways the design of a distillation column can be specified. For instance, the composition of one product and the fraction of one component that should be recovered in one product could be specified. In all cases, the overall material balance around the entire system and the material balance for one component (usually the more volatile component) around the entire system,

$$F = P + B$$

$$Fx_F = Py_P + Bx_B$$

can be used to obtain the compositions of both product streams. The important point is that the number of stages required in each section of the system is to be determined. The problem is slightly different for other problem specifications, those where the number of stages is fixed but one of the other parameters must be determined (such as one product composition or the throughput).

There are at least two cases where one may want to specify the number of stages and leave another parameter to be determined. In one case, an existing tower may be available that is believed likely to provide a fixed number of stages of separations. Then if one wanted to specify the compositions of the effluent streams, one could determine the reflux ratio needed to give the number of stages in the existing column. That is, one would want to determine the energy output required (different reboiler and reflux condenser rates) or the throughput of the system (for a specified reboiler and condenser rate). Then one would try different intersection points along the q line for the two operating lines until the point is found that gives the proper number of stages. Of course, one knows that if the number of stages required is greater than the target number, the intersection point will have to be lowered to move the operating lines farther from the equilibrium curve, and if the number of stages required is less than the target value, one can raise the intersection point. Interpolation of results usually allows reasonably accurate results to be reached in a few trials.

Another case can occur when, for a given operating condition (reflux ratio and number of stages in each section of the tower), one wishes to know the composition of the product streams. This case can arise in multicomponent systems. The number of stages may be set for the removal of one major component, but it is necessary to also determine the distribution of one or more trace components between the two distillation products. (In such cases, the equilibrium curve would have to be based upon the mixture of major components along the tower.

To determine the separation of other components once the number of stages has been determined based upon the removal of another (critical) component, one must remember to use different equilibrium curves for each component. Then the concentration of that component in either product stream should be estimated (guessed) for the first trial. The composition of either product stream can be used for each trial because the compositions are related. From the material balance on that component,

$$Fx_{Fi} = Py_{Pi} + Bx_{Bi}$$

This is just like the overall material balance on the volatile component given earlier, but an additional subscript i is used to remind the reader that this equation applies to the other independent volatile components.

Since only the compositions of the product streams are unknown, selecting one of the compositions fixes the composition of the other stream. Once the two product compositions are specified, the operating lines can be drawn for each section of the tower. Note that the slopes of these operating lines will be the same for all volatile components since the slopes are determined by the flow rates, which are the same for all components. Then the number of stages can be calculated for each component. If the number is higher than that calculated for the critical component and selected for the tower, the assumed separation is too large. The concentration of the volatile component in the overhead product is too high, and the concentration in the bottoms product is too low. Alternatively, if the number of stages is too low, the separation will be better than that assumed. Then new composition(s) can be assumed until the number of stages calculated approaches the desired value.

Note that when the number of stages is specified from the need to separate the critical component, that sets the number of stages in each section of the tower, not just the total number of stages. That means that one should shift the calculations from the operating line for the enriching section to the operating line for the stripping section after the number of stages specified for each section, not at the position that would be optimum for component i. The optimum position for the feed point will not shift greatly from component to component unless the shapes of the equilibrium curves are notably different. Although that is less likely for trace components, it could arise. An especially difficult situation would result if there is a desired distillation of the component i but that separation requires a minimum reflux ratio that is greater than that of the assumed critical component. If that occurs, it is another sign that the wrong component was selected to size the tower. Although the complexities of true multicomponent distillation are not covered in this book, this case does illustrate some of the sources of those complexities. Similar trial-and-error calculations are needed for each component in multicomponent distillation. However, when the equilibrium vapor composition for each component is dependent on the concentrations of the other components in the liquid (and thus the vapor), the calculations become even more complex.

Calculations for these specifications can be done as described by trial and error. The graphical calculations just described are so simple and quick that there may seem to be little need for using the powerful computer codes available. However, when trial-and-error calculations are required, one often develops more respect for the computer codes. Other more complicated distillation problems will be mentioned where the availability of automated calculations will be appreciated.

Analysis of Distillation Systems with Mass Transfer Coefficients

Some distillation systems are constructed of plates that resemble stages, but many towers, and most new towers, are filled with newer structured packing materials. The plates and packing materials used in distillation towers can be similar to those used in absorption and gas stripping towers, and more details on the packing materials can be found in Chapter 3. Several smaller or older towers also may be filled with random packing. Such towers can be analyzed in terms of heights equivalent to an equilibrium stage, just as discussed for absorption or stripping towers, but as with absorption and stripping it may be more desirable to use a mass-transfer-based analysis, that is, one involving transfer units rather than stages. The mass transfer analysis is essentially identical to that described for gas absorption/gas stripping. The same operating lines and equilibrium curves are used. Note, however, that the mass transfer coefficients and the interfacial surface area may be functions of the liquid and/or vapor rates and may be different in the two sections of the tower because the flow rates are different. The mass transfer coefficients will certainly be a function of the liquid velocity in the tower. The discussion of minimum reflux ratio applies equally for mass-transfer- and stage-based analyses of distillation systems. The optimum reflux ratio is also likely to fall within the same range of values for the reflux ratio, between 1.5 and 2 times the minimum reflux ratio.

Systems without Equal Molal Overflow

Mixtures of greatly different compounds are less likely to be easily approximated by the principal assumption that the liquid and vapor rates remain constant over any section (enriching and stripping) of the tower. The assumption of constant molal overflow can hold over limited regions or for a great number of systems. For instance, in many environmental systems the more volatile component may be an organic material at a relatively low concentration, and the vapor may be largely the volatile organic compound. Note that if the liquid phase is mostly water (dilute systems), there is little change in the bulk liquid composition even when the VOC is removed. In such cases, the constant molal overflow assumption may be moderately good in that dilute region, although it may not be good over a wider range of concentrations such as may be found if an attempt is made to concentrate the VOC sufficiently that the liquid in the upper portions of the enriching section is no longer dilute. One way to estimate for a separation if the constant molal assumption is adequate

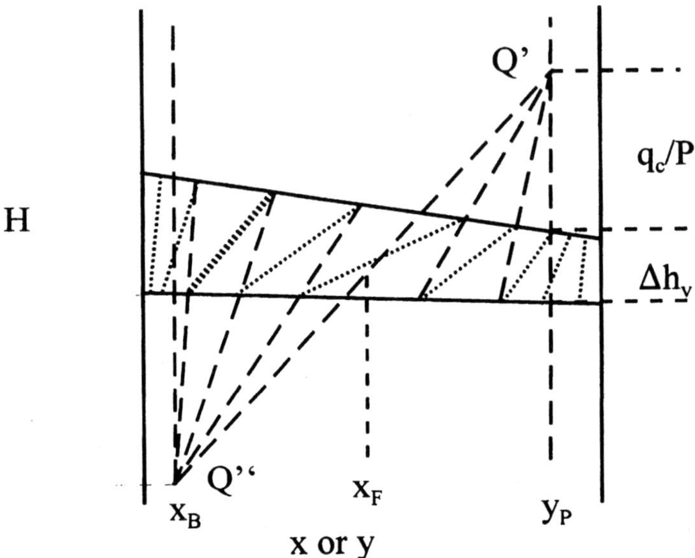

FIGURE 13 Enthalpy concentration diagrams and stage calculations.

for any concentration range is to simply look at the molar heat of vaporization of the two components in the mixture to be separated. If the two heats of vaporization are not greatly different, there is a good chance that the simplifying assumption will be adequate; if they differ greatly, it will always be desirable to look further before using the simplified expressions for linear operating lines. Note, however, that even similar molar heats of vaporization of the pure components do not ensure that the heats of vaporization for all mixtures of the two components will be similar.

The exact way to determine if the constant molar overflow assumption is valid is to look at an enthalpy concentration diagram (Figure 13). This is another equilibrium diagram, but it gives enthalpy as well as concentrations at equilibrium. Note that the enthalpy of a mixture is plotted on the vertical axis, and the compositions of the mixtures (both x and y) are plotted on the horizontal axis. The composition is x if the mixture is a liquid, and y if the mixture is a vapor. There are two curves running from one side of the diagram to the other. The upper curve represents the enthalpy of saturated vapor of the specified composition, and the other curve represents the enthalpy of saturated liquid with different compositions. Thus, any condition below the lower curve represents supercooled liquid, and any condition above the upper curve represents superheated vapor. Between the two curves two phases are represented, a liquid phase and a vapor phase. The enthalpy of the mixture is greater than that of the saturated liquid, but not sufficient to vaporize all of the liquid to vapor. This region between the saturated liquid and saturate vapor curves represents the two phase equilibrium conditions that exist throughout most of a distillation system. Distillation towers operate at "saturated" conditions, that is, with liquid and vapor in equilibrium.

There are tie lines (shown as dotted lines) that connect the saturated liquid and saturated vapor curves. Since the figures are traditionally plotted in terms of the concentration (mole fraction) of the more volatile component, these tie lines always slope toward the right (positive slope). If we had adopted the tradition of making the plots in terms of the less volatile component, the tie lines would have a negative slope. The tie lines connect the vapor composition that is in equilibrium with a given liquid composition. That is, to obtain the composition of vapor that would be in equilibrium with a given liquid concentration, one would locate the liquid composition on the lower (liquid) curve and follow the tie line that connects that point to the upper curve (saturated vapor). There is a tie line connecting every point on the liquid curve to the vapor curve. However, it is obviously not practical to draw an infinite number of tie lines without making the region between the two curves on a printed diagram totally black, so only a few representatives are shown. The user is expected to interpolate and estimate the slope of tie lines for the numerous conditions or for which they are not given. To make such an interpolation, one simply looks at the slopes of the tie lines on both sides of the point of interest and constructs a tie line with an approximately appropriate slope between those two slopes.

To know if the equal molar overflow assumption is sufficiently accurate, one need only check the heat of vaporization of mixtures over the range of interest, that is, between the distillate and bottom composition. The heat of vaporization is the difference between the enthalpy of the vapor curve and the corresponding position (connected by tie lines) on the liquid curve. What constitutes "significant" deviations in the constant molar overflow assumption, of course, depends upon the degree of accuracy needed. This usually means that the vapor and liquid equilibrium curves will be approximately parallel, that is, approximately the same distance apart over the region of interest.

When the molar heat of vaporization changes significantly over the concentration range between the distillate and bottom products, a McCabe-Thiele type analysis could still be used if the operating line were allowed to curve. Although variations in the molar heat of vaporization over the concentration range of interest will cause the operating line to curve, the curvature will not necessarily be great, even with 10–30% variation in the molar heat of vaporization. However, the most convenient way to handle a curved operating line is to actually do the calculations directly on the enthalpy-composition curve. The stage by stage calculation proceeds by alternating use of the equilibrium curve and a material balance on the more volatile component, just as in the McCabe-Thiele approach. However, in this case the material balance is expressed not as a single line but as a series of lines that resemble the tie lines that express the equilibrium conditions.

The easiest way to understand the material balance lines is to note that when two streams are mixed, there is no loss of either total enthalpy or volatile component, although both enthalpy and volatile component can be exchanged between the two phases if the mixture results in two phases. Thus, the enthalpy of the mixture and the concentration of the more volatile component will be an average of the enthalpy and concentration of the two initial streams. This average will be weighted in proportion to the amount of each stream used to form the mixture. The position of the mixture on the enthalpy-composition curve will lie along the line joining the enthalpy and composition of the two initial streams (Figure 13). The position of the mixture on that line will depend upon the amount of each stream used to form the mixture. The location of the mixture will lie closer to the point for the mixture used in the greatest excess. If equal moles of the two initial streams are used, the mixture will lie exactly midway between the positions of the two initial mixtures on the diagram. If two parts of one stream and one part of the other stream are used to form the mixture, the mixture will be located one third of the distance from the position of the mixture used in the greater quantity.

Conversely, note also that "splitting" a stream into two components is the reverse of combining two streams. That is, the position of the original stream being split will lie along a line joining the two streams produced. If the original stream is split into two equal-size streams, it will lie an equal distance from the two streams produced; otherwise, it will lie proportionally closer to the stream produced in the greater quantity.

The reason this observation of the locations of mixtures of two streams on an enthalpy-composition diagram is so important is that it is much like the material and enthalpy balance that are made around the upper stages of a distillation system that are the equated to the operating line used in the McCabe-Thiele approach. Consider the material balance around the dotted envelope shown in Figure 11. The equations for the material and enthalpy balances are

$$V_{n+1}y_{n+1} = L_n x_n + P y_P$$

and

$$V_{n+1}h_{n+1} = L_nH_{Ln} + PH_P + Q_c$$

where H denotes the enthalpy of a vapor, and h denotes the enthalpy of a liquid. The distillate product could be a liquid or a vapor, so H or h could be used. It is important to use the enthalpy of the distillate product, whether liquid or vapor. Q_c is the heat removed in the condenser. Of course, the size of Q_c will depend upon the state of the distillate product (liquid or vapor) as well as the reflux ratio. It is necessary to condense the overhead product that is to be refluxed, but the withdrawn product could be collected as either a condensed liquid or a saturated vapor. One could even supercool the liquid product if the condenser removed sufficient heat from the overhead stream for the first stage.

Note from Figure 13 and from these equations that the portion of the distillation system enclosed by the envelope appears to split the vapor coming from stage $n+1$ into two streams, the liquid from stage n and the distillate product. However, to keep the enthalpy balance correct for a simple splitting of the stream, the heat removed by the condenser would have to be added back to the distillate product. This can be expressed as

$$PH_P + Q_c = P\left[H_P + \frac{Q_P}{P}\right] = PH_Q$$

H_Q is the enthalpy of the distillate product if the heat removed by the condenser were added to the distillate stream. Remember that the condenser removes heat from the entire vapor stream coming from the first stage (the reflux stream and the overhead product), but we are adding the heat from the condenser only to the overhead product stream, the stream withdrawn from the system.

The composition of the vapor from the first stage is known because it is the same as the composition of the distillate which is usually specified. The vapor from the first stage is then located on the vapor equilibrium curve at the composition y_P. Since the liquid from the first stage is in equilibrium with the vapor leaving the first stage, its position and composition can be obtained by following a tie line from the location of the vapor to the liquid curve (Figure 13). To obtain the enthalpy and composition of the vapor coming from the next (second) stage, first locate the point (y_P, H_Q) on the diagram. This point will lie directly above the distillate product composition and will be at a point above the distillate product enthalpy a distance Q_c/P. Since one can consider the vapor from the second stage being split into this imaginary stream and the liquid from the first stage, the vapor from the second stage must lie along a line connecting the points (y_P, H_Q) and (x_1, h_1). Furthermore, the vapor from the second stage must be saturated, so it must lie on the vapor equilibrium curve. Figure 13 shows a line drawn from the liquid curve at $x = x_1$ to (y_P, H_Q). That line intersects the vapor curve at (y_2, H_2). Repeating the procedure, the point (x, h_2) is obtained by following the tie line from (y_2, H_2) to the liquid curve. Then y_3 and H_3 can be obtained by constructing another line between (x_2, h_2) and (y_P, H_Q) and looking for the intersection of that line with the vapor curve. This procedure can be repeated until the feed point is reached.

In the stripping section, a similar balance can be made by noting that the liquid stream from stage n' appears to be separated into a bottom product and the vapor stream from stage $n' + 1$:

$$L'n'x_{n'} = V'y_{n'+1} + Bx_B$$

and

$$L'h_{n'} = V'H_{n'+1} + Bh_B - Q_b = V'H_{n'+1} + BQ_B$$

Q is the heat added to the reboiler, and $Q_B = h_B - Q_b/B$. The point (x_B, h_B) is located on the liquid curve at the composition specified for the bottom product, and the point (x_B, Q_B) is located directly below that point a distance Q_b/B. Then when the feed point is passed, the calculations should start using lines from the point (x_B, Q_B) to the liquid composition of stage n' on the liquid curve and extrapolating the calculations to the vapor curve to obtain the composition and enthalpy of the vapor stream from stage $n + 1$. This is also illustrated in Figure 13.

Once the point (y_P, H_Q) is located, the point (x_B, Q_B) is usually more easily calculated graphically by noting that the overall distillation system splits the feed stream into two parts:

$$Fx_F = Py_P + Bx_B$$

$$Fh_F = PH_P + Q_c + Bh_B - Q_b = PH_Q + BQ_B$$

Thus the composition and enthalpy of the feed must lie on a line joining (y_P, H_Q) and (X_B, Q_B). This means that once the point (y_P, H_Q) is located, a line can be drawn from there through the feed point and extrapolated to find where it intersects with a vertical line at x_B. The point of intersection is the point (x_B, Q_B).

This graphical analysis is called the Ponchon-Savarit method and is only slightly more complicated than the McCabe-Thiele method, but it does require more information, the enthalpy of the saturated streams as well as the compositions. The minimum reflux ratio must also be identified here as in the McCabe-Thiele calculations. Although the reflux ratio was not specified explicitly in the discussion above, it was implied when the heat removed by the condenser was mentioned. The larger the reflux ratio, the larger Q_c/P will be. Thus the condition for total reflux is for Q_c to be infinite and the operating lines to all be vertical. The minimum reflux ratio is the lowest ratio for which the slope of the operating lines will always be greater than the tie lines going through the same liquid (of vapor) composition. Review the calculation procedure and notice that the alternative use of tie lines and operating lines must always move the calculations to lower and lower compositions.

Azeotropes

One relatively common complication in distillation that may not occur as often in environmental problems is the formation of an azeotrope. This is a "constant boiling" mixture, a mixture for which the vapor composition has the same composition as the liquid. This is usually viewed as a "nonideal" situation and separations beyond those compositions cannot be made by simply adding more stages. Azeotropes can be identified on y versus x vapor-liquid equilibrium diagrams as points where the curve crosses the diagonal. On a enthalpy-composition diagram, an azeotrope occurs at a point where a tie line is vertical. For more information on separation azeotropes, one is referred to more detailed texts on distillation. Common approaches involve producing a product close to the azeotrope and feeding this stream to another system operating at a different pressure; azeotropes can sometimes be moved to different compositions, or even eliminated, by changing pressure. Other approaches involve adding additional components to alter the vapor- liquid equilibrium and move or eliminate the azeotrope.

Multi-component Distillation

Many distillation systems must separate more than two components. These are called multi-component separations. If all of the components (except one component, such as water) are dilute, the concentrations of the different solutes may not necessarily affect the vapor-liquid equilibrium of the other components, and then the separations can be treated as several individual binary distillation problems, as discussed earlier. However, when the presence of more than two components affects the vapor-liquid equilibria of other components, the system is truly multi-component, and the treatment becomes considerably more complex. Multi-component separations are more difficult to analyze. For M components, there will be $M - 1$ material (and enthalpy) balances. The concentration of each component in the vapor can be a function of the concentration of all components in the liquid. Thus it is difficult to construct a simple graphical method for multi-component distillations. It is even difficult to describe the vapor-liquid equilibria for all of the components in a graphical form, especially when the number of components becomes large. The calculations are usually best made numerically using methods that are analogous to the stage by stage calculations described for binary distillations, alternately using equilibrium and material balance relations.

To understand multi-component equilibria, considerable more data may be required, especially if the equilibrium is far from ideal. That is, the equilibrium curves described for binary systems must be repeated not only for each component other than the least volatile component, but possibly for incremental steps in the concentrations of all of the other components over the range of compositions occurring in the tower. For numerical calculations, the multi-component equilibrium relations also need to be expressed numerically. Furthermore, to be practical, it is necessary for the data to be expressed in some form of a model that will be able to interpolate the data and reduce the number of measurements required to cover the range of concentrations needed. Suitable models are incorporated into the more advanced distillation simulation (design) codes currently available.

For more than two components, it is not possible to specify the composition of all components in the product streams. This has already been illustrated for the special case of multi-component systems that can be described as a number of binary separations. If one specifies the concentration of one component in all streams, the concentrations of the other components will be set. But those concentrations are not evident and must be estimated. This means that trial-and-error calculations are usually needed to establish the compositions of the products as well as the compositions on each stage. These trial-and-error calculations are best made with packaged computer programs. More details on multi-component distillation can be found in more specialized books on distillation, but automated calculations are recommended for systems with very many components and/or many stages.

Equipment Used in Fractional Distillation

Fractional distillation (continuous distillation with countercurrent flow of liquid and vapor, that is, with reflux) is usually carried out in "towers" with the liquid flowing down the tower by gravity and vapor flowing up the tower from the pressure drop generated by the reboiler. The entire system can be operated at atmospheric pressure, at a reduced pressure (vacuum), or at elevated pressure. The tower can consist of

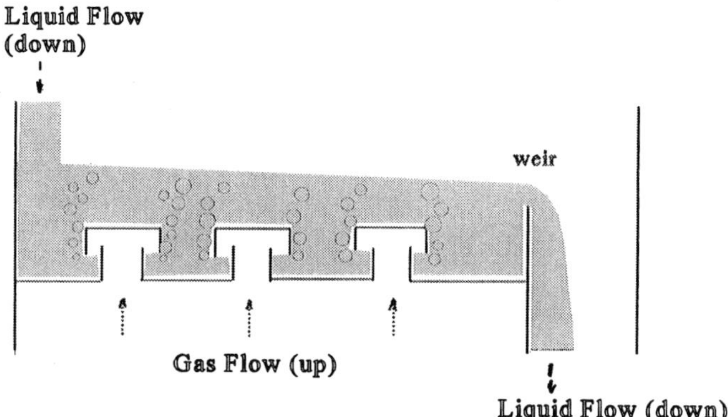

FIGURE 14 Bubble cap tray.

discrete units or be filled with a continuous packing material. The discrete units are often viewed as "stages," but they may not correspond exactly to theoretical stages. (See the discussion of stage efficiency given below.) Some of the older discrete units used in distillation towers are "bubble" trays, such as in Figure 14. Although there are many bubble cap towers in operation throughout the chemical industry, they are usually not being installed today. The principal problem with bubble cap towers is their cost, but some of the newer packing materials are also significantly more efficient (shorter stage heights, lower pressure drop, or both). Although high capital cost may offer no incentive for removing existing bubble cap systems, it severely affects the motivation for building new units. Thus newer tower internals started penetrating the market in new installations, but the performance of new packing materials can make it economical to replace the internals of some older distillation towers.

The systems that initially replaced bubble cap units for large towers were sieve plate units. These can resemble bubble cap trays, but the bubble cap is replaced by a plate with circular holes or slots which are much easier to fabricate. They must be sized to allow the vapor to flow upward through them without allowing liquid to flow down through them. This requires a knowledge of the gas and liquid flow rates, and such sieve trays are usually able to handle effectively only a significantly narrower range of flow rates than bubble cap trays. This problem was handled by incorporating lifts on the sieve tray openings that are forced open when there is sufficient vapor flow, but which are closed when there is no vapor flow or insufficient flow. This change increased the range of applicability of sieve trays with only a modest increase in the capital costs.

Packed towers have been used as long or longer than bubble cap units, but the packing material was usually a form of random packing, usually Raschig rings or Berl saddles. These packing materials can also be capital intensive, especially when they are made of relatively costly ceramic materials or metals. Nevertheless, the cost of packing materials decreases essentially linearly with the volume of the tower, and the cost of bubble cap and other tray type internals decreases more slowly with tower diameter (or volume). This made random packing more attractive for relatively small distillation systems. There have been significant advancements in the design of random packing materials, and the performance has improved significantly. The improvements usually involved incorporating more surface area into the packing material; this translates into more interfacial area and higher mass transfer rates per unit volume of tower. Some of the newer plastic packing materials also can be made relatively easily, and the capital costs of random packing materials have become more competitive. Many of the random packing material used in distillation towers are similar (or often identical) to those used in absorption and stripping towers, and they are described in slightly more detail in Chapter 3.

The most significant innovation in distillation tower design seems to be structured packing materials. These were discussed in Chapter 3, and those descriptions apply for distillation also. Because of the great expense devoted to distillation operations in the chemical and process industries, the largest motivation for developing improved packing materials, such as the structured packing materials, probably came from the need for improvements in distillation rather than in absorption/stripping. Structured packing materials are usually monoliths of metal or plastic constructed to fit within specific tower diameters. (The temperature of the operations can limit the use of some plastic materials.) The monoliths are usually constructed of layers of corrugated material, with each layer oriented at a specific angle from the layer above. The materials are also porous so liquid and gas can flow along the corrugations and pass through the corrugations. The capital cost of structured packing can be relatively high, perhaps two or three times the cost of sieve towers with the same diameter, but their additional cost can usually be justified because of their greater mass transfer performance and lower pressure drop.

Development of improved tray devices continues, and further improvements can be expected [4]. However, at the present time the most significant advances into new applications seems to be coming from the newer structured packing materials.

Equipment Size

Process design of distillation equipment usually involved determining the diameter and height of the distillation tower. As noted, distillation towers can be constructed of distinct units such as old bubble cap trays or any of the sieve tray type units. These units are usually viewed as "actual" stages. The number of actual stages may not correspond to the number of "theoretical" stages determined from stage calculations such as those described above. Thus the concept of stage efficiency has been developed. There are numerous ways to describe stage efficiencies, but for many systems it is adequate to define it as the number or stages determined by the performance (that is, the number of equilibrium stages) divided by the number of physical stage-like units (trays for a tray type tower). This definition is obviously simple and ignores the fact that the behavior of a tray could be quite different in different regions of the tower. Another approach is to define the stage efficiency as the ratio of the actual change in concentration from tray to tray to the change in concentration in a theoretical stage. This bases stage efficiency upon local conditions within the tower, that is, conditions at each stage. For packed distillation towers, the "efficiency" of the packing can be described by the "height equivalent to a theoretical stage." It is probably obvious that there is not a great deal of difference between the concept of stage efficiency and the height equivalent to a theoretical stage. Stage efficiency is a way to describe the number of mechanical stage heights that correspond to a theoretical stage.

Thus, the height of a distillation tower is likely to be approximately proportional to the number of theoretical stages needed for the separation. If the stage efficiency or the height equivalent to a theoretical stage were truly constant, this would be exactly true. However, as noted in the description of the mass-transfer-based "transfer unit" concept in Chapter 3, the number of stages and the number of transfer units required for a separation will be approximately equal or proportional to each other as long as the operating line and the equilibrium line are approximately parallel. Since the equilibrium curve is fixed, only the slope of the operating line is determined by the designer. Fortunately, the optimal (most economical) designs for distillation towers tend to make the equilibrium curve and operating line as close to parallel as practical by placing the feed point at the optimum location and by choosing the slopes of the operating lines to "parallel" the equilibrium curve in each section of the tower as closely as possible. This is usually close to the optimum slope for the operating line (or the optimum reflux ratio). It makes the contribution of each stage to the separation (the concentration change per stage) approximately the same or for all stages, and that is likely to be the optimal situation. In many cases, this corresponds to values of the reflux ratio that are 1.3 to 2 times the minimum reflux ratio.

The diameter of a distillation tower is determined by the vapor and liquid rates required. Since these rates are a function of the reflux ratio (or a function of the slope of the operating line), the tower diameter is determined by the reflux ratio as well as the throughput required for separation. Although it is possible for the liquid rate to have a significant effect on the capacity of the tower packing or trays (and thus the required tower diameter), it is very likely that the capacity of the tower internals or the required tower diameter will be determined largely by the vapor rate. This results partially from the great difference between the densities of the liquid and vapor in most distillation systems. The maximum flow capacity may be called the "flooding rates" because that is (are) the rate(s) at which the column can no longer separate the two phases, and entrainment of one phase in the other phase occurs. Flow rate capacities are usually available from manufacturers of tray units and packing materials, and those data should be used when they are available. Correlations are available in standard textbooks for the more generic packing material shapes, such as Raschig rings or Berl saddles, but there is less likely to be data available for newer proprietary packing materials. Some of the correlations are relatively old, and Perry's handbook [5] quotes the Sherwood and Hollaway correlation, which applies approximately to a number of shapes of random packing [6] and was first reported in 1938. Studies of the newer structured packing materials are becoming available, and their flow capacity is often very good.

Heights of theoretical stages and heights of transfer units or mass transfer coefficients for many standard packing materials are also available in standard textbooks and handbooks dealing with distillation, Perry's for instance. Often the data are reported for a particular distillation, and one has the option of seeking data for the system most similar to the one of current interest and minimizing extrapolation of the results. Often the dominant term in determining the height of a transfer unit or stage is the interfacial area per unit volume of tower, and this term often does not vary greatly with the systems used, providing the packing is wet with the same phase. As the flow rates increase and approach flooding condition, the performance can actually improve and then fall sharply as the flooding rate is reached. The mass transfer coefficients increase with flow rates, but usually at less than a linear rate. Thus the HTU or height of a stage usually decreases with the flow rate in the reference phase (the phase on which the mass transfer coefficient is based). However, the HTU based upon one phase can depend upon flow rates in the other phase. In estimating the HTU or height of a theoretical stage from data on a similar distillation system, it is important to compare systems that have a similar distribution of the mass transfer resistance in the liquid and vapor phases. It would not be very helpful to use information (data) from a distillation system where all of the mass transfer resistance was in the liquid phase if the resistance in your systems were principally in the vapor phase.

When using sieve trays or another tray system, the term equivilent to the HTU or height of a theoretical stage is the stage efficiency. There are several ways to express stage efficiency, but we have described the efficiency simply as the ratio of the number of theoretical stages required for a given change in concentration to the number of trays required. This is an overall efficiency, and one could use a local efficiency based upon the ratio of the change in the concentration in one phase in a tray to the change in concentration in a single stage. This is called the Murphree stage efficiency, and it can change from stage to stage. The Murphree local efficiency will change if the curvature in the equilibrium curve is very large over the region of interest.

The efficiency of a tray depends upon the geometry of the tray, the flow rates, and the physical properties of the two phases. A review of tray efficiency predictions was reported by Ognisty and Sakata [8]. There has been considerable work on improving the prediction of tray efficiencies for several years at the University of Texas [9], and this work should be considered.

MULTISTAGE STEAM STRIPPING

When the less volatile component is water and the concentration of the volatile component in the bottom product is very low, the vapor leaving the reboiler is essentially pure steam with very little volatile component. In such cases, there is little reason to expend the higher capital costs of heat transfer surfaces in a reboiler. Instead, steam can be injected into the bottom of the distillation column [10]. This makes the lower part of a conventional fractional distillation similar to a stripping column (see Chapter 3), and this type of operation is sometimes called "steam stripping."

If the steam fed to the bottom of the stripping column is saturated, the vapor flow rate in the stripper will be approximately the same as the steam rate; the rate will be exactly the same if the conditions of constant molal overflow hold. Then on a McCabe-Theile diagram, the operating line for the lower portion of a steam stripper would be straight, but rather than intersecting the diagonal at the bottom liquid product composition, as in Figure 12, it will intersect the "x-axis" at the feed composition since

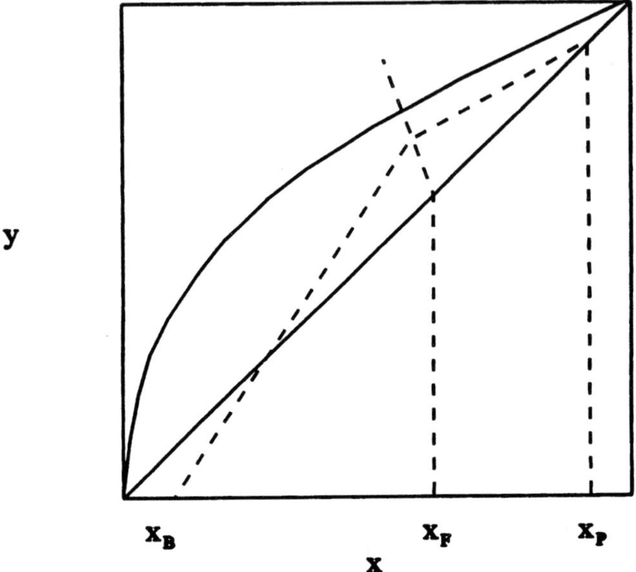

FIGURE 15 McCabe-Thiele calculations for steam stripping and distillation.

the concentration of the volatile component in the steam (bottom feed) is zero because there is none of the more volatile component in the steam fed to the tower (Figure 15).

If the steam is supersaturated, it will vaporize some of the liquid in the bottom stage, and the vapor rate in the stripping section will be greater than the steam rate. If the constant molal overflow approximation is sufficiently accurate, it is relatively easy to calculate the vapor rate in the stripping section.

Once the operating line for the lower (stripping) section of the tower is established, the required stages in the tower are easily calculated, as illustrated for other distillation columns.

EXAMPLES AND DISCUSSION OF APPLICATIONS IN ENVIRONMENTAL AND WASTE OPERATIONS

Distillation is more likely to be selected for removing relatively large concentrations of contaminants or contaminants with volatiles that are relatively close to that of water. However, one must remember that distillation, like so many other separation methods, can be used within a processing facility to separate components and prevent them from ever reaching discharge streams, and these may not be easily recognized as "waste or environmental" operations. Of course, if the volatility of the contaminant is too close to that of water, distillation may be difficult. In such cases, adsorption or other separations that do not rely upon differences in volatility can be considered.

Thus distillation is most likely to be used when there is not sufficient relative volatility to use air stripping or single stage steam stripping or evaporation without vaporizing too much water, but when there is still sufficient relative volatility to justify using distillation rather than other methods such as adsorption. Generally, one would want a relative volatility of approximately 1.3 or greater, but there is no absolute value for the application of distillation because it would have to depend upon the alternative contaminant removal processes, such as membrane processes. If the concentration of contaminant in the product water must be very low, it will often be preferable to use steam feed to the bottom of the column rather than a reboiler (steam stripping). This is likely to be the case for many wastewater treatment problems.

Distillation and steam stripping will often compete with air stripping, which usually becomes more competitive as the concentration of the contaminant decreases and or as the volatility of the contaminant increases. Bravo [10] suggests that the distillation or steam stripping will usually be more cost effective at a feed concentration above 0.1 wt% of volatile organic contaminant, but can be effective at even lower concentrations for some systems. The cost effectiveness of steam stripping relative to air stripping depends upon the scale of the operation as well as the properties of the volatile organic to be removed. Air stripping is usually less capital intensive and is more likely to be the choice for smaller systems with lower throughput.

Azeotrope formations can be a problem in some distillation processes because they can limit the purity of the products produced (purity of the water or contaminant product). However, azeotropes may not prevent the use of distillation for many environmental applications, even if they are formed. Because most environmental applications are likely to be concerned with dilute systems, it is less likely (but not impossible) that an azeotrope will be formed between the feed composition and pure water. It is more likely that an azeotrope (if formed) would lie between the feed composition and pure contaminant. An azeotrope in that region is less likely to hinder the application of distillation greatly. For instance, if an azeotrope was formed with 50% contaminant and 50% water, it can only mean that the contaminant could not be recovered at a concentration greater than 50%, but that would still be a highly contaminated substance and perhaps a concentration that could still be fed to an incinerator or other device for destruction or fixation of the contaminant. When an azeotrope is formed, higher contaminant concentrations can be reached by a second fractional distillation tower operating at a different pressure/temperature where there is no azeotrope or where the composition of the azeotrope is significantly different. One can also add a third component that eliminates the azeotrope or moves the composition significantly.

Note that it may not be optimum to use either distillation/steam stripping or air stripping to reduce the concentration of highly toxic volatile components to the extremely low concentrations that may be required. In some cases, adsorption may be the most practical method for reaching the final discharge limits. This results because stage heights for adsorption can be relative short, and it is relatively economical to add stages to obtain better removal. In such cases, one should consider the possible use of a multiple step removal scheme where distillation/steam stripping (or air stripping) removes the bulk of the volatile contaminate and another method, such as adsorption, is used to remove the last traces of contaminant before the water is discharged or reused. Adsorption often becomes more economical for treating dilute solutions because the regeneration cycle (time between regeneration) decreases with concentration, but there is no equivalent decrease in the cost of operating most other separation systems. An economic evaluation will be required to determine what concentration to reach by the distillation/steam stripping operation.

The low concentrations of contaminants in most waste and environmental streams were credited with being a factor limiting the application of distillation in environmental problems. However, one should note that some separation methods concentrate the contaminants, often by large factors, but they still do not produce a stream of pure contaminant that could be reused or stored at a minimum volume. Distillation could be the most effective way to separate the concentrated contaminant from the other components. Air stripping and adsorption regeneration can produce relatively concentrated organic contaminants (VOCs), but the product may still contain considerable water or other bulk components. However, distillation is less likely to be needed if condensation of the products from gas stripping or adsorbent regeneration results in two liquid phases, with one phase being essentially pure organic contaminant, or when the contaminant is to be destroyed rather than reused, and the presence of water or other bulk components is not important.

REFERENCES

1. Walas, S. M. *Phase Equilibrium in Chemical Engineering*. Butterworths, Boston (1995).
2. Palmer, D. A. *Handbook of Applied Thermodynamics*. CRC Press, Boca Raton, FL (1982).
3. Skogestad, S., et al. *AIChE J. 43*, 971 (1997).
4. Sasson, R. and R. Pate. *Oil Gas J.* Aug. (1993).
5. Perry, J. H. *Chemical Engineer's Handbook*, 4th ed. McGraw-Hill, New York (1963), pp. 18–28.
6. Sherwood and Hollaway. *Ind. Eng. Chem. 30*, 768 (1938).
7. Humphrey, J. L., et al. *Separation Technologies: Advances and Priorities.* U.S. DOE Report DOE/ID/12920-1 (Feb. 1991).
8. Ognisty and Sakata. *Chem. Eng. Prog. 83*, 60 (1987).
9. Prado, M. and J. R. Fair. *I&EC Res. 29*, 449 (1990).
10. Bravo, Jose. *Chem. Eng. Prog.* Dec., 54 (1994).

8
Surface Filters

As noted, surface filters will be described as filters where the particulates are collected outside a filter media, that is, on a fiber cloth, a metal frit, or other media that has a defined surface. Sometimes this type of filtration is called "cake filtration" because of the formation of a "cake" of filtered particles on the filter surface. The other major type of filter discussed in this book will be called deep bed filters, where the particles are collected within the filter media, and these will be discussed in another section. Although this definition appears to divide all of the many types of filters into two groups, the division is not so exact. There are sometimes "deep bed" properties involved in many surface filters. Small particles can penetrate the fibers of cloth and other surface filter media and become trapped by deep bed processes, and the filter cake accumulated on surface filters can aid in removal of very fine particles by deep bed mechanisms. This can cause the behavior of very small particles in surface filters to change as significant filter cake is accumulated. Nevertheless, the definitions of surface and deep bed filters will be based upon the principal mechanism for particle removal, with no requirement that this be the only mechanism.

Surface filters are particularly effective and practical for removing moderately large particles (significantly greater than 10 micrometers). Surface filters are also more effective for removing granular-shaped particles that pack in a relatively porous mass and less effective for removal of soft or fibrous particles that are more likely to pack into a dense mass or compress into a dense mass from applied pressure. That is, surface filters are most effective for removing those particles that form relatively permeable filter cakes that will allow useful filtration rates to continue even after considerable cake has accumulated. As noted, surface filters also can provide a concentrated collection of removed particles, and thus they are often preferred when recovery of the solids is a major objective of the filtration. Removed solids can be recovered from many deep bed filters,

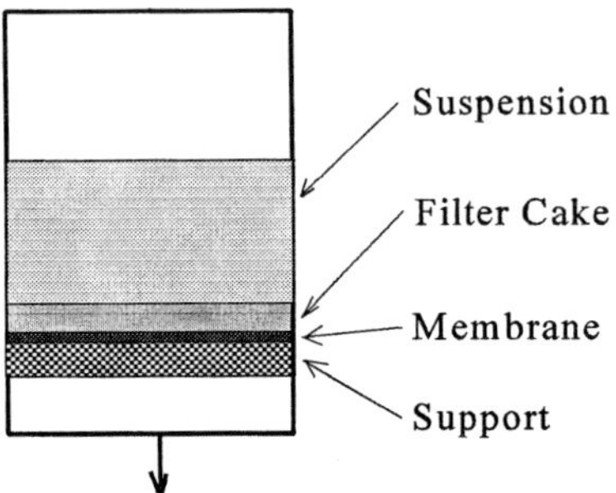

FIGURE 1 Surface filter with "caking." The filter medium (often supported on a screen or other device) under a cake of filtered particles that is under the suspension being filtered.

such as packed beds of granular materials, but they are seldom removed in the high concentrations that can be achieved with surface filters.

A schematic of a simple filter is shown in Figure 1. A filter may contain a support such as a screen or a perforated plate which supports the filter medium. The filter medium is often a cloth of cotton, glass, or ceramic, or a porous frit (of glass, ceramic, or metal), but could also be a felt (unwoven fibers) such as paper. Since the principal mode of filtration is intended to be exclusion of particles from the filter medium, it is anticipated that the openings in the medium will be smaller than most of the particles to be removed. However, a "precoat" can be placed on the filter medium that has a gradation of particle sizes to protect the medium from the smaller particles, that is, to prevent the smallest particles from passing through, or accumulating within, the filter medium. The gradation may place only the largest precoat particles nearest the filter medium, and the largest particles in the precoat will not penetrate the media. However, the upper portions of the precoat (farthest from the filter surface) may contain smaller precoat particles that will be able to retain all or most of the smallest particles to be filtered. When there is a large range of particle sizes, the retention of the smallest particles within the filter cake or precoat is likely to be by a deep bed mechanism.

As filtration proceeds, a cake of solids will accumulate on the filter medium (or on the precoat). The amount of accumulation will be approximately proportional to the volume of fluid that has been filtered per unit area of filter, since most filters remove almost all of the particles. The flow resistance of the filter will increase as cake accumulates, and the pressure drop required to maintain the desired throughput will increase. It will eventually become necessary to remove the cake to achieve acceptably high filtration rates with acceptable applied pressure. All of the cake can be removed along with the precoat, or the cake can be scraped from the medium with a sharp blade, leaving most of the precoat on the filter. Filter cakes from gas filters can often be removed, at least partially, by shaking, vibration, or sudden burst of gas in the reverse direction.

When filtering suspension in water or other liquids, it is sometimes desirable to remove as much liquid as practical from the solids before removing the cake from the filter medium. This can be done by "compressing" the cake with a press (a batch operation) or continuous rollers (a continuous operation). This is often called "expressing" the cake. The effectiveness of compression in removing excess water depends upon the compressibility of the filter cake, a property of the particles in the filter cake; some filter cakes can have little compressibility, so additional pressure can remove little more water.

TRANSIENT BEHAVIOR OF BATCH SURFACE FILTERS

Because the filter cake thickness increases during operation of a surface filter, the pressure drop and/or the throughput change with time, and the operations are transient in nature, not steady-state. The equations used to describe the effects of filter cake accumulation on surface filters are often called "design equations," but the important parameters usually cannot be determined without resorting to experiments with the materials of interest and with conditions near those to be used in the commercial equipment. Nevertheless, the equations are useful because they give a realistic basis for correlating data on a given filter medium and suspension, and the experimental data can be interpolated and even extrapolated to find optimum or preferred filter operating conditions.

Flow through both the filter medium and the filter cake is usually in the viscous region, and the flow rate through the filter is then assumed to be proportional to the pressure gradient [1]. The flow rate can be written as

$$\frac{dV}{dt} = \frac{\Delta P}{\mu(aW + r)} \tag{1}$$

where V is the fluid velocity through the filter (volumetric flow rate divided by the cross-sectional area), t is the time of filtration, ΔP is the pressure drop across the filter, μ is the fluid viscosity, a is the flow resistance of the filter cake per mass of solids removed, W is the accumulated solids on the filter, and r is the resistance of the filter medium and precoat.

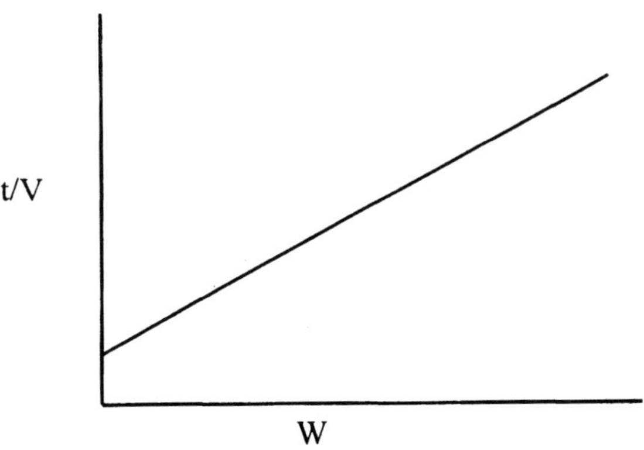

FIGURE 2 Filtration plot showing the time divided by the volume of fluid passed through the filter vs. the mass of particles accumulated in the filter cake illustrating that the filtration rate (V/t) declines linearly with the filter cake buildup with constant applied pressure. The slope and intercept can be used to evaluate the flow resistance through the filter cake and the filter media respectively.

If the filter removes essentially all of the solids, W will be approximately the product of volume of fluid filtered times the concentration of solids in the fluid. The depth of filter cake accumulated on the filter medium is proportional to W.

Two obvious ways to operate the filter would be with constant pressure (and thus decreasing filtration rates as the filter cake builds up) or with constant fluid flow rates (and thus increasing pressure drop through the filter as the filter cake thickness increases). If the filter is operated with constant pressure, Equation (1) is integrated to

$$\frac{t}{V} = \frac{\mu a W}{\Delta P} + \frac{\mu r}{\Delta P} = K_c W + C \tag{2}$$

The lumped parameters K_c and C are shown because for a given filtration system, viscosity, cake permeability, and filter medium permeability are not likely to be varied. A filtration test can produce data on volume of fluid filtered as a function of time, and a plot of t/V versus V (which is proportional to W). The resulting curve can be used to extrapolate to different filtration times (Figure 2). Thus, such a plot can describe a filtration system (medium and a given suspension) performance over a range of filtration times and allow the user to decide when to replace or "clean" the filter.

It would appear that the curve could also be used to extrapolate filtration results to different pressures. That would assume that a, or K_c, is constant and independent of pressure. However, in many cases, a or K_c can vary considerably with applied pressure. The flow resistance, r, of the precoat and filter medium can depend upon the applied pressure. This variation can be caused by "compression" of the filter cake [2]. It is common to describe the variation in a or K_c with pressure by the relation

$$a = a'P^n \qquad \text{or} \qquad K_c = K_c'P^n \tag{3}$$

The exponent n can vary from near zero to approximately 0.8. The lower exponents correspond to relatively incompressible cakes such as those formed from hard granular materials such as crystalline solids. In those cases, a or K_c will be approximately constant over a modest range of filtration pressures. The higher exponents are found for cakes of polymeric or gel-like materials which are often highly compressible. As the filter cake is compressed by higher pressure gradients, the bed compacts, becomes less porous, and develops a higher flow resistance. In some cases, a or K_c will be almost proportional to P, and the filtration rate, dV/dt, will not vary greatly with applied pressure.

There are numerous potential sources of compressibility in filter cakes. In some cases, the particles themselves can become compressed. In other cases, the particles can be rearranged physically by the applied pressure gradient. This is likely to be especially important when the particles have unusual shapes that would normally settle into cakes with high void fractions. Rigid spheres or granules are not as likely to form high void fraction cakes that can be compressed greatly. The electrical charge on the settled (filtered) particles can even be the cause of some compressibility in the cake [3]. The particles may remain separated in the cake because of their electrical charge, but sufficient applied pressure may bring the particles closer together and thus compress the cake. Once a cake has been compressed and the void fraction within the cake has been reduced, there is essentially no likelihood that the void fraction will be increased to its original value once the pressure is released. In essentially all cases, compression of filter cakes is irreversible with no significant recovery of the void fraction or permeability when the pressure is released.

When a filter is operated with a constant pressure difference, the highest pressure gradient occurs during the initial cake formation (first part of the filtration cycle), and the most compressed portion of the filter cake is that portion closest to the filter medium. This portion of the cake is most affected by the magnitude of the applied pressure. As the filter cake thickness increases, the pressure gradient within the cake decreases, so the compression of the cake is probably not uniform. That is, the portion of the cake nearest the filter medium is likely to be more highly compressed than the portion near the top of the cake (farthest from the filter medium), but the author is not aware of any experiments that demonstrate this. Equation (3) is essentially an empirical expression, but it has been useful in describing filter cakes of many materials. It describes the cake in terms of a compressibility, as if the entire cake were subject to the same compression. The equation is effective in describing the behavior of many filter cakes, but cake compression is usually highly inelastic. If the applied pressure is decreased, filter cakes usually will not return to the lower flow resistances they normally have with lower applied pressure.

Although compression of the filter cake has been a reasonable explanation for the nonlinear dependence of filtration rate on applied pressure, there are alternative explanations or phenomena that contribute to changes in filter cakes under different applied pressures. For instance, additional penetration of particles into the precoat or filter medium could play a role in this behavior. If more small particles enter the precoat or filter medium, the flow resistance of the precoat or filter medium can be reduced. When compressibility exponents are used, one should usually consider them to be empirical constants and not place too much emphasis on the exact mechanisms involved.

If variations in applied pressure are to be considered for filter operations, it will be necessary to evaluate filter performance at several pressures and determine a value for the exponent as well as the flow resistance at one pressure. The value of the exponent can be evaluated from plots of the measured values of $\log a$ or $\log K_c$ versus $\log P$ (Figure 3). Very often such plots give approximately straight lines; that is, they give an approximately constant value for the exponent n. Remember that the pressure gradient within the cake will change as the filter cake grows, and the initial portion of the cake formed nearest the filter medium will be formed with the greatest pressure gradient. This suggests that the portion of the cake nearest the filter media is likely to be the most highly compressed. Also, if compression is a transient phenomenon and compression continues with time, the greater "age" of the initial portion of the cake could contribute to its greater compression.

The constant r, or C, obtained from the intercept of the plot of t/V versus V, represents the resistance of the filter medium and any precoat applied to the medium. Thus it would appear that for cases with no precoat, the value of the intercept should correspond to the flow resistance measured for the filter medium alone or that r should be a property of the filter media and not dependent on the solids being filtered. However, one cannot measure the flow resistance of the filter media independently and obtain the same value for r or C as by extrapolating filtration data to the initial time when there was no filter cake. In many cases, values for r (or C) measured with a filter cake by extrapolating t/V versus W [Equa-

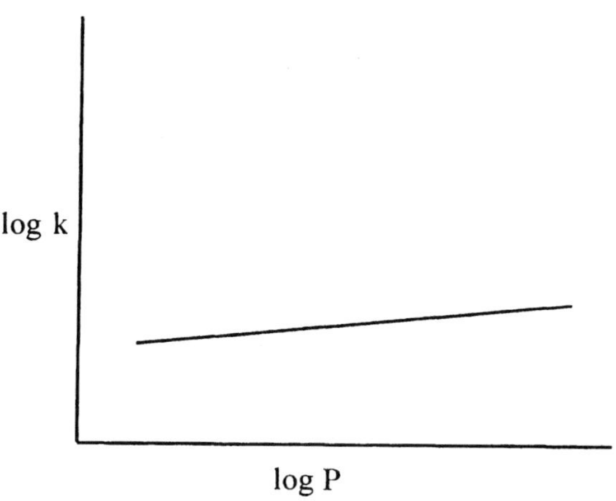

FIGURE 3 Plot of the permeability of the filter cake as a function of applied pressure across the filter. The permeability decreases (the flow resistance increases) with the applied pressure; the dependence is often approximated by a straight line on a log-log plot.

tion (2)] to the point where *W* is zero gives a significantly higher value for *r* than direct measurements of the flow resistance without a filter cake.

This probably indicates that these parameters include effects from penetration of the filter media by some of the particles, mostly penetration by the smallest particles being filtered. Penetration of the filter media probably occurs principally near the beginning of the filtration period before significant filter cake is formed which can trap fines. The "bleeding" of fine particles through surface filters has been observed to occur principally in the early stages of operations before much filter cake has accumulated. This also illustrates that deep bed effects in the filter cake can be important in surface filtration.

FILTER MEDIA

The various useful filter media can be classified in terms of materials used or the physical form of the medium. The medium could be in the form of a woven cloth, a nonwoven cloth, a frit, or a foam. The material could be cotton or other textile fibers, metals, glasses, or ceramic. The choice of material to be used depends upon the chemical properties of the fluid to be filtered, the temperature, and, sometimes, the strength needed in the medium. A fabric filter (woven or nonwoven) may be supported by a screen or other rigid porous structure, or the pressure drop across the filter could be supported by the filter medium itself.

Woven materials come in a variety of weaves, thread size, and weave tightness. Obviously the use of smaller threads and tighter weaves decreases the size of the openings in the media and thus promotes the retention of smaller particles. It also increases flow resistance. It is necessary to minimize "blinding" of the media by filtered particles that become embedded or stuck in the medium. This is likely to occur if the openings in the medium are only slightly smaller than the particles to be filtered. However, since most materials to be filtered do not consist of uniformly sized particles, there could be some fines that are smaller than the openings in the weave, perhaps considerably smaller. Note also that it is possible for small particles to penetrate the thread itself, that is, between the fibers in the threads of the woven filter medium. With a medium woven from threads which are also porous (made of numerous smaller fibers) there will be some flow through the threads as well as through openings between the threads. If the weave is very "tight" (closely spaced threads), a substantial fraction of the flow may be "through" the threads. Accumulation of particles within the threads is another reason why the flow resistance of the filter media measured without a filter cake may not be the same as that measured with the cake.

Woven filter media can also be made from "single fiber" threads, usually of a polymer, metal, or glass. In such cases, entrapment of particles within the threads will not occur, and particle entrapment will be limited to entrapment within the weave, not within the threads. Woven metal and glass filter media are more likely to be made from single fiber threads, but these can also be made from multiple fiber threads. However, cotton or wool media will invariably be made from multi-fiber threads.

Nonwoven filter media are simply pressed fibers. Perhaps the most familiar such filter media is filter paper, which is used frequently in chemical laboratories. However, nonwoven media can be prepared from essentially any fiber. Obviously, randomly oriented fibers will give a wider range of opening sizes than woven media. With such variation in pore size, it is usually necessary for nonwoven media to be many fibers thick to ensure that few particles below the desired filtration size are able to leak (or "bleed") through the media. However, nonwoven media need not be excessively thicker than woven media since the fiber thickness should be compared not with the threads in woven media but with the individual fibers used in the threads. Nonwoven media usually have relatively small pores and are thus likely to be used to filter very small particles.

Frits of metal, ceramic, or glass are often used in small filters, but they may be too expensive for frequent use in some large filters. Frits are more likely to be used for difficult filtrations under extreme conditions like high temperatures or corrosive conditions. They are usually prepared by sintering small particles under temperature and pressure. It is preferable to use relatively uniform particles in preparing frits, and particle preparation as well as frit formation can be expensive relative to the cost of manufacturing many woven materials. By using highly uniform particles for sintering into the frit, very narrow pore size distributions can be achieved.

CLEANING OF SURFACE FILTERS

One of the principal advantages of surface filters is the ability to remove the particles and reuse the filter medium. In some cases, the filter medium may be sacrificed and used only once. This is likely to be the case with paper filters. However, multiple uses of the medium as well as other parts of the filter are more likely, especially when more expensive filter cloths are used. As filter cake accumulates on the filter medium, the pressure drop increases, and it eventually becomes necessary to remove the filter cake and clean the filter medium if it is to be reused.

Cleaning the filter could occur when the filter is taken off-line. In some cases these off-line time periods can be very short, and cleaning could consist of only momentary efforts to remove loose filter cake on the medium. Of course, it is usually desirable to keep the filter in operation as long as practical by using only short momentary particle removal steps and then performing more complete cleaning only as required.

An example of momentary cleaning is illustrated in the operation of "bag" filters for gas streams (Figure 4). A bag filter often consists of a cylinder or "sock" that is open at both ends. The bag is made of a filter cloth. The gas to be filtered may flow from outside to inside or from inside to outside. The bags are often mounted vertically so that solids that are not held strongly on the filter media will be able to fall from the bag into collection chutes below the bags. Sock type filters are likely to be chosen when large gas flow rates and high solids contents are involved, and on-line cleaning of the filters is essential for economical operation of the system. Solids can also be encouraged to fall into the solids collection chutes by periodic shaking of the bags or by sudden bursts of back flow (flow in the reverse direction) through the bags. Shaking can be done while the filter is operating.

Obviously reverse flow cannot occur while the filter is operating, but with a filtration system consisting of many bags, the bursts of back flow can be imposed momentarily on only one or a few of the bags at a

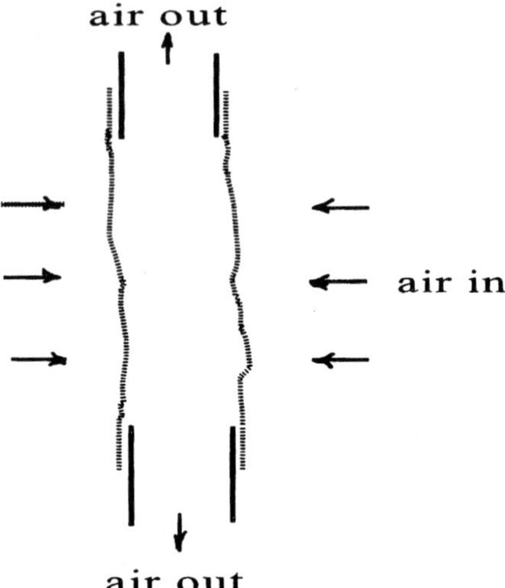

FIGURE 4 Sock filter with flow coming from the outside of the sock. Usually used in high throughput gaseous systems, these filters can be cleaned in place by short pulses of back flow or mechanical shaking.

time, leaving all other bags operating. This in effect appears to make the overall filtration system operate continuously. With bag filters, all particles removed by very short vibration or bursts of back flow may not have time to escape to the collection chute before the gas flow brings them back to the surface. However, some particles will always be reaching the collection chute, and other particles will be working their way toward the bottom of the bag and thus toward the chute. It often appears that particles that "reattach" to filter bags are not held firmly and probably reach the collection chutes quickly in subsequent cleaning cycles.

Cleaning by pulses of back flow gas involves more than simple back flow through the filter cake. These pulses can be highly effective with bag (sock) type filters even when very little back flow of gas is used. The effectiveness of cake removal probably results at least partially from acceleration of the medium (the sock) when the pulse is applied and a sudden deceleration when the medium reaches its maximum extension (constrained expansion) [4]. These limits of medium movement may result from strength of the medium fabric itself or from protective wire cages. Repeated flexing of a cloth filter medium from pulsed back flow can cause some particles to be redeposited deeper within the filter cloth. This results in a small seepage of particles through the filter. All flexible

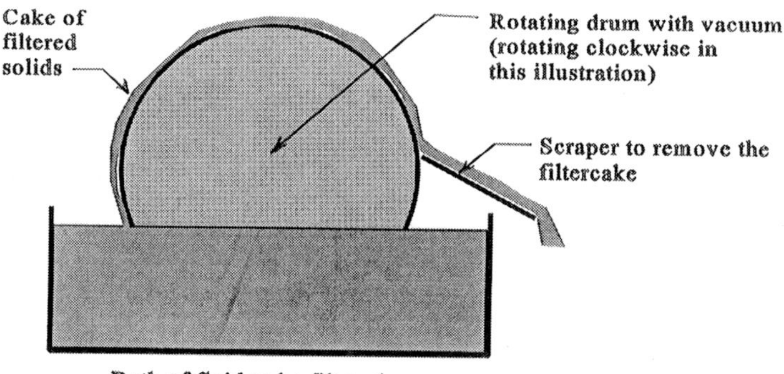

FIGURE 5 Rotary vacuum filter.

cloth filters cleaned by scraping, reversed flow, or pulses suffer some seepage, but this problem is usually more severe in filters cleaned by pulsing flow.

The solid filter cake can also be removed by physical blades or knives that strip all or part of the filter cake from the medium. This is often used in liquid filtration where the filter cake may be structurally strong and not prone to "falling" from the filter medium. In some continuous filters, the medium could be a moving belt on the surface of a rotating drum. As the medium becomes loaded with filter cake, it can be rotated out of the liquid, and the filter cake can be scraped from the medium before it is rotated back into the liquid (Figure 5). A drum filter is often a "vacuum" filter. That means a vacuum is pulled on the inside of the drum, and the difference between the outside (atmospheric) pressure and the lower pressure inside the drum is the driving force for filtration. The filtered liquid then is removed from inside the drum.

If the goal is only to concentrate the slurry of particles (not to isolate the particles), a cross-flow filter can be used to limit the buildup of filter cake and thus reduce the pressure drop. In cross-flow filtration, the filtration medium is the surface of a flow channel, and the principal flow direction is not toward the filter medium, but across its surface and along the channel (Figure 6). The channel could be cylindrical, and the filter medium would be on the surface of the cylinder; or the channel could be rectangular with the filter medium on the flat plates that make up the major axis of the channel. The distinguishing feature of cross-flow filtration is the high flow rate across the filter surface, a flow rate many times that through the filter media. Cross-flow filtration uses the fluid shear along the wall to remove particles and prevent accumulation of fil-

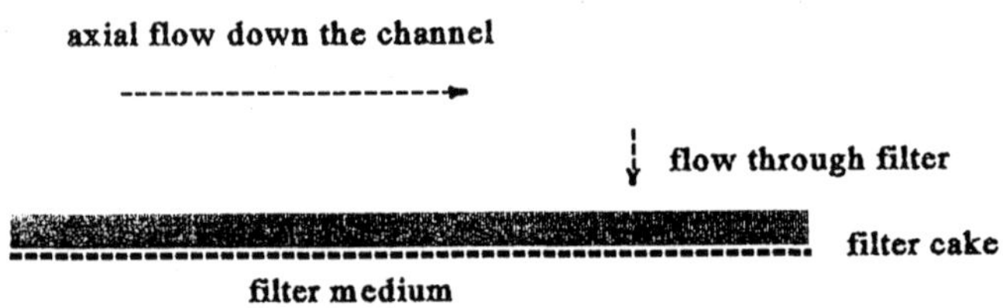

FIGURE 6 Flow through the filter forcing particles toward the filter surface and cross-flow eroding the filter cake and limiting its buildup.

ter cake on the filter medium. Ideally, a cross-flow filter would have no filter cake buildup, but the cross-flow can be beneficial even by limiting the buildup of filter cake. The cake thickness can approach a steady state where filter cake is sheared off as rapidly as it is formed, and the steady-state cake thickness may be acceptable for steady operations. The eventual (approximately) steady-state level will depend upon the cross-flow velocity. Higher velocities result in less accumulation on the filter medium but higher pumping costs for the cross-flow circulation. However, the velocity component toward the filter (usually much smaller than the velocity across the filter surface) continues to play an important role in filter cake buildup. This velocity is simply the filtration rate itself. Higher filtration rates caused by more porous filter media or higher filtration pressures will carry particles more rapidly toward the filter surface, and this will result in greater buildup of filter cake. Optimization of the flow rates in both directions (across the filter face and through the filter) in cross-flow filters is required.

Frits are particularly well suited for some cross-flow filtration operations since they can be made with strong and rigid surfaces. When the particles to be filtered carry a net electrical charge (or when a net electrical charge can be placed on the particles), application of a potential on the filter medium repels the particles (same sign as the charge on the particles) and has a similar effect as cross-flow and reduces the buildup of filter cake on the medium. Minimization of filter cake thickness by cross-flow or applied electric potential reduces the flow resistance of the filter and enhances the filtration rate.

COAGULATION AND FLOCCULATION

Coagulation and flocculation are often used interchangeably. They both involve adding a component to the suspension to enhance the filtration rate. We will take the more common approach of calling the addition of an inorganic salt to reduce the thickness of the electric double layer "coagulation" and call the addition of an organic polymer "flocculation." Actually, this is not a simple differentiation in the two terms because similar phenomena can be involved in both operations. This discussion provides only a qualitative description of the phenomena because that is likely to be all that is required for most people involved in selecting or even designing separation systems for waste and environmental facilities. These topics apply only to filtration of particles from liquids, principally from water. There are generally no equivalent phenomena for use in filtering particles from gases.

Although coagulation and flocculation may be more important in sedimentation because they improve sedimentation rates greatly, they can also be important in filtration. Filter cakes of agglomerated particles will generally be far more porous than cakes of small individual particles, and the concentration of ultrafine particles that slow filtration rates can be reduced greatly by coagulation. Larger agglomerated particles can also reduce the need for very fine openings in the filter medium and increase the filtration rate further.

When the filter cake is to be dewatered, there can be some adverse effects of agglomeration, the inclusion of more water in the filter cake. This is discussed briefly in Chapter 10. Coagulants can add slightly to the mass and volume of solids in the filter cake, but they can have greater effects by increasing the water content of the cake. Remember that the objective of coagulation was to create larger voids between the agglomerates so the flow resistance of the cake will be less. These voids contain water, and if the cake is to be dewatered, more water will have to be removed. In many cases, cakes with coagulated particles can be highly compressible, and removal of the extra water mechanically can be quite easy. However, some cakes of agglomerates can be difficult to compress. When heat is use to dewater filter cakes, all of the water in the cake adds to the energy required for drying. There may be a need to compromise between the coagulants that enhance the sedimentation and filtration rates the most and the coagulants that allow relatively inexpensive dewatering of the filter cakes [6].

As noted earlier, particles that resist agglomeration usually carry electric charges that prevent the particles from coming in contact and thus agglomerating. The electrical charges can result from properties of the particles (such as the presence of OH groups on the surfaces of many oxide particles) or from trace quantities of materials adsorbed on the particle surfaces (especially adsorption of ions). Particles in many surface waters, and even in many groundwaters flowing near the surface, may be

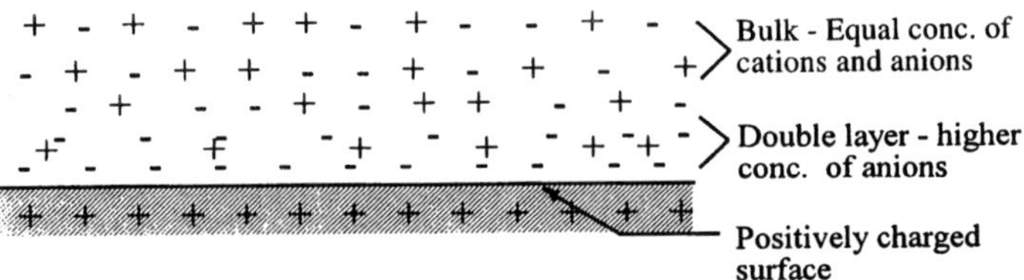

FIGURE 7 Double layer of ions at a particle surface illustrating how the thickness of the region affected by net charges on the particles is decreased as the concentration of ions in the solution increases.

stabilized by adsorbed humic acids or other organic materials [7]. The electrical charges can result from properties of the particle material itself, such as partially ionized OH groups mentioned earlier on many oxide surfaces, or from trace quantities of ions that become associated with the partially ionized OH groups. Coagulation often works by reducing the effects of these surface charges.

To understand how coagulation works, it is first necessary to examine how a surface charge on particles affects an aqueous solution around the particle. All electrical charges on the surface are neutralized by ions in the solution, but these ions are free to migrate in the solution and are not bound to remain on the surface; they must only remain in the vicinity of the charged surface (Figure 7). Diffusion forces are balanced with the electrostatic forces, and the neutralizing charge is spread over the region a short distance from the surface. When particles approach each other within this region, they no longer see neutralized charges since some of the neutralizing charges are beyond the distance separating the approaching particles. This causes the particles to be repelled, but if they could approach close enough, the strong, but short distance van der Waals forces would be able to take over and bring the particles together.

The thickness of the diffused charge around a charged surface is affected by the electrolyte concentration in the solution. Higher concentrations of electrolyte (higher ionic strength) bring this diffuse layer closer to the surface, and eventually it will be so thin that more particles will approach close enough for van der Waals forces to exert their attraction and bring the particles into contact. This is one way to cause coagulation.

Coagulation can also be initiated by changing the pH of aqueous suspensions. As noted earlier, some particles develop surface charges because molecules on their surface can ionize. A common behavior of metal oxides is hydration and the formation of hydroxyl groups at the surface. In acid, and often even in neutral solutions, the hydroxyl group can ion-

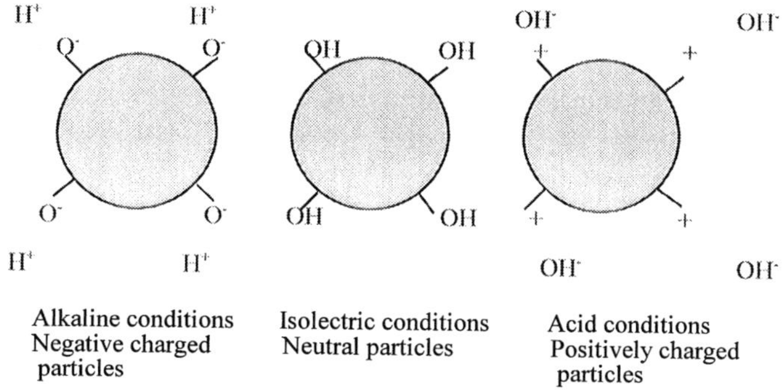

FIGURE 8 How hydroxide groups on many oxide surfaces can be negatively charged in alkaline solutions and positively charged in acid solutions. There is an immediate pH where there is zero surface charge.

ize and leave the surface or hydrogen ions can adsorb on the surface. The particle surface can then develop a net positive charge. Inversely, in basic solutions (high pH) hydroxyl ions may adsorb on the surface and give the surface a negative charge (Figure 8). At an intermediate pH, there may be no net electrical charge, and no repulsive forces to keep the particles separated. Thus pH adjustment to the value where there is no surface can be used to stimulate coagulation. The pH at which the charge on the surface becomes zero depends upon the material involved. For many common materials of interest, the neutral point may be at pHs greater than 7 or greater than the pH of the water stream to be filtered. Thus lime or another base can be helpful in stimulating coagulation; often only small additions are needed.

PRECOATS FOR FILTERS

When very small particles have to be filtered, it is often desirable or even necessary to install a "precoat" of the filter media. The precoat is essentially a filter cake of a special material that is added to the filter medium before the desired filtration operation begins. The precoat is prepared from a porous material that is easily retained by the filter medium. When the filtration operation begins, the precoat will act as a deep bed filter for removal of the smallest fraction of the suspended particles, and the larger particles will be removed from the surface of the precoat.

A precoat is likely to be necessary if the suspension to be filtered contains a significant fraction of fines, or very small particles that would normally pass through the filter medium. Often a filter will retain the fine fraction of suspended particles after a sufficient filter cake is established on the filter, but the fine fraction will "bleed" through the filter during the initial operational period. The use of a prefilter eliminates or reduces this initial bleed period of operation. A precoat may also be needed if the porosity of the filter cake would normally be too low to be practical. These are cases where the many of the particles being filtered are too small to allow a suitable flow rate through a filter cake.

The materials used in the precoat need to be porous, compatible with the fluids of interest, and economical. Diatomaceous earth is probably the most commonly used precoat material. It is obtained from deposits of skeletal remains of early organisms grown in lakes and/or oceans. Diatomaceous earth retains the structure of the early creatures but is essentially a silicate material. The higher quality diatomaceous earths have usually been processed to remove contaminants. The other most commonly used precoat material is perlite, another silicate material manufactured from volcanic ash. Other precoat materials may be used when special fluids are being filtered that are chemically incompatible with these more common silicate materials. Carbon or cellulose is sometimes used. Asbestos has been commonly used, but its use is likely to decline as the difficulty and expense of handling asbestos increases.

The filter cake may not be removed completely from the precoat, so at least some precoat material will usually be present with the solid filter product. If the filtered material is hazardous, the precoat will add at least somewhat to its volume unless it is practical to separate the two materials subsequently. If the solid product is valuable enough to be recovered, the precoat could be a contaminant that would have to be separated, usually along with other particulate contaminants. Removal of filter cakes and the precoat can involve use of scrapers or knives which mechanically take the solids from the filter medium or back flow of fluid that entrain the solids from the medium. In some operations, a relatively thick precoat is installed, and after each filtration cycle (that is, after each buildup of filter cake to its maximum desired thickness), a "knife" cuts the filter cake from the filter media at a position just below the interface with the precoat. Thus a minimal amount of precoat is removed with each cycle, and one precoat application can be used for several filtration cycles. Of course, the extraction knife must be advanced further toward the filter medium with each cycle to remove additional precoat. The precoat then becomes thinner with each cycle, and a new precoat will eventually have to be formed. The use of a relatively thick initial precoat will usually be acceptable if the principal pressure drop is through the filter cake, usually because of the small size of the particles being filtered, and addition of sufficient precoat for several filtration cycles may not add greatly to the overall pressure drop, at least not significantly to the pressure drop that developed after a significant filter cake has accumulated on the precoat.

FILTER AIDS

When a suspension forms an extremely nonporous filter cake, it may be necessary to add a filter aid to increase the cake porosity. This can occur when the cake is a slime of polymers or other long molecules or when the suspension consists of only very small particles. In these cases, a precoat may prevent initial bleeding of particles through the filter medium, but a precoat will not prevent a rapid buildup of pressure drop after relatively short filtration times. Although many suspensions contain a significant fraction of very small particles (usually those with diameters of a few micrometers or less), it is not always necessary to use a filter aid because there may be sufficient larger particles that the filter cake will be at least moderately porous. As noted, a filter cake created from larger particles can remove the smaller particles by deep bed mechanisms. However, where there are few larger particles, it may be necessary to add some larger particles. These added particles are known as filter aids. It is preferable for these materials to be porous and pack to form a porous filter cake. The most common filter aids are the same materials as those used as precoats, diatomaceous earth and perlite. Filter aids function much like precoats; they provide a deep bed in which the small particles can be retained. One can view filter aids as materials that are added to form additional precoat continuously. Of course, any filter aid added to the suspension will appear in the filter cake and will be removed with the solids that are being removed. If one wishes to recover the filtered solids, it may not be desirable to use filter aids. If the solids are a waste material, the filter aid will add somewhat to the waste volume, but that is often an acceptable cost for greatly enhanced filtration rates.

WASHING FILTER CAKES

Washing filter cake is important when the solids removed are to be used as a product and it is desirable to remove liquid from within the filter cake. Many environmental and waste management applications may not need cake washing if the purpose of the filtration is to remove a toxic solid material from a discharge stream.

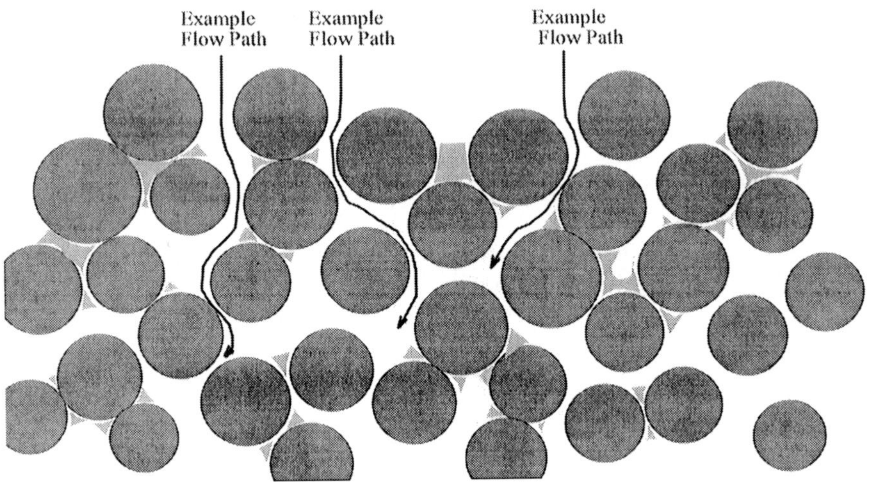

FIGURE 9 Flow through a bed of packed particles (filter cake) showing how some solution can be retained near points where particles are in contact and be difficult to wash from the filter cake.

Filter cake washing displaces the solution within the voids in the cake with fresh water. As a first approximation, the water added to the cake will displace the original solution after sufficient water is used to displace the void volume of the cake (Figure 9). If wash water passed through the filter cake in plug flow, the concentration of soluble materials in the water leaving the filter would be the same as that in the water originally being filtered until sufficient wash water is used to displace the solution in the voids. However, this is essentially never sufficient because the real concentration will vary with time, as shown in the "realistic" curve in Figure 9. Note that the clean water front still passes through the filter cake, but the front is not sharp. That means that the concentration of dissolved material in the water leaving the filter cake goes from the original concentration to zero over a finite (and usually significant) time or volume. The broad (or disperse) breakthrough front results largely from nonuniform flow within the filter cake. Dispersion and nonplug flow in filter cakes is similar to problems in packed bed adsorbers and ion exchange equipment. The wash water will flow principally through the larger channels between the particles. The velocity will be very low near each particle surface, particularly in the regions near the points where particles contact. Removal of solute from those essentially stagnant regions into the regions or channels where most of the flow occurs may be driven by diffusion as

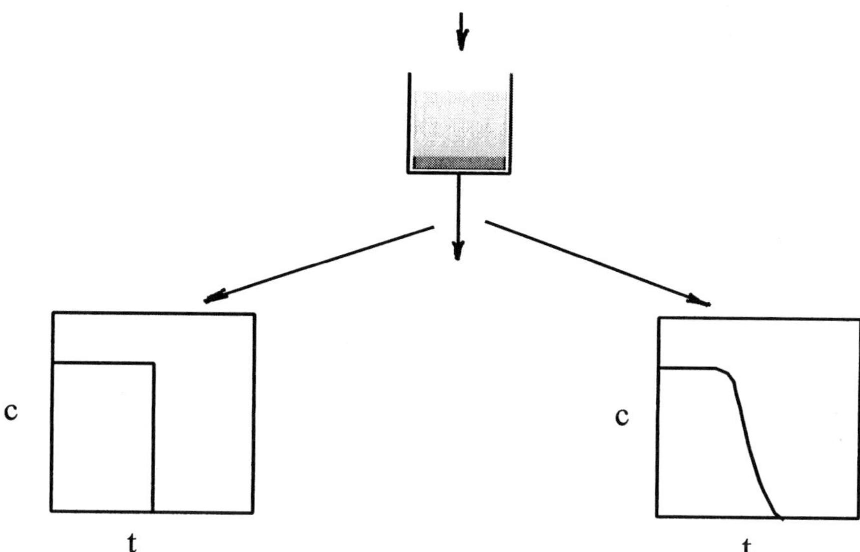

FIGURE 10 Idealized filter cake washing where residual solution is removed with essentially the use of only one void volume of wash fluid (sketch on the lower left) and the realistic case where several volumes will be required to remove essentially all of the original fluid.

much as by flow (Figure 10). The plot in the lower left portion of the figure shows the concentration of a filter cake as a function of time for an ideal case. The liquid concentration remains constant if the air or wash water is simply displacing the water in the pores. However, for the more reasonable case, shown in the lower right portion of the figure, significant fractions of the original liquid are "trapped" in the regions near points where particles in the filter cake touch, and the last portion of the original liquid must diffuse into the flow channels as suggested in Figure 9. If the diffusion rate is too slow, it may be desirable to "resuspend" the solids with clean water and filter again. This is usually not desirable if a precoat is used.

When moving belt type filter media are being used, the wash step can be carried out with a water spray on the cake. The flow through the cake and even the forces on the cake from this type of wash are complex and may need to be evaluated experimentally for each system studied.

As a first approximation [8–10], the water flow rate through the filter cake can be approximated by the flow resistance of the bed at the end of the filtration step. This simply assumes that the filter cake does not change form after the filtration step is completed. If the bed is compressible, use of a higher pressure to wash the cake would, of course, alter the porosity and thus the flow resistance of the cake. Although the flow resistance of the cake at the end of the filtration step is at its highest value, washing can still be a relatively short step because only a small volume of water in the cake voids must be displaced.

Air suction is another way to remove water from within the filter cake. Strictly speaking, this may not be considered "cake washing" but will be mentioned here because it is used for the same purpose. To force air to displace water with the voids, it is necessary to overcome the capillary (surface tension) forces within the voids. The capillary "pressure" increases with decreasing pore size, so it is impractical to remove water from filter cakes in this way when the particle size, and thus the pore size, is too small. Generally, the particle size should be greater than 10 to 15 micrometers for air suction to be effective. Note also that air suction will only remove the water from the larger pores. Capillary forces will retain the water within the crevices and smallest pores. With air suction, there is no mechanism equivalent to diffusion to remove solute from within these regions. Of course, air suction and water washing can both be used together to enhance removal of solution from the filter cake.

MECHANICAL DEWATERING

It is often desirable to remove as much water from a filter cake as practical to produce a dry product for marketing or (in environmental problems) to remove as much mass as practical from a toxic filter cake before it is sent to disposal. In many cases, the filter cake material could be dried with heat, but that usually requires more energy than mechanical dewatering, and in some cases the higher temperatures may result in some solid product degradation or, in some environmental problems, higher costs for containing toxic materials in vapors released by the drying process.

Mechanical dewatering involves applying pressure to the filter cake and squeezing the water from the cake. The pressure can be applied by solid surfaces in plate type filters or by rollers, especially when belt type filters are used. The pressure could be applied toward a single filter medium, or several filter medium surfaces in a filter press stack may be squeezed together to force fluid through all filter surfaces. The rate of compression of the cake and thus the rate of dewatering can depend upon the properties of the filter cake, and tests should always be made on realistic cake material. Applying pressure will have little effect on highly noncompressible filter cakes. (The importance of filter cake compressibility on changes in filtration rate with applied pressure difference has already been discussed.) Dewatering involves compressing the filter cake mechanically, so the amount of dewatering that can be achieved depends upon the "compressibility" of the filter cake.

The thickness of a cake under applied pressure can be measured as a function of time [11,12]. Mechanical dewatering can be carried out at constant pressure or under varying pressure. In many cases when constant pressure is applied, the volume of water displaced (or the change in cake thickness) initially will be proportional to the square root of time at which the pressure is applied. The dewatering pressure is sometimes called the "expression pressure," and dewatering by applying pressure is sometimes called "expression" of the filter cake. However, this proportionality of dewatering rate to the square root of time should not be considered a universal rule, and experimental plots of the log of water displacement versus the log of time should be made to establish that this behavior occurs. Failure to observe a slope of 1/2 on such a plot should not be considered a sign of bad data. After the initial dewatering period the rate of dewatering will decline more quickly and eventually reach impractical rates. This generally establishes the limits for practical mechanical dewatering. Removal of additional water may be best left to thermal methods.

Plate type filter presses are designed especially for dewatering by applying pressure to the cake trapped between the filter "plates." The filters consist of a series of filter surfaces with flow into and out of alternating chambers between surfaces (Figure 11). The filter surfaces are usually rectangular and are mounted on a frame, so after the filter is loaded with solids a mechanical press can be applied across the filter surfaces to dewater the solids. This is a practical operation for many applications, but it is a batch operation usually requiring significant manual work in changing the components of the filter press between batch operations. This could be an especially undesirable operation if the particles are toxic and special precaution required to protect workers from contact with the solids or with residual liquid. Opening a plate filter press to remove the filter media and the filtered particles is also a mechanical operation, and care may be required to contain the solids as well as the residual liquid.

Continuous screw press filters have been reported that discharge the filtrate through the fabric lined walls of a perforated barrel and use a screw press to move the particles (compressed cake) down the barrel for collection. Such units may be more attractive than batch type plate filters for high throughput systems such as sewage plants.

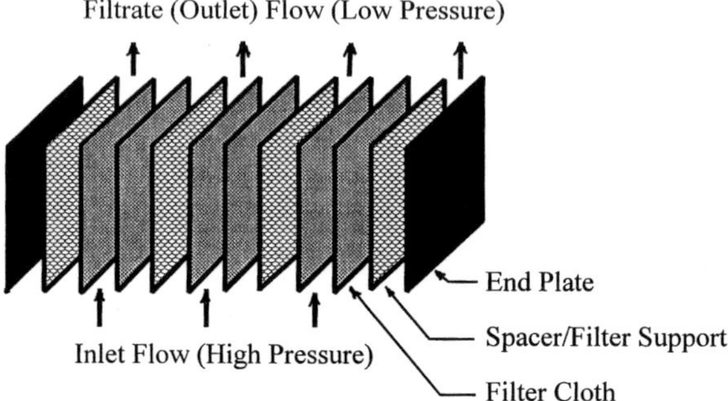

FIGURE 11 Filter press showing different filters stacked for parallel flow of the suspension through one-half of the chambers and filtered fluid removed through the other half of the chambers. Filter presses allow the resulting filter cakes to be compressed mechanically to remove much of the excess water.

CROSS-FLOW FILTRATION

When it is desirable to filter small particles without using a precoat, one can consider cross-flow filtration. Cross-flow filters use high velocities across the filter surface to prevent or minimize the buildup of filter cake on the filter medium. For instance, in a tubular frit filter, a high flow rate through the filter tubes would shear the filter cake from the filter medium surface and prevent the buildup of filter cake, but not always eliminate it. This allows the filter to maintain a relatively thin filter cake and thus an acceptable filtration rate per unit area of filter medium surface. This is illustrated in Figure 6. The steady filter cake thickness increases with pressure drop (filtration rate) because this is the effective velocity of fluid toward the filter surface, and the cake thickness decreases with the cross-flow velocity down the tube because this increases the shear forces near the wall. Both increase the filtration rate but increase the energy losses from pumping. Increases in the applied pressure will increase the filtration rate, but not linearly with pressure because of the increased thickness in the filter cake that would result from the higher flux of particles toward the filter surface. The filter cake thickness increases until the particle removal rate from cross-flow shear is again equal to the rate at which particles are carried to the filter cake. The filtration rate can also be increased by using higher cross-flow velocities, but increasing either the applied pressure or the cross-flow velocity increases the energy consumption and pumping costs.

The thickness of the filter cake on a cross-flow filter depends upon particle properties (principally size and density), fluid velocity in the channel, and flux (flow) through the filter. The flux can be controlled by the flow resistance in the filter medium itself or by the resistance in the filter cake. It is more common, especially when filtering very small particles, for the pressure drop and flux to be controlled by filter cake thickness. Once a significant filter cake is developed, the flux through the filter may be affected little or not at all by the properties of the filter medium; this is a common aspect of all surface filter operations. A high flux filter media may initially allow filter cake to build up because the high flux through the filter carries particles to the surface faster than the turbulence and cross-flow can sweep the particles back into the flowing stream. However, as the filter cake builds up, the flux through the filter declines, and when the filter cake reaches a "steady" thickness, the rate at which particles are carried back into the flowing stream equals the rate at which particles are convected toward the surface. Note that this is similar to the buildup of a concentration gradient on the surface of a pressure driven membrane (Chapter 4). Remember that the steady-state thickness of filter cake can be altered by changing the fluid (cross-flow) flow rate.

A recent paper proposed an analysis of the buildup of filter cake as well as the steady-state cakes [13]. The filter cake is usually considered to be a homogeneous media in such analyses, principally because that is the simplest way to treat them. However, the resulting analyses can sometimes be considerably in error. One recent study investigated nonuniformities in the cake and allowed the cake or part of the cake in a cross-flow filter to flow (or creep) [14]. As long as the filter cake has a solids fraction that is less than the maximum settling fraction, the cake has a finite viscosity and can flow. Even relatively modest flow of the cake can result in significant reduction in cake thickness and an increase in filtration rate. This analysis is relatively new and has not been applied to a great many sets of data.

One obvious feature of cross-flow filtration is its inability to produce pure (fluid free) particles. Cross-flow filters are useful only for concentrating the particles into a smaller volume of fluid. Because the shear forces at the wall are so much greater when the fluid is a liquid, cross-flow filtration is more likely to be used with liquid suspensions, but that should not be considered an exclusive condition. If the particles can be used as a concentrated suspension, there are obvious reasons to consider cross-flow filtration. Note that a cross-flow filter will circulate essentially all of the particles to produce the concentrated retained product. This means that the filtration rate and performance have to take into account the concentration of solids in the circulating suspension, not just the concentration in the feed stream. In many cases where the particles are concentrated several-fold, the difference in the "average" concentration of particles in the circulating solution and the feed composition could be much different. For a liquid suspension, the final liquid removal could involve drying (thermal process) if a dry solid is needed.

ULTRAFILTRATION

Ultrafiltration is simply filtration carried out with very small particles, usually in the colloidal size range (a few micrometers or less). Ultrafiltration is a relatively new concept that has become increasingly important. A number of frit and plastic ultrafilter media are now available. Ultrafiltration is particularly important in the food and biotech industries. It usually operates in the cross-flow mode and with no precoat. Thus, the filter medium must be able to separate the particles from the fluid without the help of a precoat.

It is probably obvious that ultrafiltration is in many ways a bridge between filtration and reverse osmosis. In fact, many of the "particles" removed by ultrafilter in the biotech industry may be considered molecules. Also many ultrafiltration membranes reject smaller molecules and ions, but the rejection is usually only partial. The limited rejection of dissolved salts and molecules by ultrafiltration membranes is usually not the reason why they are used, and the partial rejection of salts is not likely to be utilized.

Thus, one can consider ultrafiltration as a step toward reverse osmosis from surface filtrations, but the principle aim remains the filtration of very small particles or large molecules or molecular complexes. The behavior of an ultrafilter is more like that of cross-flow filters than like reverse osmosis systems, so the discussion of ultrafilter is given here The importance of ultrafiltration to waste and environmental problems may be limited, but there certainly is interest. Ultrafilters have been proposed for separating dilute oil and colloid suspensions from water streams [15]. They can also be used to remove metal ions from water streams if the ions are complexed with a suitably large ligand. Although cadmium can be precipitated from solution at high pH, it is not removed effectively until the pH is quite high, well above the limits for discharge and sufficiently high that lime or caustic consumption could be significant. However, by adding traces quantities of a suitable ligand such as Na-dimethyldithiocarbamate, the Cd can be removed by ultrafiltration [16]. Ligands such as this with molecular weights of 500 or greater are likely to be effective in removing metal ions by using ultrafiltration membranes with the smallest pores.

Filters with particularly small openings (usually a few angstroms) are sometimes called "nanofilters." These filters are beginning to bridge the gap between "membranes" and "filters." It is not necessary to draw a fine line between porous membranes and filters as long as the reader/user remains aware of the transition between the two closely related separation systems. Nonporous membranes that achieve their separation capabilities by "dissolution" of the permeable components in a nonporous film are more clearly different from filters. However, even they act similar to porous membranes that achieve their separation by adsorption of selected components on the pores and the subsequent transport of the adsorbed film through the pores. Such behavior resembles dissolution of the transported components in the film.

SAND FILTERS

Sand filters are widely used in water treatment and can play similar important roles in treatment of many wastewaters. It would be appropriate to discuss sand filters as surface filters or as deep bed filters because both surface filtration and deep bed phenomena can be involved. Sand filters are mentioned in the discussions of deep bed filters and of surface filters. However, a decision is made here to give the most extensive discussion to sand filters as surface filters because most of the solids are usually removed from a filter cake located on top of the sand bed.

On the other hand, sand filters have appearances much more like deep bed filters. The filter medium is relatively thick, usually 1 1/2 to 3 feet, and some deep bed particle collection mechanisms are certainly important, at least during the first phase of an operating cycle and in removal of the smallest particles. Also, cleaning sand filters is usually accomplished by back washing, much like many deep bed filters. Nevertheless, most sand filters remove most of the particles from water suspensions on filter cakes formed on the sand surface, so they will be discussed here as surface filters. The thick bed of sand can be viewed as primarily as a precoat and secondarily as a deep bed.

In most water treatment systems, use of sand filters to treat water for municipal or industrial use, the filter will follow a sedimentation step, often enhanced with coagulation using pH adjustment or added coagulants. Since sedimentation is usually not complete, small amounts of coagulates may play an important role in quickly forming a filter cake on the sand that effectively removes the smaller particles of interest. In some cases, sand filters have been known to require a brief time to reach the desired particle removal efficiency. This is similar to the behavior that

FIGURE 12 Schematic of a typical sand filter showing that the filter medium (sand) is supported on a bed of gravel, perhaps even gravel with different layers, each with different sizes of gravel particles.

one would expect from surface filters that rely upon the formation of an effective filter cake to remove the smallest particles.

Construction of Sand Filters

Sand filters are usually large devices built to handle high flow rates of water. They can be constructed in steel tanks or even in large concrete structures, usually for the largest filters. The filters usually have 1 1/2 to 3 feet of sand supported by several inches of gravel. It is desirable to have the smallest sand particles near the top surface of the bed, and that usually occurs after each back wash cycle, so no special effort is likely to be needed to grade the sand particles. However, it may be necessary to have several sizes of gravel arranged in layers below the sand. The largest gravel particles are near the bottom of the filter, and the particles must be large enough that they will not penetrate the slotted pipe or other device used in the bottom of the filter bed to remove the filtered water. The gravel just below the sand should have particles small enough that the sand does not penetrate the gravel level too deeply and eventually reach the outlet pipes. This may require two or more layers of gravel, with the larger gravel on the bottom and the smallest gravel on the top, in contact with the sand. Figure 12 is a simple sketch of a sand filter.

The feed water must be introduced into the filter through a suitable distribution system. Since sand filters are often very large devices with large surface areas, it is important, but not necessarily simple, to have effective feed distributors. The suspension could be introduced through spray systems or troughs over the filter surface from which the feed can overflow. The filter could be contained within a vertical or horizontal cylindrical tank or in a rectangular concrete system. For larger "tank" type sand filters, the horizontal systems can usually give higher filter areas for a given tank diameter. In all cases, adequate space must be allotted above the filter bed for bed expansion during back flow washing and for deintrainment of sand particles during washing. For large systems, concrete structures often give the lowest initial capital cost per unit area of filter, but tank systems can be pressurized and provide higher filtration rates per unit filter area. As noted earlier, the merits of using higher pressures (higher pressure difference driving forces) in surface filters depend upon the compressibility of the filter cake.

The filtration rate per unit cross-sectional area depends upon the flow resistance of the filter and filter cake and the pressure drop available for driving the filtration. The flow resistance of the filter cake is a function of the time between wash cycles, and the fastest overall filtration rates require that moderate to short filter cycles be used. Sand filters can operate from the pressure head above the filter bed, usually a pressure of only a few feet of water, or with higher applied pressures, usually in tank type sand filters. To increase the filtration rate further, the filter can be pressurized to a few psi. However, it is usually not practical to apply very high pressures to sand filters—that is, pressures as high as can be applied to some supported fabric or frit filters. This limitation appears to result largely from the properties of the filter cake of coagulates that are commonly formed on sand filters and the size of the sand particles. If the applied pressure become too great, the coagulate will be "forces through" the sand, and the filtration efficiency will drop drastically. Of course, this behavior results largely from the types of suspensions usually separated with sand filters and should not be considered a generalization for all suspensions.

Sand filters are usually not discussed extensively in textbooks dealing with industrial filtration; instead, one is likely to find more details in books dealing with water treatment. Some books even give sketches of sand specific filter designs. However, the principles associated with sand filter operations are no different from those discussed for other surface filters except for the differences in the size of the more common units and the wash methods. Fluidization washes are less common in applications using other surface filters.

Cleaning Sand Filters

Sand filters can be cleaned in two ways. Like all surface filters with precoats, it is possible to skim the filter cake from the top of the precoat, usually removing as little of the precoat as practical, that is, by removing as little of the sand as practical. Since some precoat (sand) is removed each time the filter cake is removed, the thickness of the sand precoat is decreased with each cake removal sequence. Eventually it is necessary to replace the sand precoat.

Alternatively, the filter cake can be removed by back flow, an operation also sometimes associated with other surface filters. The difference in this operation is the likelihood of fluidizing the sand and even some of the gravel in the sand filter. Although it would seem to be desirable to fluidize and remove the filter cake without affecting the carefully prepared sand precoat, there are at least some compensating positive results from the fluidization of the sand. As long as the back flow rate does not entrain significant quantities of sand, the sand precoat will "settle" back into place once the back flow is ended. The fluidization of the sand does limit the back flow velocity to values that will not entrain significant sand particles. However, it also removes most of the finer particles that have become trapped deeper in the sand by deep bed filter action. It can also separate any sand that may have worked its way into the gravel near the bottom of the filter. Back flow operations usually involve flow rates that will not fluidize the gravel particles. Finally, as the sand settles back to reform the sand precoat on the gravel support, there will be some segregation of the different size sand particles leaving more smaller sand particles near the top of the precoat and more of the larger sand particles near the bottom of the precoat and resting on the gravel support. This is the type of gradation of sand particles that one would prefer. Somewhat more detail on settling of binary and multisize particle suspensions is given in Chapter 10.

Note that when the particles are removed by back flow, they may be in a concentrated suspension, not in a dry or wet paste like most filter cakes. However, since the particles are concentrated, usually many-fold, from the original suspension, there are more practical options for dealing with the suspension by adding coagulants or even drying the suspension than would be practical with the original dilute suspension.

WHEN TO CONSIDER SURFACE FILTERS

Surface filters are certainly among the most common types of filter used in the process industries, and their importance is expected to extend into the growing waste and environmental treatment processes. These are relatively high pressure drop filters that are capable of removing even relatively fine particles. They are capable of accumulating relatively large quantities of particles in the filter cake before the cake must be removed. These are almost certain to be the preferred filter type when it is important to recover the filtered particles, usually as a product. Although it is often not necessary to recover toxic materials as a product, it is always desirable to recover toxic materials in a form as concentrated as possible since this reduces disposal/destruction costs.

The wide applicability of surface filters usually means that a surface filter is usually considered first unless there is a good reason for considering another type. The crucial consideration that usually decides if a surface filter is to be used is the flow permeability of the filter cake. Poor permeability of the filter cake is most likely to result when the particles are all small or when they have fibrous shapes that pack too tightly on the filter medium. Usually, granular particles with a significant number of larger particles (>20 micrometers) form moderately porous filter cakes. Even when the filter cakes are extremely nonporous, it may be possible to use small quantities of flocculation agents or filter aids to make the cake more porous. Generally one goes to deep bed filters, the principal alternatives, when there are no practical ways to use surface filtration; more details of this decision are given below.

SELECTION OF SURFACE VERSUS DEEP BED FILTERS

The decision to use a surface filter rather than a deep bed filter can depend upon several factors, and, once the decision is made to use a surface filter, there are several types of surface filters to consider. Consider first only the decision between the use of a surface or deep bed filter:

1. What is the size of the particles to be removed, and what is the flow resistance of a "cake" of particles?
2. What is the concentration of the particles to be removed?
3. Is the objective of the filtration to recover the particles or to clarify the fluid?
4. Is it acceptable to add additional foreign material to the filtered particles?

Since surface filters create a filter cake, it is important that the cake have a reasonable flow permeability and not require unacceptably high pressures to maintain the desired filtration rates. The permeability of a filter cake can be estimated by sedimentation rates since larger particles have higher sedimentation rates and form more highly permeable cakes. Fibrous particles that give low permeability cakes are also likely to have low settling rates. Sedimentation may be used in an initial assessment of the potential for surface filters. Of course, if the particles settle very rapidly, sedimentation may be used rather than filtration. On the other hand, if sedimentation rates are exceptionally slow, the particles may be so small that filter cakes will not have sufficient permeability. With very slow sedimentation rates, deep bed filters may be favored. It is in the immediate region where surface filtration is most likely to be most attractive. Fortunately, there are many applications in this range.

However, sedimentation rate, and even average particle size, is not a clear basis for selecting the surface of deep bed filters. For instance, particle density affects sedimentation rates but may have little effect on filter cake permeability. Particle size distribution can also be important since the smaller particles are the ones that most affect cake permeability, but the presence of substantial quantities of larger particles can help to keep the cake permeability high.

Particle shape can also affect filter cake permeability. Long polymer particles may form "slimy" filter cakes with very low permeability. Precoated filter cakes and filter aids can extend the use of surface filters to smaller particle sizes.

Generally, high concentrations of suspended solid favor the use of surface filters. Deep bed filters have a limited capacity for solids before the pressure drop exceeds acceptable limits, and it is not practical to clean some types of deep bed filters. Surface filters, however, can often hold considerable quantities of particles, and some surface filters are easily cleaned and operate almost continuously. On the other hand, some environmental and waste operations require removal of only trace quantities of toxic or radioactive particles, and deep bed filters may operate for long times with those dilute streams before they have to be replaced.

Surface filters are also usually preferred when the solid needs to be recovered. Particles trapped within some deep bed filters are very difficult to recover, so those particular forms of deep bed filters are not likely to be desired if the particles need to be recovered. Even those deep bed filters from which particles can be recovered usually cannot produce particles at concentrations as high as most surface filters. Particles may be recovered from deep bed granular filters as a concentrated slurry by back washing (perhaps fluidizing) the granules. In environmental and waste operations, the particles may often be a toxic material that needs to be removed from the fluid, and deep bed filters may be adequate if the particles are to be sent to disposal. In those cases, there can still be merit in surface filters if they can recover the toxic particles in higher concentration and minimize waste volume. Of course, if the particles are present in extremely low concentrations, extremely high concentration of the recovered particles may be less important.

From these few criteria, one can often quickly determine whether surface or deep bed filters will be preferred. However, in some cases, such as those with high concentrations of particles that do not form permeable filter cakes, the decision may not be as obvious. Those are often the more difficult filtration problems because there are problems or undesirable properties with both general types of filters. There is at least one "expert system" reported for selecting filtration systems [17] which discusses a more detailed list of considerations for selecting surface and deep bed filters. The reader may find such a system helpful for the more difficult problems.

CHOICE AMONG SURFACE FILTERS

When the particles form a filter cake that is sufficiently permeable, type of surface filter must still be chosen. Fabric filter media are likely to be preferred for moderate temperatures and chemical conditions that are compatible with common fabrics. Since fabric filter media can be made from a variety of materials, there are many options available even for moderately corrosive materials. Frit media are more likely to be used when high temperature strength is required. However, fabric filters can be made of ceramic fibers and operate up to 700°C or higher for short periods of time [18].

Systems needing high throughput are more likely to result in selection of continuous systems. For gas systems, this may be the bag filters. For liquid systems, this may be a belt filter such as a drum filter (as long as only modest pressure drop is required for acceptable filtration rates). Both of these filters permit high throughput and recovery of the particles in concentrated forms. For bulk water treatment, sand filters have proven to be attractive.

Another consideration in filter selection is the pressure drop required. Obviously, simple bag filters with unsupported "socks" cannot hold extremely high pressure drops. The need for such high pressure drops may indicate that other filter types should be considered. However, if a surface filter is still preferred, perhaps because it is desirable to recover the particles in high concentrations, one could consider any of the common pressure filters. Pressure filters can come in a variety of plate and cylindrical shapes, but they are all less likely to incorporate the automatic cleaning features as easily as the lower pressure drop bag (or sock) filters for gases or the belt filters for liquids. If fabrics are used as the filter medium, they are likely to be supported by porous or perforated metals.

For filtering large flow rates of water, sand filters should usually be considered. These are usually inexpensive (for the throughput) when built on a large scale, easily operated, and easily washed. Furthermore, sand filter design is well established. However, to remove very small particles, it is necessary to form a filter cake on the sand surface. If there is a coagulation/sedimentation step just prior to the filtration, no additional step may be required to form the filter cake. However, if the stream comes directly from another type of treatment step, it may be necessary or desirable to add a coagulant. For examples of sand filters, one can look at essentially any municipal or industrial water treatment system. These are standard high throughput water filtration systems.

For removing particulates from off-gas systems, fabric filters, often called bag houses, are common practice. To minimize the particulate load on fabric filters, precipitators or some other low pressure drop device may be used to remove a substantial fraction of the solids, particularly the larger particles. Fabrics have been developed that will operate from room temperature to moderately high temperatures. Fabric filters are becoming more common in the electric utility industry where large flue gas flows are involved and where increasingly restrictive regulations are limiting the quantities of fly ash that can be released. These also appear to be very popular for treatment of similar off-gases in other industries. Such systems can also be considered for many internal streams in process facilities, but some of those applications are not so obviously related to waste and environmental processing.

When it is desirable to collect and use the filtered particles, it becomes desirable to limit the quantities of precoat or filter aids to be used, since those materials will contaminate the solid product. Although surfaces filters are favored over deep bed filters when the solids are an important product, the need for excessive quantities of precoat or filter aids can be a problem. The need to remove very fine particles and recover them in a concentrated form can present difficult problems. Even when the particles are not to be recovered as a product, it may still be desirable to minimize the quantities of precoat and filter aid collected with the solids if they increase the volume of waste considerably. The increasing cost of disposing of toxic solids can increase the need to recover some filtered particles in a form that is as concentrated as practical.

Vacuum filters are commonly used for water systems which require only moderate pressure differences to achieve suitable filtration rates. Continuous belt type vacuum filters are common and reliable. There is always a choice between higher pressure driving forces and higher filter areas to achieve desired filtration rates. Obviously there are several factors, such as the relative cost of pressure vessels, the cost of additional filter aids, and the energy cost from the use of higher pressure differences that must be considered. The importance of many of these factors is highly system dependent, so only a few generalizations can be made. Perhaps the most obvious factor to consider is the compressibility of the filter cake. If the filter cake is highly compressible, there may be relatively little to be gained by increasing the pressure driving force very far, so increased filter area is more likely to be the desired choice. However, if the filter cake is essentially noncompressible, higher pressure driving forces should be considered.

When solids in highly compressible filter cakes are to be collected as product, it may be desirable to use a filter that is capable of compressing the filter cake to remove as much excess water as possible. For small systems, plate type filters may be considered, but these often require considerable labor and exposure of workers to the filter cake which could be problems if the solids are particularly toxic. For large systems continuous devices have been used which can continuously dewater compressible filter cakes using moderately high applied pressure.

ADDITIONAL EXAMPLES OF APPLICATIONS OF SURFACE FILTERS FOR ENVIRONMENTAL AND WASTE PROBLEMS

Water Streams

As noted earlier, surface filters are usually preferred for removal of relatively large particles that form porous filter cake. To remove very small (near colloidal) particles, a deep bed filter is likely to be considered, especially if there is no need to recover the filtered particles. Effective use of surface filters may require a precoat and filter aid. The presence of very small particles alone does not mean that the filter cake will not be sufficiently porous, because there may be sufficient quantities of larger particles to make the filter cake sufficiently porous.

Some Uses of Ultrafilters

As noted earlier, when very small particles are to be removed by surface filters, the pressure drop can be controlled by using cross-flow filtration. This is simply ordinary surface filtration where a high flow rate of fluid is applied across the filter surface to minimize the buildup of filter cake. A cross-flow filter can be a tube-in-shell type with the high pressure side on the inside of the tubes. A recirculating pump would then pump the slurry through the tubes at a relatively high flow rate, and the shear at the filter surface would remove particles from the surface and thus control the thickness of the cake on the tube surface. In many ways this resembles the use of high cross-flows often applied across membrane surfaces to reduce "concentration polarization" of soluble components [19].

The additional cross-flow, of course, consumes additional energy, requires stronger filter surfaces, and adds to the complexity and capital cost of the system. An optimization is needed to select the best cross-flow rate, since higher cross-flow rates increase energy consumption for the recirculating pump (and possibly capital cost), but they decrease pressure drop for flow through the filter and/or reduce the filter area needed, which would reduce capital costs.

In recent years, surface filters have been developed for removing particles in the colloidal size range. These are usually called "ultrafilters." They have found considerable use in the food industries. Ultrafilters can remove individual molecules if they are sufficiently large, e.g., many organic dyes [20]. If the dye can be concentrated sufficiently, it may be possible to recover it for reuse rather than simply for disposal. As one would suspect, ultrafiltration usually involves cross-flow to restrict the buildup of filter cake [21,22].

Ultrafilters can remove clusters of molecules such as colloidal micelles of surfactants. This has been suggested as one method for removing metal ions from waste streams by filtration. A surfactant is a molecule with a hydrophobic part and a hydrophilic part. The hydrophilic part of the molecule could be an ionizable group or a neutral group with fixed charges separated by short distances; the hydrophobic part is usually an organic part with aliphatic or aromatic groups. If one part of the surfactant associates with the metal ions, there could be a cluster of surfactant molecules that is too large to pass through a suitable ultrafilter. This is sometimes called micellular enhanced ultrafiltration.

For this technique to be attractive for environmental or waste treatment, the surfactant may have to be inexpensive and selectively attach to the toxic metal ions of interest, not all of the cations present. One study [23] reported that lecithin, an inexpensive natural and biodegradable surfactant which can be produced from a number of animal and vegetable sources, selectively binds to heavy metals such as cadmium, copper, nickel, and zinc, and can be used to remove them from some waste solutions. It is notable that lead was one heavy and toxic metal that did not seem to bind with lecithin. Another study used polystyrenesulfonate to remove chromium ions from solutions [24].

GAS STREAMS

Surface filters usually compete with centrifugal devices (cyclones) and electrostatic precipitators for treatment of gas streams for removal of particles from exhaust gases. Although surface filters do not have the capabilities to remove particles as fine as some deep bed filters, bag (sock) type filters are generally the choice for high throughput systems that require exceptionally high removal efficiencies. Centrifugal devices are usually more effective for larger particles and electrostatic devices are often useful for systems requiring only moderately high removal efficiencies. Electrostatic precipitators may have lower pressure losses than the more efficient bag filters. For large throughput problems and a wide range of particle sizes to be removed, it may prove best to use a multiple step system with a cyclone or an electrostatic precipitator backed by a bag/sock filter.

Very small gas streams are also usually suitable for surface filters because of their relatively low capital cost. The choice of surface filters over deep bed filters (usually fiber filters) is likely to depend upon the size of particles to be removed, the concentration of particles, and the need to recover the particles.

REFERENCES

1. Chase, G. G., J. Arconti, and J. Kanel. *Sep. Sci. Technol. 29*, 2179 (1994).
2. Harvey, M. A., K. Bridger, and F. M. Tiller. *Filtr. Sep. 25*, 21 (1988).
3. Koenders, M. A. and R. J. Wakeman. *AIChE J. 43*, 946 (1997).
4. Dennis, R. and J. Wilder. "Fabric Filter Cleaning Studies." *EPA-650/2-75-009* (1975).
5. Morris, W. J. *Filtration Sep. 21*, 50 (1984).
6. Liu, S. X. and L. A. Glasgow. *Sep. Technol. 5*, 139 (1995).
7. Trados, F. F. *Chem. Ind. (London)* 7, 210 (1955).
8. Michaels, A. S., W. E. Baker, H. J. Bixler, and R. W. Vieth. *I&E Chem. 6*, 29 (1967).
9. Michaels, A. S., W. E. Baker, H. J. Bixler, and R. W. Vieth. *I&E Chem. 6*, 33 (1967).
10. Han, C. D. and H. J. Bixler. *AIChE J, 13*, 1058 (1967).
11. Shirato, M., T. Murase, M. Negawa, and H. Moridera. *J. Chem. Eng. Jpn. 4*, 263 (1971).
12. Schwartzberg, H. G., J. R. Resan, and G. Richardson. *AIChE Symp. Ser. 163*, 77 (1977).
13. Chang, D-J., and S-J. Hwang. *Sep. Sci. Technol. 30*, 2917 (1995).
14. Datta, S. and J. L. Gaddis. "Dynamics and Rheology of a Fouling Cake Formed During Ultrafiltration." Paper presented at the Ninth Symposium on Separation Science and Technology for Energy Application. Gatlinburg, TN, Oct. 22–26, 1995; submitted to *Sep. Sci. Technol.*
15. Reed, B. E., et al. *Sep. Sci. Technol. 32*, 1493 (1997).
16. LeGoff, P., et al. *Sep. Sci. Technol. 32*, 1615 (1997).
17. Ernst, M., R. M. Talcott, N. C. Romans, and G. R. S. Smith. *Chem. Eng. Prog. 87*, 22 (July 1991).

18. Ergundenler, A., et al. *Sep. Purification 11*, 1 (1997).
19. Mikulasek, P. and P. Dolecek. *Sep. Sci. Technol. 29*, 1943 (1994).
20. Juang, R-S., J-F. Liang, and J-D. Jiang. *Sep. Sci. Technol. 28*, 2049 (1993).
21. Chang, D-J. and S-J. Hwang. *Sep. Sci. Technol. 29*, 1593 (1994).
22. Huang, Y-C., B. Batchelor, and S. S. Koseoglu. *Sep. Sci. Technol. 29*, 1979 (1994).
23. Ahmadi, S., L. K. Tseng, B. Batchelor, and S. S. Koseoglu. *Sep. Sci. Technol. 29*, 2435 (1994).
24. Krehbiel, D. K., et al. *Sep. Sci. Technol. 27* (1992).

9
Deep Bed Filters

In deep bed filters, the particles are removed within a bed of the filter media, not on the surface of woven or thin bed of the filter media. The media for surface filters were relatively thin sheets of woven fabrics, pressed fibers (felts), or frits, and the particles were largely prevented from entering the media but formed a cake of filtered particles on the surface of the media. As noted in the discussion of surface filters, there may not be a clear distinction between surface filtration and deep bed filtration since surface and deep bed effects may be present. One should usually classify a filter according to the principal particle removal mechanism, but the effects of the other mechanism may still be observable and even important.

In deep bed filters, there is no filter cake, and the particles become trapped in the filter media itself. The media in deep bed filters can be relatively thick, and the filter performance can depend upon the thickness (depth) of the media. In surface filtration, there may be no incentive to increase the thickness of the media beyond the minimum needed to avoid holes or provide strength to the filter media.

To enter the filter media in a deep bed filter, the particles must be smaller than the openings (void spaces) in the filter media. This obviously means that particle removal does not occur by size exclusion. Particles are removed in deep bed filters because they stick to the internal surfaces of the filter media. The retaining (or sticking) forces are usually van der Waals forces, but electrostatic forces can be applied and assist in removing and retaining particles when the charges on the filter media surfaces are opposite to those on the particles. Lint and other particularly fibrous particles can become entangled in the filter media. Most models for describing the performance of deep bed filters assume that once a particle touches an internal surface of the filter, it is retained by the filter. The focus of these models is upon the mechanisms and forces that transport the particles to the internal surfaces of the filter media because these forces must be considered in designing or estimating the performance of deep bed filters.

Deep bed filters are particularly attractive for removing ultrafine particles, since it is usually not practical to accumulate a significant filter cake of such particles on a surface filter. Filter cake of very fine particles would have a significant flow resistance that would make surface filtration impractical.

Particles trapped in many deep bed filters cannot be removed easily from the filter, and this can be a serious problem when recovery of the particles (without dilution with the filter media) is desired. However, if the objective is only removal of the particles from the fluid (and not recovery of the particles), difficulties in regenerating deep bed filters may not be so serious. Granular deep bed filters (one type of deep bed filter) can be fluidized with upflow to remove a concentrated stream of the trapped particles, but this is more likely to produce only a more concentrated suspension of particles, not a highly concentrated paste like those that can be produced by surface filters.

Two types of deep bed filters will be discussed: fiber filters and granular filters. These names refer to the type of filter medium used. Although either type of filter can be used (and is used) with both liquids and gases, it is more common to use fiber filters to remove particles from gases and granular filters to remove particles from liquids. Fiber filters can be constructed with very high void fractions that permit high flow rates through the filter with low to moderate pressure drop. This makes them especially suitable for handling high volume gas throughput. However, high void filters may not have sufficient strength to avoid compression when high density fluids like liquids are to be treated. Granular beds usually have greater structural strength and are more likely to be used when particles are to be removed from liquids. The void fraction in granular bed filters is usually not much more than 50%, and this value cannot be varied greatly. However, the void fraction in fiber filters can be very high, even significantly greater than 90% or 95%, and can often be varied considerably by changing the fiber packing density (number of fibers per unit volume of the bed).

FIBER FILTERS

Fiber filters are frequently used for removing particulates from gases. Most homeowners are familiar with the relatively thin "furnace filters" that are used with most central heating/cooling systems in homes. Although those filters may seem thin, they still may be called deep bed filters because particles are trapped throughout the filter medium, not just on the surface. Most homeowners may be familiar with the long lint particles that accumulate on the front surface of household furnace filters. Some of the lint may be removed by a "surface mechanism," but lint will penetrate the filter media and be removed deeper in the bed. As the reader will see, even with deep bed filters, the accumulation of particles will be highest near the inlet to the bed. Fiber filters are usually constructed of loose unwoven fibers; the void fraction within the filter is usually near unity (100% void volume). As noted earlier, such loosely packed fibers are often more suitable for filtering gases than liquids because the pressure drop or drag on the fibers is less with low density and low viscosity gases, and the low drag forces on the fibers are not as likely to move the fibers and "compress" the bed.

Deep bed fiber filters should be contrasted with woven cloth filters, also made from fibers. The difference is that cloth filters usually act principally as surface filters, and the particles are removed at or near the surface, principally because particles are unable to penetrate far within the small opening in the woven cloth. Of course, with filter medium with nonuniform spacing between the fibers and/or with nonuniform particle sizes in the fluid to be filtered, one can find some particles penetrating the medium surface and becoming trapped within the fibers, even with woven cloth. However, such filters will be called surface filters because most particles will be trapped on the surface. Surface filters can also use nonwoven fibers, but they are usually compacted into a mat or felt. Deep bed fiber filters are normally open or loose structures. As discussed in the preceding chapter, surface filters, such as those made of woven fibers, can also remove particles smaller than the openings between the woven threads and thus act partially like deep bed filters. Nevertheless, we usually characterize a filter on the basis of the principal mode of particle removal, but recognize that other modes of removal can also contribute.

It is often desirable to operate a filter so that only one mode dominates. Contributions from the other mode may enhance particulate removal, but at the expense of operating difficulties. Surface filtration can "blind" deep bed filters and result in more rapid increases in pressure drop. Significant internal (deep bed) filtration by surface filters may result in an accumulation of particles within the filter that resist removal in order to return the surface filter approximately to its initial condition.

Fiber filters, like all deep bed filters, contain pores or open spaces between the fibers that are much larger than the particles being removed from the fluid. Thus, geometrical entrapment makes no significant contribution to particulate removal. Instead, particles are removed because of their affinity for the fiber surface. This affinity can result from surface forces (van der Waals) or electrostatic forces if the particles are charged electrically (and many particles suspended in gas streams do carry some electric charge).

Deep bed fiber filters are particularly effective for removing long fibrous particles from gas streams. Fibrous particles are easily tangled in the fiber filter media. Fiber filters used in heating and ventilation systems (like your home furnace filter) often collect numerous fibrous solids which can behave significantly different from the near spherical granular particles which are discussed in more detail here. The discussion in this book is focused on near spherical (granular) particle shapes because they are believed to be the shapes of most interest in waste and environmental processes. Very small granular-shaped particles are usually especially difficult to filter because there are no large dimensions to become tangled in the filter media. There are, of course, some environmental applications where shapes far from spherical are important, such as filtration to remove asbestos from air- or water streams.

In most air, and even in most liquid, suspensions, particulates become electrically charged from the chemistry of the surface or from charges accumulated by contact with other electrically charged surfaces. Chemical surface charges also can result when the particles have oxides or other materials on the surface that adsorb ions or lose ions to leave a net electrical charge on the surface. Induced charges may result when the particles contact other surfaces with such electric charges or even from contact with surfaces with applied charges.

Electric charges play a major role in keeping particles suspended in air or liquids, so filters are more likely to be needed for removing particles with net electrical charges. If there are no repulsive forces, particles are more likely to stick or agglomerate once they collide. This could be much like coagulation of particles in water suspensions. Turbulent or Brownian motion can cause the particles to collide. After several collisions, the agglomerated particles may form clusters with sufficient size and mass to settle from the suspension by gravity. However, if the particles carry sufficient electrical charges (with the same sign), the electrostatic forces will repel the particles from each other, and random turbulent and Brownian motion will be less likely to bring them together close enough for van der Waals forces to take over and permit actual joining of the particles. Then agglomeration will not occur at a significant rate, and filtration will be more difficult.

It is probably obvious that electrically charged particles can be retained on the fibers of the filter media with opposite electrical charges; electrical charges can be induced on filter fibers to enhance the rate as

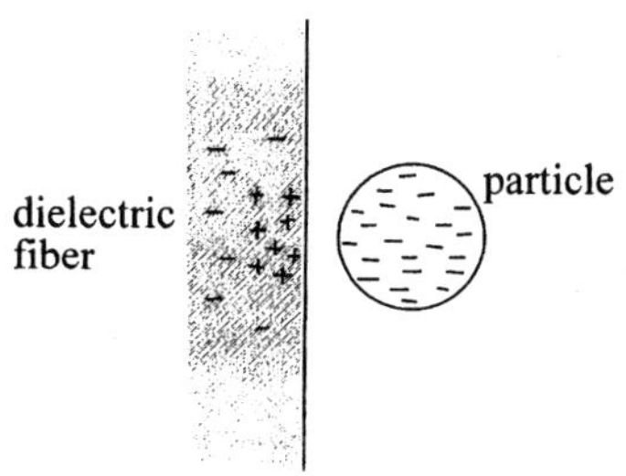

FIGURE 1 How image forces are generated when a charged particle approaches the surface of a dielectric material such as a fiber in a deep bed filter. The charge on the particle displaces charges of like sign on the filter surface and attracts charges of opposite sign. The resulting difference in the distance of the particle from the particles of like sign and opposite sign results in a net attraction to the fiber surface, although the fiber retains net electrical neutrality.

well as the retention of oppositely charged particulates. However, image forces can also contribute to retention of electrically charged particles on neutral (uncharged) fiber surfaces. As a charged particle approaches a neutral fiber surface, dielectric effects will attract opposite charges to the surface near the charged particle and repel like charges from the surface to regions further from the particle. Although the fiber itself may remain uncharged overall, this separation of charges near the fiber surface contributes to retention of charged particles (Figure 1). The electrical charge that is on the side of the particle closer to the surface is equal to the charge on the opposite side of the particle that is further from the fiber surface. However, because the charges on the fiber that are opposite to the charge on the particle are closer to the particle than the charges on the fiber that are like the charge on the particle, the attractive force from the charge closer to the particle will be greater than the repelling force from opposing charges farther from the particle.

Since particles that come sufficiently close to a fiber surface are assumed to stay on the surface by van der Waals forces or by electrostatic forces, the filter performance is determined largely by the rate at which the particles can be brought to a fiber surface.

MATHEMATICAL DESCRIPTION OF FIBER FILTER PERFORMANCE

Models of fiber filter performance are usually based upon the probability that a particle will come in contact with a fiber [1]. Retention of the particle once contact is made with a fiber is considered very likely (near unity probability), but retention coefficients can be incorporated into the models.

With fiber filters, the fibers are usually separated sufficiently that each fiber can be considered to operate independently. Furthermore, the flow near a fiber is assumed to be unaffected by the presence of other fibers. (Note that later in the discussion of deep bed granular filters, this cannot be assumed because the granules are packed much more closely than fibers in most fiber filters.) Seldom will filters be operated with closely packed fibers where the interactions of nearby fibers need to be taken into account [2]. So the probability that a particle will become attached to any fiber in a differential length of filter can be estimated by simply multiplying the probability for attachment to a single fiber by the number of fibers in a differential length of filter bed. This can be expressed in terms of a unit cross-sectional area of the filter and the depth of distance down the filter bed for a particle of a given size. The probability of capture over a differential length of filter is

$$P = mN_f\, dL \tag{1}$$

where N_f is the length of fiber per unit volume of filter, m is the probability of capture by a unit length of a single fiber, and L is the length of the filter. Of course, the length of fiber per unit volume is proportional to the volume fraction of fiber in the bed. Then the number of particles captured would be

$$\frac{dn}{dL} = -nmN_f \tag{2}$$

Integrating over the length of the filter gives

$$\frac{n}{n_0} = 1 - \exp(-mN_f L) \tag{3}$$

The initial concentration of particles is n_0, the exit concentration is n, and the length (depth) of the filter is L.

This is the key design equation for fiber filters or any deep bed filter (however, m and N_f may be combined in models for other forms of deep bed filters). If a filter is tested with one thickness or length L, the performance of a similar filter with a different thickness can be evaluated by simply applying a different length in the equation. Thus the filter thickness required for any particular particle removal efficiency can be estimated from a single test.

Unfortunately, it is not easy to predict the value of the extinction coefficients N_f and m accurately without measurements. However, there are approximate theories and correlations that are useful for estimating these parameters. Even when the predictions are not sufficiently accurate for a reliable design, they provide valuable insight into how the different design parameters affect filter performance.

The key to determining filter performance is determination of N_f and m. N_f can be easily measured; it is simply the weight of the filter per unit volume of filter divided by the weight per unit length of the filter:

$$N_f = \frac{w_v}{w_l} \tag{4}$$

The second attenuation coefficient m is more difficult to evaluate. This term includes the results of several forces or effects that bring particulates to the filter fibers. Some of these effects are listed and described below. The attenuation factor may account for any or all of the following effects.

1. Interception
2. Inertial forces
3. Gravity
4. Diffusion and Brownian motion
5. Electrostatic forces
6. Short-range forces

In many fiber filters, the void fraction is so close to unity (fibers take up little of the filter volume) that the filter performance can be estimated well by models developed from consideration of individual isolated fibers. The fluid flows around the fibers, and all streamlines pass around the fibers and thus do not impact a fiber. Although the particles tend to follow the streamlines, factors which take particles to the fiber surface can cause them to deviate from the streamlines.

"Interception" results from the finite size of the particles. Although streamlines approaching a fiber split as shown in Figure 2 and pass around both sides of the fiber, the streamlines become more compressed or crowded near the fiber surface. A particle approaching a fiber along a streamline sufficiently close to the fiber surface will intercept the fiber if its center follows a streamline that passes within one particle radius of the fiber surface. Although the streamline would pass around the fiber, a particle whose center follows the streamline would still touch the fiber surface. Thus interception occurs even for neutrally buoyant particles whose centers follow fluid streamlines essentially exactly. The effect of direct interception on the particle collection efficiency has been estimated [3] to be

$$E_{IT} = \frac{2d_p}{d_f} \tag{5}$$

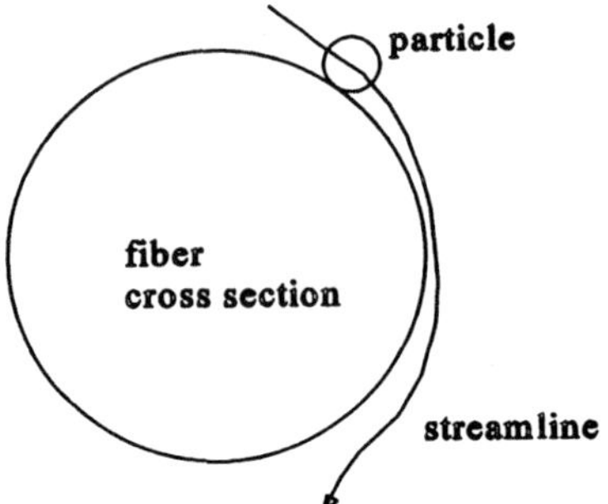

FIGURE 2 Particle following a flow streamline around a fiber, but the finite diameter of the particle causes the particle to touch the fiber surface even when the streamline does not reach the surface.

where d_p and d_f are, respectively, the diameters of the particles being filtered and the fibers. This is based upon an evaluation of the streamlines that pass within a particle radius from the fiber and the distance portion of the bed cross-section that would be bounded by those streamlines at positions far removed from the fiber. The bed cross-section bounded by the streamlines represents the probability that a particle would be within those bounds and thus the probability that a particle would intersect with the fiber by this mechanism.

Inertial effects cause the particle paths to deviate from the fluid streamlines. When the fluid streamlines turn to pass on either side of a fiber, inertial forces make dense particles tend to follow less curved paths, and this makes more particles strike the fiber surface (Figure 3). Both theoretical (often numerical) and empirical estimates have been developed for predicting the contribution of inertial forces to filtration efficiency. In theoretical calculations, the flow field is usually described as either viscous or inertial, but many fiber filters operate in the transition region. Even when the velocities suggest that the flow is in one region or the other, it may still not be accurate to use the velocity fields developed for isolated cylinders because wakes from other fibers upstream may still have an influence; they could introduce turbulence even when the Reynolds number would suggest only viscous flow.

For sufficiently large and dense particles, gravity can also make an important contribution to particle removal by causing the particle paths to deviate from the streamlines. When the filter is oriented so that the flow is downward, gravity causes the particles to be deflected less by the fluid flowing around the fibers. Qualitatively, this behavior looks much like

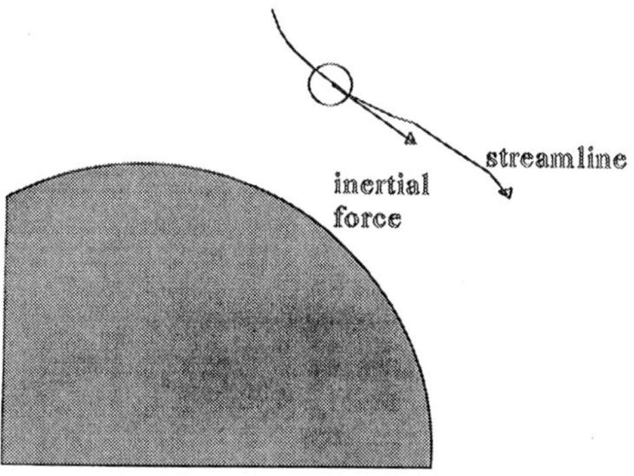

FIGURE 3 Inertial forces will cause particles to deviate from fluid streamlines. On the upstream side of the fiber, this deviation brings the particle closer to the surface than the streamline and enhance the chances of particle capture by the fiber.

the effects of inertia, and the resulting particle paths are qualitatively like those shown in Figure 4. The contribution of gravity to filter performance obviously depends upon the direction of flow. For downward flow Ranz and Wong [6] suggested that

$$E_G = \frac{g d_p \rho_p}{18 \mu v} \tag{6}$$

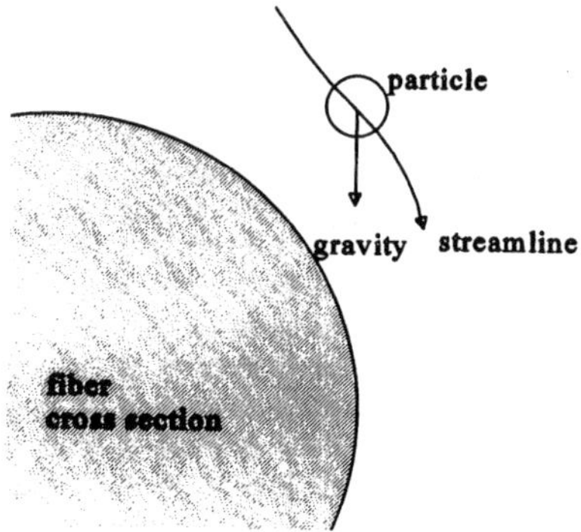

FIGURE 4 Effects of gravity to move particle from the fluid streamlines. For dense particles and downflow filtration, the effects of gravity on the upstream side of the fiber move the particles closer to the fiber and increase the chances for capture.

For beds oriented with flow in other directions, different effects of gravity could be expected. Nevertheless, for essentially any orientation, the effects of gravity usually make a positive contribution to filter performance.

Diffusion or Brownian motion acts differently from inertia and gravity forces; the deviations from fluid streamlines are random and approximately as likely to be in one direction as in another direction. Nevertheless, a fraction of these random motions causes particles near a fiber surfaces to move toward the surface and contact it. This adds to the number of particles contacting the surface and thus to filter performance.

The effects of interception, inertia, and gravity all increase as the particle size increases, and they become of less importance as particle diameter decreases. Diffusion effects, however, become more important as particle size decreases. Brownian motion results from random, uneven deflections of very small particles caused by molecules striking the particle unevenly in different directions. Smaller particles have less momentum, and their smaller size increases the probability that the results of molecular interactions will not be symmetrical and that significant Brownian motion will occur. Thus only very small particles are deflected significantly by Brownian motion. For sufficiently small particles, Brownian motion may become the dominant factor affecting filter performance because inertial and gravitational forces decrease with particle size. For Reynolds numbers less than 1, Friedlander [5] estimated that

$$E_B = \left[\frac{1.71}{2 - \ln N_{\text{Re}}}\right] N_{\text{Pe}}^{-2/3} \tag{7}$$

where $N_{\text{Pe}} = d_f v / D_B$ and D_B is the Brownian diffusion coefficient.

From only a qualitative examination of the factors just described, one can explain one important behavior of fiber filters, a minimum in the removal efficiency that often occurs for particles between approximately 0.1 and 5 micrometers. This minimum is important because it defines a range of particle sizes that are particularly difficult to remove, and this range of particles can be important to environmental problems. This minimum results because one group of mechanisms dominates for larger particles but their effects decrease with decreasing particle size. However, another mechanism increases particle removal efficiency with decreasing particle size and then dominates the removal of especially small particles. The contributions of interception, inertia, and gravity increase with particle size, while the effects of Brownian motion decrease with increasing particle size. There can then be good removal of both large and very small particles and a minimum retention of intermediate size particles.

As noted earlier, many particles that remain suspended in gases or liquids do not agglomerate because they carry electrical charges that repel particles and prevent contact and thus agglomeration. Thus many (perhaps most) difficult filtration problems involve particles with electrical charges that can also contribute to removal rates. As noted earlier, electrical charges can contribute to short-range forces that help hold particles on fibers when the fibers are uncharged. The image forces can even attract particles a slight distance from the surface, but longer range electric forces result when an electric field is applied.

Applied electric forces can be another method for moving the particles from the streamlines so that they reach fiber surfaces and become attached. Applied electric forces are not used in all fiber filter systems; they may involve unnecessary expense or be unsafe to apply at the voltages needed. Obviously, it is not practical to apply high voltages to systems with conducting fluids, such as water. However, the possible enhancement in filter performance that can be achieved with applied voltages should not be ignored, especially when improved performance is needed and is difficult to achieve by other methods [6–8].

There are two principal ways to apply electric potentials to improve filter performance. First the electric potential can be applied directly to the fibers; an oppositely charged electrode can be located upstream from the filter. In this case, the electric field lines (which denote the direction of the resulting electrostatic force) all point directly toward the fiber surfaces, so the applied electrical force acts to attract particles to the fibers. For the applied electric potential to reach all of the fiber surface, it is necessary for the fibers to be made of electrical conductors, usually a metal.

The second way to apply an electric field is with both electrodes located outside the filter. Then there is no net electrical charge on the fibers, but the fibers reside within an electric field. It may not be obvious that this will attract particles to the fibers, but note that even a uniform electric field oriented in the direction of flow would act much like a gravitational field or inertial forces and cause the particles to deviate from the fluid streamlines. This would increase filter efficiency in qualitatively the same way that inertial and gravitational forces increase the efficiency. If the field is applied in a direction normal to the direction of flow, the electric forces will attract particles to one side of the fibers and repel particles from the other side. The net effect enhances particle removal because particles on the side with repulsion would not otherwise reach the surface anyway, but more particles would reach the surface on the side where they are attracted.

Actually, an applied field across a filter usually results in electric field lines that converge into the fibers, rather than just straight field lines. This results because the dielectric constant for the solid fibers is usually greater than that of the fluid, especially when the fluid is a gas. (Of course, there is

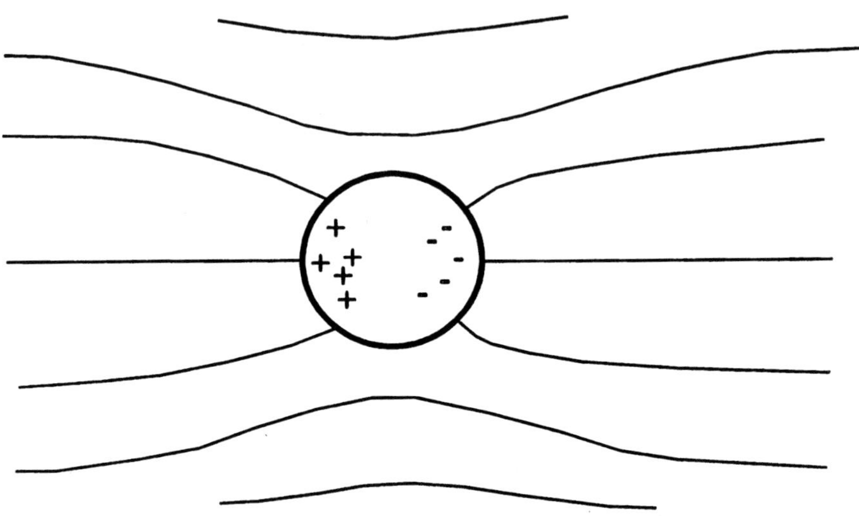

FIGURE 5 Distortions in a uniformly applied electric field caused by the presence of a dielectric fiber. These distortions result in electric field gradients that attract charged particles to fibers.

no merit in applying an electric field in this manner in a conducting fluid like water.) Then the applied field induces electric charges on both sides of the dielectric fiber, but the charges are equal in magnitude. Thus the fiber retains net neutrality but has separated charges on opposite sides (Figure 5). Note that the induced charges also create their own electrical field which must be added to the imposed field (Figure 5). Field lines converge into the fiber surface. Thus, induced charges in dielectric fibers also serve to divert the net electric field lines into the fiber surface and enhance filter performance. The particles, however, will be attracted only toward the surface that acquires an electric charge opposite to that of the particles.

Directly applied electric charges on fibers or induced charges in dielectric fibers can be relatively strong and may dominate the rate of particle removal if the charges on the particles are very high. The induced charge on dielectric fiber filters can also assist in the removal of uncharged particles if the particles also have significant dielectric coefficient (Figure 6). This can occur because the particles themselves also experience a separation of charges. The particles will remain electrically neutral, but will have charge separation, positive charges on one side of the particle and negative charges on the other side. There will be no force on such

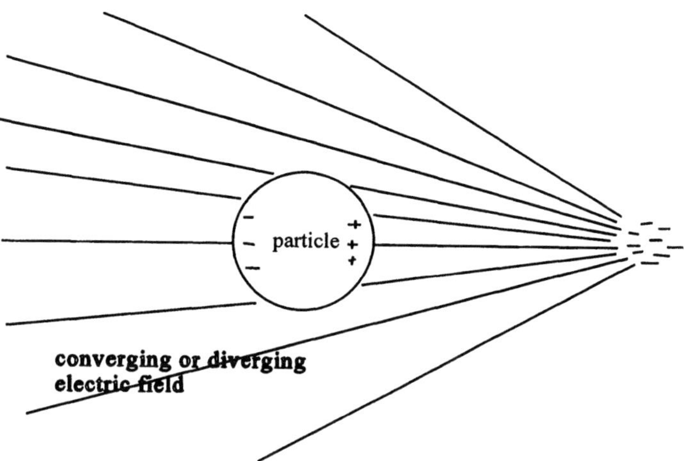

FIGURE 6 Separation of charge on a dielectric particle with no net electric charge. In the presence of an electric field gradient, the particle will be attracted to the region with the highest field gradient because the electric field intensity is slightly different on the different sides of the particles.

particles in a uniform electric field because the positive side of the particles will be attracted in one direction with the same force that the other side of the particle is attracted in the other direction. However, when the "polarized" particles are in a nonuniform electric field, there will be a net force on the particles because the force on the positively charged part of the particles will not be the same as the charge on the negative side of the particles. The difference in the forces on the two portions of the particles results from the slight separation of the charges and the fact that in a nonuniform electric field, the field intensity is slightly different at the two opposite sides of the particles. The forces will always be the highest on the side of the particles that is in the highest field, so the net force will always be toward the regions of highest electric field strength.

When electric charges are induced in a dielectric fiber bed, the electric field lines will converge toward those charges and thus to the fiber surfaces. This means that particles with induced dipole charges will be attracted to the fiber surface. The dipole charges can be significantly weaker than electrostatic forces on highly charged particles, but the dipole forces can be significant and even dominate forces when particles and fibers have high dielectric coefficients. The force on the particle increases with particle size (greater dielectric charges can be induced across larger particles).

Short-range forces are especially important because of their role in retaining particles on the fiber surfaces, but they can also contribute to the transport of particles to the surface. These forces include van der Waals forces and electrical image forces, but the effects of these forces decline rapidly with increasing distances from the fibers. They usually have less effect in transporting particles to the fiber surface than in holding the particles once they reach the surface.

In general, the transport effects that bring particles to fiber surfaces are additive, but not necessarily linearly. In many filter applications, only one or two forces are likely to contribute most to particle removal. If only one force is dominant, the filtration efficiency can be estimated from the contributions of that force alone, no combination of effects is needed. When two effects must be considered, there could be errors from simply adding the contributions of each effect if there is a coupling of those effects.

Theoretical analyses usually assume that retention of particles is essentially complete once the particle reaches the fiber surface, but this is only an approximation. A coefficient of retention can be used to account for loss of particles that strike the surface, or even release of particles that have resided on the fiber surface, but the effects of nonunity retention factors may be outside the accuracy of theoretical models for fiber performance which also have to include other significant assumptions. When unknown filtration parameters for a model are measured experimentally, the effects of a complete set of retention parameters are incorporated within those measurements. It usually is not easy to separate the effects.

Effects of Particle Loading on Fibers

Most models and descriptions of fiber filter performance are based upon the behavior of clean filters, those with no particles. Thus they describe the behavior of fresh filters when they are first put into service. Note that the description of the mechanisms for transporting particles to fiber surfaces assumed a clear fiber surface. However, after a filter has been left in service for an extended time, particles may completely coat the fibers, perhaps even several particle diameters thick. The deposits will not be uniform throughout the filter, but will be concentrated near the leading end of the filter, near the inlet. This should be evident from the approximate first order behavior of particle removal; the rate of particle removal in a unit length of filter is approximately proportional to the concentration of particles. Stated another way, the probability for a particle to be removed in a given unit length of filter is constant and independent of concentration. This means that the greatest accumulation of particles on the filter will occur near the inlet end of the filter bed where the concentration of particles in the fluid is the highest. (Recall that particles are concentrated near the front of a filter, even with deep bed filters where the particles are removed within the filter media or bed.) However, accumulation of significant quantities of particles does affect filter performance, and then the filter performance will no longer be independent of position in the filter because the particle loading is different throughout the filter.

One obvious effect of thick coatings of particles on fibers is an occasional loss of particles by fluid shear, and such lost particles will need to be caught in downstream regions of the filter. However, there are other effects that are often just as important, and these can cause the filter performance (as measured in particle removal efficiency) to increase rather than decrease with loading. These changes result from geometric changes in the fiber-particle structure. Added particles increase the effective diameter of the fibers, and that increases the ability of the fiber to intercept and catch particles. In general, filter efficiency will increase with particle loading as long as the improvements in particle removal increase more rapidly than the loss of particles from shear forces.

The qualitative effects of particle accumulation on attraction of particles by electric fields deserves some special mention. Accumulated particles can neutralize applied electric charges on fibers unless the charge of the particles can be bled to the fiber and then to the source of the applied potential. This will usually occur when electric charges are imposed on conducting fibers. The induced charge on nonconduction dielectric fibers can define the maximum filtering capacity for removal of charged particles.

On the other hand, if the collected particles are conducting or dielectric materials, they can form more localized sources of charge where the potential gradients are particularly high. Note that removal of polarized, but uncharged, particles does not neutralize charges on the fibers. These points where dielectric particles have accumulated on the fiber can then be points where other particles are attracted even more strongly [9]. This results in growth of accumulated particles in long chains or "whiskers" (dendrites) rather than in smooth coatings on the fibers. The high field gradients near the edges of whiskers can enhance the attraction of particles to the whiskers and thus enhance their growth. Note that the dendrites can also increase the effective fiber length and/or diameter and increase the rate at which particles are attracted to the fibers by other mechanisms. However, whisker growth can also occur in the absence of an applied electric field, possibly because whiskers protrude into the flow and thus intercept particles effectively. Although whiskers can be entrained in the fluid flow and thus removed from the filter, and their net contribution can be difficult to estimate, they will most likely increase removal efficiency.

Pressure Drop

The pressure drop through fiber filters is often given by the manufacturer, and when such data are available they should be used. There are a number of correlations and semitheoretical equations available for estimating the pressure drop across fiber filters. For the most common highly porous fiber filters, theoretical models are likely to be based upon drag forces on individual cylinders in the flow field, and the pressure drop is then estimated from the sum of the forces on all fibers in the bed. For the less common densely packed deep bed fiber filters, the Carman-Kozeny equation may be used as an approximation. However, data on the particular fiber filter of interest should be sought before resorting to approximate equations and correlations.

Another way to estimate the pressure drop across loosely packed fiber filters is to estimate the force on individual fibers independently and then multiply this force by the number of fibers in a unit volume of the filter. This will give an estimate of the total force on the solids (fibers) in the bed, and thus the total force on the fluid, or the pressure drop. The complications are the variety of fiber orientations with respect to the fluid flow. The fibers can have any orientation, and in some cases the orientation can be approximately random. However, the distribution of orientations may not be random, and such an assumption could be in error.

Although the particle removal efficiency may not degrade with particle accumulation, the pressure drop across the filter will certainly increase. The increase in the pressure drop required to maintain sufficient flow through the filter may be the principal reason for eventually replacing or regenerating the filter. The pressure drop across a fiber filter can usually be described reasonably as the sum of the drag on all of the fibers. As noted, since the fibers are separated far from each other in high void fraction filters, the drag on each fiber can be estimated by the drag on a cylinder in a infinite fluid.

Increases in the effective fiber diameter will increase the drag on the fiber and thus on the pressure drop, but it may not be easy to estimate the effective increase in diameter simply from the loading. If the particles form uniform coatings on the fibers, reasonable estimates probably can be made. However, as noted earlier, the particles may accumulate preferably on one side of the fibers or even form whiskers that extend from the fiber surface. The pressure drop will then be affected differently, depending upon how the particles adhere to the fibers, and estimating the change in the pressure drop can be complicated. The estimates would need to account for the arrangements of particles on the fibers.

APPLICATIONS OF FIBER FILTERS IN ENVIRONMENTAL AND WASTE MANAGEMENT

Like all deep bed filter systems, fiber filters are usually used to collect particles that are not easily handled by surface (cake) filters. These are usually applications that involve very small particles or fibrous particles that form highly nonporous filter cakes. Although fibrous particles are not as common in environmental and waste problems, fiber filters are particularly suitable for handling fibrous particles such as asbestos. As noted above, fibrous particles often will not penetrate granular deep bed filters; instead they usually form filter cakes with low porosity at the entrance to a granular bed. Although deep bed granular filters may be alternatives to fiber filters for removal of some near spherical (granular shaped) particles, they are not likely to be useful for removing fibrous particles; fiber filters are highly likely to be selected for those cases.

As noted earlier, fiber filters are more likely to be used with gaseous systems where the drag forces on the fibers are not great enough to move the fibers and compress the filter media. Because the fiber spacing can be very large relative to the fiber diameter, the pressure drop across the filter can be low, at least relative to that of comparable filters. However, if both the total pressure gradient and the particle removal coefficient are approximately proportional to the fiber packing density, there may be little advantage to reducing the fiber packing density further if the fraction of particles removed is the principal performance criteria.

Fiber filters can be used to remove particles from liquids, but significant structural strength will be required to hold the fibers in place for the required liquid flow rates. Fiber filters are less common for removing particles from liquids; such applications are likely to be filled by granular bed filters (for small particles) or by surface filters (for larger particles that form porous cakes).

The most common application of fiber filters in environmental and waste management operations are likely to be in controlling particulate emissions in off-gases. Systems to filter large gas flow rates are likely to consider cyclones or electric precipitators to remove the bulk of the particles, especially the larger particles in the stream. Filters are likely to be needed only if the particles are sufficiently toxic and small, and if large fractions of the smallest particles must be removed. Even in such cases where filters are needed, deep bed filters will compete with other filter systems such as large sock (bag) filters described in Chapter 8. Bag filters have the important advantage that they can use automatic cleaning systems (with shaking or burst of reverse flow to remove the particles from the sock). However, deep bed fiber filters do have the potential to remove particles that are too small to remove efficiently by any type of surface filter.

Some of the advantages of bag filters and precipitators disappear as the size of the gas stream decreases. The capital cost of precipitators and automatic bag cleaning systems may not be justified for very small gas streams. The low capital cost of small fiber filters make them especially attractive for small gas streams. This is illustrated in the extensive use of fiber filters in small consumer devices such as household furnace filters or intake air filters for automobiles. (However, these applications often involve removal of a significant quantity of fibrous particles, another condition that favors use of deep bed fiber filters.) Similarly, fiber filters are likely to be both effective and economical for removing very fine particles from many small off-gas streams in environmental and waste processing.

When extremely effective filtration of off-gas streams is required, fiber filters are likely to be attractive choices. The high efficiency HEPA filters that are so common in the nuclear industry are examples of the choice of fiber filters for maximum removal of even fine particulates.

GRANULAR DEEP BED FILTERS

Deep bed filters can be filled with granular materials as well as fibers. The granular material in the filter can be sand, glass, or any other nonfibrous packing. Granular filters are deep bed filters in the same sense as fiber filters and retain the removed particles within the filter bed (the filter medium), not just on the forward surface of the bed. The advantage of a granular bed over a fiber filter bed can be the greater structural strength of the packed bed. The loosely filled fiber filters offer low flow resistance, but they usually have limited structural strength and are subject to compaction of the widely spaced fibers under significant forces from the fluid being filtered. A packed bed has high structural strength with only limited compressibility, and packed beds usually will maintain a similar flow resistance over a wide range of imposed pressure gradients. A loosely packed fiber filter (one with high void fraction), however, may compress (become compacted) and lose most of its permeability if sufficiently compressive forces are applied.

Because of their greater ability to function with higher pressure forces, granular filters are more likely to be the choice to remove small particles from liquids, especially dense or viscous liquids. In describing granular filters, there is an opportunity for confusion because two different groups of solid particles must be discussed, the small particles being filtered from the fluid and the larger particles in the bed. In this discussion, the larger particles which make up the bed will be called "granules," and the smaller particles being removed from the fluid will be called "particles." An excellent book summarizing much of the work and information available on the behavior of granular filters is Tien [10], and readers interested in more detail, especially on modeling and design of deep bed granular filters, should consult it.

The granular material used in the filters can be of any shape, but it is desirable that the packed bed have sufficiently large pore (void) regions that the pressure losses will not be excessive and that the pores be as uniform as practical so all of the pores will be used approximately equally. It would be undesirable for most of the flow to go through only a small fraction of the pores in the bed. The need for uniform flow through deep bed filters is similar to the need for uniform flow through packed adsorption or ion exchange beds. Uniform flow is usually achieved when the granules are of approximately uniform size and are allowed to settle randomly into the bed. The most common shapes of granules are irregular convex shapes much like those of common sand. These shapes can usually be approximated as near spherical, but that term can be assumed to apply to many convex shapes.

To keep the granular sizes as nearly uniform as practical, a natural material is screened, or a synthetic product is manufactured with approximately uniform particle sizes. The flow resistance of the bed is controlled principally by the diameter of the granules used in the bed, but uniform granule sizes generally result in less flow resistance than highly nonuniform size mixtures. Mixtures of nonuniform size granules usually result in lower void volumes because smaller particles are likely to fill in some of the pores (void volume) between larger particles and/or result in pore sizes that are associated with the smaller size particles in the mixture, not with the larger or even the average particle size. If a high throughput is needed or if the fluid is very viscous, it will be necessary to use relatively large granules. For water, other low viscosity liquids, or gases, it may be preferable to use smaller granule particles that are likely to give higher filtration efficiencies per length of filter bed.

Recall that the principal merits of deep bed filters are their ability to remove very small particles without developing the extremely high pressure drops from the buildup of a filter cake of very small particles at

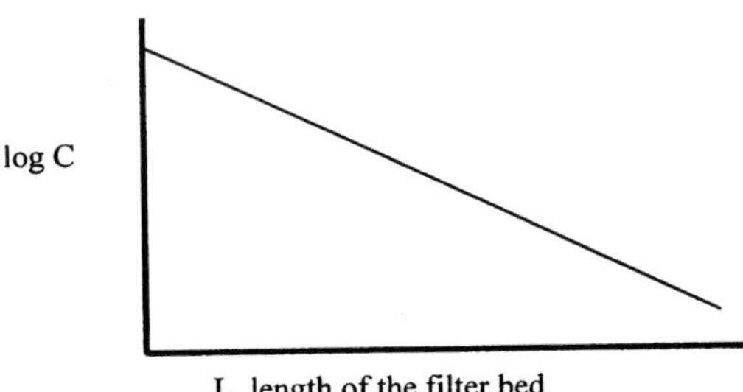

FIGURE 7 Decrease in particle concentration with position in a granular (or other deep bed) filter.

the entrance of the filter. In all types of deep bed filters, the particles are removed throughout the deep bed, not just on the forward surface of the filter.

Forces Bringing Particles to Granule Surfaces

The forces that affect particle motion and that attract or retain particles on granular filters are the same as those that are important in fiber filters. These are interception, inertial forces, gravity, diffusion or Brownian motion, electrostatic forces, and short-range van der Waals forces. In most ways the analysis of granular filters is similar to the analysis of fiber filters. Any or all of these forces that transport particles to the medium surface in fibers can contribute to the rate of particle removal in granular deep bed filters. If the concentration of particles is relatively low, one can assume that in a fresh or unloaded bed any particle acts independently of the presence of other particles, and the rate of removal is proportional to the concentration of particles. This is a first order process and results in an exponential decay of particle concentration with increasing bed depth (Figure 7). That is,

$$\frac{C}{C_0} = 1 - \exp(-m_g L) \tag{8}$$

where C is the concentration of particles leaving the bed, C_0 is the concentration of particles at the inlet to the bed, L is the length of the bed, and m_g is the extinction coefficient or the fractional rate of particle removal per unit length of bed. Note that this expression is essentially the same as that used for fiber filters. The principal difference is the incorporation of the surface area (or the number of fibers or granules) into the extinction coefficient, eliminating the surface area as an independent variable. This is done because with a packed bed of granules, the number of particles per unit volume of bed is not usually an independent parameter; it depends upon the diameter and shape of the particles. In contrast, deep bed fiber filters usually use very loose fiber spacing, and the spacing can be adjusted to have more or fewer fibers per unit volume. The filter surface area per unit volume is then an adjustable parameter, and it is then more appropriate to include the specific surface area or the concentration of fibers as a separate variable. Since the surface area per unit volume is not an adjustable parameter for granular filters, there is usually little justification for including the surface area as a separate parameter.

The analysis or prediction of filter performance without experiments is more difficult for granular filters than for fiber filters principally because the internal structure and the flow patterns with the bed are more difficult to predict. However, this may not be of major importance to many users if experiments are performed to evaluate the extinction coefficient. The complexities of the flow patterns in granular beds are included in any measured extinction coefficients.

In contrast to granular bed filters, the flow around each fiber in a fiber bed could be approximated relatively well by assuming that each fiber was sufficiently far from other fibers that it acts independently of the other fibers. (There were, however, questions of the range or spectra of fiber orientation which is set by the filter construction.) Also, flow through loosely packed fiber beds, as noted, is more likely to be in the transition or inertial flow regime, but flow through granular filters is more likely to be in the laminar (viscous) flow regime, especially when relatively viscous liquids are to be filtered.

It is often easier to model viscous flow than transition flow, even if it is necessary to use numerical techniques, but the geometry of the flow channels of granular beds is so complex that mathematical modeling is not likely to be practical even for viscous flow without considerable simplification. The particles are usually dumped into the bed and their positions involve considerable randomness. The flow channels within the bed are largely random, contain many constrictions, and are interconnected with other flow channels. Thus, there is no exact way to evaluate accurately the flow patterns within such a system, so idealized flow patterns which approximate the patterns have been studied.

Developing a useful estimate of the flow pattern within a granular bed is probably the most difficult conceptual problem in predicting the behavior of granular filters, and one should not consider any of the current flow patterns suggested for randomly packed beds more than models or approximations of the real case. Calculating particle motion within the assumed flow field is often called "trajectory analysis." The calculations are conceptually simple but can be mathematically complex. A review of analyses for granular filters prior to 1979 was prepared by Tien and Payatakes [11]. Important publications describing particle motion within granular beds are Windk et al. [12], Rajagopolan and Tien [13], and Viadyanathan and Tien [14].

One commonly used model for approximating the flow patterns in randomly packed beds was developed by Happel. This is a "cell" based model [15], which means that the flow through the bed is assumed to be approximated by flow through a series of cells. This model assumes that each cell is a sphere containing a granule surrounded by a spherical shell of fluid. The ratio of fluid volume to granule volume is the same in the cell as in the bed. Thus, the void (fluid) fraction in the bed defines the volume (or thickness) of the shell of fluid in the cell. In packed beds of near spherical granules, the void fraction is likely to be near 0.5 and not vary greatly from this value (0.55 is approximately as high as the void fraction becomes for beds of very uniform spheres), so the fluid volume is not likely to be far from the granule volume.

Happel assumed that the fluid velocity is zero at the granule surface and that there is no shear at the outer surface of the spherical cell. With this "free" flow at the outer boundary of the cell, fluid flows into the cell from the upstream side of the cell and flows out from the outer surfaces of the cell on the downstream side of the cell.

With the hydrodynamics of the Happel model, the trajectories of particles through the cell can be estimated as described for flow around individual fibers in a deep bed fiber filter. The same effects (inertia, gravity, etc.) described for fiber filters can influence the particle motion and divert particles from streamlines so they impact the granule surface where van der Waal forces can hold them to the granule. The principal difference between granular filter and fiber filter analysis is that the streamlines for flow in granular beds (as illustrated in a Happel cell) differ from flow around a cylindrical fiber used to model fiber filters.

The analysis then indicates that a fraction of the particles entering a Happel cell will touch the granule surface and be retained in the cell. In terms of the overall granular bed, the model suggests that this is the fraction of particles removed in a volume of bed corresponding to the Happel cell, one granule and the corresponding void volume. This type of analysis estimates the fraction of particles removed in each volume of a granular bed filter equivalent to the Happell cell volume. This first order removal rate results in the exponential decrease of particle concentration with increasing bed length, as indicated in Equation (9). This behavior is generally verified qualitatively by experiments. The exact value of the extinction coefficient, however, is more difficult to estimate accurately because the models for flow patterns around the particles are only estimates.

Similar conclusions could be reached using other cell models for flow within a granular bed. Since essentially all models predict a constant extinction coefficient, the relative merits of different models of this type would be their ability to predict the extinction coefficient for the exponential decay of concentration with bed depth. Although the Happel cell is one of the best such models, it is advisable to use experimental data on particle retention rather than predictions from any model when data are available. The exponential decrease in particle concentration with bed depth (Figure 7) is more likely to be correct than values of the concentration reduction parameter predicted by any model. Besides the rather significant simplifications in the cell models, many real problems involve the use of nonspherical granules, nonuniform size granules, nonspherical particles, and nonuniform mixtures of particles. All or any of these factors can add to the unreliability of applying theoretical extinction coefficients and make it more desirable to use the theories only as guidelines and rely upon experimental data for estimating the exact and quantitative performance of a granular filter. The principal merits of the different models may be their ability to predict how different parameters such as granular size or fluid properties affect the extinction coefficient rather than the ability to predict the value of the extinction coefficient accurately and reliably.

Like fiber filters, the captured particles will not be distributed evenly throughout the bed, but they will be concentrated near the entrance of the bed. This is the consequence of a constant extinction coefficient, and the behavior is just like that described for fiber filters. The concentration of retained particles will decrease with distance down the bed in the same manner that the concentration of free (unfiltered) particles decreases with distance down the bed. As long as viscous shear forces within the bed do not remove significant numbers of particles that have been collected on the filter, the effectiveness of the granular bed in removing particles will not change significantly. Like fiber filters, the particle removal efficiency could actually increase as the flow channels become smaller. However, the flow resistance of the bed will increase, and the buildup of pressure drop is more likely to be the reason for replacing a granular bed filter. With the smaller flow channels in granular filter beds, fewer particles can be retained in the bed before the pressure drop starts to increase significantly. Since the void volume of a granular bed is only approximately 50%, the capacity of granular beds for retaining collected particles is usually quite limited, often significantly less than the capacity of fiber filter beds.

Granular deep bed filters, like fiber filters, can be discarded after use. However, the randomly packed granular filters often can be regenerated by reverse flow. This is a significant difference between granular bed filters and fiber bed filters. If the reverse flow velocity is sufficiently high, the granules will fluidize, and many particles may be stripped from the granules as the granules tumble vigorously in the fluidized stream. Of course, the particles that are removed by fluidizing a granular bed are then recovered in a (usually) more concentrated fluid stream, not in a concentrated paste or a dry form that is relatively free of the fluid.

The performance of granular filter beds, like the performance of fiber filters, can be enhanced by applied electric fields [16], but electric fields are not used when water is the fluid being filtered because of its high conductivity. This is unfortunate because water is probably the fluid that is most likely to need filtration in environmental applications. Nevertheless electric fields can be used with gases or nonconducting liquids (less likely to be used in environmental problems). Magnetic fields can be used in a similar way for particles that are paramagnetic. Although there is a great deal of similarity in some magnetic filters and electrostatic dipole filters, magnetic filters with granular or fiber beds are discussed in Chapter 12, where the behavior of such filters can be combined with other magnetic phenomena to separate particles from particles and particles from fluids.

APPLICATIONS OF GRANULAR DEEP BED FILTERS IN ENVIRONMENTAL AND WASTE TREATMENT

Granular deep bed filters, like all deep bed filters, are likely to be used when the particles to be removed form filter cakes with very high flow resistances. Unlike fiber filters, granular bed filters are usually limited to the removal of granular (approximately spherically shaped) particles. Although fiber filters can remove fibrous contaminants by entangling them on the fibers of the filter, fibrous contaminants usually will not enter a granular filter bed but will collect at the surface (entrance) of the bed and build up a nonporous filter cake with high flow resistance. The openings to granular filters (between granules) are likely be relatively small and less able to allow fibrous particles of comparable size to enter the bed; larger openings are obtained more easily in fiber filters.

As noted earlier, granular beds can offer much greater structural strength, so they are more likely to be used for removing particles from liquids than fiber filters. Since water and air are probably the most common streams to need filtration for environmental and waste treatment, the role of granular bed filters in these applications is largely for water treatment.

Water treatment is well-developed technology, and there is much experience in the use of granular beds with water. Although there is far more experience devoted to treatment of raw (natural) water, the same technologies usually can be easily adapted to treatment of discharge streams or other aqueous waste streams in environmental or waste treatment. Because the water treatment industry, like much of the environmental and waste industry, often needs to treat large volumes of water, the granules in the bed need to be readily available in the quantities needed, to be of low to moderate cost, and to introduce no additional contaminant to the water. These conditions are usually satisfied by common sand, and there is enough experience with sand filters (granular filters using sand as the granules) that they will be discussed in detail.

Sand Filters as Deep Bed Filters

Sand filters are commonly used in water treatment and are especially common for large-scale treatment systems. It is difficult to know exactly where to place sand filters in this book. A more detailed discussion of sand filters is given in Chapter 8 because surface filtration (cake filtration) is usually the most important aspect of sand filter operation. However, a brief discussion of sand filters is placed here principally because they appear to look much like deep bed granular filters, and the deep bed filtration action often is an important aspect of many sand filters since the smallest particles are often removed by deep bed mechanisms during the initial period of an operating cycle. Sand filters usually accumulate more particles on the surface of the filter, which would make it more appropriate to discuss sand filters along with surface filters. Once a substantial filter cake is built up on a sand filter, the portion of particle retention from surface filtration will increase even further. This discussion will focus on the behavior of sand filters as deep bed filters. In many cases, a sand filter may remove a significant fraction of the particles by each mechanism.

As the name implies, sand filters are usually constructed of beds of sand, usually with mean particle sizes near or less than 0.5 mm in diameter. Actually other types of particles can be used, but the cost and chemical inertness of silica are usually more suitable for most applications. The attractiveness results largely from two features, the ability to construct relatively large filters at very reasonable cost and the ability to regenerate the filter relatively easily. Particles removed by surface filtration can be removed by back flow (usually by fluidizing the bed) or by removing the upper few inches of the bed, and particles caught deeper within the bed are removed by fluidizing the bed (back flow).

A sand filter consists of a layer of sand, usually 18 to 28 inches deep. The bed of sand is usually supported by a thinner layer of coarser particles, often some form of gravel. The gravel layer usually covers a series of perforated horizontal pipes that drain the filtrate from the bed. The filter bed may be enclosed in metal or concrete structures, and the filtration usually is carried out under pressure or under a head of only a few feet of water. Very large systems that are to be operated without overpressure are likely to be constructed in concrete, but smaller filters and pressurized filters are usually constructed more easily of metal. Sand filters usually have characteristics of surface filters and deep bed filters, and most particles usually are collected on the sand surface. Usually only a fraction of the smaller particles penetrated the sand deeply, and that penetration probably occurs during the initial part of an operation cycle before significant filter cake is formed. Thus, sand filters are discussed as surface filters or as deep bed filters. In this book, the major discussion of sand filters is as surface filters since that is generally believed to be the principal mechanism of particle removal, and only a brief mention of sand filters is given here.

After a sand filter is loaded and needs to be regenerated, any filter cake accumulated on the surface of the bed can be skimmed off with the top few inches of sand. That removes the particles that are removed by surface filtration and many particles removed by deep bed mechanisms that are also concentrated near the surface but within the filter medium. In addition, or alternatively, the solids can be removed by upflow of water with sufficient velocity to fluidize the sand granules and elutriate the filter cake and captured particles. This operation will remove only particles that were trapped within the filter medium by deep bed mechanisms. If the upflow rate is controlled sufficiently, there will be little or no fluidization of the gravel granules that support the sand granule, so sand granules that happen to migrate into the gravel will be washed from the gravel. Even if some gravel is fluidized, the larger gravel granules will settle faster than the sand and thus reform a support for the sand, which settles later.

In water treatment, sand filters play a major role in removing the bulk of the bacteria and other organic solids from the water. This is not effective enough to remove all biological hazards from the water, but it can remove enough material that the demand for chemical disinfection agents is reduced greatly. Bacteria can also grow on/in the sand bed and can play a role in particle removal. More details on sand filters and the surface filtration of sand filters are given in Chapter 8.

MAGNETICALLY STABILIZED GRANULAR FILTERS

Another special case of a deep bed granular filter is the magnetic stabilized granular filter first commercialized by Westinghouse and described by Rosenweig [17,18]. This filter concept offers the advantage of low pressure drop with high gas throughput. The void fraction in a bed of paramagnetic granules is fluidized by an upflow of gas. The pressure drop across fluidized beds is limited to values that are usually not much greater than the weight of the bed divided by the cross section of the bed. The pressure drop across a fluidized granular bed may be greater than the pressure drop across many fiber beds, but at high fluid rates the pressure drop can be significantly less than the pressure drop across a packed bed. Although the pressure drop is moderately low, a fluidized bed is usually not an effective filter because motion of the granules transports any particles collected on the granules to the outlet end of the bed, and attrition of particles from the granules can degrade filter performance further.

However, when a sufficiently strong magnetic field is applied with the direction of the field parallel to the axis of a bed of paramagnetic granules, the motion of the particles stops. Then the "magnetically stabilized" bed will act much like a granular filter with a higher void volume than a packed bed of the same granules. Recall that it is more difficult to alter the void fraction and (for a given granule size) the size of the flow passages in a granular bed. The application of a magnetic field after the bed is fluidized is one way to do that. The pressure drop across such an expanded and magnetically stabilized bed would then be less than the pressure drop across a packed bed. There is probably some nonrandom orientation of the particles in a magnetically stabilized bed into longitudinal "strings," but the beds still perform much like random beds of particles. The principal difference is the higher bed void fraction. Apparently work on magnetically stabilized granular filters continues [19]. One reason to stabilize a bed is to obtain higher void fractions so the beds can be operated at high flow rates and reduce the pressure drop at the higher rates. However, one can also operate at lower flow rates those beds that have been stabilized and expanded at higher flow rates (without turning off the magnetic field). Since the expansion of the bed is retained at the lower flow rates (lower than the minimum fluidization velocity), lower pressure drops can also be maintained at the lower flow rates as well.

The magnetic field can also enhance the removal of particles of paramagnetic materials. The uniformly applied magnetic field that stabilizes the motion of paramagnetic granules in the bed is distorted by the paramagnetic granules. The distortion causes the magnetic field lines to converge into the granules in the same way that electric fields cause electric field lines to converge into fibers of dielectric materials in fiber filters. Paramagnetic particles then are drawn toward the granules in the presence of the magnetic field in the same way that dielectric particles are drawn toward fibers in the presence of an electric field. This enhancement can be significant but depends upon the paramagnetic properties of the particles to be removed.

Magnetic fields can also be used to separate different particles from each other based upon their paramagnetic properties, but separations of different particles from each other will be discussed in another chapter.

Electrically Stabilized Granular Filters

Electrostatic forces can be used to "stabilize" expended granular beds in a manner much like stabilization with magnetic fields. Electrostatic stabilization of fluidized beds and granular filtration in such beds was evaluated and discussed by Barker et al. [20]. Electrostatic forces on granular particles follow equations similar to those used to describe magnetic stabilization. The forces depend upon the dielectric constant of the granules instead of the paramagnetic properties.

There are, however, two important differences in electrostatically stabilized beds. The first is that strong electric fields cannot be applied in conducting fluids such as water. Electrostatic granular filters are potentially useful in nonconducting liquids and essentially all gases. For environmental applications, the most likely applications are with gases.

The second important difference is the effect of the electric field in enhancing the filtration of electrically charged particles or dielectric particles; many (perhaps even most) stable suspensions of particles in nonconducting fluids contain electrically charged particles because the electric charges prevent agglomeration of the particles. As described earlier, the electric field enhances filtration efficiency even without expansion and stabilization of the bed. The same forces and the same enhancement are achieved (qualitatively) with stabilized expanded beds as with packed beds when the electric field is applied. However, because the flow paths are altered significantly by the fluidization and stabilization, the filter efficiencies cannot be predicted quantitatively from data on packed granular beds.

Although enhanced filtration of uncharged dielectric particles in the presence of an electric field is similar to enhanced filtration of paramagnetic particles in the presence of magnetic fields, there is no magnetic equivalence to the enhanced removal of electrically charged particles. As mentioned in the section on fiber filters, the applied electric field generates a stronger force on particles with significant electrical charge than it generates because of the dielectric effects; so for charged particles the electrostatic effects are likely to be significantly more important than the dielectric effects. Another significant difference in the effects of electric fields is the greater likelihood that particles will have significant dielectric constants than that they will have significant paramagnetic properties. Note that when considering the effects of electric or magnetic fields, it is important to know the properties of the particles as well as their size.

REFERENCES

1. C. J. Guzy, E. J. Bonano, and E. J. Davis. *J. Colloidal Sci. 95*, 523–543 (1983).
2. M. Beizaie. *Sep. Technol. 1*, 132–141 (1991).
3. T. Gillespie. *J. Colloid Sci. 10*, 299 (1955).
4. W. Ranz and J. Wong. *I&E Chem. 44*, 1371 (1951).
5. S. K. Friedlander. *AIChE J. 3*, 43 (1957).
6. F. S. Henry and T. Airman. *Fluid Filtration: Gas 1*, 13 (1986).
7. S. J. Judd and G. S. Solt. *Colloid Surf. 39*, 189 (1989).
8. G. S. Solt and S. J. Judd. *Chem. Eng. Symp. 113*, 105 (1989).
9. S. Tousi, F. S. Henry, and T. Ariman. *Sep. Technol. 1*, 66–72 (1991).
10. C. Tien. *Granular Filtration of Aerosols and Hydrosols*. Butterworths, New York (1989).
11. C. Tien and A. C. Payatakes. *AIChE J. 25*, 737 (1979).
12. W. J. Windk, D. Gidaspow, and D. Wasan. *Chem. Eng. Sci. 30*, 1035 (1975).
13. R. Rajagopolan and C. Tien. *AIChE J. 22*, 523 (1976).
14. R. Viadyanathan and C. Tien. *Chem. Eng. Sci. 43*, 289–302 (1988).
15. J. Happel. *AIChE J. 4*, 197 (1958).
16. R. E. Barker, R. R. Brunson, S. D. Clinton, and J. S. Watson. *Sep. Technol. 1*, 166–174 (1991).
17. R. E. Rosensweig. *Science 204*, 57 (1979).
18. R. E. Rosensweig. *Ind. Eng. Chem Process Des. Dev. 18*, 260 (1979).
19. C.-S. Luo and S.-D. Huang. *Sep. Sci. Technol. 28*, 1253 (1993).
20. Barker, R. E., R. R. Brunson, S. D. Clinton, and J. S. Watson. *Sep. Technol. 1*, 166 (1991).

10
Sedimentation and Settling

Sedimentation is the separation of solids from (usually) liquids based upon gravitational forces. There is no difference between sedimentation and settling except that some may use "settling" to describe relatively rapid separation and "sedimentation" to describe slower and more difficult separations. Other terms often discussed with sedimentation are coagulation and flocculation. These terms have much different meanings and refer to approaches to increase the settling (sedimentation) rates. Gravity which drives the separation could be either the normal gravity experienced on earth or similarly acting forces from centrifugal fields. In this discussion, centrifugation and sedimentation will be discussed together because the principles are so similar, although the equipment used is considerably different. Although sedimentation could refer to separation of solids from any fluid, the discussion will focus on separation from water. This is the most common difficult environmental sedimentation problem, and the reader is likely to have little difficulty in applying the principles to separations from other fluids.

Separation of solids from liquids by sedimentation depends upon the differences in densities of the solid and liquid; however, the rate of separation depends upon particle size and shape and the fluid viscosity as well as the difference between particle density and fluid density. It is not practical to consider all particle shapes that one may want to separate; instead, we consider only spherical and fibrous particles, and most of our discussion is devoted to spherical particles. Spherical particles are simply a "model" shape to describe granular and other near spherical particles, that is, most convex-shaped particles.

Sedimentation plays a major role in wastewater treatment either for discharge water or intake water to industrial facilities. It is usually the most practical way to remove the largest particles in a wastewater, and it may be used in conjunction with coagulation and flocculation processes to remove smaller particles. Sedimentation is an important companion to precipitation operations that are used to remove a variety of undesirable dissolved materials, especially the heavier metals. Precipitation operations are discussed in another chapter, but the subsequent separation of the solids and liquid is discussed here and in Chapter 11.

Coagulation and flocculation are widely used to assist sedimentation in treating well water or surface water for use as municipal or process water. These operations assemble the fine individual particles into much larger bunches of particles (agglomerates) that can settle more rapidly.

COAGULATION AND FLOCCULATION

These operations are discussed together simply because they are usually used together. In fact, they seem to often be confused with each other. There will seldom be reason to separate the two operations. They are usually used together when an important fraction of the particles is so small that sedimentation rates would be undesirably slow without coagulation and flocculation. The objective of these operations is to cause the particles to stick to each other or to other particles to form larger masses that will settle much more rapidly.

"Coagulation" refers to changing the water composition to reduce the repulsive forces (electrical charges) that hold the particles apart. "Flocculation" refers to efforts to increase interparticle contact. It should be obvious why these operations are usually carried out together.

Most particles suspended in water carry a net electrical charge. This charge can result from imperfections on a crystalline surface, from charged ions that have left the surface, from ions that adsorb on the surface, or from materials that have otherwise become attached to the particle surface. The charges need not be uniformly spread over the particle, at least if the particle is not an electrical conductor. Clays such as those found suspended in groundwater usually carry a net positive charge at pH values below 6.

The pH of the water can greatly affect the electric charge on the surface. As noted earlier, clays carry a positive charge in low pH (acid) solutions. However, in higher pH solutions, the surface charge will go to zero and eventually become negatively charged at higher pH values. This can be thought of as resulting from the addition of negatively charged hydroxyl groups to the surface, but the surface chemistry can be more complex than that. Most suspended solid particle surfaces that have an oxide or hydrous oxide surface will reverse electrical charge at some pH, often near neutrality (near pH 7). This means that if the pH is raised (or possibly lowered) to the correct value, there will be no net electrical repulsive force keeping the particles suspended. Then the particles can become attached and eventually settle as larger clusters of particles far more rapidly from the suspension than individual particles. Raising the pH to reach the conditions for zero surface charge is called coagulation, and enhancing interparticle contact so that larger clusters of particles can form is called flocculation. The net result is the production of agglomerates of numerous particles. Nonuniform surfaces may make it difficult to find an ideal pH for zero surface charge on all surfaces. This is a complication, but the pH for zero charge on many oxide surfaces is near enough to neutral that the surfaces charges on most surfaces can be reduced greatly if not eliminated entirely.

Other materials also may be added to enhance coagulation. In the United States, aluminum sulfate is probably the most commonly used coagulant, but other soluble aluminum compounds such as aluminum chloride or soluble ferric iron compounds could be used. At high, or even neutral pH, these compounds hydrolyze and form insoluble amorphous precipitates. These precipitates entrap or otherwise coagulate other particles and promote sedimentation. The exact mechanism for coagulating common suspended particles is complicated and not always fully understood; several mechanisms may be involved. The surfaces of suspended particles may act as nucleation sites for the growing precipitate (during precipitation operations), and/or the suspended particles could simply become entangled in the growing polymeric hydrous oxide precipitate. Organic "polyelectrolytes" can also be used to coagulate suspended particles. Organic coagulants may be more expensive and/or less desirable for treating waters that are to be discharged to the environment.

It is important to add the proper amount of coagulant, just as it is important to coagulate at the proper pH. It is probably apparent that sufficient coagulant should be added to coagulate the suspended particles adequately. On the other hand, adding excessive amounts of coagulant would be wasteful. Perhaps more importantly in waste management when the solids are to be sent to disposal operations, excessive coagulant would add unnecessary waste volume, and even result in less effective coagulation. The reasons are not all evident, but during precipitation they could involve factors such as nucleation of precipitate within the solution, not principally on the suspended particles where they could enhance coagulation most effectively.

Once the conditions are set for reducing the repulsive forces that prevent agglomeration of the particles, time is required for the particles to agglomerate; this agglomeration of particles is called flocculation. In many cases, the flocculation process will occur in two steps. If the initial suspended particles are sufficiently small, that is in the colloidal range (less than 1 micrometer), motion of the particles relative to each other will be governed by Brownian motion, and interactions between particles will be random. This is usually a second order process; that is, the rate of contact between particles within a given volume of solution is proportional to the square of the particle concentration. If the particles are larger, or once the agglomerated groups of particles become large enough that Brownian motion no longer controls the rate of interaction, viscous shear forces begin to play a greater role. In this regime, the rate of interaction is related to the particle diameter, their spacial separation, and the shear field in the water. The rate of interaction is then first order, proportional to the concentration of particles. Nonuniformity in particle sizes can also affect interaction rates, especially as larger particles or flocs begin to settle through the smaller particles or when larger and smaller particles are swept past each other by convection. Thus any force that moves the particles of different size past each other at different velocities can enhance interaction and flocculation.

One may instinctively feel that stirring would tear agglomerates apart, and with sufficiently vigorous stirring that can happen. However, low to moderate stirring is more likely to enhance agglomeration rather than impede it. This discussion has implied that particle interaction (touching) produces a high probability for the particles to become attached and thus to flocculate. It also pointed out that the rate of flocculation is improved by shear (stirring). However, it is true that extremely high shear stresses can tear flocs apart and result in smaller, not larger, flocs. Deflocculation will occur at sufficiently high shear stresses in rather turbulent shear. Some deflocculation may occur even at low shear stresses, but the deflocculation rate may be small compared with the flocculation rate leaving a net growth in floc size. Thus, it is desirable to stir, but not stir with excessive vigor, the flocculating suspension to promote contact between particles.

Coagulation/flocculation equipment usually has two stages: an initial mixing stage when the coagulant is added and a second stage when most flocculation occurs. The initial mixing stage is carried out with rather vigorous mixing. When precipitating the solids, it is desirable to distribute the precipitation agent and the coagulant uniformly throughout the mixture before precipitation begins to occur. As noted earlier, concentrations of flocculating agents that are too high can actually decrease flocculation rates, perhaps by promoting nucleation of the flocculent in solution rather than on the particles where it can be most effective. Good mixing lowers (minimizes) the time for which local concentrations of the flocculent exceed the eventual average concentration.

This initial stage of vigorous mixing is followed with a longer period of gentler mixing. Some shear is desirable during this period to promote flocculation, but excessive shear will break up the floc. This second period is when the floc grows to its maximum or final size; thus, this is probably the period when the floc is most susceptible to breakdown by high shear forces. These two flocculation steps could be carried out in separate chambers; the rapid mixing is usually in a very small volume (short residence time). Alternatively the small rapidly mixed volume could be a region of a larger vessel in which the flocculation occurs. The rapid mixing could occur near the edge of mixing blades, and the second flocculation step could be carried out in the much larger remaining volume of the vessel where there is gentler mixing. If the two steps are carried out in the same vessel, it is necessary to add the agents in the region where there are high mixing rates.

Need for Testing Coagulation/Flocculation Systems

It is wise to test coagulants and flocculation conditions with a particular suspension before designing a coagulation-flocculation system. The surface properties of particles can be complex and poorly understood, and even the simplest tests may provide helpful information to design a coagulation-flocculation system more cheaply than extensive characterization studies. There may be soluble contaminants in the suspension that can affect coagulation or flocculation, and the simple tests with real suspensions may detect the presence of such materials more easily than chemical analysis. Such contaminants could include organic phosphate compounds like those used in some older detergents which hinder many coagulation processes. Any trace contaminant that adsorbs on the particle surfaces and affects the flocculation process can be important.

Tests for selecting coagulation and flocculation conditions often use simple equipment, little more than sealed flasks (or jars) in which specific quantities of coagulants are added to the suspension, and the settling rate is observed by measuring the time for the suspension to settle a specified distance. It is usually necessary to test a range of coagulant concentrations and (perhaps) pH. It is also important to examine the supernatant (fluid remaining after flocculation and settling are complete) to see what suspended particles remain. This is a check of the effectiveness of both flocculation and coagulation. If the flocculation has not allowed time for all particles to interact with others, some will remain free in the suspension after the agglomerates have settled. If the coagulant has not interacted with all of the particles, some may still retain repelling electrical charges. Remember that all particles in a suspension may not be the same;

some may be affected more strongly by the coagulant and others may be affected less or not at all.

The properties of the settled bed of solids may be affected by the use of some flocculants. Generally flocculants are likely to lower the density of the bed of settled particles. At a minimum, any solid flocculants will add to the volume of the settled solids simply because additional solids are involved, but the mass or volume of flocculants required can be a very small portion of the settled volume. However, the increase in volume of settled solids can be even greater than one would guess from the mass of flocculent used. Since flocculants increase settling rates by aggregating the small particles into larger clusters of particles, the settled bed is likely to retain much of the structure of the settled agglomerates. This means that there is likely to be larger void volumes between the aggregates as well as void space between the particles within the aggregates. Some flocculents produce relatively "soft" or compressible settled cakes whose volume is relatively easily decreased by compression, but other flocculents produce relatively incompressible settled cakes of solids that tend to retain water. Thus the use of flocculents may have an undesirable effect of making it more difficult to dewater settled solids, but this affect is often a minor problem when compared with the difficulty of settling many suspensions of small particles.

SEDIMENTATION OF UNIFORM SPHERES

Sedimentation can take place either in a flocculation vessel or in a special sedimentation vessel and can involve sedimentation of flocs or individual particles as they originally appeared in the stream. Sedimentation of uniform spherical particles has been studied for years, and this discussion focuses on settling of spherically shaped particles because so many of the more difficult environmental sedimentation problems are likely to involve very small granular particles that are approximated closely by the spherical shapes. Although other particle shapes can be important, there are too many alternative shapes to cover them all, and much of the behavior of settling spherical particles is similar to the behavior of many other settling particles, at least qualitatively similar. As expected, the most significant differences are likely to occur when the particles become greatly nonspherical; such cases are fibrous and plate-shaped particles.

The rate at which particles settle depends upon particle size, particle density, fluid density, fluid viscosity, and particle concentration. Quantitative study of settling rates usually starts with very dilute suspensions in viscous flow (that is, at very low settling rates). For these conditions, an exact analytical solution for the settling rate has been developed and is usually referred to as Stokes' law.

$$v_S = \frac{2r^2 g(\rho_p - \rho_f)}{9\mu_f} \tag{1}$$

Here v_S is the settling velocity, d_p is the diameter of the suspended particle, ρ_p is the density of the particle, ρ_f is the density of the fluid, and μ_f is the viscosity of the fluid.

As the settling velocity or the particle diameter is increased, the settling rate will eventually be in the transition region, and with even larger particles in the turbulent zone. Under these conditions, analytical expressions for the settling rate cannot be obtained because more terms in the equations of motion would have to be taken into account. Flow around the particle is no longer symmetrical, but a complex wake will form at the top of the settling particle. Most accurate information on settling rates in this region come from experimental observations. The information is often expressed in terms of the "drag coefficient"

$$C_D = \frac{F}{\pi r^2 (1/2 \rho_f v_s^2)} \tag{2}$$

where F is the force on the particle, r is the radius of the particle, ρ_f is the density of the fluid, and v_S is the settling velocity.

The behavior of the drag coefficient over a wide range of Reynolds numbers is shown in Figure 1. At low Reynolds numbers, less than approximately 1.0, the drag coefficient is inversely proportional to the Reynolds number. That is exactly as predicted by Stokes' law. However, as the Reynolds number increases and enters the transition regime, the drag coefficient begins to level off. Eventually after the Reynolds number enters the turbulent regime, the drag coefficient becomes more nearly constant and has a value near 4. Most sedimentation problems are likely to involve Reynolds numbers in the viscous or transition range; sedimentation in the inertial range is likely to be sufficiently rapid that the problem will require little concern.

The relationship in Figure 1 can give excellent predictions for the settling rates of individual particles in large bodies or fluids and even for settling of very dilute suspensions of uniform spherical particles. However as the concentration of suspended particles increases, there will be significant deviations between these relations and the observed settling rates. This is usually called "hindered settling." There are several reasons for these deviations. First the particle may be affected by the wake from

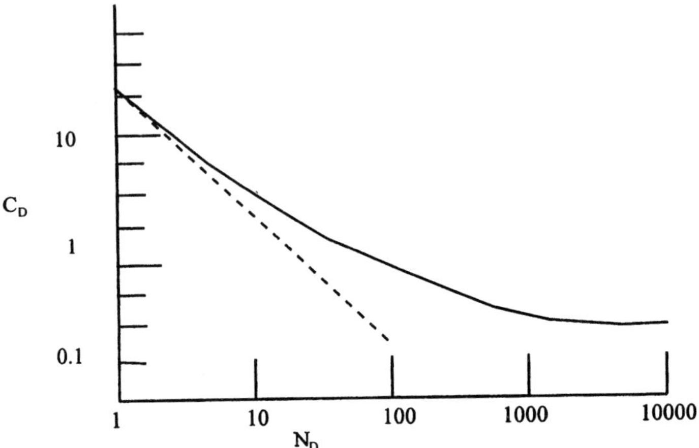

FIGURE 1 Drag coefficient for a single sphere settling in an infinite medium.

other particles below and beside it. For dilute suspensions, the velocity field around a settling particle can be assumed to approach zero at essentially an infinite distance from the particle. However, in a concentrated suspension, nearby particles are also affecting the fluid motion. This compresses the velocity patterns and results in higher velocity gradients and, thus, higher drag and lower settling velocities.

The second effect of concentrated settling is a net flow of fluid upward in the opposite direction to the settling. This upward flow of fluid results from displacement of fluid by the settling particles. This should be evident from a simple examination of the net flow across any horizontal cross-section of the sedimentation vessel. If the bottom of the vessel is sealed, there can be no flow out of it. Since the sedimentation of particles is always downward (assuming that the particles are denser than the fluid), there must be an equivalent upward flow of the same volume of fluid. This is the fluid that is being displaced from the region of the vessel below the reference horizontal cross-section by the particles passing through that cross-section.

The arrangement of settling particles is largely random, and the flow pattern around the particles is complex. Thus it is very difficult to estimate accurately how settling rates will be affected by particle concentrations. A very simple and yet effective equation for predicting settling rates in concentrated suspensions was disclosed several years ago by Richardson and Zaki [1]. They used some theoretical arguments but principally empirical results to say that

$$\frac{v_b}{v_S} = e^n \tag{3}$$

Here v_b is the hindered settling rate, the rate in concentrated solutions, v_S is Stokes' settling velocity observed in infinitely dilute suspensions, e is the void (or fluid) volume fraction in the suspension, and n is an empirical constant. Of course, when the void fraction is essentially unity as in very dilute suspensions, the observed settling velocity should be approximately Stokes' settling velocity. As the void fraction decreases and becomes less than unity, the hindered settling velocity will be come increasingly less than Stokes' velocity. The exponent n will be greater than unity. One can show that a power of 1 in the exponent results from the upflow of displaced fluid, and the additional changes in the velocity patterns around the particles increases the exponent further.

The constant n is independent of particle concentration, but it is dependent on the particles themselves. Richardson and Zaki found n to depend upon the Reynolds number when the particle is settling in a dilute suspension, that is, the Reynolds number for Stokes' settling. However, n is not a notable function of void fraction, and that is the principal value of the equation.

There have been numerous studies to check the Richardson-Zaki expression, and improvements have been suggested. However, there has been little improvement in the general equation itself. Perhaps, the most significant improvements [2,3] have been in the expressions for n.

Richardson and Zaki approached the problem as a semitheoretical modification of the Stokes settling problem, but other investigators have attempted to approach the problem in a more fundamental way and calculate the expected settling rates in different arrays of particles or groups of particles. Eventually, those approaches may yield more rational theoretical results with less empirical input. They could also provide a far better understanding of the behavior of settling particles. However, such approaches were not covered in this discussion because the author does not believe that they have yet produced results that can be as readily used by most practicing engineers as the semitheoretical methods that have evolved from the Richardson-Zaki approach.

SEDIMENTATION WITH PARTICLE SIZE DISTRIBUTIONS

It is relatively rare to find a sedimentation problem with a waste or environmental suspension that has uniform size particles, or even uniform densities. If the size distribution is relatively sharp, one can use the Richardson-Zaki expression to estimate settling rates and ignore the mod-

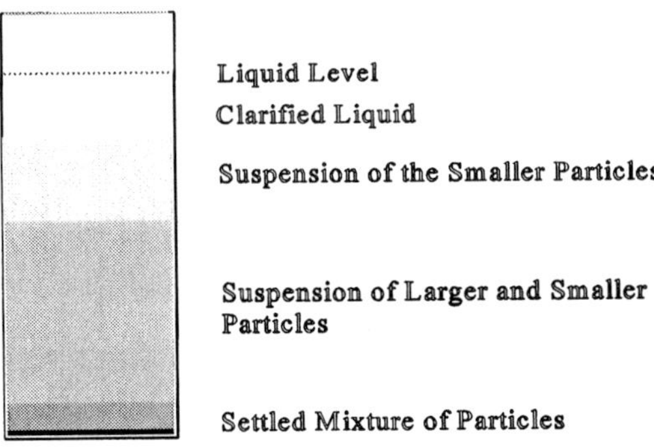

FIGURE 2 Settling of a binary mixture of two different particle sizes.

est variation in particle sizes in the suspension. However, some problems involve widely varying particle sizes, and describing the sedimentation rate based upon a single particle size could be inaccurate.

To examine sedimentation of particle mixtures, look first at the settling of a binary mixture, one with only two particle sizes. When a binary mixture begins to settle, the larger particles obviously settle faster than the smaller particles. This leads to a settling pattern illustrated in Figure 2. The suspension initially consists of a single region where all of the particles are suspended uniformly. As settling progresses, an interface will move down the vessel. The interface represents the movement of the larger particles through the suspension. There will be essentially no larger particles above that interface. Another interface will also move downward from the top of the initial suspension, but this interface will move more slowly; it will be the interface between the upper region which contains only smaller particles (larger particles having already descended below this interface) and the solids free zone above the smaller particle region. That front will be the top of the suspension. At the bottom of the vessel, a region of "settled" particles will begin to accumulate. This region will first contain a mixture of large and small particles. After all of the larger particles have settled (when the lower settling interface reaches the more slowly rising level of settled particles) there will be no more large particles in the suspension. The settled region will then rise even more slowly and will accumulate only smaller particles.

Qualitatively, this behavior is probably as expected, but some experimental observations should be noted that are not necessarily obvious. First the rate of movement of both descending interfaces remains essen-

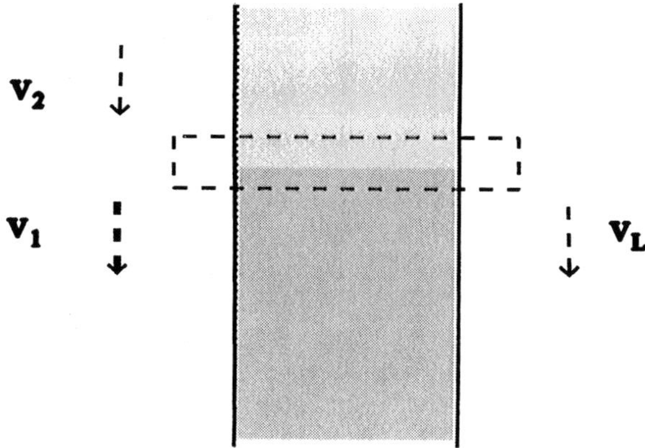

FIGURE 3 Particle balance around the interface between larger and smaller particles in settling binary mixture.

tially constant throughout the settling process. That observation can be sufficient to infer other observations. For instance, the concentrations of both larger and smaller particles in the region below the interface of settling larger particles is essentially the same as the composition of the initial suspension. This can be inferred because that interface is determined by the motion of the larger particles. If the particles settle at the same rate as they did in the initial suspension, it is not surprising that the composition in this region with larger particles is essentially the same as that of the original suspension. If the composition in this region were changing, one would expect the settling rate to also change as the larger particles and this front move downward.

The concentration of smaller particles in the region above the interface of settling larger particles also remains constant. Since we understand settling of uniform suspension, this should follow from the observation that the settling velocity was constant. The concentration of smaller particles in the upper region, and thus the settling rate in that region, can be estimated from a material balance around the lower interface of the settling larger particles (Figure 3). The accumulation of smaller particles in a volume that includes both regions and the interface of falling larger particles is the settling velocity of the larger particles (the rate at which the interface falls) and the difference in the concentration of smaller particles in the two regions. This is equal to the rate at which smaller particles enter the upper region (region 2) and the rate at which smaller particles leave by settling from the lower region (region 1) with the larger particles.

$$v_2 C_2 - v_1 C_1 = v_L (C_2 - C_1) \tag{4}$$

The v's refer to velocity of smaller particles, and C's refer to concentrations of smaller particles. The subscript 1 refers to the lower region which contains the larger particles as well as smaller particles, subscript 2 refers to the upper region where there are only smaller particles, and v_L refers to the settling velocity of the larger particles. Since the presence of the larger particles defines the lower region, the settling velocity of the larger particles is also the velocity of the interface between the lower and upper regions with only smaller particles (regions 1 and 2).

Note that C_1 is known since it is the concentration of smaller particles in the original mixture, and that v_2 and C_2 are related by the Richardson-Zaki equation. Then this relation can be solved for all terms if v_L were known. That is the key to predicting binary sedimentation rates. However, applying the Richardson-Zaki sedimentation equation to this problem does not give results that are in good agreement with observations. To fit the data for binary mixtures, Richardson and Zaki had to select a different empirical value for n to describe settling of the larger particles. Other equations have been suggested [4] with limited success. Perhaps the first equation that seems to fit all of the available data was proposed by Selim et al. [5], but it only treats the two particles as different and does not take into account how much difference there is between the two particle sizes. This results in a serious problem when the correlation attempts to predict the behavior of binary mixtures as the size of the particles change and approach each other. That limitation does not affect the equation's ability to predict the available data because there are no data on binary mixtures with similar size particles. (It is difficult to obtain particles with sufficiently narrow size distributions for such measurements.) However, if the solution for a binary system is to be extrapolated to handle complex mixtures such as continuous distributions, this limitation is serious.

There is one suggested equation that has not been widely used, but which can predict the behavior of binary mixtures with any separation of particles and which can, in principle, be used with any number of particle sizes and thus approximate continuous mixtures. The author may be biased in mentioning this paper by Shor and Watson [6]. This development assumes that the larger particles descending through smaller particles experience a fluid with a viscosity similar to that expected from a slurry of smaller particles, ignoring the presence of other large particles. There is the implication that the large particles are moving vertically downward and interact with each other only as they would interact in a settling slurry. If the larger particle settling rate is much greater than the settling rate for the smaller particles, the effective slurry composition would be essentially the same as the concentration of smaller particles. Any suitable correlation for effective viscosities of slurries could probably be used, but one should remember that the use of a single number for the viscosity is only an approximation because slurries are often not Newtonian.

However, as binary mixtures are used with the two groups of particles approaching the same size, the groups would be settling with more similar velocities; in the limit of identically sized particles they would be settling at the same velocities. This should result in less interactions between the larger and smaller particles. Shor and Watson suggested using an apparent concentration of smaller particles linearly related to the difference in the settling rates of the two particle sizes. These assumptions gave very good results and excellent agreement with the available data. The agreement with available data were essentially as good as the agreement of the Selim equation, and the result could be extrapolated to predict settling of mixtures with similar as well as greatly different particle sizes. The result of this approach is

$$v_L = v_{L0}e^{n-1}(1 - C_1)\frac{\mu_0}{\mu_s} - v_{S0}e^{n-1}C_S \tag{5}$$

The settling velocity of the larger particles in the suspension is v_L, and their settling velocity at infinite dilution (Stokes' settling) is v_{L0}. The concentrations of the larger and smaller particles in volume fraction are C_L and C_S, respectively. The viscosity of the fluid is μ_0, and the effective viscosity of the suspension is μ_s.

However, one should remember that the available data do not include studies with binary mixtures with sizes that are not far apart. This means that the linear relationship assumed by Shor and Watson has only been tested over a relatively narrow range of differences in settling velocities; that narrow range reflects the limitations of the available data.

As noted earlier, this equation can, in principle, be used to predict the performance of mixtures of many particle sizes, even continuous distributions of particle sizes. To predict settling in mixtures with several (or many) different size articles, the analysis described above is repeated for each settling zone (n zones for a mixture of *n* different particles). Similarly the same approach could be used to predict settling in different zones when the densities of the particles are different. However, there have been no experiments with a large number of particle fractions that can test these predictions, so their validity is only speculative.

SETTLING TESTS

There are a number of ways to test the settling rates of suspensions. Only one method will be discussed here, a method that closely follows the description of settling that has been presented. Descriptions of other methods can be found in more detailed discussions of sedimentation and in standard handbooks [7]. Details of the test methods will depend upon the suspension properties.

As described above, if the suspension is fairly uniform, or, if it has a minimum particle size, we would expect the particles to form a settling interface just as the binary mixture did. The simplest settling test will only follow that interface. The test can be carried out in any tall glass or clear plastic container such as a measuring cylinder. The suspension is placed in the container and stirred sufficiently to suspend the particles uniformly. Then the stirring is stopped and the position of the upper settling interface is measured as a function of time. For some slurries, it may be desirable to maintain very slow stirring with a bent wire impeller during the sedimentation to minimize wall effects [8]. Of course, any stirring during the tests must be slow enough to avoid resuspending particles or otherwise slowing the effective settling rate. The result of such tests will be a settling curve much like that shown in Figure 4. At first the interface settles with a uniform velocity. Actually, in many cases, the uniform velocity continues over much of the settling time. This is the behavior just described in the theoretical discussion. However, things will not necessarily remain so ideal forever. There is often a nonconstant settling rate as the interface approaches the settled solids near the bottom of the vessel. Figure 4 illustrates a relatively simple settling curve such as one could

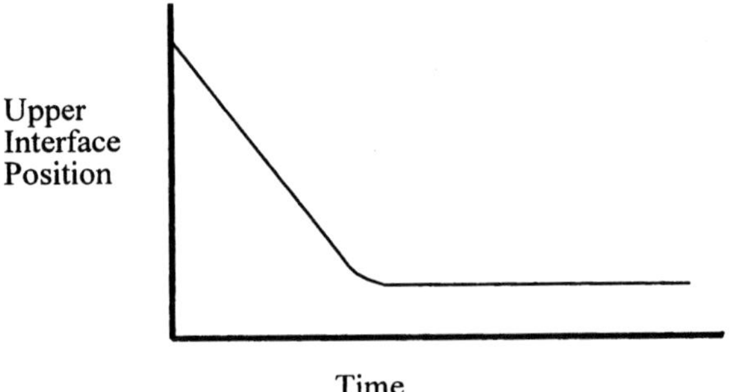

FIGURE 4 Position of the upper interface in a typical sedimentation operation (settling curve).

see with rigid and uniform spheres settling with little flocculation taking place. Only a small transition curve is shown between the initial settling rate and the zero settling rate (constant interface position). In many systems, this immediate region can be much broader than that shown in Figure 4. This is often called the compression zone, since "compression" of the settled sludge is one reason for a broad compression zone.

Flocculation often occurs during sedimentation, and it can be an important factor in settling rates, even when little or no special effort is made to promote flocculation. As the previous discussions indicate, the probability of flocculation being important can be estimated from the pH and the surface conditions of the particles.

If one needs to know the conditions within the settling particles, it will be necessary to measure the concentration of solids within the settling region as well as the position of the interface. This is usually done by taking samples from within the settling region as a function of time. Usually the only measurement is the total solids concentration, but one could even measure the concentration of different size particles if there is a wide range of particle sizes in the slurry. If flocculation is occurring, the problem can be complex, and it may be difficult to interpret anything more than empirical measurements of the total solids concentration.

There are several reasons why deviation from constant settling velocity could result, but we will not try to resolve that question here. The problem probably results at least in part from compression of the settled region of the vessel. This effect can be especially pronounced with fibrous particles. For such suspensions, sedimentation of the smaller fractions may be too slow for simple sedimentation to be practical or the concentration of the smallest size fraction may be too low to be evident visually. In those cases, it may not be possible to see an upper interface, and sedimentation tests may have to include periodic withdrawal of suspension samples from various heights in the test vessel.

Although sedimentation operations are carried out in large variety of different vessels, the principles involved are the same, and the results from these relatively simple tests can often be applied to a wide range of equipment types. The results can often even be used to predict the performance of centrifugal systems, but the sedimentation rate should be increased by a factor equal to the ratio of the centrifugal force to the gravity force used in the simple batch settling tests. There are two reasons why the performance of larger equipment could deviate from that based upon the simple tests. First, the geometry of the sedimentation vessel could be significantly different; that is, the vessel may not have straight vertical walls like those in the tests. To a degree, these effects can be estimated, and some sloping walls can even be helpful by reducing the distance for settling if the settled material quickly flows down the sloped surface. The other reason for difference in performance is small flow patterns or circulation in the larger vessels. The tests are usually run with little or no circulation in the slurry once the settling period starts after mixing. The relatively small vessels used in the tests make it relatively easy for circulation patterns to damp quickly. In larger vessels, circulation may dampen more slowly, and in continuous settlers some motion is necessary to continuously supply feed and remove settled solids and clarified liquid. Also if flocculation is part of the settling operation, it is likely to be desirable to maintain some gentle circulation.

SEDIMENTATION EQUIPMENT

Sedimentation can be carried out in batch or continuous systems. A batch system is little more than a tank filled with the suspension. As the particles settle, the interface of settling particles descends through the tank just as they do in small batch tests. Since a batch sedimentation with low compressibility systems is so similar to simple settling tests, the time required for the interface to settle is easily determined from similar batch tests. After the upper interface has passed a withdrawal point, the clear (or more nearly clear) liquid can be decanted from the vessel. The degree of concentration of particles depends upon the position of the withdrawal or decantation point. One may want to have it extend completely to the settled solid interface. However, it is usually not possible to remove liquid from too close to the settled solids without resuspending significant numbers of particles. Thus a good concentration of particles in the settled product may be all that one should expect.

Sedimentation operations can be used to concentrate the solids, remove particles from the liquid, or both. If the principal purpose of the sedimentation is to concentrate the solids, the equipment used is likely to be called a "thickener," but if the principal purpose is to produce a clear liquid free of particles, it is likely to be called a "clarifier." There are not necessarily major differences in the equipment used for either purpose, and both goals could be essentially equally important.

Clarifiers can be operated in continuous or semicontinuous as well as batch manner. In a continuous operation, clarified liquid and solids concentrated sludge would be withdrawn continuously from the top and bottom of the device. In semicontinuous operations, the feed would be fed continuously, and the clarified liquid would also be withdrawn continuously. However, the sludge could be withdrawn periodically after it accumulates.

From both the description of batch settling tests and the theory discussed, the relation between settling time and the settling distance has been emphasized. The settling time for any noncompressible suspension interface is, over most of the range of interest, directly proportional to the settling distance. Only when the upper settling interface approaches the settled solids does the settling curve deviate greatly from unity. This means that to have short settling times, the settling distance needs to be short. This would favor settling vessels with large diameters (or cross-sections) and short heights. Of course, going too far in this direction will make it difficult to decant the liquid from the settled solids without entrainment. This is the optimization that a designer of a batch settling vessel would have to consider.

Most high throughput settling equipment is operated continuously. There is no major difference in the principles for operation of continuous and batch settling equipment. Perhaps the simplest form of a continuous settler is a rectangular tank where the suspension enters one end of the tank and slowly flows to the other end. In its ideal form, this type of settler would act just like a batch settler, but the horizontal distance along the tank would be much like time in the batch settler. Fluid that has moved a certain distance down the tank will have been in the settler for a time that is equal to that distance divided by the fluid's velocity. Of course, obtaining uniform (plug) flow down a tank and avoiding turbulence and mixing are problems in such settlers. Vanes and other devices can be built into the settler to direct the flow and limit the turbulence and nonuniform velocities. These continuous devices often make the decantation step relatively easy since simple weirs or removal blades can take off the upper portion of the decanted liquid, and an underflow weir can remove the concentrated suspension from the bottom of the settler.

In cylindrical-shaped settling vessels the feed may be introduced at a position below the overflow position where the clarified liquid is withdrawn, and that can result in a new upflow of liquid. The particle settling rate must exceed any upflow velocity, and that will impose a minimum cross-section (diameter for cylindrical vessels) on the device.

When designing continuous systems, it is necessary to realize that the solids at the bottom of the device may not be allowed to reach the maximum concentration suggested in the batch settling curve in Figure 4. It may not even be possible to remove sludge from the bottom of the device if the solids concentration is allowed to reach its maximum value. Some clarifiers and thickeners have mechanical devices to "scrape" the sludge from the bottom of the device to aid in solids removal. This can be a slowly rotating blade that moves the settled sludge toward a central removal point in a cylindrical settler. Other cylindrical devices

FIGURE 5 Overall material balance.

may use a conical bottom to assist the sludge in "flowing" toward the removal port.

When the concentrated slurry of solids is withdrawn at an immediate concentration (less than the maximum concentration, it is desirable to have data on the concentration in the slurry as a function of time and position. Remember that in a continuous operation, the feed is divided into two streams, and if the flow rate and compositions of the feed and either of the other two streams is known, a material balance gives the flow rate and composition of the other stream.

Continuous settlers can be constructed in other geometries as well. For instance, the settler could be cylindrical with flow either inward toward the center or outward. Some cylindrical settlers involve angular flow around the cylinder. Thickeners can be operated in cylindrical vessels with vertical (axial) flow. This is likely to be the geometry of choice for thickeners that need some mild stirring to promote flocculation/coagulation. The behavior of such systems can be described in terms of settling rates (velocities) and a mass balance over the length of the settler [8]. Consider the volumetric feed rate, Q_f, with a solid concentration x_f, as shown in Figure 5. In the thickener this feed is split into two streams: Q_c is the stream depleted in particles, and Q_b is the stream leaving the bottom of the thickener that is concentrated in particles. A material balance on the solid particles gives

$$Q_f = Q_c + Q_b$$

and

$$x_f Q_f = x_c Q_c + x_b Q_b$$

Solving for x_b gives

$$x_b = \frac{x_f Q_f - x_c Q_c}{Q_b} \cong \frac{x_f Q_f}{Q_b}$$

The total solids flux, $Q_b x_b$, is simply the product of the concentration of particles in the bottom (concentrated solids stream) and the rate at

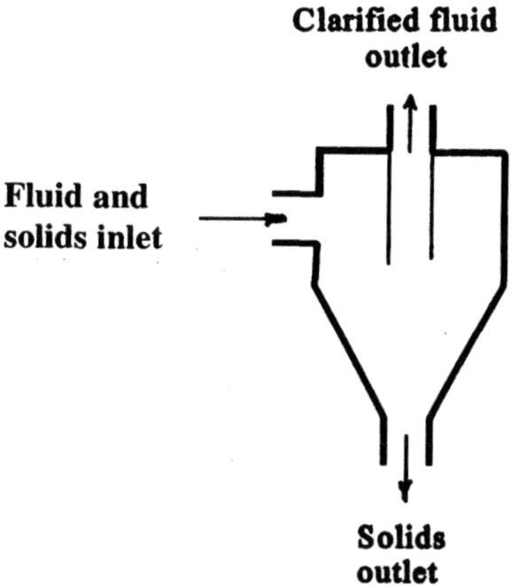

FIGURE 6 Cyclone or hydroclone.

which fluid is removed with that stream. Note, however, that the concentration of particles in the bottom stream is likely to a strong function of the residence time of the fluid in the system, so the flow rate and the concentration are not independent. However, if the bottom flow rate is small compared to the flow rate of clarified water, the product of the concentration and bottoms withdrawal rate may not change the clarified product flow rate greatly.

Continuous settling equipment is usually most effective in preparing a concentrated stream of liquid with particles that can be handled by other means. The continuous equipment could consist of tank type settlers like those described above, or the equipment could be a continuous centrifuge, gas cyclone, or a hydroclone (Figure 6).

APPLICATIONS OF SEDIMENTATION IN ENVIRONMENTAL AND WASTE MANAGEMENT

The most common need for sedimentation in environmental and waste treatment is for removing solids from water, often wastewater or groundwater and in pretreatment for other processes such as membrane or adsorption operations. Although removal of solids from air is also an important and common problem, particulate removal from air is less likely to involve sedimentation; filtration is more common. Cyclone centrifugal devices are the principle sedimentation devices for air streams. There are so many potential applications for solids removal that it would not be practical to attempt to name them all. Instead, it is probably more fruitful to describe a few classes and examples of sedimentation operations in environmental and waste processing.

Sedimentation will usually be used prior to filtration operations if a significant portion of the solids is large enough to be removed easily by sedimentation. Of course, this implies that there is a substantial fraction of particles large enough to be removed easily by sedimentation. If there is a high concentration of solids, it may be desirable to use sedimentation, but flocculation/coagulation may be necessary to achieve practical sedimentation rates unless the particle size is sufficient to give satisfactory sedimentation rates. If sedimentation rates are moderately high or if filtration rates are sufficiently slow, sedimentation can affect the filter cake on surface filters and thus the filtration rates. For instance, if the filter surface were mounted horizontally, as is common in some filter tests, the early filter cake could have a higher concentration of larger particles at the bottom of the filter cake than near the top of the filter cake. If the filter surfaces were mounted vertically, as is common in some plate filters, the larger particles may be concentrated near the bottom of the filter. It is not uncommon to find part of a filter cake appearing to flow or otherwise accumulate more toward the bottom of vertical plate filters.

Both clarification and thickening operations can be important. Toxic solids in discharge streams or particles that are troublesome for subsequent treatment (membranes or packed adsorbent beds) need to be removed from liquid streams, and this is a clarification operation. However, if the particles removed have to be concentrated for treatment or packaging for disposal, that is a thickening operation.

One notable need for thickening and clarification operations is after biological treatment of aqueous streams. Although it is becoming more common for enzymes of biological organisms to be used for destroying organic pollutants in waste streams while they are attached to packing materials or other surfaces within a bioreactor, many bioreactors still use suspended organisms. This is usually called "activated sludge," and the excess sludge must be continuously removed from the reactor. (Bioreactors are often called fermenters, although some laymen may take that term to be restricted to specific bioreactors that produce alcohol from carbohydrates.) Although solids, including biological materials, may be released from any bioreactor, those with fixed organisms or fixed enzymes are likely to release less solid material. Also those bioreactors that use only enzymes and "dead" organisms are less likely to produce excess biomass that must be removed. In the activated sludge reactor, solid-liquid separation is usually incorporated directly in the reactor design, and an activated sludge reactor may appear much like a sedimentation device, usually like a slowly stirred sedimentation system. Anyone interested in wastewater treatment should become familiar with activated sludge systems and the sedimentation operations that take place within such reactors.

Although the biological operations of activated sludge systems are not within the domain of this book, the sedimentation aspects are. The size of an activated sludge system must depend upon the ability to remove solids (including biological organisms that have participated in the desired reaction), produce new organisms to replace those that die or are swept from the reactor with the concentrated solids at the bottom of the reactor, as well as carry out the desired reaction to the degree needed, usually destroy enough of the organic pollutants. In this case, many of the solids are produced within the reactor rather than introduced into the device with the feed stream. For solids to not leave with the treated water, the sedimentation rate must be sufficient to keep clarified and treated water near the top of the reactor for withdrawal. Settling tests of sludge can be made in the same manner described above for other solids and used to predict settling characteristics of an activated sludge reactor. It is usually desirable to remove the solids at the bottom of the reactor in a concentrated form with a small liquid underflow rate. This puts most of the liquid in the clarified and treated overflow stream, leaves a high concentration of solids in the reactors to serve as a biocatalyst, and removes the solids in a concentrated form more suitable for further treatment or disposal.

Sedimentation operations are also closely associated with most precipitation operations since separation of the precipitates from the liquid are obviously necessary. As noted in the section on precipitation processes, compounds such as aluminum hydroxide or ferric hydroxide can be precipitated with one aim, to remove toxic or other undesirable components that will adsorb on the hydroxide precipitates. A number of metal ions are effectively adsorbed on the surface of these precipitates. Note that hydroxides of aluminum and ferric iron are also materials that are commonly used as flocculates to remove ultrasmall particles of other materials from suspensions. These hydroxides form polymers that agglomerate and settle as clusters of moderate size. Sedimentation of such materials in precipitation processes can be similar to removal of flocs of these materials.

Particulate removal is usually necessary when extremely high removal of contaminants in necessary, even when the contaminant is normally believed to be dissolved in the water. Because so many contaminants adsorb on some of the smallest particles, such as suspended clays or humus materials, it is necessary to remove the solids if the concentration of contaminant is to be reduced to concentrations below that of the adsorbed materials. Since the surface area of the smallest particles can be so high, it is likely to be necessary to remove small colloidal particles, and coagulation and flocculation are usually necessary to remove such particles by sedimentation.

Since particle size determines sedimentation rate, it is usually the key issue in determining if sedimentation or filtration is to be used. Although both filtration and sedimentation rates can become very slow as the particle size becomes smaller, the sedimentation rates can be affected more significantly. In fact, once the particles become as small as colloidal sizes, sedimentation can stop completely because Brownian motion can keep the particles suspended. Coagulation and flocculation are helpful for sedimentation and filtration rates, are there are other filtration methods (such as deep bed filtration) that can handle colloidal and even subcolloidal size particles. There is no practical approach to sedimentation for small particles, but high-G centrifuges can be used for some high value applications.

The next parameter likely to affect the decision to try sedimentation or filtration is particle concentration. Although high particle concentrations were shown to hinder settling rates, the hindering is not likely to be as important as the need for rapidly changing or cleaning filters when the particle concentration is very high. Coagulation and flocculation can also be particularly effective because there are larger concentrations of particles to interact and agglomerate. Once sedimentation (with or without agglomeration) removes the bulk of a high concentration of particles, filters can be used to "polish" the stream and remove the remaining particles.

Particle density can also affect sedimentation and have relatively little or no effect on filtration rates. Thus, one would not want to attempt sedimentation on particles that were exactly buoyant in the fluid. Since the densities of inorganic particles are not likely to vary greatly, density is not often a critical consideration. However, the densities of some organic materials can be relatively close to the density of water, and that can make sedimentation difficult.

REFERENCES

1. J. F. Richardson and W. N. Zaki. *Trans. Inst. Chem. Eng. 32*, 38 (1954).
2. J. F. Richardson and R. A. Meikle. *Trans. Inst. Chem. Eng. 39*, 348 (1961).
3. J. Garside and M. R. Al-Dibouni. *I&E Chem. Process Des. Dev. 16*, 206–214 (1977).

4. S. Mirsa and J. F. Richardson. *Chem. Eng. Sci. 34*, 447 (1979).
5. S. Selim, A. L. Kothari, and R. M. Turian. *AIChE J. 29*, 1029 (1983).
6. J. T. Shor and J. S. Watson. *Sep. Sci. Technol. 25*, 2157 (1990).
7. J. H. Perry. *Chemical Engineer's Handbook*. McGraw-Hill, New York (1963), pp. 19–42.
8. J. B. Christian. *Chem. Eng. Prog. 90*, 50–56 (1994).

11 Precipitation

Precipitation, long a common method for removing components from solutions, has played a major role in removing contaminants from wastewater. Precipitation plays a major part in most water treatment operations for both industrial and residential use. Strictly speaking, precipitation should be considered a chemical or molecular separation method rather than a physical separation method because it involves removal of molecules (or ions) from solution by forming a solid. Note also that many precipitation processes involve removal of one contaminant by precipitating another material that may be called a "carrier" precipitate. In those cases, precipitation could involve adsorption of the contaminant on the precipitate as well as incorporation of the contaminant into the precipitate; it is often difficult to know the exact mechanism of the removal process.

Nevertheless, precipitation is covered here in the section of the book that discusses physical separation methods. It is discussed along with particulate removal, which is usually the most visible part of the operation. Thus, the reason for placing the discussion of precipitation at this place in the book is the close association of precipitation with the subsequent step, the removal of the precipitated solids from the liquid. There are several important aspects of precipitation, such as crystal nucleation and growth, that will not be discussed extensively because, in environmental applications, the solid-liquid separation is often the most difficult, and thus the most critical, step in the operation. This difficulty can be in contrast to other related precipitation-like processes that aim to recover major components from concentrated solutions. When attempting to recover a valuable component from a concentrated solution, some precipitation processes may include careful efforts to control the crystallization process to ensure that the crystal growth rate is both sufficiently fast for good recovery rates and sufficiently slow to avoid excessive nucleation and to permit formation of crystals with desirable sizes and minimum entrapment of undesirable impurities. The objective of such operations could be the preparation of good quality and high purity solids, not the purification of the solution. Subsequently, when moderately large crystals are formed, separation of the crystals from the liquid may involve little difficulty.

Although such slow processes may prove to be important in some environmental and waste operations, the more common problem in waste and environmental applications is believed to be the removal of a trace contaminant from an aqueous solution, and the objective is purification of the water, not recovery of the precipitant/crystal. In these applications, it is more desirable to create a compound that is highly insoluble in the water (that is, to achieve high removal efficiency). These conditions are also likely to make precipitation sufficiently fast that it is practical to have only limited control over the rate of nucleation and crystal growth. The problem with rapid precipitation is the formation of very fine particles that can be difficult to separate from the solution, and that case is considered most extensively.

Removal of trace components and high removal efficiencies usually mean that precipitation must involve the formation of an especially insoluble form of the contaminant or of a material that will carry (incorporate or adsorb) the contaminant. In such cases, the precipitate is likely to form quickly, and control of the precipitate properties may be limited. Furthermore, the volume of precipitate per volume of liquid will be small if no carrier is used. These are the cases that need to be considered more carefully for most environmental applications. However, there can be other situations, even if they are less likely in environmental and waste processing where it is both easier and desirable to grow carefully sized and relatively pure crystals when the solution is not excessively supersaturated with the component being precipitated.

To precipitate a trace component and to remove large fractions of the component, it is necessary to convert the component from the original (more) soluble form to a form that is far less soluble. When a component is near its saturation concentrations, even if that is a trace concentration, it is possible to evaporate some liquid and leave the solution supersaturated in the component. Precipitation will then occur, but it will probably be called "crystallization" because evaporation will bring the solution to supersaturation relatively slowly. Then, once a modest fraction of the solid is precipitated, the concentration will decrease to the saturation concentration, and precipitation will cease. It is obvious that evaporation of substantial fractions of the liquid is likely to be necessary to remove large fractions of the component in this manner. In a similar manner it is sometimes possible to change (usually lower) the temperature of the solution so that the concentration of the component of interest becomes supersaturated. Again this is likely to allow removal of only a modest fraction of the component because there is only limited opportunity for the temperature to be lowered, and this is often not sufficient to lower the solubility greatly.

For accurate prediction of the solubility of components in a mixture, it is especially important to know the form of the contaminant in the solution and in the final precipitated product. For instance, many trace components are adsorbed on existing precipitants of other materials, although the trace component would normally be soluble. Thus, contaminants that appear to be dissolved in the water could be partially or largely adsorbed or otherwise incorporated in colloidal particles suspended in the solution. Obviously, such components will not behave as soluble compounds in adsorption, ion exchange, or other molecular separation method. In some cases, one may want to create a precipitate that will adsorb or incorporate a contaminant.

One could argue that such behavior is adsorption rather than precipitation and technically be correct. However, since the system for carrying out such adsorption operations involves a precipitation step, they are mentioned here. In other cases, the contaminant can be incorporated into the growing crystals of another precipitating component and form a double salt. Again the removal of the contaminant can be much greater than one would expect from considerations of the contaminant solubility of pure components in solutions. These operations can be called "carrier precipitation" since the removal of the contaminant is enhanced (or even made possible) by the precipitation of another bulk component. This type of behavior may be necessary for highly efficient removal of trace components whose total mass in the solution would produce such a small mass of precipitate that, if the precipitate were formed, would be so small that removal would be difficult.

Carrier precipitation is especially helpful when it is necessary to remove small traces of a contaminant. One can precipitate a component that is similar to the contaminant, and the contaminant may actually be incorporated within the precipitate of the carrier. Radioactive contaminants are examples of cases where carrier precipitation sometimes is a practical approach. Radium behaves much like barium and calcium and is effectively incorporated in barium sulfate precipitates. Strontium also behaves much like calcium and can be incorporated in most precipitates of insoluble calcium compounds.

In the cases just described, the effectiveness of the precipitation is enhanced by the formation of a different solid form of the precipitate.

The effectiveness of precipitation operations can also be affected by the form of the contaminant in the original solution. To predict the performance of the precipitation, it is necessary to know the form of the target contaminant in the solution. For instance, many toxic metals are highly insoluble in high pH solutions, but they may not precipitate effectively if they do not exist in the solution as free metal ions. A number of organic ligands can bind to metal ions and form complexes from which the metal ions are not easily precipitated. Such ligands are likely to be present in metal cleaning solutions (decontamination solutions) where ligands such as EDTA (ethylenediaminetetraacetic acid) are used to enhance metal solubilities. See Chapter 5 for more discussion of the role of ligands in increasing the effective solubilities of metal ions in solution. Important, but less effective, ligands may be found in natural groundwaters that contain humus materials. Effective precipitation of metals from solutions with strong ligands may require destruction (oxidation) of the organic ligand [1]. Ligands can alter the behavior of other separation methods for metal ions, so their effects are not limited to precipitation processes.

In many important cases for precipitation of environmentally important contaminants, even modest changes in the solution pH, the redox state of the contaminant, or the presence of a suitable anion (or cation) can lower or raise greatly the solubility of certain toxic metals. These are some cases of primary interest for wastewater processing. Perhaps the contaminants of primary interest for precipitation processes are the transition metals and heavy metals such as mercury, lead, or cadmium. In other cases, a carrier precipitate can be formed by adding a component that precipitates after such a change is made in the solution, and the precipitate can carry a number of metal ions and sometimes even some organic contaminants.

PRECIPITATION OF TOXIC METALS FROM WASTE WATER BY RAISING THE pH

Many toxic metals have insoluble hydroxides and thus have very low solubilities in alkaline media. In such cases, even a small rise in the pH of the wastewater can reduce the solubility greatly. The solubility of these metals usually continues to decrease with increasing pH, and in some cases the solubility of the metals can usually be estimated if the solubility product of the metal is known.

$$K_{\text{solubility}} = [M^{+n}][OH^-]^n$$

The hydroxide concentration is usually expressed as the pH of the solution (wastewater). Note that increasing the pH (or the hydroxide concentration) reduces the solubility of the metal ion, and the decrease can be very significant. For instance, an increase in the pH from a value of 7 (neutral) to 8 involves an increase in hydroxide concentration only from 10^{-7} to 10^{-6}, a very small change but it is a 10-fold increase, and thus would result in a 10^{n}-fold decrease in the solubility of a metal ion with a valance n. It is often practical to increase the pH to values of 9 or 10, and the resulting decrease in solubility can be several orders of magnitude. Nevertheless, this only requires a hydroxide ion concentration of 10^{-5} to 10^{-4} M. However, when treating groundwaters, it is necessary to note that EPA regulations limit the pH of waters that can be returned to an aquifer, so it may be necessary to readjust the pH of the water again to a value close to neutral if it is necessary to return it to the ground.

This simple solubility product is a simplification of many overall processes for several reasons. Most notably, the precipitation product could be different from the simple hydroxide assumed in the equation. As noted, in many cases different solids can be formed. These are usually mixed salt and hydroxide compounds or dehydrated hydroxides. The hydroxide could lose water molecules and leave a hydrous oxide with a variable composition. Hydrous oxides can form polymers with both different compositions and structures.

Even in cases where the solubility product would be appropriately constant in terms of the real concentrations of hydroxide and metal ion concentrations used in the solubility product, it may not be very helpful if little of the metal contaminant is in the simple ionic form. The solubility product may be appropriate when used in terms of the concentration of actual metal ions, but if only small fractions are present as simple unassociated metal ions, it may be difficult to estimate the concentrations of ions from the total concentration of the metal. The ions may become partially complexed with anions, and little of the metal ions may be as free metal ions. (Complexation with anions also may be associated with mixed hydroxide and anion compounds instead of the simple hydroxides.) Even the hydroxide ions may become associated with polyvalent metal ions, leaving a cation complex, but one with a lower electrical charge than the unassociated metal ion. Other anions may be present that can form tighter coordination complexes with the metal ions; in some cases, the metal ion-anion complex can even become an anion because the number of anions coordinated with the metal ion more than neutralizes the positive charge of the metal ion. In the cases of common anions, a high number of anions can be incorporated in the complex producing anion complexes. Complexation certainly can affect the effective solubility of the metal ions.

These complex anions are more likely to occur at relatively high anion concentrations and, thus, may be less likely to occur in dilute wastewaters or groundwater.

Although identified ligands may not be added to the water, several components of the humus materials in soils and groundwater can coordinate with metal ions of environmental interest. Metals can also be adsorbed or otherwise associated with colloidal solids in the soils, for instance with colloidal clay particles or colloidal humus particles. All of these interactions may mean that the concentration of actual metal ions and hydroxide ions in the wastewater or groundwater may be significantly less than the total concentration of the metal or (potential) hydroxide in the water.

A second reason why the solubility may not follow the simple solubility product closely is the formation of a different solid phase. This effect can assist or hinder precipitation, but it is more likely to improve the fraction of the metal ion precipitated. The simple solubility product assumes that the solid formed is the hydroxide, but the solid could be considerably more complex than that. The precipitate could be crystalline or an amorphous solid, perhaps even a hydrated polymer. Such cases would result in different effective solubility products because the solid product is different. If the nature of the solid precipitate changes with pH or with the presence of other components in the wastewater, the effective solubility product could change. An even greater change would occur if the solid coprecipitates with another metal ion, that is, with a carrier precipitate. Then the activity of the solid could be greatly different. In most cases, this is likely to enhance the precipitation, perhaps even greatly.

One study of the solubility of lead in basic solutions suggested several different solid compounds that precipitate [2]. Precipitates from lead chloride can contain such compounds as PbOHCl or $Pb_4(OH)_6Cl_2$. The composition of the solid phase may depend upon the concentration of anions (Cl) and the pH. If the precipitation is accomplished by adding lime to the wastewater, the pH will be a function of the calcium ion concentration. Of course, all or any of the other electrolytes in the wastewater could affect the pH as well by providing either an initial pH different from neutrality or a buffering effect. Since lead can also be precipitated from sulfate or carbonate solutions, there can be a collection of more complex solid phases formed, such as $Pb_4(OH)_6SO_4$ (which can be dehydrated to $Pb_4O_3SO_4$). Similar complex solids can be found when precipitating from a carbonate system.

The importance of ligands to precipitation processes is likely to depend upon the concentration of the metal ions as well as the concentration of the ligands present. If the contaminant is present in only trace concentrations, it takes very little ligand activity to affect essentially all of the contaminant ions, whereas it may take far more ligand to associate with a significant fraction of the metal ions at much higher concentrations. Even when a contaminating metal is present at relatively high concentrations, the process requirement may make it necessary to reduce the concentration of ligands to a very low value, and the importance of trace ligands may become more evident as the concentration of the metal ions is reduced. Because other metals ions and even hydrogen ions can compete with the contaminant metal for the ligands, the importance of trace concentrations of ligands can be affected by the presence of other metal ions and the pH, that is, by the concentration of hydrogen ions.

PRECIPITATION OF METAL IONS BY SELECTED ANIONS

Precipitation methods other than hydroxide precipitation also can sometimes be described by similar solubility products. Another group of precipitates with wide importance to environmental and waste treatment is the sulfide group. Several toxic metals have especially insoluble sulfides, often even less soluble than the hydroxides. Thus, adding even a trace of sulfide ions to the water can remove several toxic metals that may be present in the water. The effectiveness of the sulfide precipitation can also be affected by the solution pH, but, of course, the pH does not specifically describe the sulfide ion concentration like it does the hydroxide ion concentration.

When the metal to be precipitated is present only in trace concentrations, it may be impractical to remove (settle or filter) a trace precipitate. Precipitations from extremely dilute solutions may form extremely small particles, and there may not be enough solids to form an effective filter cake. (Note that in some filtration systems, the filter cake may play a major role in the filtration process, in both the filtration rate and the actual removal of particles.) In such cases, it may be necessary to add a carrier precipitate. This is a precipitate of another material on/in which the target component can be precipitated. With the carrier precipitate, there will be sufficient solids present to be removed effectively. In many pH controlled precipitation processes, there will be other components in the water that can be removed by the addition of quick lime or other material to raise the pH. The target precipitate may become entangled or incorporated into the larger precipitate particles and thus be easily removed. Even when the particles are not incorporated into the precipitated particles, the larger volume of precipitate may form a filter cake that is helpful in removing trace quantities of the contaminant precipitate.

ADSORPTION OR INCORPORATION OF TRACE METALS ON PRECIPITANTS

In some cases, it may not even be necessary for a trace contaminant to be insoluble to be removed by precipitation. Hydrous metal oxides (or hydroxides) can be effective adsorbents for a number of other metal ions. However, the precipitating hydroxides may be especially effective. Normal adsorbents adsorb components directly on the particle surface, and utilization of the remainder of the adsorbent requires that the adsorbate (the contaminant in this case) diffuse into the particles by any of several mechanisms. However, in precipitation processes, there is a continuous creation of new "surface area" as the precipitate particles grow and new precipitate particles are formed. This creates an opportunity for the contaminant to "adsorb" on the particles as they are formed and grow. This can be a very rapid process, and higher contaminant loadings are likely to be achieved than would be practical if a preformed precipitate were contacted with a solution of the contaminant. Note that adsorption of components on solid precipitates as they are formed is much like incorporating the components into the solid precipitates.

This phenomenon could be described as adsorption or as precipitation since the contaminant could be adsorbed on, or incorporated into, the carrier precipitant. In many cases, both phenomena occur, and the topic is mentioned here and in Chapter 2. The most common precipitants used to adsorb or incorporate common toxic metals are ferric hydroxide and aluminum hydroxide. When iron is used, it is usually added as ferric chloride, and aluminum is usually added as aluminum sulfate, or alum. These are acid salts of metals that form insoluble hydroxide precipitates when the pH is raised to values only slightly above neutral. The hydroxides formed are very hydrous and usually have large surface areas, just the conditions needed for adsorption or incorporation of other metal ions in the precipitates. These are also usually long polymer-shaped precipitates that often coagulate well and make settling and/or filtration less difficult.

A precipitation vessel is often stirred gently to promote agglomeration of the precipitated particles, and it is usually possible to have most of the solids settle to the bottom of the vessel so that much of the water near the top of the vessel can be decanted. Further filtration will be required if high removal efficiencies are needed. The settled solids can resemble a sludge and require dewatering to minimize the waste volume. More detailed descriptions of coagulation, agglomeration, and settling are provided in Chapter 10.

One recent study [3] compared direct precipitation of Cd and Cu with coprecipitation using ferric chloride or aluminum sulfate (to form

the hydrous hydroxides). In each case, the coprecipitation improved the removal of Cd and Cu greatly with the addition of as little as 100 ppm of the coprecipitate. In all cases, the removal could be carried out at significantly lower pH values than would be possible without the coprecipitate, and in the limit the pH range for effective removal of the toxic metal was moved from the pH range where the metal became insoluble toward the range where the coprecipitant became insoluble. The effective pH range can even reach values slightly below neutral. Cd is somewhat more difficult to precipitate at the lower range of pH values than Cu because its solubility does not decrease until higher pH values are reached. Effective removal of both Cd and Cu (initially at concentrations of 84 ppm) can be achieved at near neutral pH with as little as 100 ppm of ferric chloride, but somewhat larger concentrations of aluminum sulfate are required to lower the pH for effective Cd removal near neutral pH.

WHEN TO CONSIDER THE USE OF PRECIPITATION

Precipitation has played a major role in water treatment for a long time, and it is widely used for the treatment of many wastewaters before discharge. Precipitation is especially useful in the removal of heavy metals and other contaminants that can be "carried" by precipitates. One is likely to think of precipitation for removal of lead, cadmium, copper, mercury, and transition metals. One or more forms of precipitation can usually remove all of these metals effectively.

The principal disadvantage of precipitation processes can be the volume of solid waste produced. This is the least attractive aspect of precipitation, but its importance, of course, depends upon the volume of waste produced by competing separation methods. Waste volume has become an especially important consideration for selecting waste and environmental treatment systems only in recent years, and now further increases in solid waste disposal costs would make this an even more serious question for precipitation processes. Often the difference in the waste volume will be a function of the concentration of the contaminant metal(s). If the metals are present at only trace concentrations and carrier precipitates are needed to remove the precipitant particles effectively, the waste produced may contain the contaminant at only low concentrations and one may need to consider another separation process such as adsorption or ion exchange if an adsorbent or ion exchange material can be found with sufficient selectivity to concentrate the contaminant better than the carrier precipitation. Thus, the waste volume from carrier precipitation can be relatively insensitive to the contaminant concentration. Then there is more likely to be an alternative separation process to precipitation that produces less solid waste when the concentration of the contaminant is very low.

For the limiting case where no carrier precipitate is needed, the waste volume from precipitation can be approximately as low as would be possible from any alternative process. In those cases, the waste volume still can depend upon the selectivity of the precipitation process. Of course, precipitation processes initiated by raising the pH (and even by adding sulfide ions) are not particularly selective and may carry down a number of metals, leaving a precipitate relatively dilute in the contaminant of interest. If there is a need to remove only a single toxic metal, a more selective removal method may be considered. However, when there are a number of heavy metals in the wastewater, it may be necessary to remove all of these metals, so there can be considerable merit in removing several contaminants in one step. In many cases, it is fortunate that so many toxic metals can be removed by single precipitation processes.

EXAMPLES OF PRECIPITATION IN WASTE AND ENVIRONMENTAL TREATMENT

Although there are numerous applications of precipitation for water treatment (wastewater as well as process water), this discussion will focus largely on studies that have evaluated the precipitation processes. In 1980, the EPA issued a report evaluating precipitation for removing the heavy metals such as cadmium, copper, chromium, nickel, and zinc that are present in plating wastes [4]. This study was specifically evaluating sulfide precipitation, but it also included evaluation of lime-based precipitation by pH control because the sulfide precipitation was applied after lime precipitation. The study includes moderately detailed cost information, but the early date of the study requires considerable extrapolation of the costs to apply them today. The feed solutions tested were acid solutions containing a range of concentrations of cadmium, copper, chromium, nickel, and zinc, but the zinc, chromium, and nickel concentrations were the highest. Lime was added to increase the pH to a value between 8 and 11. Tests were made with lime treatment alone and with both lime and sulfide treatments. The sulfide treatment was made both with the lime treatment and as a polishing step after lime treatment. The precipitated particles were removed by settling or filtration. The different combinations of these treatment steps covers a wide variation in treatment schemes.

Lime precipitation reduced the concentration by approximately 80%, but filtration (rather than settling) was required to reach that level of copper removal. As expected, filtration gave at least somewhat better removal of all of the metals than settling alone. Better than 95% of the cadmium, chromium, and zinc were removed by the lime precipitation.

The sulfide precipitation always increased the removal efficiency of all the metals, and the removal was better when the sulfide precipitation was a separate step rather than incorporated with the lime precipitation. With separate precipitation steps and filtration of the precipitate, the removal of all of metals was greater than 95%, and the removal of zinc was greater than 99%. Sulfide precipitation did not have a major effect on chromium removal, which should be expected because chromium is probably present largely as an anion (chromate) and would not be expected to interact with sulfide ions. There was a significant difference between chromium removal with settling and filtration, and this may be viewed as evidence that the chromium precipitates are relatively fine.

Costs for the precipitation processes were estimated for facilities with throughput capacities of 10,000 gal/day, 200,000 gal/day, and 500,000 gal/day. Capital costs as of 1980 ranged from approximately $233,000 for the smallest lime treatment facility using settling to approximately $1,400,000 for the lime-sulfide treatment facility with filtration. The operating costs ranged from approximately $24/1000 gal for the smallest facility using both lime and sulfide precipitation and filtration to approximately $2.10/1000 gal for the simpler lime precipitation and settling. All of these costs would have to be extrapolated considerably to apply today.

A more recent study [5] at the Lawrence Berkeley National Laboratory also gives capital and operating cost estimates as well as performance information. This paper also compares the cost of different separation processes for this application. The contaminant of interest was copper, which was present at concentrations from 16 to 94 mg/L. The wastewater was from a copper plating bath, but there was also believed to be nickel, lead, and chromium present in the wastewaters since these are usually associated with plating wastes. The precipitation treatment involved adding ferric chloride or ferric sulfate and sulfuric acid to free the copper ions from ligands believed to be present in the wastewaters. At sufficiently low pH values, the hydrogen ions could compete with copper ions for the ligands, such as ethylenediamine, citrate, carboxylate, ethylenediamine tetracetate, nitrilotriacetate, and phosphate. The pH was then raised to neutral or above (6.6 to 10) with sodium hydroxide to precipitate the ferric ion as hydroxide or hydrous oxide. The effluent concentration of copper was reduced to values between 0.24 and 5 mg/L. The removal efficiency appeared to be a stronger function of the amount of ferric ion precipitated and the form of the ferric iron than the final pH. The ferric sulfate and sulfuric acid system appeared to be more effective than the chloride system.

The costs were estimated for a relatively small treatment unit treating only about 25,000 gal/week, and the results showed precipitation to be more costly than reverse osmosis, ion exchange, or evaporation, but the range of uncertainties in the cost estimates was approximately as large as the differences between the costs. Thus, one cannot clearly state from the costs given which separation method would be least costly. It is informative to see different separation methods compared. However, care must be used in comparing cost estimates for different separation methods. Even when care is taken to ensure that all operating and capital costs are generated with the same basis (interest rates, contingency costs, etc.), it is desirable to consider the entire waste treatment costs, including final disposal of the solid wastes. Since waste disposal costs can be different at different sites (usually because of different distances to the disposal sites and different charges from operators of the disposal site), the relative cost of separation methods may change with location of the facility.

REFERENCES

1. Allen, J. S., et al. "Treatment of Metal-EDTA Wastes Using Electrochemical Reduction, Chemical Oxidation, and Metal Precipitation Techniques." *Proceedings of the 51st Industrial Waste Conference*. Purdue University, May 6–8, 1996, p. 601.
2. Baltpurvins, K., et al. "The Use of the Solubility Domain Approach for the Modeling of the Hydroxide Precipitation of Lead from Wastewater." *Proceedings of the 50th Industrial Waste Conference*. Purdue University, 1995, p. 237.
3. Karthikeyan, K., H. Elliott, and F. Cannon. "Enhanced Metal Removal from Wastewater by Coagulant Addition." *Proceedings of the 50th Industrial Waste Conference*. Purdue University, 1995, p. 259.
4. Robinson, A. K. and J. C. Sum. *Sulfide Precipitation of Heavy Metals. EPA-600/2-80-139*, June 1980.
5. Chang, Li-Yang. *Environ. Prog. 15*, 28 (1996).

12
Other Physical Separation Methods

This chapter contains descriptions of several physical separation methods that are not covered in as much detail as filtration and sedimentation methods. Some of these methods could be of importance to specific readers and their specific problems. The material on these methods is largely descriptive and does not include design methods. However, an attempt is made to include a description of the basis for the separations and some of the key issues that affect the performance of the methods. Suggestions are also made on the types of problems that are mostly likely to be attractive for the use of these methods, but the suggestions are briefer than those given in some of the lengthier chapters.

FLOTATION

Flotation is a physical process to separate particles based upon their surface properties. The technique separates hydrophobic from hydrophilic particles. That is, flotation separates particles based upon their wettability by water. It has its principal and most traditional use in mineral treatment, but additional uses in other fields appear likely to grow, especially in environmental and waste processing. When they can be used, physical separations such as floatation have important advantages in terms of cost and energy consumption. When the alternative to a physical separation is chemical dissolution, the differences in energy consumption, reagent consumption, and waste production can be especially important. The only reagents required for small quantities of surfactants and, possibly, small quantities of materials to adsorb on the particle surfaces to alter the surface properties. Since only the surface of the particles needs to be affected by additives, the quantities of additives needed are usually much less than would be required to dissolve or even leach components from the particles. There may be no wastes other than the particles themselves.

Flotation could prove to be useful in environmental and waste management separations if toxic and nontoxic particles in a waste or soil have different surface properties. For instance, if a solid waste has been spilled onto soil or become combined with nontoxic materials in a waste stream, it may be possible to remove the toxic solids by floatation if the toxic particles have sufficiently different wettability from the nontoxic bulk of the solids. In other cases, a liquid contaminant may adsorb preferably on only one type of particle in soils or waste, and removal of those particles would remove much of the contamination and concentrate the contaminant. In such cases, it may be desirable to separate the contaminant bearing particles by flotation before attempting to leach the contaminant from the particles. That could reduce the quantities of leach liquor required and thus the volume of waste produced.

Probably the most crucial factor in flotation is the ability to separate the contaminant from uncontaminated particles sufficiently that the uncontaminated particles can be released or reused. That requires, first, that some (preferably most) of the particles be sufficiently free of contaminants and, second, that flotation be able to separate contaminated particles sufficiently from the uncontaminated particles. Since flotation only separates individual particles, it is necessary for the different materials (such as toxic and nontoxic components) to be in separate particles. In many high temperature processes, particles of different materials may become fused, or one solid material may coat another material. Such particles could not be separated by flotation, and usually not by any physical separation method. Likewise, when a contaminant is adsorbed on the particles, flotation will only be helpful when the contaminant is adsorbed selectively only on a portion of the particles; the remaining particles need to be sufficiently free of contaminant that they can be used, released, or treated differently.

It is obviously necessary for the particles to be dispersed in the suspension. Agglomerated particles will act as a single larger particle, usually with average surface properties with contributions from all of its components. If the particles in a suspension tend to agglomerate, it may be necessary to add a "dispersant" to separate the particles. This is likely to be an acid or (more likely) a base which changes the pH. This changes the ionization of hydroxyl and other groups on some particle surfaces and can result in electrical charges that repel individual particles and keep them dispersed. The chemistry involved in dispersion is related to agglomeration. For further discussion of dispersion/agglomeration methods, see Chapter 10 on sedimentation. Agglomeration also is often a critical step in obtaining satisfactory filtration rates with practical pressure driving forces.

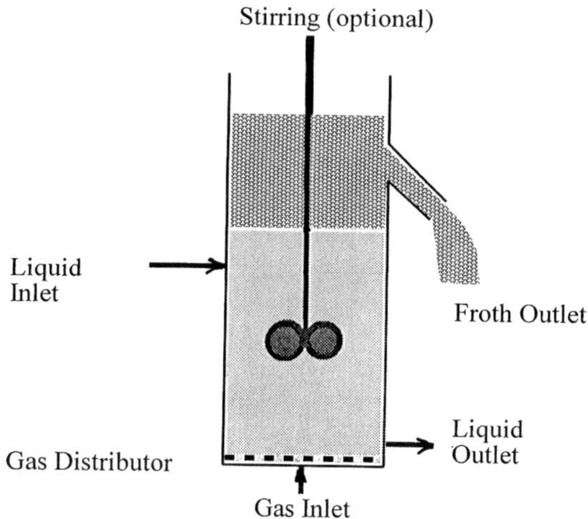

FIGURE 1 Schematic drawing of a flotation cell.

When air is bubbled through a suspension of particles, the particles that prefer to not be wet by the water are more likely to be wet by the air and collect at the air-water interface. These particles are then concentrated at the air-water interface, carried from the suspension by the rising air bubbles, and separated from the suspension. A typical flotation system is illustrated in Figure 1. The contact time between the air bubbles and the suspension must be sufficient for the particles to reach the bubbles by Brownian and turbulent diffusion and become attached to the bubbles. Agitation is usually applied in the flotation device to increase the rate at which particles are transported to the bubbles.

Because the forces that hold the particles to the bubble and, thus, remove the particles from the suspension are not strong, it is necessary to have very small particles; flotation is not likely to be effective for large particles. It is probably obvious that the particles need to be sufficiently small that the buoyant forces of the bubbles are able to lift the particles from the liquid. In most cases, the particles should be less than a millimeter, preferably much smaller. This may mean that the material has to be ground into very small particles. Grinding also may be desirable if it breaks larger agglomerates that contained several solid phases and, thus, separates the multiple phase particle into several smaller particles with fewer phases, preferably single phases. However, grinding can be a relatively expensive and energy intensive operation for some solids and can hinder the application of flotation.

A simple flotation device is shown in Figure 1. Note that air is injected into the bottom of the unit, and a large number of bubbles are generated. Agitation is also usually provided by stirrers. The bubbles collect the particles that are wet by air and "float" them to the top of the device. The froth (or foam) of bubbles overflows the unit carrying away the floated particles. The slurry can flow through the unit, or, particularly in laboratory units, the slurry could remain in the cell, and flotation could be a batch process. Especially effective separations can often be achieved by using ultrafine particles and very fine and approximately uniformly sized bubbles. The fine bubbles can be formed by dissolving the gas in the slurry at high pressure; fine bubbles are formed as the pressure is reduced and the gas comes out of solution. Electric fields at the gas inlet can also be used to assist in generating uniform gas bubbles, preferably small bubbles [1].

A typical flotation cell is principally a single stage device, but with a very stable froth it is possible to "wash" the froth with a countercurrent flow of water (or solution) and measure the performance. However, at least one multi-stage device has been reported which uses a series of draft tubes which allows local recirculation of the froth for increased mass transfer, and the use of several draft tubes in series provides the effects of multiple stages [2]. The addition of multiple draft tubes was shown to increase the effectiveness of flotation.

It is desirable for the froth to be stable enough to be removed from the cell with the attached particles attached. If the bubbles should "break" before the froth overflows from the cell, the particles attached to that bubble will be returned to the suspension. To create a relatively stable froth, it is often necessary to add a surfactant, usually called a "frother."

A number of other chemicals may be added to assist in removing or separating particles. Although flotation of coal particles may not be a common operation in environmental and waste management operations, it provides a simple example of a "promoter" (sometimes called an "activator") that enhances the separation of hydrophobic coal particles from hydrophilic mineral particles. Small amounts of oils added to a suspension of coal will selectively wet the coal particles, even if the coal particles contain local hydrophilic surfaces, perhaps because of traces of mineral matter on the surface. Once the coal particles are covered with oil, they become even more hydrophobic and more easily separated from the mineral content of the suspension. This "beneficiates" the coal by separating much of the mineral matter. Activators can also be materials that adsorb on the components to be removed and make them hydrophobic.

By using an activator, it is possible even to separate different particles that are normally both hydrophilic or hydrophobic. For instance, a material can sometimes be selectively adsorbed on only some of the particle surfaces and alter the hydrophobicity. A compound with an active group on one end of the molecule that binds with one of the mineral surfaces can have an organic (hydrophobic) group on the other end. If the particle surface becomes covered with these "collector" molecules, the effective surface made up of the organic ends of molecules can be hydrophobic.

When a particle surface does not accept a "collector" readily, it may be possible to adsorb a small amount of metal ions or other materials that then provide sites for collector molecules to become attached. Such materials may be called "activators." "Collectors" may make too many different particles hydrophobic and many different types particles will be floated from the cell. Then it may be possible to reduce the flotation of the other particles by adding a "depressant." Thus the objective of the use of any additive is to enhance the difference in the surfaces of the different particles to be separated by adsorbing the additive on one of the particles.

Surfaces of natural materials are often difficult to characterize accurately because the materials are impure and because additional impurities may already be adsorbed on the surface. Adsorbed materials on the surface can have major effects on the surface properties and the behavior of particles during flotation just as the activators and collectors that are intentionally added to affect flotation, but these adsorbed materials may represent little of the bulk mass of particles. Just as only small quantities of activators or collectors are needed to enhance flotation, it takes only a very small amount of originally adsorbed material to cover particle surfaces and greatly alter the behavior of the particle mixture in flotation units. One cannot predict flotation behavior from the bulk composition of the particles alone, so it is usually necessary to test particle mixtures to determine if flotation behavior is essentially the same as expected or if adsorbed impurities are important.

Careful experimentation is often necessary to assess how effective a flotation operation will be or even to select the best frother, collector, activator, etc. This may make the design of flotation cells and operations appear to be mostly art and only part science. However, the principles of flotation operations are well understood. The uncertainties result more from lack of information on the properties of the impure materials to be separated than from lack of understanding of the principles of flotation. Trace impurities in particles may be of little importance if they are distributed throughout the particles, but they can be of great importance if they are concentrated on the surface of the particles.

The capacity of flotation devices is set by the time required for the particles to attach to bubbles and be carried from the cell. This requires sufficient residence time for the slurry in the cell. The time required can depend upon the particular separation and the degree of separation required, and could be several minutes. Flotation cells can be operated in series to achieve longer residence times. Details on particular flotation cell performance for the more common applications are available from the manufacturer.

Environmental and Waste Applications of Flotation

Flotation has also been studied for the removal of traces of oil from water [3,4]. This study focused upon dissolved air flotation and the economics of the system. Bubble size is an important parameter in air utilization during flotation, and sudden reductions of pressure above water saturated with air can produce small and relatively uniform bubbles that are especially effective for flotation. Bubble sizes below 100 micrometers can be achieved in such systems [5]. In such operations flotation competes with other liquid-liquid separation methods such as coagulation/sedimentation, but coagulation can occur during flotation as well. Dissolved air flotation (also called "pressure flotation" was studied for removal of fine nitrocellulose from propellant manufacturing [6].

One does not usually think of flotation as a method for removing dissolved contaminants from water, but at least one paper has described preliminary studies of removal of heavy metals by combined precipitation and flotation [7]. In that study, the heavy metal ions were precipitated as sulfides by adding sulfide ions, and the particles were removed by flotation. The advantages of flotation over some other particulate removal methods may be the possible selectivity of the flotation for the sulfide precipitants and the ability of flotation to handle small particles less than 10 micrometers which are not easily removed by most filtration methods. Adding sulfide ions to some waters may not be attractive for all applications, but some of the sulfides of toxic heavy metals are so insoluble that only trace concentrations of sulfide ions may be needed. If the sulfide ions could be introduced in a sufficiently controlled manner, there could be little residual sulfide ion left in the product, but for many applications the concentrations of metals to be removed may not be constant or even accurately known.

In any case, flotation may contribute to the removal of toxic metals by removing particles on which the metals have been adsorbed (or held by ion exchange). When one must be concerned with extremely low concentrations, adsorption of ions on colloidal clay, silica, or other particles may be the most important form of the toxic metals. The effects of adsorption of radioactive contaminants can be especially important because the acceptable limits of contamination are often so low. As the concentration of the contaminant increases, the importance of the bulk solution will increase, and the chemistry of removal processes is likely to appear to be more standard and resemble the behavior of such systems at higher concentrations.

It is probably obvious that bubbling air through a liquid can also remove volatile organic compounds (VOCs) as well as solids. Although flotation systems are not designed for maximum removal of VOCs, a substantial fraction of the more volatile VOCs can be removed in flotation operations [8]. In some cases, the removal of significant VOCs could be an advantage, especially if it is needed and sufficient. However, the presence of significant toxic VOCs can require that the effluent gas from a flotation operations be treated to remove the VOCs. If the removal of VOCs is not sufficient, an additional gas stripping or other operation may be needed to remove the remaining VOCs. In such cases, it may be unfortunate to have to recover the VOCs in two ways. The VOCs removed in a flotation operation can usually be recovered from the air with an activated carbon adsorption bed. In the case of acid or basic gases, one could consider absorption operations if the volume of gas flow is sufficient to justify absorption equipment. Adsorption is likely to be the method of choice for small gas streams. If gas stripping should be used to remove the remaining VOCs from the liquid stream, it will probably be desirable to send both the flotation gas and the strip gas to the same treatment system.

Chemical Separations During Flotation

Flotation can also be used to remove soluble materials and, especially, surfactants. Although it is certainly not completely appropriate to discuss separation of molecules in a chapter largely devoted to separation of particles, there is adequate reason to include this technique with foam flotation because both dissolved molecules and particles can, and will, be separated during such operations. Thus, we will discuss chemical separations that occur during particle separations, and identical chemical separations that can be carried out even when there are no suspended particles to separate. The author is not aware of wide use of flotation for chemical separations, but the method has received, and continues to receive, attention, and the reader may want to be aware of it.

First, note that surfactants are collected at air-water interfaces, and one should expect surfactants to be concentrated in the foam from a flotation system. Thus, when the concentration of surfactants in a wastewater is too high, it would be possible to reduce the concentration of the surfactant by "foaming" the surfactant from the water. Then by breaking the foam, the surfactant would be concentrated. This also occurs in all flotation systems, and much of the surfactant added as a frothing agent is usually removed with the froth.

It is possible to produce surfactants with active groups (ligands) that complex with metal ions. Then when the surfactant is removed, the metal ions are removed. If the ligand group on the surfactant is specific for a toxic metal, the metal would be removed selectively. Note that this approach is related to the removal of complexed metal ions by nanofiltration. The difference is that the metal-ligand complex is separated by the size of the complex rather than by the surfactant properties of the complex.

Organic contaminants can also be removed by flotation through a similar mechanism. Soluble organic contaminants can become associated with the hydrophobic portion of surfactants, and when the concentration of surfactant is high enough to form micelles, the organic contaminants being less hydrophilic than water are likely to be incorporated inside the micelles. Although the micelles have their hydrophilic groups facing the water phase, they are still likely to be less hydrophilic than the surrounding water. Furthermore, additional surfactant molecules may associate with the micelles forming double layers of surfactants and make the surface of the micelles even more hydrophobic. In any case, the micelles are likely to concentrate at air-water interfaces and be carried from the solution with foam. The removal of *t*-butylphenol using foam flotation with cetyl pyridinium or dodecyl sulfate as surfactants was studied recently by John Scamehorn's group at the University of Oklahoma [9]. The removal efficiency peaked at the critical micelle concentration. This suggests that micelle formation is important to contaminant removal, probably hindering flotation of contaminant by competing for the interfacial area. This qualitative observation was strengthened by the addition of salts that lowered the critical micelle concentration and also lowered the contaminant removal rate. These interfacial separations of molecules are interesting and may have significant future applications. However, anyone dealing with flotation of particles from complex solutions that also contain soluble materials should be aware of the other processes that may be taking place while particles are being separated. Remember also that any component that adsorbs on the particles can also affect their surface properties, so the presence of other components, contaminants or not, can affect the primary flotation operation.

MAGNETIC SEPARATIONS

There are two general classes of magnetic separation operations: magnetic devices that separate solid particles from fluids, and magnetic devices that separate solid particles from each other on the basis of their magnetic properties. The first class of devices can be considered "magnetic filters." The magnetic filters can be small and relatively inexpensive because they only have to generate magnetic forces sufficient to drive the particles from a fluid to a solid. It is not necessary to have a precise operation that depends upon small differences in the particles. Magnetic filters are likely to remove essentially any particle that is ferromagnetic or moderately paramagnetic. If a filter cake of the magnetically susceptible particles is formed, the filter is likely to remove a great many of the nonmagnetic particles from the actions of the filter cake. If one wanted to remove only the magnetic particles, it would be necessary to avoid the formation of a filter cake or other severe flow restrictions that would contribute to removal of the nonmagnetic particles.

Magnetic Properties of Materials

Magnetic separation methods depend upon the magnetic susceptibility of solid materials (particles). Materials can be conveniently classed as ferromagnetic, paramagnetic, or diamagnetic. All of these groups can be described in terms of their magnetic susceptibility:

$$M = m\chi H$$

M is the magnetization of the material (particle), χ is the magnetic susceptibility (a property of the material), m is the mass of the particle, and H is the magnetic field intensity. The three groups of materials can be defined by their magnetic susceptibility.

Ferromagnetic materials have very high magnetic susceptibilities. These materials include metals such as iron, nickel, cobalt, and many of their alloys. The susceptibilities of these materials are so much higher than those of other materials that it is convenient to classify them separately. There is no continuum of materials from the ferromagnetic to paramagnetic materials (those with lower susceptibilities), but a "valley" or "gap." These susceptibilities are usually significantly lower than those of the ferromagnetic materials. Most (but not all) materials can be thought of as paramagnetic if one includes those materials that have essentially no magnetic susceptibility. However, the term has practical application only to those with significant susceptibilities. These include a variety of minerals, metals, and alloys.

Diamagnetic materials have negative susceptibilities. This means that they will magnetize to oppose the applied magnetic field. Of course, since the negative susceptibility can have any magnitude, including zero, one could also include that vast number of materials with no measurable magnetic susceptibility in this class or in the paramagnetic class.

Behavior of Particles in Magnetic Fields

As the definition of the three classes of magnetic materials indicates, the first response of a material when placed in a magnetic field is to become "magnetized," and the extent, and even the direction, of the magnetization depends upon the magnetic susceptibility of the material. This is very important, but for magnetic separations it is the interaction of this "induced" magnetization with gradients in the magnetic field that generates the forces that is usually utilized in magnetic separations. The behavior of magnetically susceptible particles in magnetic fields is similar to the behavior of dielectric particles in an electric field, and the reader may want to look at Chapter 9, especially the section on electrofiltration, for a similar discussion.

The behavior of a magnetized particle in a nonuniform magnetic field is illustrated in Figure 2. The particle is magnetized by the magnetic field, and the extent of the magnetization depends upon the strength (intensity) of the field and the magnetic susceptibility of the material in the particle, as described in Equation (1). Although magnetizing the particle does not induce a force to move the particle, there will be a net force on the particle if the magnetic field is not uniform. That results because in a nonuniform magnetic field and a finite size particle, the different regions of the particle will not be in exactly the same magnetic field. This means that the attraction of the pole at one end of the particle toward the magnetic field will be stronger than the attraction of the other end of the particle in the opposite direction along the magnetic field. This results in a net attraction of the paramagnetic particle toward regions of high magnetic field intensity. In Figure 2, the highest magnetic field gradient occurs at the wire (shown in cross-section), and the particle is drawn toward the wire. Note that this is essentially the same behavior seen for dielectric particles in an electric field, and Figure 2 is similar to Figure 6 in Chapter 9. Diamagnetic particles behave in an opposite

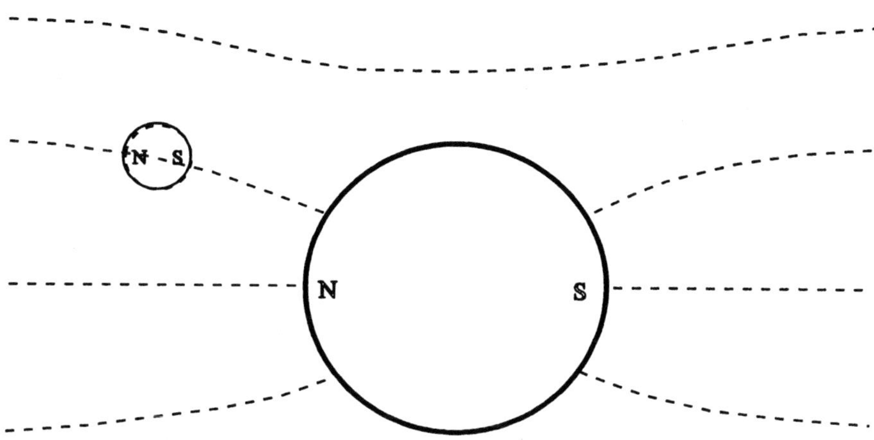

FIGURE 2 Magnetically polarized wire and a particle showing how the magnetic field lines converge toward the polarized wire.

manner to paramagnetic particles and are attracted toward regions of low magnetic field intensity.

The force on the particle is obviously dependent upon the degree of induced magnetization, and that is proportional to the magnetic field intensity and the paramagnetic susceptibility of the particle. The difference in the magnetic field intensity at the extreme ends of the particle is proportional to the size of the particle and to the gradient in the magnetic field. There is also an effect from the particle shape, but for particles with any given shape the force on the particle will be proportional to the particle size. Since the force on the particle is proportional to both the magnetic field intensity and the gradient of the intensity (and thus to the product of the intensity and the gradient of the intensity), it is often convenient to describe the force as proportional to the gradient of the square of the field intensity.

Magnetic Removal of Particles From Fluids

The most common use of magnetic separations is probably magnetic filters. That is, a magnetic field gradient is used to attract ferromagnetic or paramagnetic particles to solid surfaces for the purpose of removing them from suspension in a liquid or gas. As noted earlier, technically this is not as challenging a task as removing individual groups of particles according to their magnetic susceptibility.

Although there are several potential applications for magnetic filters, the separations are likely to be easiest if the particles to be removed all have ferromagnetic or moderately high paramagnetic properties. It may be difficult to remove all of the solids from a fluid if the solids are highly heterogeneous and some of the particles have very low magnetic susceptibilities. However, there are numerous applications where one is concerned principally with the removal of only a single solid material or only a relatively few solid materials. An obvious potential application of magnetic separations could be for the removal of metal particles from machine cutting liquids or even engine cooling oils. Should such applications be considered environmental separations? If such separations allow reuse of the fluids or even longer use of the fluids, there will certainly be important environmental benefits since less oil would have to be sent to disposal. Metal particles can also be serious problems in food and paper products where even traces of metals could degrade the product greatly.

The advantages of a magnetic filter over conventional filters can be the low pressure drop and/or the potentially easy removal of the particles from the filter. The low pressure drop can result when relatively high magnetic forces can remove the particles without requiring very narrow flow channels, such as the flow through very fine filter media or precoats. In many cases, the pressure drop across a magnetic filter can be very low. The potential ease of regeneration results when electromagnetics are used to generate the magnetic field gradients and neither the particles nor the ferromagnetic mesh or other device retains significant magnetization after the electric current is turned off. Alternatively, the particles can be recovered by removing the magnetic pieces and physically detaching the particles from them.

There are several ways to generate a magnetic field gradient to remove ferromagnetic or paramagnetic particles, and two arrangements will be discussed here. Perhaps the most obvious way to generate a magnetic field gradient involves placing a pole piece in the flow stream and surrounding it by one or more poles of opposite orientation and a greater area [10]. This causes the magnetic field to converge toward the pole piece. This can be done one time, or there could be a series of such pole pieces arranged with a filter chamber or in a series of chambers. Such systems offer the potential for very open and large flow channels, so the pressure drop through such devices can be very low. This is likely to be a candidate for removing ferromagnetic particles and relatively large particles with moderate magnetic susceptibilities. Of course, the smaller the particle and the lower the magnetic susceptibility, the higher the magnetic field and the magnetic field gradient that must be applied in such a system.

To remove small particles with low to moderate magnetic susceptibilities, it is necessary to generate higher magnetic field gradients than can be generated conveniently in single pole devices such as those just described. One convenient way to develop very high magnetic field gradients is to apply a magnetic field across a ferromagnetic mesh, usually an iron mesh. The magnetic field lines converge toward the iron (wires in the mesh), creating local very high gradients in the magnetic field that would normally be essentially uniform without the iron mesh (Figure 2). The gradient in the field will be inversely proportional to the diameter of the wires in the mesh, and for the highest field gradients fine wire mesh should be used. However, it is also necessary for the mesh to have sufficient structural strength to withstand the hydrodynamic drag of flow through the mesh and any magnetic forces that are present. It is also usually desirable for the wire diameter to be greater than the particle diameter, especially if the magnetic susceptibility of the wire is not significantly greater than that of the particles. These considerations place restraints on both the size of the wires used in the mesh and the mechanical structure of the mesh. In fluids that are even slightly corrosive, it may be difficult to use extremely fine wires in the mesh.

Again, the forces on the particles will be proportional to the susceptibility of the particles and the square of the applied magnetic field. (The magnetic field gradient and the magnetization of the particles will be proportional to the magnetic field strength.) The probability of a particle becoming attached to a wire of the mesh will be proportional to the number of wires in the mesh. That is, the probability of a particle being removed in a given volume of the filter will be proportional to the number of wires per unit volume of the filter. Thus, one would like to have the wire density (volume fraction occupied by the wire) to be as high as practical for the greatest removal efficiency (per unit volume of filter). However, the pressure drop across the filter (per unit length of filter) will also be approximately proportional to the density of wires in the mesh. Thus, there is an optimization required to select the most practical wire mesh density and the diameter of the wires in the mesh. This optimization covers the required strength of the mesh (the wire diameter), the removal efficiency, and the pressure drop.

There are also hydrodynamic forces causing the particles to reach the packing surface, as noted in the behavior of other deep bed filter systems. High gradient magnetic filtrations have been discussed by several authors [11,12]. As noted in the discussion of other deep bed filters and electrofilters, the analyses are largely directed at the initial loading of the bed. Often the bed performance improves as the loading increases, and the operating time may be controlled by the increased pressure drop across the filter as the loading increases. One paper has addressed the effects of loading on magnetic filter performance [13]. The analysis assumes rate constants for the particle removal rate, release rate, and surface area changes. There are limited data to verify the model, and several parameters are required to use the model, so it may take considerable data and comparisons with it to determine the usefulness of the model.

A mesh filled magnetic filter operates qualitatively like any deep bed filter with the probability of a particle being removed approximately the same for each unit length of the filter. That means that the removal process is "first order," and the concentration of particles decreases logarithmically with position down the filter length. Then if a given length of filter removes 80% of the particles, doubling the length of the filter (but retaining the same applied field strength) will result in 96% removal of the particles ($100 \times 0.2 \times 0.2$ or 4% of the particles will not be removed).

Mesh filled magnetic filters resemble mesh filled electrostatic filters, but there are significant differences. Electropolarization and paramagnetic susceptibility are similar, and the equations governing the behavior of the filters are essentially the same once one substitutes magnetic field gradients for electric potential gradients and susceptibility for dielectric constant. There are two points, however, that are different for the two systems. First, magnetic systems can be used in fluids that are electrically conducting (that is, in water). That eliminates an important limitation for electrofilters. Second, there will be no magnetic "monopole," which is similar to an electric charge on a particle. That is, electrically charged particles can be attracted to an oppositely charged surface even with a uniform field, and those attractive forces can be larger than induced dielectric forces when the charge on the particles is sufficiently high. However, there is nothing similar to an electric charge in a magnetic system, so all magnetic filters operate on induced magnetic forces and field gradients rather than the field itself.

Separation of Solid Particles from Other Solid Particles

Magnetic separation is also used to separate solid particles from other solid particles. The solids are separated based upon their magnetic properties, and this can be a particularly attractive method to use when the material to be separated is divided into individual particles with the components. Since there are no chemical reactions involved, the energy required for magnetic separations can be relatively modest, except when exceptionally strong fields are required. Magnetic separations also introduce no other materials to the system that can add to the waste volume.

The separations are based upon differences in the magnetic susceptibility of the individual particles in the mixture. This is a separation of particles, not necessarily a separation of different components or materials in the mixture. As with all solid-solid separation methods, if one is trying to separate the different materials (components) in the mixture, the success of magnetic separation requires that the different components to be separated occur in separate particles. Composite particles that contain mixtures of the components of interest, make essentially any physical solid-solid separation ineffective. Nevertheless, there are numerous opportunities for using magnetic separations in environmental and waste separations where composite particles are not serious problems, and such applications are expected to grow.

This discussion of magnetic separations of particles will deal with separation of ferromagnetic materials from nonmagnetic (or only slightly paramagnetic) solids. These separations utilize relatively modest magnetic fields to separate materials with vastly different magnetic properties. Under these modest fields, only a few materials like iron and highly magnetic susceptible alloys are attracted sufficiently to the applied magnets to be removed from the mixture of solids, and this type of magnetic separation is generally limited to removal of such materials. However, iron is one of the high volume materials produced in modern economies, and essentially all of those materials eventually become wastes.

The general concept of this form of magnetic separation is simple and familiar to any reader who has used a hand-held magnet to remove iron filings or machining wastes or to find a needle among nonmetallic materials. The principal difference between using a hand-held magnetic to go through a stack of hay and industrial magnetic separation is that with most industrial scale magnets it is easier to bring the waste to the magnet than to bring the magnet to the waste. In an industrial system to find the needle, one would be more likely to convey the hay past the magnet, and the needle would be captured and retained by the magnet. Two similar systems for separating ferromagnetic materials are described briefly (Figures 3 and 4). In both of these examples, the solids to be separated are carried to the magnet by a conveyer belt.

In the first illustration, the mixture of particles is fed to a belt that moves over the surface of a magnet. The belt carries the mixture across the magnet until gravity allows the nonferromagnetic materials to fall from the belt. The magnet only needs to hold the ferromagnetic materials long enough for them to fall at a different place. Generally the ferromagnetic materials will fall from the belt once they are carried beyond the edge of the magnet.

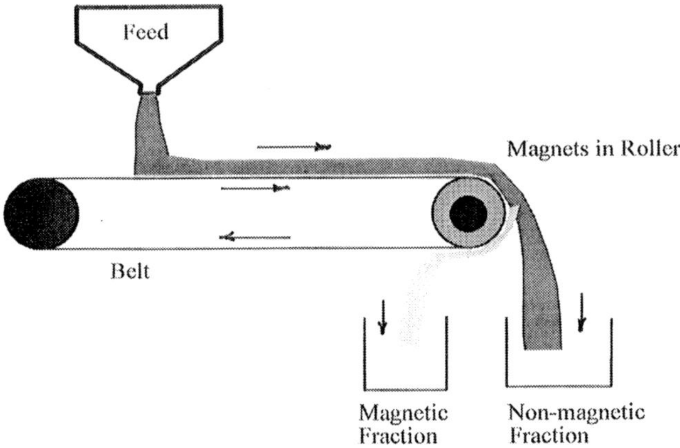

FIGURE 3 Magnetic separator that holds the magnetic particles on a moving belt no longer than nonmagnetic particles.

For effective separation of ferromagnetic and nonferromagnetic materials, it is necessary for the mixed materials to be laid on the revolving belt evenly and in a sufficiently thin layer that the particles interact with the magnetic field individually. This means that the layer of materials on the belt is sufficiently thin that the ferromagnetic materials held on the belt by the magnetic field do not trap nonferromagnetic materials and carry them over to the discharge of ferromagnetic materials. It is also necessary that the layer of solids be sufficiently thin that ferromagnetic

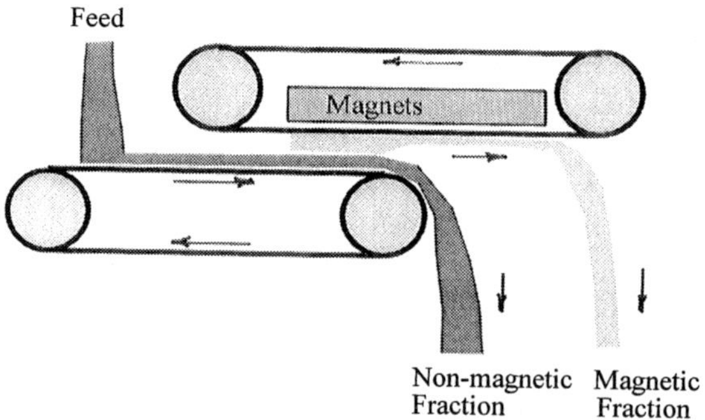

FIGURE 4 Magnetic separator that lifts magnetic particles from a moving belt and leaves the magnetic particles on the belt.

materials on the top of the layer are also retained by the magnet but do not trap nonmagnetic material and carry it into the magnetic product. Furthermore, the rotation of the belt should not be sufficiently fast that centrifugal effects force ferromagnetic materials from the belt with the nonferromagnetic materials. Thus, the limits on the thickness of the layer of feed, the rotation rate of the belt, and the size of the belt (and magnet) set the maximum throughput of the device.

Figure 4 is slightly different. In the first case, the nonferromagnetic material are removed from the ferromagnetic material retained on the belt. Here the ferromagnetic material is removed from the nonferromagnetic material. Again the mixture of solids is conveyed to the magnet, but the magnet is located above the conveyer belt and lifts the ferromagnetic material from the feed belt. After the ferromagnetic material is removed, the remaining materials are discharged from the belt. For this to be a continuous operation, the ferromagnetic material taken from the feed mixture needs to be carried from the magnet continuously. In the illustration, this is accomplished by a second conveyer belt that moves around a fixed magnet, to carry the ferromagnetic material beyond the magnet where it falls by gravity to another collector. Again, for efficient separation, it is necessary that the ferromagnetic materials not be "trapped" by the other materials that stay on the original conveyer belt and that they not carry other materials to the magnet. Since the ferromagnetic material must be lifted slightly to reach the magnet, it is not as likely to "carry" other materials to the magnet in this type of system as it is to trap other material to the magnet in the first illustration. Thus, this second arrangement is more likely to produce a ferromagnetic (iron) product free of other particles.

Environmental Applications of Magnetic Separation of Ferrous Metals from Other Solids

The principal use of magnetic separation of ferrous metals is to remove iron or a few iron alloys from most other materials. In a batch mode it can be used to remove large iron pieces from a variety of municipal and industrial wastes. In the continuous forms illustrated in Figures 3 and 4, it can be used to remove ferrous metals from such solid wastes. Automatic sorting of municipal wastes has been attempted by several communities, and magnetic separations are almost universally an important part of any such system because the ferrous metal content of such wastes is one of the more valuable components (although the value varies considerably), and the continuous separations illustrated here are relatively reliable and economical. Difficulties in obtaining a good separation may result from difficulties in other parts of the system. For instance, it is necessary to shred the solids into pieces small enough to be handled by the magnetic separation system. For reasonably sized equipment, this may require very large components of the waste to be sheared into much smaller pieces. Of course, as with any solid-solid separation method, efficient separations are possible only when the solids can be shredded into particles that are sufficiently small that the each particle contains only one component; that is, it either has ferrous metal or it does not. For many wastes shredding or grinding of the waste to uniform particle size can be the most energy intensive and costly of the physical separation system, and when grinding to extremely small particle sizes is necessary, the cost can become unacceptable. Particles with some ferrous metals as well as other material are likely to have magnetic properties between those of the ferrous metal and the other material and could be collected in either fraction.

The future of continuous separation of municipal waste and the demand for magnetic separators for that purpose depends upon the demand for products from municipal waste, and the degree to which source separation is accepted by the public as well as the technical ability to design and operate economic and effective municipal waste systems, including magnetic separators. Mass incineration has also become a competitor for separation of municipal wastes, but incineration is having major difficulties in receiving public acceptance. Although incineration does not make magnetic recovery of ferrous metals impossible, incinerator wastes are often prepared only for landfills. During incineration, significant melting of metals and mixing ferrous and nonferrous metals could make magnetic separation of ferrous metals from the ash less effective by entrapping ferromagnetic and nonferromagnetic components into individual particles.

Besides municipal wastes, there should be a growing desire to use magnetic separation of ferrous metals from other wastes and debris. These metals have significant value, although not as high as that of nonferrous metals such as aluminum or copper, and there are large volumes of ferrous metals produced that ultimately will appear in wastes. Currently approximately 50% of the steel used in the United States is recycled [14]. Thus, there is room for improving iron and steel recycling, but the technology and infrastructure for using good quality recovered steel is at least partially in place. Although burial is an option for such material, recovery will become economical for some, and probably for a growing number of, cases.

One general requirement of economic magnetic separation, and usually of any waste separation system, that is not always appreciated is a large and reliable source of wastes to treat. The second requirement is a need to shred or otherwise break the waste into pieces small enough that the ferrous metals can be separated from the other components. The source of the waste is most reliable when the organization generating the waste also does the separation/recovery of the valuable components. If a different organization is generating the waste, the separation/recovery operation depends on continuity of the supplier to maintain its operations. If the recovered product proves to be of sufficient value, the facility performing the separation/recovery operation may find itself competing with other facilities for the source of waste materials. (However, many facilities would like for the market for their separated materials to become and remain high enough for that to become a problem.)

It is especially important that the ferrous metal not contain too much of certain components that devalue the iron content of the metal. Although copper is generally a more valuable metal than iron, even modest concentrations of copper in iron devalue the iron because copper is relatively difficult to remove from iron alloys and degrades the properties of steels. It is a general observation that materials recovered for recycling have less value than the original metal because of the impurities in the recycled material. This is a special problem for metals that are alloyed with selected components, and separation or segregation of similar alloys (such as different types of steels or aluminum) is difficult or impractical. In addition, the separation may not be perfect; steel may not be separated completely from other metals such as copper or aluminum. Most steel operations can accept only a specified fraction of its iron from recycled materials. If all of the iron in different steels were recycled without additional separation steps, the impurities in the recycled material would accumulate and degrade the properties of the steel product(s). This is not necessarily an inherent problem since one could consider additional separation steps to remove impurities, but the current recycling systems are able to handle the accumulation by limiting the amount of recycled steel in their products.

Even when the ferrous metal has relatively little value, it may be a contaminant in more valuable metal wastes. Then the incentive for removing the ferrous metal could then be to obtain a less contaminated and more valuable source of the nonferrous metal.

One interesting application reported recently was the use of a magnetic belt separator to remove spent (poisoned) catalytic cracker catalyst [15]. Cracking catalyst is normally treated continuously to "burn off" carbon and recycled. However, there is also an accumulation of iron, vanadium, and nickel on the catalyst which eventually poisons it in a way that simple burning will not help. A small bleed stream is used to remove catalyst at a rate sufficient to maintain a sufficiently low concentrated of highly poisoned catalyst. However, since the fluidized bed catalytic cracker and the regenerator act approximately like a well-mixed system, the poisoned catalyst is mixed with good catalyst, and the bleed stream removes good as well as poisoned catalyst. The accumulation of all three poisons contributes to the magnetic susceptibility of the catalyst particles, but the major contribution comes from the iron (oxide). By separating the catalyst bleed stream with a magnetic separator, the catalyst with less poison (less magnetic susceptibility) can be recycled. This is one example of treatment of a waste stream to minimize the quantity of waste produced. Alternatively, the plant operator can use this approach to maintain a higher catalyst activity for the same rate of catalyst discharge; that would increase the plant throughput and contribute to less catalyst consumption per volume of product and reduce the amount of heavy tar (lower grade product or waste) produced.

Other Possible Applications of Magnetic Separation in Environmental and Waste Problems

Magnetic separations have the advantages that they consume little energy, require no reagents, and add nothing to the waste stream. The disadvantages include the limited ability to make a "clean" separation based upon relatively small differences in magnetic susceptibility. Often a complete separation is not possible because the components to be separated are not completely separate and in different particles. Any particles that contain both of the components to be separated will carry one of the components to the stream that nominally should contain only the other particles. This is an inherent limitation of all physical separation methods. In some cases, grinding or shredding the material into smaller particles will separate the components better, but it may be impractical to shred some solid-solid mixtures to sizes sufficiently to place each component completely in different particles. There are practical limits for shredding or grinding particles because smaller particles eventually become more difficult to handle and energy costs increase as grinding proceeds to smaller and smaller particles.

Nevertheless, there may often be benefits to magnetic separations, even when the separation is not complete. If the contaminant can be removed sufficiently from some of the particles, magnetic separations can at least concentrate the contaminant into a portion of the particles. It can also be used to remove particles that would interfere with subsequent operations, even if the particles removed do not contain the contaminant. The low energy consumption and the need for no addition of reagents can make the magnetic separations attractive, so one may want to explore how magnetic separations can be used to reduce the difficulties or cost of subsequent separations. Capital costs for the magnetics and asso-

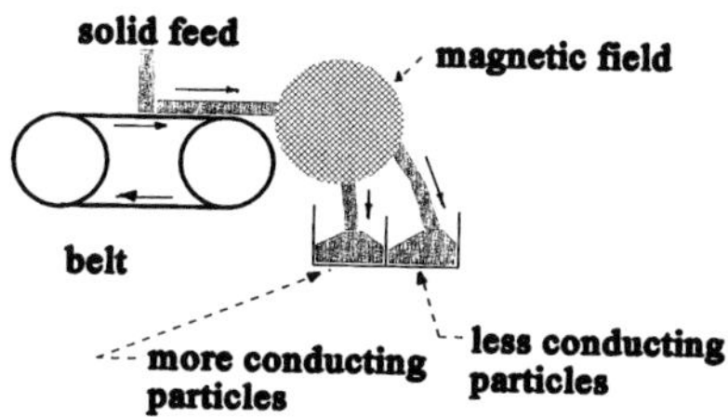

FIGURE 5 Eddy current separator.

ciated control systems may involve significant cost, but the costs will not be excessive unless extremely high fields are needed that require costly magnetics.

One can speculate about applications such as enhanced separation of components from buildings and equipment with high ferrous metal content. These are likely to be industrial buildings and other buildings that are largely steel structures. Although magnetic separation units probably could be made small enough to be portable and taken to large demolition sites, the other important components required, such as shredding equipment, may be less portable. If the material requires difficult shredding, central treatment facilities may have to become practical.

Eddy Current Separations

A mixture of solid particles can be separated on the basis of the electrical conductivity of the particles. The objective here may be to separate metal particles (particularly nonferrous metals) from glass particles or stones which have comparable sizes and are not easily separated by density or surface properties. Although there is no absolute limitation, eddy current separators are more likely to operate with larger particles because the forces increase with particle size. Eddy current separators have been incorporated into municipal waste separation systems to remove aluminum, copper, and other nonferrous metal particles from glass particles. The ferrous metals are likely to be removed in an earlier step by magnetic separation.

The principles of eddy current separations are simple and illustrated in Figure 5. In one version of a separator, the particles to be separated are projected across a magnetic field. The particles can be projected across the field by ejection from a conveyor (Figure 5). As the particles cross the magnetic field lines, eddy currents are generated within the particles that generate local magnetic fields that oppose the motion of the particles.

This is similar to the response of the rotor of some electric motors moving across magnetic fields from the stator: the higher the electrical conductivity of the particle and the greater the size of the particle, the greater the eddy current and the greater the induced force opposing the motion of the particle. That means that motion of the more conducting particles will be slowed, and in the device in Figure 5 the particles with the highest conductivity will not move as far horizontally as the less conducting particles.

If all other factors, such as particle size, shape, and density, were the same, relatively good separations could be achieved, even with moderate differences in electrical conductivity. However, in most particle mixtures, other factors affect the particle paths and thus the separation. As noted earlier, the opposing magnetic forces are greater for larger particles of the same material because of the larger paths for induced currents. The particles are also subject to aerodynamic forces. That means that smaller and less dense particles are likely to be slowed more by air drag than larger and denser particles; and larger particles (of conducting materials) are likely to be slowed more by the magnetic field. If the particles are not of uniform size and shape, the separation will be less sharp. Interactions between particles can also degrade the separation. The simple description just described the motion of individual particles. If particles strike each other or even affect the air motion around another particle, one or both of the particles will be deflected from the normal path and thus not fall where expected.

These limitations are some of the reasons why eddy current separations are usually used to separate particles with large differences in electrical conductivities; separation of aluminum from glass is an example. However, if one could prepare a mixed waste media into particles of relatively uniform size and shape, it would be possible to increase the selectivity of these devices.

The usefulness of eddy current separations in future waste and environmental processing will depend upon the need for such separation and the ability to prepare mixtures of particles with sufficient uniformity to make a useful separation. Separation of municipal solid waste has had difficulties. Physical separation of components at municipal waste collection centers has to compete with consumer separation of materials prior to collection. Consumers have generally been considered unlikely to support segregation of trash, but there may be evidence from the last few years that some degree of segregation may be acceptable to a sizable fraction of the public. Segregation of trash by consumers is more widespread in some other countries, such as Germany. However, eddy current separation may have a promising future for special separations if not for large-scale municipal wastes, for instance, in removal of metal particles from recycled glass bottles. If more and more manufactured products are shredded to separate the multiple components incorporated in the products, there may be more applications for eddy currents, particularly if shredding can make particles of single components and with approximately uniform sizes.

SCREENING

Screening is similar in many ways to surface filtration. Both screening and filtration operations involve retaining particles on a porous surface. The difference is that in filtration the aim is to retain all of the particles on the porous surface, but in screening the aim is to retain only a fraction of the particles on the porous surface, i.e., particles that are too large to pass through the screen. Screen operations can be done, in principle, in either liquids or gases, but practical considerations usually make screening in gas-solid operations more common. The goal of screening operations is to use the screen openings to separate the larger particles from the smaller, and any other forces, such as viscous forces, that are greater in liquids than in gases are likely to hinder screening operations. Screening operations can, in principle, be used on any size range of particles, but, in practice, screening operations are limited to relatively large particles, at least 100 micrometers to centimeters or even much larger.

Mesh Size

The screening surface can be a "screen," as the name suggests, a sieve plate, or even rolls of bars separated by the appropriate distance. Screens are usually made from wire mesh and are specified by the number of parallel wires per inch, i.e., the mesh size. Since woven screens have wires in two directions, usually normal to each other, the spacing between wires in the different directions does not have to be the same. That is, a woven wire screen could have one mesh size in one direction and a different mesh size in the other direction. When dealing with granular shapes (near convex shapes that are approximated crudely as spheres), the smaller spacing (the large mesh number) is usually the only size that is important. Using a larger spacing (smaller mesh number) in the other direction allows a somewhat larger open area with the same effective mesh number.

The mesh number is a crude estimate of the opening size in the screen only for the larger openings (lower mesh numbers). The actual size of screen openings can only be calculated by knowing the diameter of the wires as well as the number of wires per inch (mesh number). That is, the effective open width for particles to pass through a screen is the space (distance) between the wire centerlines minus the diameter of one wire (two radii of the wires on both sides of the opening). Thus, the opening in a screen is always less than the spacing between centerlines of the wires in the screen, and the importance of the wire thickness usually becomes more significant as the mesh number increases (as the spacing between wires decreases). Finite wire thickness is necessary to maintain sufficient structural strength in the screen, and it is not practical to reduce the wire thickness proportionally as the mesh size is increased. For fine screens with high mesh numbers, the opening size between wires can be strongly affected by the wire size, and the opening size can be considerably smaller than the spacing between wires.

Free Area of Screens

The finite wire thickness also decreases the "free" area of the screen, that is, the fraction of the screen area that is open for particles to pass. This is an important parameter for productivity of the screen. For relatively large screen openings (low mesh numbers) it is possible to maintain relatively large fractions of the screen "open," but for finer screens where the wire thickness becomes a more important factor in determining the size of the screen openings, the wire diameter also lowers the "open" fraction of the screen, often lowers it considerably.

As noted earlier, by making the wire spacing in one direction greater than the spacing in the other direction, it is possible to increase the "free" area fraction over that possible with "square" mesh screens, the same wire spacing in both directions. If very strong screens, are required, it is possible to use bars rather than woven wire. Bars are more likely to be used when relatively large particles are to be screened, and large particles can exert relatively strong forces on the screen. In some cases, bar type screens can be constructed so the spacing between bars can be adjusted. That can add flexibility to equipment that is to serve different purposes, but it is likely to be an unnecessary expense for continuous operations for long periods on the same separation.

Blinding of Screens

Surface filters can (and do) build up a filter cake, but that would greatly reduce the performance of a screening operation. For a screen to separate particles according to size, all particles must reach the screen surface and thus be given the "chance" to pass through the screen or to be rejected depending upon their size. However, particles accumulated on a filter cake are not likely to ever reach the screen surface. This means that in screening operations the particles must be spread over the screen surface, and the larger particles that are retained by the screen must not block access of smaller particles to the screen. This is usually not a problem in filtration, and the filter cake can even play an important role in the performance of a surface filter.

To prevent smaller particles from being blocked from the screen, the particles are usually spread in a relatively thin layer on the screen, and vibration is applied to mix the particles and keep giving the smaller particles opportunities to reach the screen. The particles must be removed from the screen after the smaller particles are given sufficient time to pass through the screen. In small laboratory units, batch screening can be used with the larger particles simply removed manually after screening for a specified time, but in production operations the larger particles are more likely to be removed continuously. The screen could be part of a moving belt or on a slope so that the larger particles move down the screen by gravity. Steep slopes can be used with gravity preventing particles from resting on the screen, but with vibration even a relatively gentle slope can keep particles from moving down the screen. That is, with each vibration, particles are thrown slightly above the screen and statistically move in the downward direction as they settle back on the screen.

Vibration can be imposed on the screens in several ways, and different manufactures have their own methods. Lower frequency vibrations are likely to be imposed mechanically. An "off-center" rotor can be used, and the rotor can be shaped to give different "shapes" to a vibration cycle. For higher frequencies, electromagnetic vibrators are more likely to be used. These could be frequencies in the audible range, or even beyond. Lower frequencies are likely to be used when screening relatively large particles, and relatively large screen displacements may be necessary to more the particles sufficiently. However, when screening very fine particles, very high frequencies with small displacement are more likely to be preferred.

Screening Equipment

Screening equipment can be mechanically complex, and is purchased from manufacturers as essentially complete units. It is obviously necessary to specify the particle sizes to be screened (the size of the screen opening needed), but it is also wise to consider some properties of the entire mixture to be screened. For instance, a few very large particles could damage a relatively fine screen and probably should be removed by a separate screening operation before trying to screen the smaller particles. (See the next subsection.) Different types of equipment are likely to be most economical in different ranges of solids throughput. That is, the equipment that is most economical for small-scale operations may look quite different from the equipment that is most economical for large throughput operations.

As noted above, the frequency and amplitude of the vibration needed depend upon the particle size range. For screening toxic materials, one may also want to consider how easy it is to "seal" a given manufacturer's equipment and thus minimize loss of toxic fine dust.

Classification

Classification is the separation of a range of particles according to their size. A series of screens is one way to "classify" a mixture of particles. In such operations, the "fine" particles that pass through one screen go to a smaller screen, and the "fine" particles passing that screen go to an even smaller screen, etc. The resulting product is a series of particle "fractions." The largest fraction consists of particles that were rejected by the first screen, and the smallest fraction contains the particles that pass through the finest screen. The other fractions include particles that pass through one screen and are rejected by the next finest screen. The particle fractions can be specified as greater than the screen that rejected them and smaller than the finest screen they passed through. This type of classification is simply a series of screening operations, but there are problems often seen in classification operations. Classification can be done on a production basis, but it is also commonly used to characterize mixtures of particles.

Screens are not the only way particles can be separated into size fractions. Other methods include elutriation, which utilizes the difference in hydrodynamic and gravity forces to separate the particles. Since the difference in hydrodynamic and gravity forces on a particle depend upon the density of the particle as well as the size, classification by elutriation can give confusing results when the particles have different densities. Smaller particles with higher densities can appear in the same fraction as larger particles with lower densities.

Uses for Screening in Waste and Environmental Problems

Screening is likely to be useful in treating any heterogeneous mixture of solids, especially when the different size fractions need to be treated in different ways. For instance, in soil washing operations, it may be helpful to screen the larger "rock" from soil, especially if the contamination is only in the much smaller clay particles. Even if the larger rocks are contaminated, it may be necessary or desirable to treat them differently than the smaller clay particles. For instance, if even slightly acid solutions are required for a soil washing operation, it would be desirable to remove any limestone rocks from the soil before applying acid. Otherwise, it could become necessary to dissolve the limestone to maintain even a slightly acid washing liquor.

Heterogeneous mixtures of solid waste can resemble "trash," and screening can be helpful when such materials require treatment. Of course, other treatment including methods discussed earlier can separate several of the materials in such mixtures according to properties of the material (magnetic properties, etc.), but when the different materials are in different size particles, screening can be a particularly cost effective approach to use, at least for some of the separation that may be needed.

When solids must be ground to smaller sizes, for effective treatment using methods such as leaching, screening is usually necessary to remove particles that have not been ground or shredded to the required size range. Grinding equipment is not likely to always produce uniform size particles, but more nearly uniform size particles can sometimes be obtained by recycling the larger particles to the grinder/shredder for further reduction.

SORTING

Once solid wastes from several sources have become well mixed, the physical methods described above that are based on differences in a single property of the materials may not be adequate for separating multiple components of the waste. For larger particles, usually an inch to several inches in diameter, the separation can be called "sorting." Sorting can be done manually by allowing workers to select different types of solids as the mixed waste moves down a conveyer. Workers can then sort such things as glass, steel, plastic, and nonferrous metals from complex mixtures, including municipal wastes as well as industrial wastes. Workers can sort materials on the basis of color, shape, texture, or any other visual property. Manual sorting is routinely used to sort foodstuff (that is, to remove damaged vegetables from good ones in premium products), but that is not typically considered waste treatment, even if the rejected vegetables are later treated as waste.

Manual sorting is obviously labor intensive, but workers can develop skills that improve sorting rates over those achieved by beginning workers. Manual sorting can also be dangerous to workers when hazardous materials are being sorted, but sorting can be performed to recover useful materials and reduce nonhazardous waste volume as well as to remove toxic components. Sorting toxic materials would require worker protection, usually protective clothing, and this is likely to reduce worker efficiency.

Automated Sorting

In recent decades, there have been significant advances in automated sorting, and this is a common practice for some waste materials. However, it has not always been practical. Failure can result from limitations in the sorting equipment or from unexpected variations in the wastes to be sorted. An unexpected component in a waste stream can report unexpectingly with a component and thus contaminate the product. Automated sorting equipment can be mechanically complicated and often requires some customization of the equipment unless the waste stream is very much like that used in other sorting operations. Users are advised to work closely with the manufacturer to ensure that the equipment is suitable for the waste stream to be treated and that the material is treated (sized, washed, etc.) in a manner suitable for the automated sorting equipment.

Any solids sorting operation requires three operations:

1. Spreading the materials in a dilute form (usually on a conveyer belt) where each particle can be "seen" individually
2. Sensing methods to determine which particles are to be rejected or, more generally, are to go into each classification
3. Removing the selected particles and transporting them to the proper collection bin(s)

With manual sorting, all operations can be performed manually; in some cases, it may be desirable to use mechanical methods for removing the particles, and sensing can be the only manual part of the operation if it is undesirable to handle the particles.

Particles must be spread over the conveyer belt or other transport device sufficiently dilute (separated from each other) so that each particle can be seen by the human or electronic sensing device and so that the particles can be removed individually from the others. The throughput of the operation depends upon the concentration that can be tolerated of the particles on the conveyer and the rate at which the conveyer moves. The speed of the conveyer must be sufficiently slow that the sensing and removal devices can identify and remove particles before they move beyond the range of the active part of the sorter. The limits of human sorters are probably more easily perceived by the reader, but there are also similar limitations for automatic sorting systems. However, in most applications where automatic sorting has been successful, the allowable conveyer speed for automatic sorting is usually much higher than for manual sorting.

Sensing devices for automatic sorters can utilize essentially any measurement of physical property, and a number of sensing devices have been tried for different applications. Two of the most successful automatic sorting operations (and two of the most interesting to this author) are for separating different plastics and for separating different color glass shards. In these systems, optical (spectroscopic) measurements are likely to be the most practical. The volumetric throughput of such separations is very impressive to this author. The need for "quick" sensing of the objects is probably obvious. Although considerable computations may be needed to examine spectra or other measurements to decide whether to remove the particle, any such calculations must be completed before the particle is transported beyond the removal mechanism.

Although, in principle, essentially any removal method can be used, it is necessary for the removal device to complete the particle removal quickly and be ready to remove the next particle soon if the sensing device decides that the next particle has to be removed. This quick recovery of the removal device is necessary and could limit the throughput of the system. The need for quick "recovery" of the removal device hampers the use of large and heavy components whose inertia would slow recovery of the device. "Air puffs" are common removal devices that avoid many of the limitations mentioned above. In such systems, a puff of air blows selected particles from the conveyer. The inertial limitations of this approach are associated with the relatively small valve that controls the airflow and the inertia of the particles themselves. The inertia of the particles themselves is essentially inherent in any removal method, and these devices can operate relatively quickly. Some limitations of air puffs are the difficulties in removing very dense and large particles and the possible difficulty (cost) in containing/treating air from the puff if it contains dust of toxic materials.

When Sorting Operations Can Be Helpful in Waste and Environmental Applications

Some of the conditions for which automatic sorting are most attractive are illustrated in sorting of plastic bottles and glass shards. Both applications involve relatively simple mixtures of only glass or plastics (often only plastic bottles). In both cases, there is a single and practical sensing method capable of identifying the components of interest. Plastic and glass are usually collected separately and are available relatively free of other components, or they can be treated to remove other components such as any remaining metal caps or rings. Note that both materials are also usually available with only one material in each particle. That is, a sensing element that makes a measurement on one part of the particle/object would find the same measurement on any other part of the particle. The exceptions could include metal rings that may remain on some plastic or glass bottles. Also the size of the particles/objects is convenient for sorting. Sorting of very fine particles would require more (even far more) decisions and removal steps for a given throughput of material. Thus, the throughput is usually the greatest when the particles/objects are as large as possible and still maintain homogeneous individual particles and convenient removal methods. Note that plastic bottles are relatively light and are thus convenient for removal by air jets.

Although the kind of highly automated sorting is not common in waste and environmental processing other than the recovery of selected components for recycling, such as the cases just mentioned, some degree of sorting is often used, even if informally. It is generally desirable to limit the volume of any waste that must undergo extensive and costly treatment; any effort to limit the volume of additional materials that are treated can be minimized. That can involve sorting type operations while waste solids are being collected or stored as well as sorting the material after it is mixed. A general rule of waste minimization (or at least hazardous waste minimization) is to avoid mixing toxic waste materials from nonhazardous wastes whenever practical.

REFERENCES

1. Barkat, O. and Z. Merah. "Electro-Flotation in Waste Water Treatment." Paper presented at the Ninth Symposium on Separation Science and Technology for Energy Applications, Gatlinburg, TN, Oct. 22–26, 1995; submitted to *Sep. Sci. Technol.*
2. He, D. X., et al. *Sep. Technol. 5*, 133 (1995).
3. Pascual, B., B. Tansel, and R. Shalewitz. "Economic Sensitivity of the Dissolved Air Flotation Process with Respect to Operational Variables." *Proceedings of the 49th Industrial Waste Conference*, Purdue University, May 9–11, 1994.
4. He, D. X., et al. *Sep. Technol. 5*, 133 (1995).
5. Edswald, J. K., J. Wash, G. Kaminski, and H. Dunn. *J. Am. Water Works Assoc. 83*, 92 (1992).
6. Grasso, D., et al. "Pressure Flotation of Nitrocellulose Fines: Hydrodynamics and Interfacial Chemistry." *Proceedings of the 50th Industrial Waste Conference*, Purdue University, May 1995.

7. Ghazy, S. *Sep. Sci. Technol. 30*, 933 (1995).
8. Parker, W. J. and H. D. Monteith. *Environ. Prog. 15*, 73 (1996).
9. Wungrattanasopon, P., et al. *Sep. Sci. Technol. 31*, 1523 (1996).
10. Norrgran, D. *Chem. Eng. Prog. May*, 56, (1996).
11. Kolm, H., J. Oberteuffer, and D. Kelland. *Sci. Am. 233*, 46 (1975).
12. Scott, T. C. *AIChE J. 35*, 2058 (1989).
13. Cuellar, J. and A. Alvaro. *Sep. Sci. Technol. 30*, 141 (1995).
14. *Steel: A National Resource for the Future*. Report prepared for the National Academy of Science and the U.S. Department of Energy, May 2, 1995.
15. Goolsby, T. L. and H. F. Moore. "Development of FCC Catalytic Magnetic Separator." Presented at the Ninth Symposium on Separation Science and Technology for Energy Applications, Gatlinburg, TN, Oct. 22–26, 1995; submitted to *Sep. Sci. Technol.*

Index